Plant Bioinformatics

METHODS IN MOLECULAR BIOLOGY™

John M. Walker, SERIES EDITOR

METHODS IN MOLECULAR BIOLOGY™

Plant Bioinformatics

Methods and Protocols

Edited by

David Edwards

*Australian Centre for Plant Functional Genomics,
Institute for Molecular Biosciences and School of Land,
Crop and Food Sciences, University of Queensland,
Brisbane, Australia*

HUMANA PRESS ✳ TOTOWA, NEW JERSEY

Production Editor: Christina M. Thomas
Cover design by Karen Schulz

For additional copies, pricing for bulk purchases, and/or information about other Humana titles, contact Humana at the above address or at any of the following numbers: Tel.: 973-256-1699; Fax: 973-256-8341; E-mail: humana@humanapr.com; or visit our Website: www.humanapress.com

Library of Congress Control Number: 2007933141.

Preface

Bioinformatics is a new and rapidly evolving field of research that is being driven by the development and application of various omics technologies. The term bioinformatics means different things to different people, and following the theme of this series, this volume focuses on applied bioinformatics with specific applications to crops and model plants. This volume is aimed at plant biologists who have an interest in or requirement for accessing the huge amount of data being generated by high-throughput technologies. The volume would also be of interest to bioinformaticians and computer scientists who would benefit from an introduction to the different tools and systems available for plant research.

Bioinformatics was initially driven by the need to understand and exploit the information being generated by gene and genome sequencing projects, the scope of bioinformatics now extends from the genome to the phenome. It is the integration of information relating to heritable agronomic traits, including important metabolic profiles, with the emerging genome and transcriptome data that will drive plant research and bioinformatics developments into the future. One observation during the production of this volume is the increasing integration of both related data from different plants and also the integration of multiple diverse forms of data. I expect this trend to continue as this field of research continues to develop in the future.

Scientific research by its nature progresses and changes, and this is especially true for bioinformatics. The rapid evolution of the tools and systems described in this volume may change during the lifetime of this edition, and it is suggested that the reader consults the relevant web pages directly as they frequently host detailed list of updates and changes.

David Edwards

Contents

Contributors

ROLF APWEILER • *EMBL-Outstation The European Bioinformatics Institute, Wellcome Trust Genome Campus, Hinxton, Cambridge, UK*

ERIC ARCHULETA • *National Center for Genome Resources, Santa Fe, NM*

AMOS BAIROCH • *Department of Structural Biology and Bioinformatics, Centre Medical Universitaire, Swiss Institute of Bioinformatics and University of Geneva, Geneva, Switzerland*

RANJAN BALACHANDRA • *Department of Primary Industries, Australian Temperate Field Crops Collection, Horsham Victoria, Australia*

JACQUELINE BATLEY • *Australian Centre for Plant Functional Genomics, School of Land, Crop and Food Sciences and ARC Centre of Excellence for Integrative Legume Research, CILR, The University of Queensland, Brisbane, Australia*

WILLIAM D. BEAVIS • *National Center for Genome Resources, Santa Fe, NM*

TANYA Z. BERARDINI • *Department of Plant Biology, The Arabidopsis Information Resource, Carnegie Institution of Washington, Stanford, CA*

EMMANUEL BOUTET • *Swiss Institute of Bioinformatics, Centre Medical Universitaire, Geneva, Switzerland*

VOLKER BRENDEL • *Department of Genetics, Development and Cell Biology, Iowa State University, Ames, IA*

COREY D. BROECKLING • *The Samuel Roberts Noble Foundation, Plant Biology Division, Sam Noble Parkway, Ardmore, OK, and Department of Horticulture and Landscape Architecture, Colorado State University, Fort Collins, CO*

RICHARD BRUSKIEWICH • *International Rice Research Institute, Metro Manila, Philippines*

C. ROBIN BUELL • *The Institute for Genomic Research, Rockville, MD*

RICO A. CALDO • *Department of Plant Pathology and Center for Plant Responses to Environmental Stresses, Iowa State University, Ames, IA*

EVELYN CAMON • *European Bioinformatics Institute, Wellcome Trust Genome Campus, Hinxton, Cambridge, UK*

AGNES P. CHAN • *The Institute for Genomic Research, Rockville, MD*

DEBBIE CLEMENTS • *Nottingham Arabidopsis Stock Centre, School of Biosciences, University of Nottingham, Loughborough, UK*

TIMOTHY J. CLOSE • *University of California, Riverside, CA*

JOHN A. CROW • *Center for Computational Genomics and Bioinformatics, University of Minnesota, Minneapolis, MN*

JULIE A. DICKERSON • *Virtual Reality Applications Center and Department of Electrical and Computer Engineering, Iowa State University, Ames, IA*

EMILY DIMMER • *European Bioinformatics Institute, Wellcome Trust Genome Campus, Hinxton, Cambridge, UK*

DAVID EDWARDS • *Australian Centre for Plant Functional Genomics, Institute for Molecular Biosciences and School of Land, Crop and Food Sciences, University of Queensland, Brisbane, Australia*

TAKASHI R. ENDO • *Laboratory of Bioknowledge Systems, Bioinformatics Center, Institute for Chemical Research, Kyoto University, Gokasho Uji, Kyoto, Japan*

REBECCA ERNST • *MIPS Institute for Bioinformatics, GSF Research Center for Environment and Health, Neuherberg, Germany*

ANDREW D. FARMER • *National Center for Genome Resources, Santa Fe, NM*

KAMAL GAJENDRAN • *National Center for Genome Resources, Santa Fe, NM*

MICHAEL D. GONZALES • *National Center for Genome Resources, Santa Fe, NM*

SUSUMU GOTO • *Laboratory of Plant Genetics, Division of Applied Bioscience, Kyoto University, Kitashirakawa Oiwake-cho, Kyoto, Japan*

NEIL GRAHAM • *Nottingham Arabidopsis Stock Centre, School of Biosciences, University of Nottingham, Loughborough, UK*

JUSTIN D. GUENTHER • *Agriculture and Agri-Food, Saskatoon Research Centre, Saskatoon, Saskatchewan, Canada*

GEORG HABERER • *MIPS Institute for Bioinformatics, GSF Research Center for Environment and Health, Neuherberg, Germany*

XIMIAO HE • *Beijing Genomics Institute, Chinese Academy of Sciences, Beijing, China*

LU HONG • *Interdepartmental Bioinformatics & Computational Biology, Iowa State University, Ames, IA*

DAVID V. HUHMAN • *The Samuel Roberts Noble Foundation, Plant Biology Division, Sam Noble Parkway, Ardmore, OK*

NICK JAMES • *Nottingham Arabidopsis Stock Centre, School of Biosciences, University of Nottingham, Loughborough, UK*

ERICA JEWELL • *Department of Primary Industries, Primary Industries Research Victoria, Animal Genetics and Genomics, Attwood, Victoria, Australia*

JAMES E. JOHNSON • *Center for Computational Genomics and Bioinformatics, University of Minnesota, Minneapolis, MN*

MINORU KANEHISA • *Laboratory of Plant Genetics, Division of Applied Bioscience, Kyoto University, Kitashirakawa Oiwake-cho, Kyoto, Japan*

PAUL KERSEY • *EMBL-Outstation The European Bioinformatics Institute, Wellcome Trust Genome Campus, Hinxton, Cambridge, UK*

TAMARA KULIKOVA • *EMBL-Outstation The European Bioinformatics Institute, Wellcome Trust Genome Campus, Hinxton, Cambridge, UK*

ANNE F. LAMBLIN • *Center for Computational Genomics and Bioinformatics, University of Minnesota, Minneapolis, MN*

CAROLYN J. LAWRENCE • *Genetics Building, Iowa State University, Ames, IA*

CHRISTOPHER T. LEWIS • *Agriculture and Agri-Food, Saskatoon Research Centre, Saskatoon, Saskatchewan, Canada*

DAMIEN LIEBERHERR • *Swiss Institute of Bioinformatics, Centre Medical Universitaire, Geneva, Switzerland*

CHRISTOPHER G. LOVE • *Primary Industries Research Victoria, Plant Genetics and Genomics, Victorian AgriBiosciences Centre, La Trobe University, Bundoora, Victoria, Australia*

MATTHEW LYON • *University of California, Riverside, CA*

ALI MASOUDI-NEJAD • *Laboratory of Bioknowledge Systems, Bioinformatics Center, Institute for Chemical Research, Kyoto University, Gokasho Uji, Kyoto, Japan, and Department of Bioinformatics, Institute of Biochemistry and Biophysics, University of Tehran, Tehran, Iran*

SEAN MAY • *Nottingham Arabidopsis Stock Centre, School of Biosciences, University of Nottingham, Loughborough, UK*

KLAUS F. X. MAYER • *MIPS Institute for Bioinformatics, GSF Research Center for Environment and Health, Neuherberg, Germany*

GRAHAM MCLAREN • *International Rice Research Institute, Metro Manila, Philippines*

THOMAS METZ • *International Rice Research Institute, Metro Manila, Philippines*

BLAKE C. MEYERS • *Delaware Biotechnology Institute, University of Delaware, Newark, DE*

KAN NOBUTA • *Delaware Biotechnology Institute, University of Delaware, Newark, DE*

HELEN O'SULLIVAN • *ACPFG Australian Centre for Plant Functional Genomics, University of Adelaide, Urrbrae, Australia and Department of Biological Sciences, University of Bristol, Bristol, UK*

JAMES OSTELL • *National Center for Biotechnology Information, National Institutes of Health, Bethesda, MD*

ISOBEL A. P. PARKIN • *Agriculture and Agri-Food, Saskatoon Research Centre, Saskatoon, Saskatchewan, Canada*

CHARLES E. PAULE • *Center for Computational Genomics and Bioinformatics, University of Minnesota, Minneapolis, MN*

REBECCA L. POOLE • *Biological Sciences Building, Clifton, Bristol, UK*

ARLLET PORTUGAL • *International Rice Research Institute, Metro Manila, Philippines*

JOHN QUACKENBUSH • *The Institute for Genomic Research, Rockville, MD*

PABLO D. RABINOWICZ • *The Institute for Genomic Research, Rockville, MD*

ERNEST F. RETZEL • *Center for Computational Genomics and Bioinformatics, University of Minnesota, Minneapolis, MN*

STEPHEN J. ROBINSON • *Agriculture and Agri-Food, Saskatoon Research Centre, Saskatoon, Saskatchewan, Canada*

MIKEAL L. ROOSE • *University of California, Riverside, CA*

BEATRICE SCHILDKNECHT • *Nottingham Arabidopsis Stock Centre, School of Biosciences, University of Nottingham, Loughborough, UK*

MICHEL SCHNEIDER • *Swiss Institute of Bioinformatics, Centre Medical Universitaire, Geneva, Switzerland*

HEIKO SCHOOF • *Max Planck Institute for Plant Breeding Research, Plant Computational Biology, Koeln, Germany*

LISHUANG SHEN • *Virtual Reality Applications Center, Iowa State University, Ames, IA*

BRIAN SMITH-WHITE • *National Center for Biotechnology Information, National Institutes of Health, Bethesda, MD*

MANUEL SPANNAGL • *MIPS Institute for Bioinformatics, GSF Research Center for Environment and Health, Neuherberg, Germany*

JASON E. STAJICH • *Department of Molecular Genetics and Microbiology, Duke University, Durham, NC*

PETER STERK • *EMBL-Outstation The European Bioinformatics Institute, Wellcome Trust Genome Campus, Hinxton, Cambridge, UK*

LLOYD W. SUMNER • *The Samuel Roberts Noble Foundation, Plant Biology Division, Ardmore, OK*

TATIANA TATUSOVA • *National Center for Biotechnology Information, National Institutes of Health, Bethesda, MD*

MICHAEL TOGNOLLI • *Swiss Institute of Bioinformatics, Centre Medical Universitaire, Geneva, Switzerland*

CHRIS D. TOWN • *The Institute for Genomic Research, Rockville, MD*

EWA URBANCZYK-WOCHNIAK • *The Samuel Roberts Noble Foundation, Plant Biology Division, Ardmore, OK*

KALYAN VEMARAJU • *Delaware Biotechnology Institute, University of Delaware, Newark, DE*

STEVE WANAMAKER • *University of California, Riverside, CA*

JUN WANG • *Beijing Genomics Institute, Chinese Academy of Sciences, Beijing, China*

DOREEN WARE • *Gramene, Cold Spring Harbor, NY*

DAVID WHEELER • *National Center for Biotechnology Information, National Institutes of Health, Bethesda, MD*

ROGER P. WISE • *Corn Insects and Crop Genetics Research, USDA-ARS and Department of Plant Pathology and Center for Plant Responses to Environmental Stresses, Iowa State University, Ames, IA*

1

The EMBL Nucleotide Sequence and Genome Reviews Databases

Peter Sterk, Tamara Kulikova, Paul Kersey, and Rolf Apweiler

Summary

Nucleotide and protein sequence databases are major resources for biological and medical research. This chapter introduces the European Molecular Biology Laboratory (EMBL) Nucleotide Sequence Database, a comprehensive primary data archive for nucleic acid sequences, and Genome Reviews, a secondary database that provides an up-to-date, standardized and comprehensively annotated view of the genomic sequence of selected organisms with completely deciphered genomes. Focusing on plant nucleotide sequences, we demonstrate how these data are accessed, how sequence similarity searches are performed and how we can obtain a wealth of additional information relating to genome sequences using Integr8.

Key Words: Database; EMBL; Genome Reviews; Integr8; SRS; sequence analysis; BLAST; FASTA.

1. Introduction

The European Bioinformatics Institute (EBI), part of the European Molecular Biology Laboratory (EMBL), is a non-profit organization and a centre for research and services in bioinformatics. The Institute manages many databases containing biological information including nucleic acid and protein sequences, and macromolecular structures. Data and bioinformatics services are made freely available to the scientific community.

Among the databases provided, the most widely known are the EMBL Nucleotide Sequence Database (1), further referred to as EMBL database,

From: *Methods in Molecular Biology, vol. 406: Plant Bioinformatics: Methods and Protocols*
Edited by: D. Edwards © Humana Press Inc., Totowa, NJ

and the protein database that many still refer to as Swiss-Prot, but which has in recent years become part of the UniProt Knowledgebase (UniProtKB), together with TrEMBL and PIR-PSD; UniProtKB in turn is part of the universal protein resource UniProt *(2)*. The EMBL database is a typical primary (archival) database: it contains the results from sequencing experiments in laboratories together with the interpretation provided by the submitters, but with no or very limited review. By contrast, in the UniProtKB the vast majority of its protein sequences have been derived from nucleic acid sequences from the EMBL database. The data resource is curated by many experts, assuring a very high quality of data. Because of this, and the fact that the UniProtKB curators maintain for each database entry cross-links to a great number of other databases, UniProtKB is an excellent data resource to consult during the analysis of new and existing sequence data.

In this chapter, we concentrate on nucleic acid sequences, as the first step into the bioinformatics field for most "wet-lab" researchers is the analysis of their own nucleic acid sequence(s). We explain what an EMBL database record looks like, how the EMBL database can be accessed and how a sequence similarity search is done. In addition, we have a look at Genome Reviews, a database that aims to provide a standardized and up-to-date view of complete genomic sequences, and Integr8, a resource that integrates genomic and proteomic data sources *(3)*.

2. Materials

2.1. Nucleotide Sequence Databases

2.1.1. Description of the EMBL Database

The EMBL Nucleotide Sequence Database (http://www.ebi.ac.uk/embl/), maintained at the EBI, is a comprehensive collection of nucleotide sequences and annotation *(1)*. The database is the European part of the International Nucleotide Sequence Database Collaboration (INSDC) (http://www.insdc.org/), an international collaboration with DNA Databank of Japan (DDBJ) at The National Institute of Genetics in Mishima, Japan, and GenBank at the National Center of Biotechnology Information (NCBI) in Bethesda, USA. The data in the EMBL database originate from large-scale genome sequencing projects, the European Patent Office and direct submissions from individual scientists. It is important to note that the editorial rights to an entry in the EMBL database remain with the original submitter(s). This means that apart from the addition of cross-references, the data are not updated by EMBL database curators, unless explicitly instructed by the submitter(s). When

EMBL entries are updated, older versions of the same entry can still be retrieved from the EMBL Sequence Version Archive (EMBL-SVA) (*4*; also *see* **Note 1**).

There is a quarterly release of the EMBL database, and new and updated records are distributed on a daily basis. Data are also exchanged daily among the three collaborating institutes to ensure that all data are available from all three sites.

Current statistics for the database can be found at http://www3.ebi.ac.uk/Services/DBStats/. At the time of writing, the database contained over 110 billion nucleotides in nearly 61 million entries.

Data in the EMBL Nucleotide Sequence Database are grouped into divisions, according to either the methodology used in their generation, such as expressed sequence tag (EST) and high-throughput genome (HTG) sequencing, or taxonomic origin of the sequence source, such as the PLN division for plants. This is done to create subsets of the database which separate sequence of lower quality such as EST and HTG sequences from high-quality sequence, and group high-quality sequences into groups that reflect the main areas of interest. The advantage becomes obvious when you wish to limit your database search or sequence similarity analysis to plant entries rather than the whole database; at the time of writing, only a relatively small fraction of the database, 1.5 million nucleotides or 300,000 entries, need to be searched when using the PLN division.

The EMBL database also contains entries that are grouped into separate data sets. Two of those are:

1. Whole Genome Shotgun (WGS) data: Methods using WGS data are used to gain a large amount of genome coverage for an organism. The sequences of all contigs originating from one experiment are grouped into a set. The accession numbers of all entries in each WGS set share the same prefix. The current listings of all WGS sets available can be found at http://www.ebi.ac.uk/genomes/wgs.html. At the time of writing, there were 255 such sets available, among them large data sets for *Oryza sativa*, both indica and japonica cultivars.
2. Third Party Annotation (TPA) data: The TPA data set was launched in response to requests from the research community to submit entries that include either re-annotation of existing data or combinations of novel sequence, existing primary sequence, trace archive and WGS data; this is a relatively small subset of data, but growing rapidly.

For users who are interested in the sequences that encode proteins, a subset of the EMBL database, EMBL CDS (coding sequence), a database of nucleotide sequences of the CDS features as annotated in the EMBL database and additional annotation derived from UniProtKB, is available, too (http://www.ebi.ac.uk/embl/cds/).

2.1.2. EMBL Flat File Format

The EMBL flat file is a commonly used distribution format for information in the EMBL database. EMBL flat files, in contrast to GenBank and DDBJ flat files (which in essence contain the same information), use line-type prefixes, which indicate the type of information present on each line. The flat file can be divided into three parts:

1. the header, which contains information applicable to the whole record,
2. the feature table, which contains the annotation, i.e., the features contained in the sequence and their corresponding locations, and
3. the sequence itself.

An example of an EMBL flat file is presented in **Fig. 1**.

All the line types are described in detail in the EMBL user manual (http://www.ebi.ac.uk/embl/Documentation/User_manual/usrman.html). Here, we describe the most important line types in the three sections of the EMBL flat file.

1. Header

- The AC line contains the primary accession number of the database entry, which is the primary means of identifying a sequence. This primary accession number can be followed by one or more secondary accession numbers. These are included to allow tracking of data when entries are merged or split.
- The DE line is a description of the database record.
- The OS and OC lines specify the organism and its taxonomic classification, respectively.
- Lines starting with R provide references to publications and the submitter.

2. Feature table

- The FH lines are the feature table header. The actual features and corresponding qualifiers are on the FT lines. In **Fig. 1**, there are three features: the source feature, describing the biological source of the nucleic acid sequenced, a CDS feature and an mRNA feature. Each feature may have one or more qualifiers that provide additional information about the feature.

3. Sequence

- The SQ line marks the beginning of the sequence section.
- The "//" line marks the end of the database record.

```
ID   TRBG361     standard; mRNA; PLN; 1859 BP.
XX
AC   X56734; S46826;
XX
SV   X56734.1
XX
DT   12-SEP-1991 (Rel. 29, Created)
DT   15-MAR-1999 (Rel. 59, Last updated, Version 9)
XX
DE   Trifolium repens mRNA for non-cyanogenic beta-glucosidase
XX
KW   beta-glucosidase.
XX
OS   Trifolium repens (white clover)
OC   Eukaryota; Viridiplantae; Streptophyta; Embryophyta; Tracheophyta;
OC   Spermatophyta; Magnoliophyta; eudicotyledons; core eudicots; rosids;
OC   eurosids I; Fabales; Fabaceae; Papilionoideae; Trifolieae; Trifolium.
XX
RN   [5]
RP   1-1859
RX   PUBMED; 1907511.
RA   Oxtoby E., Dunn M.A., Pancoro A., Hughes M.A.;
RT   "Nucleotide and derived amino acid sequence of the cyanogenic
RT   beta-glucosidase (linamarase) from white clover (Trifolium repens L.).";
RL   Plant Mol. Biol. 17(2):209-219(1991).
XX
RN   [6]
RP   1-1859
RA   Hughes M.A.;
RT   ;
RL   Submitted (19-NOV-1990) to the EMBL/GenBank/DDBJ databases.
RL   M.A. Hughes, UNIVERSITY OF NEWCASTLE UPON TYNE, MEDICAL SCHOOL, NEWCASTLE
RL   UPON TYNE, NE2 4HH, UK
XX
FH   Key             Location/Qualifiers
FH
FT   source          1..1859
FT                   /db_xref="taxon:3899"
FT                   /mol_type="mRNA"
FT                   /organism="Trifolium repens"
FT                   /tissue_type="leaves"
FT                   /clone_lib="lambda gt10"
FT                   /clone="TRE361"
FT   CDS             14..1495
FT                   /db_xref="GOA:P26204"
FT                   /db_xref="UniProtKB/Swiss-Prot:P26204"
FT                   /note="non-cyanogenic"
FT                   /EC_number="3.2.1.21"
FT                   /product="beta-glucosidase"
FT                   /protein_id="CAA40058.1"
FT                   /translation="MDFIVAIFALFVISSFTITSTNAVEASTLLDIGNLSRSSFPRGFI
FT                   FGAGSSAYQFEGAVNEGGRGPSIWDTFTHKYPEKIRDGSNADITVDQYHRYKEDVGIMK
FT                   DQNMDSYRFSISWPRILPKGKLSGGINHEGIKYYNNLINELLANGIQPFVTLFHWDLPQ
FT                   VLEDEYGGFLNSGVINDFRDYTDLCFKEFGDRVRYWSTLNEPWVFSNSGYALGTNAPGR
FT                   CSASNVAKPGDSGTGPYIVTHNQILAHAEAVHVYKTKYQAYQKGKIGITLVSNWLMPLD
FT                   DNSIPDIKAAERSLDFQFGLFMEQLTTGDYSKSMRRIVKNRLPKFSKFESSLVNGSFDF
FT                   IGINYYSSSYISNAPSHGNAKPSYSTNPMTNISFEKHGIPLGPRAASIWIYVYPYMFIQ
FT                   EDFEIFCYILKINITILQFSITENGMNEFNDATLPVEEALLNTYRIDYYYRHLYYIRSA
FT                   IRAGSNVKGFYAWSFLDCNEWFAGFTVRFGLNFVD"
FT   mRNA            1..1859
FT                   /evidence=EXPERIMENTAL
XX
SQ   Sequence 1859 BP; 609 A; 314 C; 355 G; 581 T; 0 other;
     aaacaaacca aatatggatt ttattgtagc catatttgct ctgtttgtta ttagctcatt        60
     cacaattact tccacaaatg cagttgaagc ttctactctt cttgacatag gtaacctgag       120
     .
     agaagctatg atcataacta taggttgatc cttcatgtat cagtttgatg ttgagaatac      1800
     tttgaattaa aagtcttttt ttattttttt aaaaaaaaaa aaaaaaaaaa aaaaaaaaa       1859
//
```

Fig. 1. EMBL flat file representation of database entry X56734. Note that the nucleotide sequence is only shown in part.

2.2. The Genome Reviews Database

2.2.1. Description of the Genome Reviews Database

Genome Reviews (http://www.ebi.ac.uk/GenomeReviews/) is a database that provides an up-to-date, standardized and comprehensively annotated view of the genomic sequence of selected organisms with completely deciphered genomes *(3)*. Currently, these include species from the superregna archaea and bacteria, including a number of plant parasites and symbionts. In addition, the eukaryote baker's yeast is available and *Arabidopsis* is scheduled for inclusion in early 2006. The primary data source is the EMBL database for prokaryotic genomes and the *Saccharomyces* Genome Database (SGD) *(5)* for baker's yeast. Genome Reviews is a secondary database and as such adds value to the archival data by offering curated and more up-to-date information. Additional data are incorporated from many data sources including the UniProtKB, the EMBL Nucleotide Sequence Database, the Gene Ontology (GO) Annotation (GOA) project *(6)* and InterPro *(7)*. In addition, annotations used inconsistently among the original submissions have been standardized, and deleted when the coverage is low; for example, this is the case when UniProtKB curators have identified a region that is annotated in the primary entry as a coding region but is unlikely to encode a real protein. Genome Reviews also introduces new features. Currently, these include tRNA features, identified with tRNA-Scan-SE *(8)*, and added when tRNA features are missing from the original submission; new CDS features when a UniProt protein entry can be mapped to a region in the genome that has not been annotated as a CDS. The latter procedure is also used to correct CDS features whose translation does not agree with the protein sequence in the corresponding UniProtKB entry. Furthermore, where a UniProt entry suggests that mature peptides are produced after cleavage, the features sig_peptide, trans_peptide, pro_peptide, peptide and mat_peptide are added below the corresponding CDS feature where appropriate describing functional chains and signal sequences.

In Genome Reviews, evidence tags are attached to most of the feature qualifiers. An evidence tag contains a cross-reference to the database and identifier within that database that specifies the source of the information that was used for that qualifier. The use of evidence tags is illustrated in **Fig. 2**.

2.2.2. Accessing the Genome Reviews Database

Genome Reviews are accessible through the Genome Reviews browser and Ensembl-style graphical interface *(9)* providing a zoomable graphical view of all chromosomes and plasmids represented in the database. The location and

```
FT    CDS              complement(2438999..2439730)
FT                     /codon_start=1
FT                     /gene="zipA {UniProtKB/Swiss-Prot:Q9PAG1}"
FT                     /locus_tag="Xf2557 {UniProtKB/Swiss-Prot:Q9PAG1}"
FT                     /product="Cell division protein zipA homolog
FT                     {UniProtKB/Swiss-Prot:Q9PAG1}"
FT                     /function="protein binding {GO:0005515}"
FT                     /biological_process="barrier septum formation {GO:0000917}"
FT                     /biological_process="cell cycle {GO:0007049}"
FT                     /cellular_component="inner membrane {GO:0019866}"
FT                     /cellular_component="integral to membrane {GO:0016021}"
FT                     /protein_id="AAF85354.1 {EMBL:AE003849}"
FT                     /db_xref="GO:0000917 {GOA:Q9PAG1}"
FT                     /db_xref="GO:0005515 {GOA:Q9PAG1}"
FT                     /db_xref="GO:0007049 {GOA:Q9PAG1}"
FT                     /db_xref="GO:0016021 {GOA:Q9PAG1}"
FT                     /db_xref="GO:0019866 {GOA:Q9PAG1}"
FT                     /db_xref="HOGENOM:HBG405294 {HogenProt:Q9PAG1}"
FT                     /db_xref="HSSP:1F46 {UniProtKB/Swiss-Prot:Q9PAG1}"
FT                     /db_xref="InterPro:IPR007449 {UniProtKB/Swiss-Prot:Q9PAG1}"
FT                     /db_xref="InterPro:IPR011919 {UniProtKB/Swiss-Prot:Q9PAG1}"
FT                     /db_xref="UniParc:UPI000013C3EB {EMBL:AAF85354}"
FT                     /db_xref="UniProtKB/Swiss-Prot:Q9PAG1 {EMBL:AE003849}"
FT                     /transl_table=11
FT                     /translation="MSDVTLLRIGIAIVGILFVAAVFFFSTPKTSAHRVRTKKEEPPRE
FT                     RREPMLSTEADNSPPQGVDEVPASVSQQQVNPEANKPGEVQLGKRPTNHFDKIILLFVA
FT                     AKAEHTLRGEDIVVAAEKTGMIFGYMNVFHRLVEGYPEHGPIFSMASILKPGSFDMANI
FT                     REMQIPAISFFLTLPAPMTALDAWEKMLPTVQRMAELLDGVVLDESRNALGRQRIAHIR
FT                     DELRAYDRQQQVPPLIKNSRW"
```

Fig. 2. The coding sequence (CDS) feature for the *zip*A gene of *Xylella fastidiosa* strain 9a5c. Evidence tags (here in bold typeface) are added in curly brackets to most feature qualifiers and reference the databases and the identifiers within those databases from which the information was derived.

structure of all genes are shown, and the distribution of features throughout the sequence is displayed and hyperlinked to pages displaying information about the gene, gene structure, transcript and protein. Searches can be performed using a query interface on the Genome Reviews homepage or in Integr8 (*see* **Subsection 2.3.**). On the Genome Reviews homepage, you will also find a link to the Genome Reviews flat files, which can be downloaded from the EBI ftp server. The database can also be searched in Sequence Retrieval System (SRS, explained in **Subsection 3.1.**), which gives access to the flat file representation of the Genome Reviews data, and external databases cross-referenced.

2.3. The Integr8 Web Portal

2.3.1. Description of Integr8

The Integr8 Web portal *(3)* (http://www.ebi.ac.uk/integr8) has been developed to offer a consistent view of genes, their transcripts and their proteins in their genomic context, and provide a focal point through which relevant data can be identified and downloaded. The focus of Integr8 is on species whose genomes have been completely sequenced. For each species, Integr8 provides an overview of its biology and a detailed statistical analysis of its genome and proteome, represented using both textual and graphical displays. Comparative analysis between genomes is supported, while additional tools allow users to customize their own analyses. For the genes in each genome, a clearly structured view of gene products, homologues and neighbours is provided. The current coverage of Integr8 in the plant kingdom is restricted to *Arabidopsis thaliana*, although data concerning plant parasites such as *Agrobacterium tumerfaciens*, *Xanothomans campestris* and *Xylella fastidiosa*, and symbionts such as those from the genus *Rhizobium*, are also available.

Information available about each species includes a description, a list of recent publications and information about the components of the genome (such as their length, average GC content, codon usage and length of the CDSs they contain). In addition, a proteome set has been constructed for each species, using information derived from the UniProtKB. These sets may contain additional/altered protein sequences (compared with those in the original submission of the annotated genome to the EMBL/Genbank/DDBJ databases), but redundant sequences (that may have been submitted by other scientists working on this species) are filtered out. The sets are available for download and are also used to prepare a statistical analysis of each proteome available on the Integr8 site. These analyses are performed through the application resources that can be used to classify individual proteins, such as InterPro *(7)*, CluSTr *(10)* and GO *(11)*.

InterPro is a database of protein domains, families and functional sites (and of computational methods used to identify these in previously unclassified sequence). CluSTr is a database of hierarchically arranged clusters of proteins defined by the degree of mutual sequence similarity among their members; it can be used to co-classify proteins containing no recognized protein domains. The GO is a controlled vocabulary for the annotation of gene products. GO is a useful tool for the analysis of proteomes, as different groups can agree to annotate to the GO standard, certain tools exist for automatically converting other types of annotation to GO, and specific terms can be replaced by more general terms to allow the integration of annotations made with varying levels of precision. Intetgr8 uses these (and other resources) to describe and compare complete proteomes in terms of the proteins they contain and to allow the user

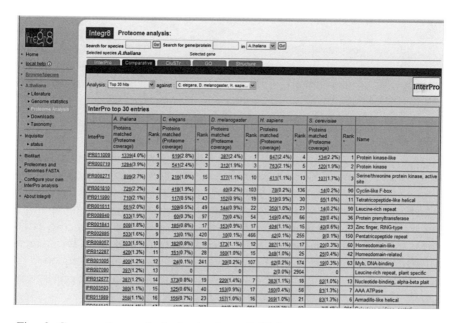

Fig. 3. Comparative analysis in the Integr8 Web portal, showing the frequency of certain well-represented InterPro domains in the complete proteomes of several eukaryotic species.

to specify customized data sets for download. An example of the type of results that can be accessed is shown in **Fig. 3**.

2.3.2. Integr8 Components

2.3.2.1. INTEGR8OR

Information about individual genes can be accessed through the "Integr8or", component of the Integr8 portal, which provides a clear view of the products of each gene, and the annotations (and cross-referencing entries in external resources) that have been associated with each. Moreover, the Integr8or provides a schematic visual representation of the location of each gene in the context of its neighbours and allows for the detection of syntenic regions by supporting the customizable expansion of this view into a comparative display of multiple genomes centred on putative orthologues in each species. In this display, each gene is colour-coded according to the domain architecture of its protein product, allowing the quick comparison of genes not yet assigned a descriptive name. A typical view will be generated in **Subsection 3.3**.

2.3.2.2. INQUISITOR

Another component of Integr8, the "Inquisitor", provides an expert system for protein sequence analysis, running a number of different sequence analysis tools to identify each sequence in Integr8 or, if the sequence is novel, to classify it as fully as possible.

3. Methods

In this section, we show a number of ways to access information from the EMBL and the Genome Reviews databases and how we can obtain additional information from a host of other databases through cross-references in EMBL and Genome Reviews. All databases and tools are available from the main EBI Services Web page (*see* **Note 2**). Here, we will explain the use of SRS *(12)* to retrieve information from databases (*see* **Subsection 3.1.**), FASTA sequence similarity searching *(13)* to find sequences similar to a query sequence (*see* **Subsection 3.2.**) and Integr8 to explore gene and gene order conservation in a number of related species (*see* **Subsection 3.3.**).

3.1. Database Browsing

SRS is a powerful tool for retrieving molecular biology data at the EBI. It has two main functions: a data retrieval system (more than 200 databases are currently incorporated) and a data analysis applications server providing applications for the analysis of proteins and nucleic acids. The system is available through a uniform query interface on the Web (http://srs.ebi.ac.uk). This interface allows you to:

- perform simple and complex queries across one or several databases,
- view your results in different formats,
- create your own views for your results,
- save results to file or to a browser,
- launch applications on results, and
- link results to different databases.

The tasks that can be performed include database searching against nucleic acid or protein databases, similarity searches (BLAST *(14)* and FASTA), and generation of multiple alignments (ClustalW) *(15)*.

In this tutorial, we will search SRS for EMBL database entry X56734. To demonstrate some of the power of SRS, we will not only search the EMBL database but include UniProtKB in the search as well. To achieve that, we follow the following steps:

- Start up an Internet browser and point it to URL http://srs.ebi.ac.uk/.
- At the top of the page, there are a number of tabs. Choose library page, which will take you to the page that lists all available databanks. Tick EMBL and UniProtKB.
- In the "Quick Search" box at the top, enter accession number X56734 and click on the "Quick Search" button next to it.

The above steps are shown in **Fig. 4**. The search returns two entries, one from EMBL (TRBG361) and one from UniProtKB (BGLS_TRIRP). Each entry can be viewed by clicking on the hyperlink. When you click on the EMBL:TRBG361 link, you are taken to the EMBL entry. You can display the entry in two formats, the default is the most user-friendly, but you may also wish to view the entry in EMBL flat file format as shown in **Fig. 1** by clicking on the Text Entry link at the top of the page. To go to the UniProtKB entry, you can either click on the "Back" button in your browser and select the UniProtKB:BGLS_TRIRP link or select the "Next Entry" button at the

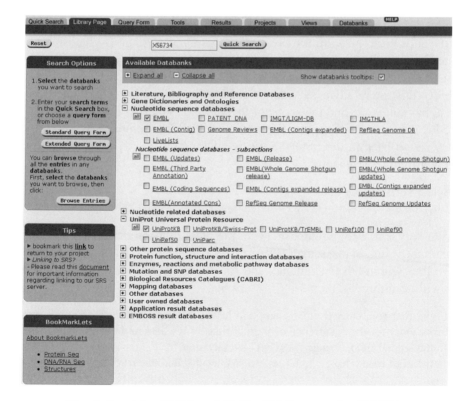

Fig. 4. Searching EMBL and UniProtKB for accession X56734.

top of the page. Let us concentrate on the EMBL entry for now. When you compare the flat file in **Fig. 1** with the SRS page, apart from the differences in format, you will notice that there are many hyperlinks. These provide direct links to many of the databases present in SRS. For example, you can retrieve the abstract of the associated publication by clicking on the Pubmed link, view the taxonomic classification of the organism by clicking on the taxon link, and find functional information on the protein encoded by the CDS in a range of databases. For example, if you click on the InterPro link, it will take you to the InterPro page for entry IPR001360, where you will find further information on the glycoside hydrolase family 1, including hyperlinks to a number of other databases, including protein domain database. At the top left of the page, you will find a link to the InterPro homepage, which enables you to find out more about that particular database. This example shows that a quick search for an EMBL database entry in SRS opens the way to a lot of additional information stored in other databases. You can achieve this by specifying multiple databases before you start your search, but even if you only search in one database, you can access other databases through hyperlinks in the result page.

If you want to perform a more general search, e.g., plant β-glucosidases in general, rather than a specific one, SRS provides two query forms that allow you to do more advanced searches. We will use the standard query form to achieve this; more advanced queries can be formulated in the extended query form, but that is beyond the scope of this tutorial. You will find extensive help in SRS if you wish to explore this and any other feature not covered here. To start, we go back to the library page by clicking on the appropriate tab at the top and follow these steps:

- Select EMBL in the library page.
- Click on the "Standard Query Form" button in the "Search Options" box on the left.

We will search for β-glucosidase in the plant division of EMBL, PLN, and we are interested in messenger RNAs rather than genomic DNA sequences. We have four search boxes available, and for the search criteria we have defined, we will need three of them.

- In the first box, enter β-glucosidase and leave "AllText" selected. This will search every entry for the occurrence of β-glucosidase anywhere in the entry.
- In the second field, change "AllText" to "Division" and enter PLN.
- In the third field, change "AllText" to "Molecule" and enter mRNA.

Figure 5 shows the form after we have entered our search criteria. Click on "Search" to perform the query in SRS. The search returns (at the time of

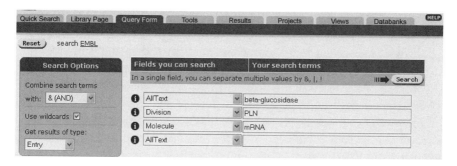

Fig. 5. Using the SRS Standard Query Form to search EMBL for plant mRNA's encoding β-glucosidases.

writing) 93 results. If we had not restricted our search to the EMBL PLN division and mRNA, the query would have found over 1300 entries. Using the search forms with the appropriate search terms is the best way to get a sensible number of results. Despite that, the number of results are still too high, you should think of additional ways to refine your search.

3.2. FASTA Sequence Similarity Search

Sequence similarity search tools allow the user search an input nucleotide or protein sequence against a nucleotide or protein sequence database and find sequences similar to the input sequence. In this way, the identity of anonymous sequences, e.g., newly sequenced cDNA clones or newly predicted proteins, can be established if one or more close matches to known sequences are found. The most popular are the families of BLAST and FASTA tools. The algorithms used in BLAST and FASTA are different; in general, BLAST tends to produce its results faster, but FASTA produces the best local alignments. Both BLAST and FASTA similarity searches are available at the EBI. Both BLAST and FASTA have very similar Web interfaces with extensive help pages, but we will only cover FASTA in this chapter.

FASTA (pronounced FAST-Aye) stands for FAST-All, reflecting the fact that it can be used for a fast protein comparison or a fast nucleotide comparison. This program achieves a high level of sensitivity for similarity searching at high speed. This is achieved by performing optimized searches for local alignments using a substitution matrix, in this case a DNA identity matrix.

We will consider the sequence from our previous example, EMBL entry X56734, and look for sequences that are similar in the EMBL database. As X56734 is present in the EMBL database, we will expect to find a sequence that is a perfect match to our test sequence. The FASTA Web interface can be found at http://www.ebi.ac.k/fasta33/. The form consists of two parts, the top part allows us to select the program parameters, and in the bottom part, we can either paste our test sequence or upload it from a file. Let us start with the sequence. We can copy this from SRS by searching EMBL for X56734 as explained in the SRS tutorial, scroll down to the sequence, select the sequence with the mouse and select "Copy" from the "Edit" menu in your browser. As this particular sequence has a poly(A) tail, it is advisable to exclude this from your selection. This is particularly important if the query sequence does not have many close hits to sequences in the database. Excluding the poly(A) tail will remove any bias towards sequences containing stretches of A's, or T's if a match is on the complementary strand. Next, paste the sequence in the appropriate FASTA field and change PROTEIN to DNA/RNA at the top of the box.

In the top box, we will need to make a few changes to the default selections. The most important one is the database selection. Because we will be searching in EMBL, under DATABASES, "Protein" should be changed to "Nucleic Acid," and we will also limit our search to plant sequences by changing the "EMBL" to "PLANT". We can choose whether we want to get the result by email, in which case we have to supply our email address, or in a browser window, in which case we change email to interactive under RESULTS. We leave all other parameters as is (concerning parameters *see* **Note 2**). **Figure 6** shows the completed FASTA form. Press the "Run Fasta3" button, which will enter your submission to the EBI. If you opted for your results to be sent by email, you will normally receive an email within a few minutes. Otherwise, your results will appear in your browser.

Figure 7 shows the part of the results received by email and lists the 20 best hits. These results are read as follows: each line starts with the database name and identifier in that database, followed by the accession number (with the version number appended with a dot). Next, we have the first 20 characters of the description of that entry followed by the length of its sequence and whether the hit is on the forward (f) or reverse (r) strand and finally the optimized score, bits score and the E-value, which determine the quality of the hit. Not surprisingly, the best hit – the one with the lowest E-value – is our input sequence. The E-value is given at the end of each line. A perfect match is represented by a value of zero; close matches have E-values very close to zero, which is the case for all of the other matches in the list. It is not always easy to decide whether larger E-values represent a true match, and in these cases, it is best to examine the alignments.

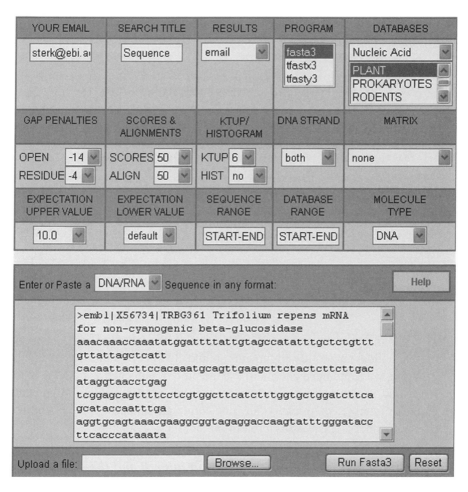

Fig. 6. FASTA search of EMBL sequence X56734 in the EMBL plant division.

In our analysis, the best score is a perfect match, which we can easily verify by looking at the first alignment. The alignment over the first 60 nucleotides is shown in **Fig. 8A**. In the space between the query sequence and the database sequence, perfect matches are represented by a colon (:), non-matching nucleotides by a space. To produce optimal local alignments, FASTA may introduce gaps in either sequence represented by one of more dash characters (-). You see this in the second alignment, partly shown in **Fig. 8B**. In coding sequence, the more gaps that are introduced, in particular when they are not a multiple of three (codon length) and the more mismatches, the more

```
The best scores are:                                       opt bits E(318592)
EM_PL:TRBG361 X56734.1 Trifolium repens mRNA f  (1859)  [f]  9150 1435.3        0
EM_PL:AJ630653 AJ630653.1 Cicer arietinum mRNA  (1769)  [f]  3735  592.9 9.4e-167
EM_PL:CAR5950 AJ005950.3 Cicer arietinum parti  (1580)  [f]  3068  489.1 1.7e-135
EM_PL:PS50201 U50201.2 Prunus serotina prunasi  (2055)  [f]  2603  416.8 9.5e-114
EM_PL:AF411131 AF411131.1 Prunus serotina prun  (1819)  [f]  2596  415.7 2.1e-113
EM_PL:PA39228 U39228.1 Prunus avium beta-gluco  (1731)  [f]  2595  415.5 2.3e-113
EM_PL:AF411009 AF411009.1 Prunus serotina prun  (1911)  [f]  2575  412.4   2e-112
EM_PL:AF221527 AF221527.2 Prunus serotina puta  (2059)  [f]  2567  411.2 4.6e-112
EM_PL:TRBG104 X56733.1 T.repens mRNA for cyano  (1690)  [f]  2467  395.6 2.3e-107
EM_PL:AY766303 AY766303.1 Dalbergia nigrescens  (1964)  [f]  2364  379.6 1.5e-102
EM_PL:AF221526 AF221526.1 Prunus serotina prun  (1960)  [f]  2281  366.7 1.2e-98
EM_PL:AF163097 AF163097.1 Dalbergia cochinchin  (1957)  [f]  2234  359.4 1.8e-96
EM_PL:PS26025 U26025.2 Prunus serotina amygdal  (1915)  [f]  2207  355.2 3.4e-95
EM_PL:AB088027 AB088027.1 Camellia sinensis mR  (1729)  [f]  1884  304.9 4.6e-80
EM_PL:AK070962 AK070962.1 Oryza sativa (japoni  (1853)  [f]  1828  296.2 1.9e-77
EM_PL:AF170087 AF170087.1 Cucurbita pepo silve  (1669)  [f]  1756  285.0 4.6e-74
EM_PL:AF480476 AF480476.1 Hevea brasiliensis P  (1831)  [f]  1661  270.2 1.3e-69
EM_PL:BT013386 BT013386.1 Lycopersicon esculen  (1763)  [f]  1627  264.9    5e-68
EM_PL:AB024024 AB024024.1 Arabidopsis thaliana (73921)  [f]  1592  260.3 1.2e-66
EM_PL:AK117809 AK117809.1 Arabidopsis thaliana  (1687)  [f]  1575  256.8 1.4e-65
```

Fig. 7. The 20 top hits from a FASTA similarity search with X56734 in the plant division of the EMBL database.

unlikely it is that the two sequences are related. Although there are exceptions to this, nucleotide–nucleotide alignments should always be examined closely. In our example, we could have chosen to use the protein sequence as FASTA input sequence and search against UniProt. In general, protein–protein alignments are easier to interpret than nucleotide–nucleotide alignments. This depends of course on whether a trustworthy protein prediction is available.

3.3. Graphical Representations of Genes in Integr8 and the Genome Reviews Browser

Sequence similarity search tools such as BLAST and FASTA are useful gene analysis tools, but analyses are usually done on a per gene basis. Now that many complete genome sequences have become available, we have the opportunity to make genome comparisons, look for similarities beyond a single gene, e.g., conservation of genes (orthologues) and gene order (synteny), and postulate preservation of function where this occurs.

To illustrate this, we will use Integr8or to compare a section of the genomic sequence around gene *zip* A from two strains (strain 9a5c and strain Temecula1/ATCC 700964) of *X. fastidiosa* (a plant parasite that causes a range of economically important plant diseases), *Xanthomonas axonopodis citri* (which causes citrus canker) and *Xanthomonas campestris campestris* (which

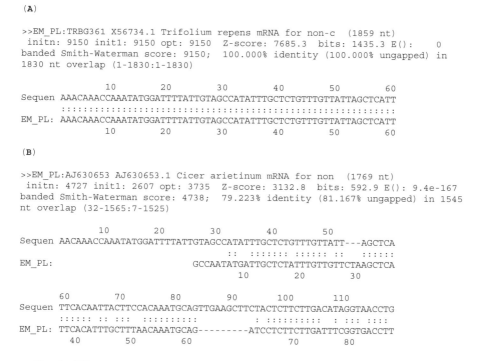

Fig. 8. **(A)** Alignment of query sequence X56734 with itself results in a perfect alignment. **(B)** Alignment of query sequence X56734 with EMBL entry AJ630653 (the second best match). The FASTA output clearly shows matches and mismatches and has introduced gaps in both sequences to produce the optimal local alignment.

causes black rot in certain crucifers). In addition, we will have a look at this region in one of these species in the Genome Reviews browser.

- Start by opening URL http://www.ebi.ac.uk/integr8/.
- In the "Quick Search" box, leave the first field blank, enter zipA in the gene/protein field and select "specify…" in the third field. Click on "Go!"
- On the next screen, we have to specify the species. Click on the letter "X" and select both *X. fastidiosa* strains by clicking on the name. Click on the word "continue" at the top.

The search returns two results, as shown in **Fig. 9**. There are hyperlinks to a description of the organism, Integr8or (i8 icon) and the Genome Reviews Browser (GR icon), and links to the UniProtKB pages for the *zip*A-encoded proteins.

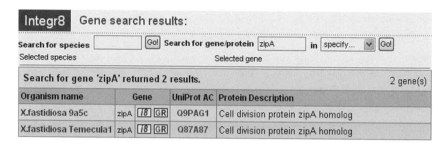

Fig. 9. Integr8 results for search of zipA in *Xylella* strains.

We will use Integr8or to view the neighbourhood of putative *zip*A ortho-logues in the species represented in Integr8.

- Click on one of the Integr8or (i8) icons under Gene, which should take you to the Integr8or page with the Gene tab selected.
- Click on the Protein tab; you will see a table containing information related to the *zip*A-encoded protein and hyperlinks to external databases. At the top of the table, next to the Protein name, you will see two icons, "O" (orthologue) and "P" (paralogue). Click on the orthologue icon, which will return a list of similar sequences in other species.

Integr8	Integr8or

Search for species [ ] [Go!] Search for gene/protein [ ] in [X.fastidiosa 9a5c ▾] [Go!]
Selected species *X.fastidiosa 9a5c*   Selected gene *zipA* [GR]

| Gene | Transcript | Protein | Results | Context | History |

Comparative genome view ▽ scroll △

X.fastidiosa 9a5c	X.fastidiosa Temecula1	X.campestris ATCC 33913	X.axonopodis
Xf2550	PD1934	asnS	XAC1617
Xf2551	gyrA	XCC1562	asnS
Xf2552	mtnA	rpsF	XAC1619
mtnA	PD1937	rpsR	rpsF
Xf2554	lysS	rpll	rpsR
Xf2555	PD1939	XCC1566	rpll
Xf2556	ligA	smc	smc
zipA	zipA	zipA	zipA
Xf2558	smc	hisC	XAC1625
rpll	rpll	lig1	hisC
rpsR	rpsR	lysS	lig1
rpsF	rpsF	XCC1572	lysS
Xf2562	PD1946	XCC1573	XAC1629
asnS	asnS	gyrA	XAC1630
Xf2564	PD1948	gcd	gyrA

The table cells represent genes. Colours represent InterPro domain architectures.

Fig. 10. A comparative view of regions around the *zip*A gene in two *Xylella fastidiosa* strains and two *Xanthomonas* species. Observe the strong gene and gene order preservation (note that the gene order is reversed for the *Xanthomonas* species).

- Select the top three species in the table (those with the highest Z-score) in the "Select" column and click on "Comparative view" at the top of the table.

The results are displayed in **Fig. 10**. Examining the results show that *zip*A and many of the surrounding genes have been conserved in all of the strains. Colours represent protein architecture according to InterPro and help identify potential synteny for genes not yet assigned a descriptive name.

We can get a detailed view of chromosomes in the Genome Reviews browser. From the page displayed in **Fig. 11**, we click on the "GR" icon (next to Selected gene zipA), which launches a zoomable view of the chromosome of *X. fastidiosa* strain 9a5c centred around *zip*A. Moving the mouse over genes will display menus that allow the user to explore the gene, transcript and encoded protein in more detail. Context-based access to external databases is available, and whenever needed, help is available on every page in the Genome Reviews browser.

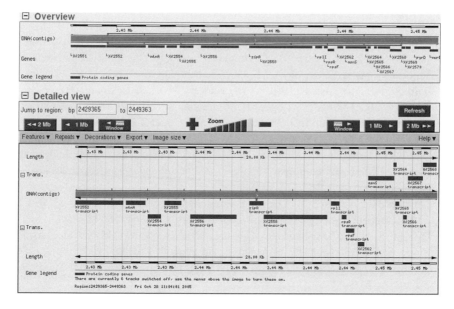

Fig. 11. The Genome Reviews browser displaying the neighbourhood of gene *zip*A in *Xylella fastidiosa* strain 9a5c.

Notes

1. EMBL-SVA *(4)* server is a repository of all entries that have been made public since release 1 of the EMBL database. It comprises more than 100 million entries and includes entries pre-dating the first electronic release of the database in 1982.
2. The main EBI Services Web page (http://www.ebi.ac.uk/services/) provides access to all available databases and tools. The 2can Web site (http://www.ebi.ac.uk/2can/) is the EBI bioinformatics educational resource, which provides information about tutorials on many of the databases and services hosted by the EBI. When performing a FASTA search, certain parameters affect the score that is calculated for an alignment. For example, you can raise the value for the gap penalty to favour alignments with a lower number of gaps. Have a look at the FASTA help pages and the 2can tutorial on FASTA for a detailed explanation on the use of the FASTA parameters.

Acknowledgements

Integr8 and Genome Reviews are funded by the European Commission as the TEMBLOR, contract-no. QLRI-CT-2001000015 under the RTD programme "Quality of Life and Management of Living Resources".

References

1. Kanz, C., Aldebert, P., Althorpe, N., Baker, W., Baldwin, A., Bates, K., Browne, P., van den Broek, A., Castro, M., Cochrane, G., et al. (2005) The EMBL Nucleotide Sequence Database. *Nucleic Acids Res.* 33, D29–D33.
2. Bairoch, A., Apweiler, R., Wu, C. H., Barker, W. C., Boeckmann, B., Ferro, S., Gasteiger, E., Huang, H., Lopez, R., Magrane, M., et al. (2005) The Universal Protein Resource (UniProt). *Nucleic Acids Res.* 33, D154–D159.
3. Kersey, P., Bower, L., Morris, L., Horne, A., Petryszak, R., Kanz, C., Kanapin, A., Das, U., Michoud, K., Phan, I., et al. (2005) Integr8 and Genome Reviews: integrated views of complete genomes and proteomes. *Nucleic Acids Res.* 33, D297–D302.
4. Leinonen, R., Nardone, F., Oyewole, O., Redaschi, N., and Stoehr, P. (2003) The EMBL sequence version archive. *Bioinformatics* 19, 1861–1862.
5. Dwight, S. S., Balakrishnan, R., Christie, K. R., Costanzo, M. C., Dolinski, K., Engel, S. R., Feierbach, B., Fisk, D. G., Hirschman, J., Hong, E. L., et al. (2004) Saccharomyces genome database: underlying principles and organisation. *Brief Bioinform.* 5, 9–22.
6. Camon, E., Barrell, D., Lee, V., Dimmer, E., and Apweiler, R. (2003) Gene Ontology Annotation Database – an integrated resource of GO annotations to UniProt Knowledgebase. *In Silico Biol.* 4, 0002.

7. Mulder, N. J., Apweiler, R., Attwood, T. K., Bairoch, A., Bateman, A., Binns, D., Bradley, P., Bork, P., Bucher, P., Cerutti, L., et al. (2005) InterPro, progress and status in 2005. *Nucleic Acids Res.* 33, D201–D205.

8. Lowe, T. M., and Eddy, S. R. (1997) tRNAscan-SE: a program for improved detection of transfer RNA genes in genomic sequence. *Nucleic Acids Res.* 25, 955–964.

9. Hubbard, T., Andrews, D., Caccamo, M., Cameron, G., Chen, Y., Clamp, M., Clarke, L., Coates, G., Cox, T., Cunningham, F., et al. (2005) Ensembl 2005. *Nucleic Acids Res.* 33, D447–D453.

10. Petryszak, R., Kretschmann, E., Wieser, D., and Apweiler, R. (2005) The predictive power of the CluSTr database. *Bioinformatics* 21, 3604–3609.

11. Gene Ontology Consortium (2004) The Gene Ontology (GO) database and informatics resource. *Nucleic Acids Res.* 32, D258–D261.

12. Zdobnov, E. M., Lopez, R., Apweiler, R., and Etzold, T. (2002) The EBI SRS server – recent developments. *Bioinformatics* 18, 368–373.

13. Pearson, W. R. (1994) Using the FASTA program to search protein and DNA sequence databases. *Methods Mol. Biol.* 24, 307–331.

14. Altschul, S. F., Madden, T. L., Schaffer, A. A., Zhang, J., Zhang, Z., Miller, W., and Lipman, D. J. (1997) Gapped BLAST and PSI-BLAST: a new generation of protein database search programs. *Nucleic Acids Res.* 25, 3389–3402.

15. Thompson, J. D., Higgins, D. G., and Gibson, T. J. (1994) CLUSTAL W: improving the sensitivity of progressive multiple sequence alignment through sequence weighting, position-specific gap penalties and weight matrix choice. *Nucleic Acids Res.* 22, 4673–4680.

2

Using GenBank

David Wheeler

Summary

GenBank(R) is a comprehensive database of publicly available DNA sequences for more than 205,000 named organisms and for more than 60,000 within the embryophyta, obtained through submissions from individual laboratories and batch submissions from large-scale sequencing projects. Daily data exchange with the European Molecular Biology Laboratory (EMBL) in Europe and the DNA Data Bank of Japan ensures worldwide coverage. GenBank is accessible through the National Center for Biotechnology Information (NCBI) retrieval system, Entrez, which integrates data from the major DNA and protein sequence databases with taxonomy, genome, mapping, protein structure, and domain information and the biomedical journal literature through PubMed. BLAST provides sequence similarity searches of GenBank and other sequence databases. Complete bimonthly releases and daily updates of the GenBank database are available through FTP. GenBank usage scenarios ranging from local analyses of the data available through FTP to online analyses supported by the NCBI Web-based tools are discussed. To access GenBank and its related retrieval and analysis services, go to the NCBI Homepage at http://www.ncbi.nlm.nih.gov.

Key Words: NCBI; Entrez; DNA sequence; BLAST; MegaBLAST.

1. Introduction

This chapter is designed to serve as a practical guide to using the GenBank nucleotide sequence database in biological research. The chapter is divided into five subsections including *Introduction*. **Subsection 2** provides a summary of the content of the database and describes various methods of access. **Subsection 3** presents several common 'methods' that can be applied to

From: *Methods in Molecular Biology, vol. 406: Plant Bioinformatics: Methods and Protocols*
Edited by: D. Edwards © Humana Press Inc., Totowa, NJ

the GenBank data. The notes in **Subsection 4** provide details ranging from summary tables and tips to computer code for data parsing and downloads. **Subsection 5** provides email addresses to use when submitting data to GenBank and getting help on using the data.

2. Materials

The sections below describe the GenBank database, the release formats and release cycle, composition, methods of access, and integration with other biological resources.

2.1. The GenBank Database

GenBank *(1,2)* is a comprehensive public database of nucleotide sequences and supporting bibliographic and biological annotation. GenBank is maintained and distributed by the National Center for Biotechnology Information (NCBI), a division of the National Library of Medicine (NLM), at the US National Institutes of Health (NIH) in Bethesda, MD. NCBI builds GenBank from several sources including the submission of sequence data from authors and from the bulk submission of expressed sequence tag (EST), genome survey sequence (GSS), whole-genome shotgun (WGS), and other high-throughput data from sequencing centers. The US Office of Patents and Trademarks also contributes sequences from issued patents. GenBank, the European Molecular Biology Laboratory (EMBL) *(3)*, and the DNA Databank of Japan (DDBJ) *(4)* comprise the International Nucleotide Sequence Database Collaboration (INSDC) (http://www.insdc.org/) whose members exchange data daily to ensure a uniform and comprehensive collection of sequence information.

2.1.1. The GenBank Release Formats and Release Cycle

NCBI makes GenBank available at no cost through FTP and the Web-based Entrez search and retrieval system *(5)*. The FTP release consists of a mixture of compressed and uncompressed ASCII text files, 944 in release 153, containing sequence data and indices that cross-reference author names, journal citations, gene names, and keywords to individual GenBank records. For convenience, the GenBank records are partitioned into 18 divisions (*see* **Note 1**) according to source organism or type of sequence. Records within the same division are packaged as a set of numbered files, so that records from a single division may be contained in a series of many files – 512 in the case of the EST division in release 153. The full GenBank release is offered in two formats: the GenBank 'flatfile' format (*see* **Note 2**) and the more structured and compact Abstract Syntax Notation One

(ASN.1) (http://www.ncbi.nlm.nih.gov/Sitemap/Summary/asn1.html) format used by NCBI for internal maintenance. Full releases of GenBank are made every 2 months beginning in the middle of February each year. Between full releases, daily updates are provided on the NCBI FTP site (ftp://ftp.ncbi.nih.gov/genbank/ and ftp://ftp.ncbi.nih.gov/ncbi-asn1/). The Entrez system always provides access to the latest version of GenBank including the daily updates.

2.1.2. The Composition of GenBank

From its inception, GenBank has doubled in size about every 18 months. Release 153, in April 2006, contained over 61 billion nucleotide bases from more than 56 million individual sequences. Contributions from WGS projects supplement the data in the traditional divisions to bring the total beyond 130 GB. The number of eukaryote genomes for which coverage and assembly are significant continues to increase as well, with over 125 such assemblies now available. Database sequences are classified and can be queried using a comprehensive sequence-based taxonomy *(6)* developed by NCBI in collaboration with EMBL and DDBJ with the assistance of external advisers and curators. Over 225,000 named species are now represented in GenBank, and new species are being added at the rate of over 3000 per month. Detailed statistics for the current release may always be found in the GenBank release notes (ftp://ftp.ncbi.nih.gov/genbank/gbrel.txt).

2.1.3. Sources of Plant Sequences

GenBank release 153 contains over 16 million plant sequence records from some 67,000 plant species. Although these plant sequences are of many types such as the sequence-tagged site (STS) sequences, found in the STS division of GenBank, high-throughput genomic sequences, found in the high-throughput genomic (HTG) division, and high-throughput cDNA sequences, found in the high-throughput cDNA (HTC) division, more than 99% of the plant sequences in GenBank fall into one of three divisions: PLN, EST, or GSS (*see* **Note 3**); the bulk of these data comprises genomic assemblies or batches of ESTs.

Genomic assemblies include that of the *Arabidopsis thaliana* genome produced by the Arabidopsis Genome Initiative *(7)*, the assembly of *Oryza sativa* (japonica cultivar group), resulting from the International Rice Genome Sequencing Project *(8)*, and that of *O. sativa* (indica cultivar group) produced by the Chinese Academy of Sciences in Beijing, China *(9)*.

Data resulting from large-scale EST *(10,11,12)* sequencing projects for over 70 plants have been deposited into GenBank. A list of large-scale plant EST sequencing projects with links is available at http://www.ncbi.nlm.nih.gov/ genomes/PLANTS/PlantList.html.

NCBI encourages the submission of sequencing data ranging in complexity from a transcript sequence annotated with a single-coding region to sets of aligned sequences supporting population or phylogenetic studies or large-scale genomic assemblies with detailed annotations. For a description of some submission methods and details of the submission process, see **Note 4**.

2.1.4. Annotations Found in GenBank Records

Each GenBank entry includes a concise description of the sequence, the scientific name, and taxonomy of the source organism, bibliographic references, and a table of biological features (*see* **Note 5**). Annotation is best for GenBank records in the PLN division, whereas records in divisions such as EST and GSS contain minimal annotation or may contain no annotation at all.

2.1.5. Integration of GenBank Data with Other Resources

More than 25 plant species for which more than 70,000 EST sequences have been deposited into GenBank have been incorporated into the UniGene database *(13)*, where they are combined with other transcript sequences in GenBank and partitioned into over 200,000 gene-oriented clusters. The UniGene clusters are subjected to further analysis to link them to the sequence-tagged sites in UniSTS, a non-redundant derivative of the data in the GenBank STS division. Links are also made to the genes in Entrez Gene, gene homologs in HomoloGene and proteins in Entrez Protein wherever possible. As a consequence, Entrez (*see* **Subsection 2.2.1.**) can be used to match a GenBank EST accession number (*see* **Note 6**) to a gene location, a protein sequence, and homologous genes in many organisms.

ESTs arising from several plant species, as well as the UniGene clusters derived from them, are aligned to the well-assembled *A. thaliana* and *O. sativa* genomes. These alignments make an important link between EST data and assembled, well-annotated genomic sequence. In addition to these ESTs, almost four million single-nucleotide polymorphisms *(14)* have been mapped to the *O. sativa* genome at NCBI.

2.2. Accessing GenBank

GenBank data can be downloaded and put to use on any conventional computer platform. For interactive use through Entrez on the NCBI Web pages, a Web browser is the only requirement. Entrez may also be accessed through scripts using the Entrez programming utilities (E-utilities). Bulk downloads from the NCBI FTP site may be made using a Web browser or an FTP program.

2.2.1. Interactive Access with Entrez

The sequence records in GenBank are accessible through Entrez (http://www.ncbi.nlm.nih.gov/Entrez/) *(5,15)*, a robust and flexible database retrieval system that covers over 30 biological databases containing, in addition to millions of individual DNA and protein sequences, genome maps, sequences, and alignments from population, and phylogenetic studies, sequence sets from environmental (ENV) samples, gene expression data, the NCBI taxonomy *(6)*, protein domain information, protein structures from the Molecular Modeling Database, MMDB *(16,17)*, and MEDLINE references through PubMed and PubMed Central *(18,19)*. The Entrez Nucleotide sequence database contains the GenBank data but includes data from various sources other than GenBank – the GenBank data may be selectively accessed within Entrez using query limitations (*see* **Note 7**).

Records within the Entrez system are linked to other records both within and across databases. An example of a simple linkage is that between a GenBank sequence record and the PubMed abstract for the paper listed in the 'Journal' section (*see* **Note 2**) of the GenBank record. Computational linkages are also made between nucleotide and protein sequences, such as those based on sequence similarities discovered using BLAST *(15,20)*. In addition, records available in Entrez may offer LinkOuts *(21)* that lead to external plant-specific databases *(22–25)* or to other resources such as the Gene Ontology *(26)*. Queries on the Entrez databases are made with simple text words, combined using Boolean logic, and either limited to a particular record field (*see* **Note 7**) or applied to all fields. A Web-based service called Batch Entrez allows bulk sequence downloads specified by an arbitrary set of GenBank identifiers supplied within a local file.

2.2.2. Scripted Access Through Entrez with the Entrez Programming Utilities

Entrez queries of GenBank and downloads of individual records or sets of records may be made through the Entrez system from scripts through a set of server-side utilities called the Entrez Programming Utilities *(27)* (*see* **Note 8**).

2.2.3. Bulk Downloads of GenBank Through FTP

The full bimonthly GenBank release and the daily updates, which also incorporate sequence data from EMBL and DDBJ, are available by anonymous

FTP from NCBI in both flatfile and ASN.1 formats (ftp://ftp.ncbi.nih.gov/ genbank/ and ftp://ftp.ncbi.nih.gov/ncbi-asn1/). A mirror site exists at the University of Indiana (ftp://bio-mirror.net/biomirror/genbank/). As described in **Subsection 2.1.1**, the full release in flatfile format is distributed as a set of compressed files in the directory 'genbank'. To download a single compressed GenBank file using a command line FTP client, 'ftp' to ftp.ncbi.nih.gov, log in as 'anonymous' and give your email address as password. If there are problems connecting, check the client's 'active/passive' mode settings (*see* **Note 9**). For a sample FTP session, (*see* **Note 10**).

For most purposes, a download of the entire GenBank databases is not required, as standard sequence-similarity searches, such as BLAST, may be performed remotely on the GenBank data at NCBI. However, if a local copy of GenBank is required, there are two major considerations: local storage space and download time.

As of GenBank release 153, the combined size of the compressed flatfile format files is about 33 GB, and the compressed ASN.1 version of GenBank requires only 25 GB of space. A local copy of the uncompressed GenBank flatfiles will require somewhere on the order of 220 GB of disk space, and the uncompressed ASN.1 version will require about 189 GB. The second consideration, download time, depends on Internet connection speed. Given a standard 10 Mb/s ethernet connection, it is possible to download the release 153 flatfiles in about 7.5 h (5 h for the compressed ASN.1 files). A typical broadband connection (DSL or Cable), at 1 Mb/s, will take on the order of 75 h and is, therefore, of dubious practicality. Internet access with transfer rates in excess of 100 Mb/s allows the download of the database in less than an hour.

Once a full release of GenBank had been saved locally, it can be kept current using incremental updates. For this purpose, NCBI provides a non-cumulative set of updates in the 'daily-nc' subdirectory of the 'genbank' directory. A Perl (http://www.perl.org/) script is provided in the 'tools' directory of the GenBank FTP site to convert a set of daily updates into a cumulative update (ftp://ftp.ncbi.nih.gov/genbank/tools/).

3. Methods

Four general methods that are central to making use of GenBank include downloads of all or a subset of the database to support local analysis, the construction and use of local GenBank-derived databases for sequence similarity searches and the execution of remote searches of the database using Web or command line network clients. These methods are discussed in the sections that follow.

3.1. Download All GenBank PLN Division Sequences

3.1.1. Strategy

A download of all the GenBank PLN division sequences may be required to take advantage of their high level of annotation during a local analysis. Because the annotation in the PAT, WGS, EST, HTG, HTC, and GSS divisions is not as plentiful and uniform as that in the PLN division, sequences in these divisions can be omitted to build a compact, information-rich local database. For simple, division-oriented bulk downloads, FTP transfers are the most convenient.

To affect the download, it is simply necessary to connect to the NCBI FTP site and specify for download files of the name 'gbpln*.seq.gz', the asterisk is a wild card that matches the entire series of numbered PLN division files (e.g., gbpln1.seq.gz....gbpln17.seq.gz). The download will result in a local set of compressed files for sequence records in the GenBank flatfile format. To view and use the records, they must be uncompressed.

3.1.2. Execution

To begin, perform an anonymous FTP login to the NCBI FTP server using either a command line FTP program or a Web browser. See **Note 11** for information on reaching a command prompt in various operating systems. After connecting using an FTP program, issue the following commands:

1. get gbpln*.seq.gz.
2. Following the completion of the transfers, type 'quit' at the FTP prompt.

To perform the download using a Web browser, simply navigate to

1. ftp://ftp.ncbi.nih.gov/genbank and
2. click on each gbpln*.seq.gz file in turn to download it.

The files are compressed to save storage space and to increase FTP transfer rates and must be uncompressed locally prior to use (*see* **Note 12**).

As described in **Subsection 2.1.1**, the GenBank data are also available in the compact and versatile ASN.1 format. Besides their significantly smaller size, the ASN.1 format files offer other advantages over the compressed flatfiles. Using a suite of command line tools available from NCBI (*see* **Note 13**), ASN.1 files can be used to generate records in various other formats. These formats include the GenBank flatfile (*see* **Note 2**), FASTA (*see* **Note 14**), five-column feature table (*see* **Fig. 1**), and INSD XML. In addition, both the nucleotide sequences and the protein sequences derived from their coding sequence (CDS)

```
>Feature gb|M81884.1|EPFCPCG
69        1        gene
70028    70023
                        gene    His-tRNA
69        1        tRNA
70028    70023
                        product  tRNA-His
                        anticodon            (pos:complement(36..38),aa:His)
                        note     anticodon gtg
                        label    anticodon_gtg
1        19799    misc_feature
                        note     large single copy region (LSC)
569       234     gene
                        gene    pseudo-psbA
                        pseudo
569       234     CDS
                        pseudo
```

Fig. 1. Initial portion of the output of 'asn2all' for a remote fetch from GenBank at National Center for Biotechnology Information (NCBI) of the *Epifagus virginiana* chloroplast genome in record, accession M81884, formatted as a five-column feature table. The definition line, beginning with a '>', gives the GenBank accession number for the sequence. Feature coordinates are given in the body of the table in columns 1 and 2 with the corresponding feature names in column 3. Columns 4 and 5 give feature qualifiers and their values, respectively. There may be multiple qualifiers for each feature.

annotations are readily accessible. ASN.1-formatted GenBank records can also be used to generate databases, both nucleotide and protein, for local analysis using BLAST (*see* **Subsection 3.3.**).

Although not as structured as the ASN.1 format files, the GenBank flatfiles are also easily parsable using scripts that can be written in various languages (*see* **Note 15**). For a simple example illustrating the parsing of a GenBank flatfile written in Perl, (see **Note 16**). Using scripts, one can also easily extract portions of GenBank flatfiles for insertion into spreadsheet software. For an example, (see **Note 17**).

3.2. Download Set of GenBank Sequences for a Single Plant Species

3.2.1. Strategy

A method to download a complete set of sequences for a single plant species, such as *Vitis vinifera*, must retrieve all *V. vinifera* sequences, regardless of GenBank division. Several GenBank divisions, besides the PLN division, such as the EST, STS, and GSS divisions, contain plant sequences. However, to avoid the download of each division in its entirety, just to get the *V. vinifera* data, a more flexible method than a simple bulk FTP download of the GenBank division files is needed. The Entrez search and retrieval system is ideally suited for this sort of selective download. Using Entrez, one can specify the subset of GenBank to download, choose a download format, and then download the sequence records as a single batch.

3.2.2. Execution

The GenBank sequences are a subset of the Entrez Nucleotide database; however, this database also contains nucleotide sequence records derived from other sources, for example from the NCBI RefSeq database *(28,29)*. RefSeqs are curated sequence records that are derived from sequences in GenBank and are extremely valuable as aids to genomic annotation and analysis. However, because they are a derivative of GenBank and do not constitute primary data, they will be excluded in this download. The following Entrez query, when entered into the Entrez query box as shown in **Fig. 2**, will limit the retrieval to sequences within GenBank:

vitis vinifera[orgn] AND srcdb ddbj/embl/genbank[prop]

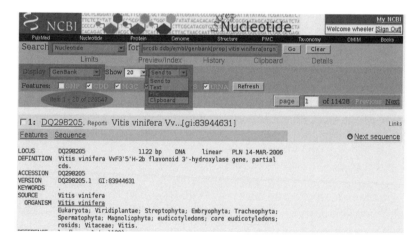

Fig. 2. An Entrez screen at the point of beginning a download of 228,547 *Vitis vinifera* sequences from GenBank. The Entrez Nucleotide query, shown in the top right shaded area, limits the download to GenBank sequences from the organism *Vitis vinifera*. The sequences will be downloaded in the GenBank flatfile format, as indicated in the 'Display' pulldown, also shaded. The 'Send to' pulldown indicates that the output will be written to a local file.

Records must satisfy both terms to match. Terms may also be linked using the boolean 'OR' or 'NOT' operators. For instance, the following query will capture the *V. vinifera* sequences that are not a part of GenBank:

vitis vinifera[orgn] NOT srcdb ddbj/embl/genbank[prop]

As of early 2006, the 'AND'ed query yields more than 200,000 records – note the count of 228,547 highlighted in **Fig. 2** within the shaded oval. To download these records, choose the appropriate format using the Entrez 'Display' pulldown, 'GenBank', for instance, and then choose 'File' from the 'Send to' pulldown as shown within the lower shaded rectangles in **Fig. 2**. Entrez will then offer to save the sequences using the browser's file dialog. Alternatively, a download of a subset of GenBank sequences may be performed within a script through the Entrez Programming Utilities (*see* **Note 8**).

3.3. Establish and Perform Sequence Similarity Searches on a Local Database of PLN Division Sequences

3.3.1. Strategy

If GenBank sequences are to be used in local sequence analysis, a program for rapid sequence similarity searches is required. The Basic Local Alignment Search Tool, or BLAST, is the most widely used program for such searches. When searching a nucleotide database, BLAST has the ability to translate the database sequences on the fly to compare coding regions. BLAST binaries for standard computing platforms are available for download on the NCBI FTP site (http://www.ncbi.nlm.nih.gov/BLAST/download.shtml). A comprehensive discussion of the BLAST algorithm and the interpretation of results are beyond the scope of this chapter; however, details can be found in **ref. *20,30,31***.

3.3.2. Execution

To prepare a local database of GenBank PLN division sequences for sequence-similarity searches through BLAST requires the BLAST executable package, available for the most common computer platforms at (http://www.ncbi.nlm.nih.gov/BLAST/download.shtml). The executable packages must be 'uncompressed' to produce the program files; the method used to unpack them will vary from platform to platform (*see* **Note 12**). Three programs from the BLAST package will be discussed: formatdb, blastall, and megablast. The program 'formatdb' is used to create a local BLAST database from a file of concatenated FASTA format sequences (*see* **Note 14**) or from an

ASN.1 format GenBank file. The programs 'blastall' and 'megablast' are used to search the databases created by formatdb.

The following command line is used to create a local nucleotide database from a file of concatenated FASTA format sequences contained within the file 'myfasta':

formatdb -i myfasta_file -p F -o T

The '-p' parameter indicates whether the sequences to be formatted are protein, 'T', or nucleotide, 'F'. The '-o' parameter, when set to 'T', causes formatdb to create indices that allow individual sequence records to be retrieved by another program in the BLAST package, 'fastacmd' (*see* **Note 18**), on the basis of sequence identifiers found in the definition lines of the records.

To create a nucleotide sequence database with a set of ASN.1 format GenBank sequence files, use

formatdb -a T -b T -p F -o T -i gbpln1.aso

The '-a' parameter is set to 'T', indicating that the input file is in ASN.1 format, and the default value is 'F', indicating the FASTA format. The '-b' parameter is set to 'T' to indicate that the ASN.1 files are of the 'binary' rather than 'text' type. Success or failure messages are recorded in a file called 'formatdb.log'. For the run above, the following appears at the end of 'formatdb.log':

Formatted 85705 sequences in volume 0

This indicates that the run was successful and that a local database called 'gbpln1.aso' has been created consisting of several files with various extensions that are summarized in **Table 1**. Although the BLAST database comprises many files, the database name used on the BLAST command line when the database is searched is the same as the name given to formatdb when the database was formatted.

To create a protein sequence database from the corresponding translations of annotated CDS features on the nucleotide sequences contained in gbpln1.aso, use

formatdb -a T -b T -p T -o T -i gbpln1.aso

Success is indicated by the following entry in the 'formatdb.log' file:

Formatted 42334 sequences in volume 0

Table 1
Nucleotide Sequence Database Files Created by
'formatdb' Comprising a BLAST Database

BLAST database file	Size (bytes)	Content
gbpln1.aso.nhr	13,686,977	Deflines
gbpln1.aso.nni[a]	2724	GI indices
gbpln1.aso.nsq	36,723,724	Sequence data
gbpln1.aso.nin	1,028,536	Indices
gbpln1.aso.nsd[a]	16,487,735	Non-GI data
gbpln1.aso.nnd[a]	68,5640	GI data
gbpln1.aso.nsi[a]	383,032	Non-GI indices

[a] Files created only when '-o T' is used.

The database files created are shown in **Table 2**.

To search the nucleotide database formatted above using the 'blastn' program, with a FASTA-formatted nucleotide sequence within a file named 'myfile', type

blastall -i myfile -d gbpln1.aso -p blastn -F T -e 1e-5 -o myoutfile

To search the protein translations arising from the CDS features on the records in gbpln1.aso, assuming that the protein sequence version of gbpln1.aso has been created using formatdb as described above, use

blastall -i myfile -d gbpln1.aso -p blastp -F T -e 1e-5 -o myoutfile

Table 2
Protein Sequence Database Files Created by 'formatdb'
Comprising a BLAST Database

BLAST database file	Size (bytes)	Content
gbpln1.aso.phr	5,507,439	Deflines
gbpln1.aso.pni[a]	1372	GI indices
gbpln1.aso.psq	15,209,797	Sequence data
gbpln1.aso.pin	338,744	Indices
gbpln1.aso.psd[a]	4,852,260	Non-GI data
gbpln1.aso.pnd[a]	338,672	GI data
gbpln1.aso.psi[a]	108,095	Non-GI indices

[a]Files created only when '-o T' is used.

In command lines above, the parameter '-i' is followed by the name of a file containing one or more FASTA-formatted sequences to be used as queries. The '-d' parameter is followed by the name of a database formatted using formatdb. The parameter '-p' is followed by the name of a BLAST program and may be one of 'blastn', 'blastp', 'blastx', 'tblastn' or 'tblastx'. The '-F' switch is set to 'T' to indicate that the query sequence or sequences should be filtered for 'low complexity' (*see* **Note 19**). The quality of the alignments returned by BLAST can be controlled using the '-e' parameter to set an 'expect value'. In this case, an expect value of 0.00001 has been specified, indicating that alignments expected to occur by chance more than 0.00001 times in a database of the size of gbpln1.aso should be excluded. The '-o' parameter is followed by the name of a file into which the output will be placed. An example of the output generated by 'blastall' is given in **Fig. 3**.

Local databases allow computationally intensive searches that may be impractical elsewhere. For instance, to compare a large batch of EST sequences to the nucleotide sequences in gbpln1.aso with an intermediate six-frame translation of both the EST sequences and the database sequences from gbpln1.aso, use

blastall -i myfile -d gbpln1.aso -p tblastx -F T -e 1e-5 -o myoutfile

Such a doubly translated search requires the equivalent of 36 protein to protein sequence comparisons (36 separate blastp searches) on the translations of the full-length EST and database sequences. A 'tblastx' search is a flexible and sentitive way to compare nucleotide sequences that are believed to encode proteins, because it operates at the level of protein sequences that are better conserved than their corresponding nucleotide CDSs, and it is not dependent on or limited to the CDS annotations on the records. However, this sort of search can be impractical against a large database on a public Web server because of its computational intensity and is therefore a good candidate for local implementation.

A large number of parameters may be specified on the 'blastall' command line that affect the number, format, and quality of the alignments returned. To see them all, type 'blastall –'.For detailed documentation on the parameters, see the 'readme' files that come with the package.

The MegaBLAST program is useful for extremely rapid scans of a nucleotide database with a large batch of nucleotide queries, such as EST sequences. MegaBLAST is described in more detail in **Subsection 3.4**. The following command line can be used to perform a MegaBLAST scan of

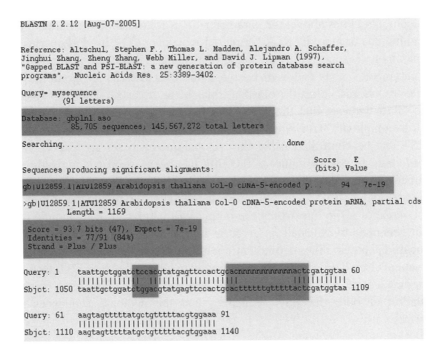

Fig. 3. Output from 'blastall' for a query against a local database created from GenBank FTP file 'gbpln1.aso', as downloaded from the GenBank FTP site at the National Center for Biotechnology Information (NCBI) and formatted using 'formatdb'. The 91-base query aligns to one sequence in the database with an Expect value of 7e-19, as seen in the one-line summary section, second rectangle. The actual alignment is shown below, with statistics indicating 77 identical residues between the two sequences in the aligned region, third rectangle. Two mismatches are shown in the fourth rectangle, followed by region in which the query has been filtered (n) as 'low-complexity' sequence, visible within the fifth rectangle.

a local database with a batch of FASTA-formatted nucleotide sequences contained in the file 'batchseq':

megablast -i batchseq -d gbpln1.aso -D3 -o batchout

The new option, '-D', appearing in this command line specifies one of three tabular output formats that are commonly used in batch searches. These outputs are easily parsable and are therefore amenable to automated analysis through scripts. The analogous tabular MegaBLAST output for the search of **Fig. 3** is shown in **Fig. 4**.

```
# BLASTN 2.2.12 [Aug-07-2005]
# Query: mysequence
# Database: gbpln1.aso
# Fields: Query id, Subject id, % identity, alignment length, mismatches, gap openings, q. start, q. end, s. start, s. end, e-value, bit score
mysequence    U12859.1    100.00 43    0    0    49    91    1098  1140  2e-16  85.7
```

Fig. 4. Tabular output format for a MegaBLAST search. The query sequence has the identifier 'mysequence'; the database sequence is 'U12859.1'. An alignment between the two of 43 bases at 100% identity is reported. Fields are separated by tabs for ease of parsing and are identified in the 'Fields' comment line.

3.4. Identify ESTs Using Annotated GenBank Records

3.4.1. Strategy: Find Good Matches to Well-Annotated Sequences in GenBank

To assign a gene name to each member of a set of EST sequences derived from a single plant species, it is desirable to find a nearly perfect match to a well-annotated nucleotide sequence from the same organism that provided the EST sequence. Failing this, a good match to a well-annotated nucleotide sequence in another organism may suffice. Finally, failing a match at the nucleotide level, one may attempt to match one of the six possible translations of the EST to the conceptual protein translation annotated on a GenBank record. Note that, because the latter option is computationally intensive, it is wise to begin with one of the two nucleotide to nucleotide comparisons.

3.4.2. Execution

3.4.2.1. Best Case: Finding a Match to Nearly Identical Sequences from the Same Organism

To find a nearly perfect match between EST sequences and annotated GenBank sequences derived from the same organism, a MegaBLAST *(33)* search can be run against the 'nr', or 'non-redundant', nucleotide database using the BLAST Web interface at NCBI *(31)*. The 'nr' database includes the well-annotated GenBank sequences found in the traditional PLN division. Poorly annotated sequence types such as EST and GSS sequences are not included in 'nr' but are available as separate BLAST databases. MegaBLAST is a good choice for the search, because it is optimized to rapidly scan nucleotide databases for close matches. In addition, the Web interface to MegaBLAST at NCBI supports the upload of batches of sequences.

The EST sequences for MegaBLAST searches are submitted as a concatenated set of FASTA format sequences (*see* **Note 14**). It is best to perform any editing of the data using a text editor such as one of those described in **Note 20**. Save the input file as ASCII text if a word processor is used.

A batch of EST sequences may be submitted for a MegaBLAST search using the 'megablast' link on the BLAST Homepage (http://www.ncbi.nlm.nih.gov/BLAST). **Figure 5** shows the MegaBLAST page just prior to launching a search of a single *O. sativa* EST sequence against *O. sativa* sequences in the traditional GenBank divisions. If a large batch of ESTs is to be used as a query, it is possible to upload them from a local file using the file upload box directly below. The second shaded box from the top highlights the database to be searched, 'nr'. The next two shaded areas highlight limitations to be applied to

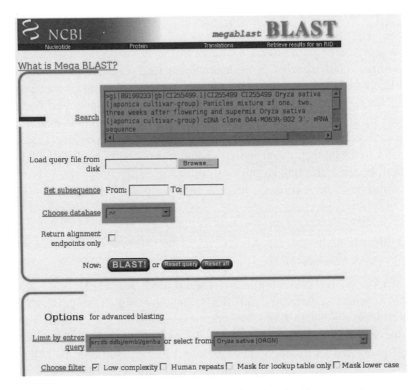

Fig. 5. The BLAST Web form used to submit a batch of expressed sequence tags (ESTs) for a MegaBLAST search against sequences in GenBank. Batches of sequences are submitted by pasting concatenated FASTA-formatted sequences into the query box, shaded at the top of the figure, or by specifying a file containing such a batch using the 'Browse' button, visible just beneath the query box. In the query box, the definition line for one *Oryza sativa* EST sequence is shown. The search will be run against the nucleotide 'nr' database, limited to *O. sativa* sequences found in GenBank using the Entrez limitations highlighted within the two lower shaded boxes.

the 'nr' database when searching. The search is limited to GenBank sequences using the Entrez query 'srcdb ddbj/embl/genbank[prop]' and to sequences from *O. sativa*, using the adjoining pulldown. BLAST searches against GenBank at NCBI using the BLAST Web interface or the BLAST network client (*see* **Note 21**) may be limited to virtually any subset of GenBank sequences using Entrez queries (*see* **Note 7**).

The result of the search, shown in **Fig. 6**, indicates strong hits to *O. sativa* sequences belonging to a UniGene cluster representing the lectin gene.

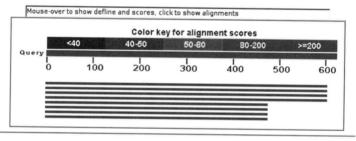

Query= gi|89199233|gb|CI255499.1|CI255499 CI255499 Oryza sativa (japonica cultivar-group)
Panicles mixture of one, two, three weeks after flowering
and supermix Oryza sativa (japonica cultivar-group) cDNA clone 044-M063R-B02
3', mRNA sequence
Length=625

Distribution of 8 Blast Hits on the Query Sequence

Sequences producing significant alignments:	Score (Bits)	E Value
gi|3248319|emb|AL662990.3|OSJN00194 Oryza sativa genomic DNA...	1109	0.0
gi|58530790|dbj|AP008210.1| Oryza sativa (japonica cultivar-g...	1109	0.0
gi|32988415|dbj|AK103206.1| Oryza sativa (japonica cultivar-g...	1109	0.0
gi|530789|gb|M24504.1|RICLECTIN Oryza sativa lectin mRNA, comple	1109	0.0
gi|4689405|gb|AF140682.1|AF140682 Oryza sativa subsp. japonic...	872	0.0
gi|4689403|gb|AF140681.1|AF140681 Oryza sativa subsp. indica lec	867	0.0
gi|4689401|gb|AF140680.1|AF140680 Oryza rufipogon lectin gene, p	867	0.0
gi|4689399|gb|AF140679.1|AF140679 Oryza sativa subsp. japonic...	861	0.0

Fig. 6. One-line summary section of the MegaBLAST results for the search of **Fig. 5**. A number of BLAST database hits with very low Expect values (E-value = 0.0) are returned (S.F. Altschul, The statistics of sequence similarity scores, http://www.ncbi.nlm.nih.gov/BLAST/tutorial/Altschul-1.html), indicating very significant alignments. The 'U' graphics highlighted by the shaded ovals are links to the *Oryza sativa* UniGene cluster for Lectin. Good alignments to these sequences strengthen the evidence for the lectin gene as the source of the query-expressed sequence tag (EST).

3.4.2.2. First Fallback: Finding a Match Across Species with Discontiguous MegaBLAST

If MegaBLAST is unable to find a match between the EST and the annotated GenBank transcripts from the same species, one can attempt to match across species using Discontiguous MegaBLAST (http://www.ncbi.nlm.nih.gov/blast/discontiguous.shtml). Discontiguous MegaBLAST is optimized to detect the weaker similarities between nucleotide sequences expected across species and is especially useful when comparing CDSs. The procedure is analogous to that described above; choose the 'discontiguous megablast' link rather than the 'megablast' link on the BLAST page to reach the corresponding form.

3.4.2.3. Second Fallback: Performing a Translated BLAST Search (blastx) Against Plant Proteins

If Discontiguous MegaBLAST fails to find a match, one may perform a translated search against the plant proteins annotated on GenBank records. The 'blastx' program translates a nucleotide query in all possible reading frames (three forward and three reverse frames) and compares these translations with the sequences in a protein sequence database. In this case, one chooses the 'blastx' link from the BLAST page and searches the default 'nr' protein database with the limitation to *O. sativa* sequences derived from the translations of annotated CDSs in GenBank in the same manner described above. Because translated searches are computationally intensive, it is best to avoid them when processing large batches of sequences unless nucleotide-level comparisons fail.

4. Notes

1. The files in the GenBank releases are partitioned into 'divisions' that correspond roughly to taxonomic groups such as bacteria (BCT), viruses (VRL), primates (PRI), and rodents (ROD). In recent years, divisions have been added to support specific sequencing strategies. These include divisions for EST, GSS, HTG, HTC, and ENV sequences, making a total of 18 divisions. To facilitate downloads, most divisions are partitioned into multiple files for the bimonthly GenBank releases on the NCBI FTP site. In addition, two special classes of record exist that do not appear within the usual 18 divisions of GenBank: WGS and Third Party Annotation (TPA) records. Over 60 million bases of WGS sequence appear in GenBank as sets of WGS contigs, many of them bearing annotations, originating from a single sequencing project. TPA (http://www.ncbi.nih.gov/Genbank/TPA.html) records support the reporting of published, experimentally confirmed sequence annotation by a scientist other than the original submitter of the primary sequence. The content of the GenBank divisions is summarized in **Table 3**.

Table 3
Division Codes and Content of the 18 GenBank Divisions

Code	Description
Traditional GenBank Divisions	
BCT	Bacterial sequences
PRI	Primate sequences
MAM	Other mammalian sequences
VRT	Other vertebrate sequences
INV	Invertebrate sequences
PAT	Patent sequences
PLN	Plant, fungal, and algal sequences
VRL	Viral sequences
PHG	Bacteriophage sequences
SYN	Synthetic and chimeric sequences
UNA	Unannotated sequences, including some WGS sequences obtained through environmental sampling methods
Non-traditional GenBank Divisions	
EST	EST division sequences, or expressed sequence tags, are short single pass reads of transcribed sequence. About 9.2 million ESTs are derived from almost 50,000 plant species
STS	STS division sequences include anonymous STSs based on genomic sequence as well as gene-based STSs derived from the 3'-ends of genes and ESTs. STS records usually include primer sequences, annotations, and PCR conditions. About 100,000 of the STS sequences in GenBank are of plant origin
GSS	GSS records are predominantly single reads from bacterial artificial chromosomes ('BAC ends') used in various genome sequencing projects. GSS records for plant species number over 6.3 million
ENV	The ENV division of GenBank, for non-WGS sequences obtained thorugh environmental sampling methods in which the source organism is unknown, debuted with release 147 in April 2005
HTG	The HTG division of GenBank contains unfinished large-scale genomic records that are in transition to a finished state. These records are designated as phase 0–3 depending on the quality of the data. Upon reaching phase 3, the finished state, HTG records are moved into the appropriate taxonomic division of GenBank
HTC	The HTC division of GenBank accommodates high-throughput cDNA sequences. HTCs are of draft quality but may contain 5'-UTRs and 3'-UTRs, partial coding regions, and introns

(*Continued*)

Table 3 *(Continued)*

Code	Description
CON	Large records that are assembled from smaller records, such as eukaryotic chromosomal sequences, are represented in the GenBank 'CON' division. CON records contain sets of assembly instructions to allow the transparent display and download of the full record using tools such as NCBI's Entrez

BCT, bacteria; ENV, environment; EST, expressed sequence tag; GSS, genome survey sequence; HTC, high-throughput cDNA; HTG, high-throughput genomic; PCR, polymerase chain reaction; PRI, primates; STS, sequence-tagged site; VRL, viruses.

2. **Figure 7** shows a complete GenBank flatfile, and a portion of the succeeding flatfile, as found within the FTP file 'gbpln15.seq.gz' of GenBank release 153. The '//' symbol, circled, marks the boundary between successive records. Record field labels appear in the first twelve columns, shaded for the first record. Each record begins with a LOCUS line followed by a header containing the database identifiers, the title of the record, references, and submitter information. The header is followed by the feature table (*see* **Note 5**) and the sequence itself on the line following the 'origin' field. The GenBank flatfile is described in detail in the GenBank release notes (ftp://ftp.ncbi.nih.gov/genbank/gbrel.txt). In the Entrez system, the GenBank format is the default display for records in the traditional divisions.

3. **Table 4** shows the number of records in various divisions for some prominently represented plant species in GenBank release 153.

4. Virtually, all records enter GenBank as direct electronic submissions, with the majority of authors using the BankIt or Sequin programs described on the GenBank submission Web page (http://www.ncbi.nlm.nih.gov/Genbank/submit.htm). Most journals require authors with sequence data to submit the data to a public database as a condition of publication. GenBank staff can usually assign an accession number (*see* **Note 6**) to a sequence submission within two working days of receipt. The accession number serves as confirmation that the sequence has been submitted and can be used to retrieve the data when it appears in the database. Direct submissions receive a quality assurance review that includes checks for vector contamination, proper translation of coding regions, correct taxonomy, and correct bibliographic citations. A draft of the GenBank record is passed back to the author for review before it enters the database, and authors may ask that their sequences be kept confidential until the time of publication. As GenBank policy requires that deposited sequence data be made public when the sequence or accession number is published, authors are instructed to inform GenBank staff of the publication date of the article in which the sequence is cited to ensure a timely release of the data. Although only the submitting scientist is permitted to modify sequence data or annotations, all users are encouraged to report lags in releasing data or possible errors or omissions to GenBank at update@ncbi.nlm.nih.gov. Submissions

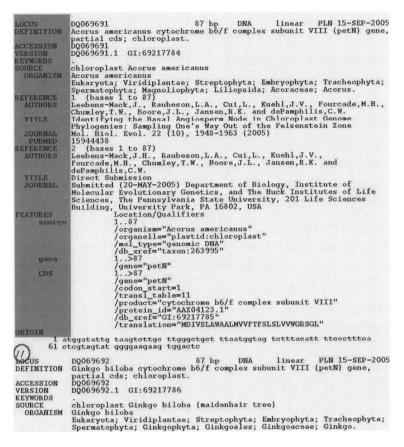

```
LOCUS          DQ069691                87 bp    DNA     linear   PLN 15-SEP-2005
DEFINITION     Acorus americanus cytochrome b6/f complex subunit VIII (petN) gene,
               partial cds; chloroplast.
ACCESSION      DQ069691
VERSION        DQ069691.1  GI:69217784
KEYWORDS       .
SOURCE         chloroplast Acorus americanus
  ORGANISM     Acorus americanus
               Eukaryota; Viridiplantae; Streptophyta; Embryophyta; Tracheophyta;
               Spermatophyta; Magnoliophyta; Liliopsida; Acoraceae; Acorus.
REFERENCE      1  (bases 1 to 87)
  AUTHORS      Leebens-Mack,J., Raubeson,L.A., Cui,L., Kuehl,J.V., Fourcade,M.H.,
               Chumley,T.W., Boore,J.L., Jansen,R.K. and dePamphilis,C.W.
  TITLE        Identifying the Basal Angiosperm Node in Chloroplast Genome
               Phylogenies: Sampling One's Way Out of the Felsenstein Zone
  JOURNAL      Mol. Biol. Evol. 22 (10), 1948-1963 (2005)
   PUBMED      15944438
REFERENCE      2  (bases 1 to 87)
  AUTHORS      Leebens-Mack,J.H., Raubeson,L.A., Cui,L., Kuehl,J.V.,
               Fourcade,M.H., Chumley,T.W., Boore,J.L., Jansen,R.K. and
               dePamphilis,C.W.
  TITLE        Direct Submission
  JOURNAL      Submitted (20-MAY-2005) Department of Biology, Institute of
               Molecular Evolutionary Genetics, and The Huck Institutes of Life
               Sciences, The Pennsylvania State University, 201 Life Sciences
               Building, University Park, PA 16802, USA
FEATURES             Location/Qualifiers
     source          1..87
                     /organism="Acorus americanus"
                     /organelle="plastid:chloroplast"
                     /mol_type="genomic DNA"
                     /db_xref="taxon:263995"
     gene            1..>87
                     /gene="petN"
     CDS             1..>87
                     /gene="petN"
                     /codon_start=1
                     /transl_table=11
                     /product="cytochrome b6/f complex subunit VIII"
                     /protein_id="AAZ04123.1"
                     /db_xref="GI:69217785"
                     /translation="MDIVSLAWAALMVVFTFSLSLVVWGRSGL"
ORIGIN
        1 atggatattg taagtcttgc ttgggctgct ttaatggtag tctttacatt ttccttttca
       61 ctcgtagtat ggggaagaag tggactc
//
LOCUS          DQ069692                87 bp    DNA     linear   PLN 15-SEP-2005
DEFINITION     Ginkgo biloba cytochrome b6/f complex subunit VIII (petN) gene,
               partial cds; chloroplast.
ACCESSION      DQ069692
VERSION        DQ069692.1  GI:69217786
KEYWORDS       .
SOURCE         chloroplast Ginkgo biloba (maidenhair tree)
  ORGANISM     Ginkgo biloba
               Eukaryota; Viridiplantae; Streptophyta; Embryophyta; Tracheophyta;
               Spermatophyta; Ginkgophyta; Ginkgoales; Ginkgoaceae; Ginkgo.
```

Fig. 7. A GenBank flatfile and a portion of the next as found within the GenBank release FTP files. The GenBank flatfiles are packaged for release as concatenated sets of records belonging to the same GenBank division. Shaded are the first 12 columns of the record, which contain the field names. The record separator, '//', is circled. For a short script that can be used to parse this file, see **Fig. 11**.

may be prepared through a Web form called BankIt, using a standalone program called Sequin or using specialized procedures for large-scale projects or batches of sequences.

BankIt

Using BankIt (http://www.ncbi.nlm.nih.gov/BankIt), authors enter sequence information directly into a form and add biological annotation such as coding regions or mRNA features. Free-form text boxes, list boxes and pulldown menus allow the submitter to describe the sequence. BankIt validates submissions, flagging many

Table 4
Prominent Plant Species in GenBank Release 153

PLN non-WGS (~0.60 million)	EST (9.2 million total)	GSS (6.3 million total)
Arabidopsis thaliana (58649)	*Z. mays* (724,343)	*Z. mays* (1,946,817)
Oryza sativa[a] (38,799)	*Triticum aestivum* (605,747)	*Brassica oleracea* (595,420)
Oryza sativa (9770)	*A. thaliana* (622,788)	*S. bicolor* (573,724)
Zea mays (13,356)	*O. sativa* (1,182,558)	*A. thaliana* (438,887)
Sorghum bicolor (5910)	*H. vulgare* (407,521)	*L. esculentum* (320,398)
Hordeum vulgare (3133)	*Glycine max* (356,920)	*O. sativa* (250,417)
Nicotiana tabacum (4476)	*P. taeda* (329,469)	*O. ridleyi* (204,729)
Lycopersicon esculentum (4233)	*Saccharum officinarum* (246,301)	*O. coarctata* (195,285)
Triticum aestivum (4134)	*Malus x domestica* (253,581)	*O. minuta* (169,651)
Solanum tuberosum (2221)	*Medicago truncatula* (225,129)	*M. truncatula* (168,679)

EST, expressed sequence tag and GSS, genome survey sequence.
[a] Japonica cultivar group.
[b] Whole genome shotgun in PLN; Oryza (japonica cultivar group), 50,231 contigs; *O. sativa*, 35,047 contigs.

common errors, and checks for vector contamination using a variant of BLAST called Vecscreen, before creating a draft record in GenBank flat file format for the submitter to review. BankIt is the tool of choice for simple submissions, especially when only one or a small number of records are to be submitted. BankIt can also be used by submitters to update their existing GenBank records. Help for using BankIt, as well as example submission scenarios, is available (http://www.ncbi.nlm.nih.gov/BankIt/help.html).

Sequin

NCBI offers a standalone multiplatform submission program called Sequin (http://www.ncbi.nlm.nih.gov/Sequin/index.html) that can be used interactively with other NCBI sequence retrieval and analysis tools. Sequin handles simple sequences such as cDNA, as well as segmented entries, phylogenetic studies, population studies, mutation studies, environmental samples, and alignments for which BankIt and other Web-based submission tools are not well suited. Sequin offers complex annotation capabilities and contains a number of built-in validation functions for quality assurance. In addition, Sequin is able to accommodate large chromosome-scale sequences and read in a full complement of annotations through 'five-column feature' tables such as that shown in **Fig. 1**. Once a submission is completed, submitters can e-mail the Sequin file to the address gb-sub@ncbi.nlm.nih.gov. Versions of Sequin for common computer platforms are available through anonymous FTP (ftp://ftp.ncbi.nih.gov/sequin).

NCBI works closely with sequencing centers to ensure timely incorporation of bulk data into GenBank for public release. Submitters of large, heavily annotated genomes may find it convenient to use 'tbl2asn' (http://www.ncbi.nlm.nih.gov/Sequin/table.html) to convert a table of annotations generated through an annotation pipeline into an ASN.1 record suitable for submission to GenBank. Special procedures for the batch submission of EST, GSS, and STS sequences are described at http://www.ncbi.nlm.nih.gov/Genbank/submit.htm.

5. The feature table is the portion of the GenBank record that provides information about the biological features annotated on the nucleotide sequence. These features include coding regions and their protein translations, non-coding regions, genes, variations, sequence-tagged sites, transcription units, repeat regions, and sites of mutations or modifications. The International Sequence Database Collaboration (http://www.insdc.org/) produces a document describing and identifying the features allowed on GenBank, DDBJ, and EMBL records (http://www.ncbi.nlm.nih.gov/collab/FT/index.html).

6. Each GenBank record, consisting of both a sequence and its annotations, is assigned a stable and unique identifier, termed an 'accession number', which is shared across the three INSDC members (GenBank, DDBJ, and EMBL). The DNA sequence within a GenBank record is assigned a unique NCBI identifier, called a 'gi', that appears on the VERSION line of GenBank flatfile records. A third identifier of the form 'Accession.version', also displayed on the VERSION line of flatfile records, combines the information present in both the gi and the accession numbers. An entry appearing in the database for the first time has an 'Accession.version' identifier equivalent to the ACCESSION number of the GenBank record followed by '.1' to indicate the first version of the sequence for the record, for example,

ACCESSION AF000001
VERSION AF000001.1 GI: 987654321

A similar system tracks changes in the corresponding protein translations. These identifiers appear as qualifiers for CDS features in the FEATURES portion of a GenBank entry, for example, /protein_id='AAA00001.1'. Protein sequence translations also receive their own unique gi number, which appears as a second qualifier on the CDS feature, for example,

/db_xref='GI:1233445'

The one- or two-letter accession prefixes (e.g. 'AF' in the example above) that are used for non-WGS GenBank records indicate the INSDC member to which the sequence was submitted and often, but not always, indicate the type of sequence. A table of prefixes with their associated INSDC member is given in (http://www.ncbi.nlm.nih.gov/Sequin/acc.html).

7. The GenBank database and protein sequences arising from CDS annotations on GenBank records can be searched at NCBI using BLAST thorugh a Web interface or through a command line network client. In either case, subsets of the data may be selected for searching using Entrez (*see* **Subsection 2.2.1.**) queries. A query used to limit a search to sequences from a particular organism has the form 'organism[orgn]', where 'organism' is an organism name, and the search is limited to terms indexed within the Entrez 'organism' field by specifying 'orgn' within square brackets. For example, to specify only sequences from *A. thaliana*, use 'Arabidopsis thaliana[orgn]'. Some Entrez queries involving terms indexed in the Entrez 'properties' field are summarized in **Table 5**. Entrez queries can be combined using Boolean operators such as 'AND', 'OR', and

Table 5
Entrez Queries that are Useful in Limiting BLAST Searches

Terms in the Entrez 'properties' field	Effect	Example of Use
'gbdiv X', where X is one of pri, rod, mam, vrt, inv, pln, bct, vrl, phg, syn, una, est, pat, sts, gss, htg, htc, env	Limit to sequences within a GenBank division	gbdiv pln[prop]
'biomol X', where X is one of crna, genomic, genomic mrna, other, pre mrna, rna,rrna, scrna, snorna, snrna, transcribed rna, trna	Limit to a molecule type	Biomol mrna[prop]
'srcdb X', where X is one of ddbj, ddbj/embl/genbank, embl, genbank, pdb, tpa ddbj, tpa ddbj/embl/genbank, tpa embl, tpa genbank, refseq[a]	Limit to source database. To limit to GenBank, use 'ddbj/embl/genbank'[b]	srcdb ddbj/embl/ genbank[prop]

[a]Additional 'refseq' terms are available but are not shown.
[b]The term 'srcdb genbank[prop]' limits to records within GenBank that entered by submission to GenBank at National Center for Biotechnology Information (NCBI). Because the data in GenBank also includes data submitted to the DNA Databank of Japan and European Molecular Biology Laboratory partners of the International Nucleotide Sequence Database Collaboration, one must specify 'srcdb ddbj/embl/genbank[prop]' to include all the records in GenBank.

'NOT' (*see* **Subsection 3.2.2.**). On the BLAST Web pages, Entrez queries are typed into the box labeled 'Limit by Entrez query'. Using the BLAST network client, 'blastcl3' (*see* **Note 21**), the '-u' option is used to specify the Entrez query. A knowledge of the content of the BLAST databases is important in the construction of effective Entrez query limitations. For instance, a query such as 'gbdiv est[prop]' that limits the search to sequences from the GenBank EST division results in a zero-length database when used in conjunction with the 'nr' BLAST database, as this database contains no ESTs. Similarly, a query such as 'gbdiv pln[prop]' when applied to the BLAST 'EST' database will yield no sequences as this database contains only sequences from the GenBank EST division. For descriptions of the standard BLAST databases, click on the 'Choose database' links that appear on each of the BLAST Web forms accessible from the BLAST Homepage (http://www.ncbi.nlm.nih.gov/BLAST).

8. The Entrez Programming Utilities are a set of nine server-side scripts that allow automated access to the Entrez search and retrieval system. The utilities accept a set of parameters that may be URL-encoded or transferred through the SOAP protocol. Searches of Entrez are performed using the 'esearch' E-utility; short record summaries are retrieved using 'esummary'; full records may be downloaded using 'efetch'; and linking between records may be performed using 'elink'. Other E-utilities for specialized roles exist. The E-utilities may be used from within any programming language that supports the posting of a URL. Results are returned in XML for all E-utilities except 'efetch', which supports return modes of XML, HTML, text, and ASN.1 as well as return formats such as GenBank flatfile, FASTA, and the INSDC XML format. The E-utilitiy calls shown in **Table 6** will find all plant sequences in GenBank, returning sequence counts, and NCBI sequence ids in XML; retrieve a brief sequence description for an *A. thaliana* chromosome-4 contig record (GenBank accession AJ270058 and NCBI gi 42494965) in XML format; retrieve the entire contig record with sequence and annotations; and return ids for all protein records derived from conceptual translations of CDSs annotated on the *A. thaliana* chloroplast genome (GenBank accession AP000423 and NCBI gi 5881673).

Table 6
Representative URLs for Entrez Programming Utility Calls

esearch.fcgi? db=nucleotide&term=srcdb+ddbj/embl/genbank[Properties]+plants[orgn]
esummary.fcgi? db=nucleotide&id=42494965
efetch.fcgi? db=nucleotide&id=42494965&retmode=text&rettype=gbwithparts
elink.fcgi? db=protein&db_from=nucleotide&id=5881673

These URLs should be prefixed with the E-utility base (http://eutils.ncbi.nlm.nih.gov/entrez/eutils/).

E-utility URLs may be tested using any Web browser; however, they are designed to be used from within a script. A simple script, written in Perl (see scripting languages in **Note 15**) that will use 'esearch' to find all plant sequences in GenBank of molecule type 'trna' (for other molecule types, *see* **Note 7**) and download them in GenBank flatfile format is shown in **Fig. 8**.

For more information, see **ref. 27**. Four-day workshops focusing on advanced E-utility use are given quarterly at NCBI (http://www.ncbi.nlm.nih.gov/Class/PowerTools/eutils/course.html).

9. Using passive FTP mode, the local FTP program connects to the server to establish a data connection, whereas using the active FTP mode, it is the server that connects first to establish the connection. If the local machine is connected to the Internet from behind a firewall, then passive FTP is usually required. However, in some firewall configurations, active FTP works, whereas passive FTP does not, so try both and use the mode that works.

10. **Figure 9** shows the output to the computer screen resulting from a typical Linux command line FTP download of the first GenBank file for 'PLN' division sequences. Commands issued by the user are highlighted. The FTP command line uses two parameters that may not be valid for every FTP client. The first parameter '-a' instructs the ftp program to use an anonymous login protocol; in this case, the user does not have to respond to a 'login' or 'password' prompt. The second parameter, '-A', forces 'active mode'. Once logged on to the NCBI ftp site, it is convenient to 'change directory' or 'cd' into the 'genbank' subdirectory where the GenBank files reside

```
1   use LWP::Simple;
2   $file="outfile";
3   $rettype="gb";$retmode="text";
4   $term="srcdb+ddbj/embl/genbank[prop]+plants[orgn]+biomol+trna[prop]";
5   $ebase="http://eutils.ncbi.nlm.nih.gov/entrez/eutils/";
6
7   $ts=time;
8
9   $url="$ebase"."esearch.fcgi?"."usehistory=y&db=nucleotide&term=$term";
10  $results=get($url);
11  $results=~/<Count>(\d+)<\/Count>.*<QueryKey>(\d+).*<WebEnv>(.+)<\/WebEnv>/s;
12  ($nrecords,$qkey,$webenv)=($1,$2,$3);
13
14  print STDERR "The url posted was: $url\n";
15  print STDERR "Search results returned: $results\n";
16
17
18  $results="";
19  $url="$ebase"."efetch.fcgi?"."db=nucleotide&tool=gbplant&retmax=$nrecords";
20  $url.="&rettype=$rettype&WebEnv=$webenv&query_key=$qkey&retmode=$retmode";
21
22  print STDERR $url;
23
24  $results=get($url);
25
26  open (OUT,">$file");print OUT $results;close OUT;
27  $t=time-$ts;
28  print STDERR "Elapsed time for search and download: $t seconds\n";
29
```

Fig. 8. Perl script for a search of GenBank using the E-utilities, followed by a batch download. The search and subsequent download of over 100 sequences in GenBank format requires only 2 s using a cable modem connection.

```
ftp -a -A ftp.ncbi.nih.gov
Connected to ftp.wip.ncbi.nlm.nih.gov.
220-
Warning Notice!

This is a U.S. Government computer system, which may be accessed and used
only for authorized Government business by authorized personnel.
Unauthorized access or use of this computer system may subject violators to
criminal, civil, and/or administrative action.

All information on this computer system may be intercepted, recorded, read,
copied, and disclosed by and to authorized personnel for official purposes,
including criminal investigations. Such information includes sensitive data
encrypted to comply with confidentiality and privacy requirements. Access
or use of this computer system by any person, whether authorized or
unauthorized, constitutes consent to these terms. There is no right of
privacy in this system.
---
Welcome to the NCBI ftp server. The official anonymous access URL for NCBI
ftp server is "ftp://ftp.ncbi.nih.gov", please use it.

Public data may be downloaded by logging in as "anonymous" using your E-mail
address as a password.
220 FTP Server ready.
331 Anonymous login ok, send your complete email address as your password.
230 Anonymous access granted, restrictions apply.
Remote system type is UNIX.
Using binary mode to transfer files.
ftp> cd genbank
250 CWD command successful
ftp> get gbpln1.seq.gz
local: gbpln1.seq.gz remote: gbpln1.seq.gz
500 Illegal EPRT command
200 PORT command successful
150 Opening BINARY mode data connection for gbpln1.seq.gz (62582012 bytes)
100%
|*********************************************************************************|
61115 KB   564.47 KB/s   00:00 ETA
226 Transfer complete.
62582012 bytes received in 01:48 (564.38 KB/s)
ftp>
```

Fig. 9. The output of a sample command line FTP session. The login is made as user 'anonymous', after which the user moves into the 'genbank' directory through the 'cd' command. The message following the row of asterisks indicates that the 65 Mb compressed GenBank flatfile, 'gbpln1.seq.gz', has been downloaded in a little under 2 min. The file is uncompressed (*see* **Note 12**) locally prior to use.

before beginning the download. In the case shown, a 65-Mb file is downloaded in 1 min and 48 s at about 5.6 Mb/s – typical of a cable modem connection.

11. On Unix, Linux, or Mac OSX systems, use an 'xterm' window to reach a command prompt. On DOS and Windows, use a 'DOS prompt' window.

12. GenBank is distributed on the NCBI FTP site as a set of text files that have been compressed for speedier downloads. These files must be decompressed before use. There are a number of free decompression programs available for common computer platforms (http://www.thefreesite.com/).

13. NCBI offers command line utilities for working with ASN.1-formatted data. Each utility program accepts a number of command line arguments, specified using a dash and a single-letter option code followed by an option value. Some values are Boolean and are given as either 'T', true, or 'F', false. Others are specified using one-letter codes, such as format specifiers, or strings, such as file names or GenBank accession numbers. To see a complete list of command line parameters for any of the programs, run the program with a trailing dash and no parameter. A list of the seven programs with brief descriptions is given in **Table 7**, whereas a detailed description of one of the most versatile programs, 'asn2all', follows. The

Table 7
**National Center for Biotechnology Information Utility Programs for Conversion
of Data from and to the Abstract Syntax Notation One (ASN.1) Format**

ASN.1 Converter	Function
asn2all	Converts GenBank release files in ASN.1 format to a variety of other formats
asn2fsa	Converts binary or text ASN.1 sequence files to FASTA format
asn2gb	Converts binary or text ASN.1 sequence files to GenBank or GenPept flatfile formats
asn2idx	Generates accession/file offset indices for Bioseq-set release files
asn2xml	Converts binary or text ASN.1 sequence files to XML format
asnval	Validates ASN.1 release files
tbl2asn	Automates the creation of sequence records for submission to GenBank by reading feature annotations given in the five-column feature table format and generating an ASN.1 file

program asn2all is primarily intended to generate reports from the binary ASN.1 Bioseq-set GenBank release files (ftp://ftp.ncbi.nih.gov/ncbi-asn1/).

Depending on the value of the '-f' argument, the program can produce GenBank and GenPept flatfiles (*see* **Note 2**), FASTA sequence files (*see* **Note 14**), INSDSet-structured XML, TinySeq XML, and five-column feature table (*see* **Fig. 1**) formats. Prior to running asn2all, the GenBank release files, which have an '.aso.gz' suffix, should be uncompressed (*see* **Note 12**), resulting in files with suffix '.aso'.

Using asn2all, the name of the file to process is specified with the '-i' command line argument. Use '-a t' to indicate batch processing of a GenBank release file and '-b T' to indicate that it is binary ASN.1. A text ASN.1 record, such as one obtained on the Web from Entrez, can be processed by using '-a a -b F' instead of '-a t -b T'.

Nucleotide and protein sequence records within ASN.1 files can be processed simultaneously. Use the '-o' argument to indicate the nucleotide output file and the '-v' argument for the protein output file.

The '-f' argument determines the format to be generated. Legal values of '-f' and the resulting formats are given in **Table 8**.

The command

     asn2all -i gbpln1.aso -a t -b T -f g -o gbpln1.nuc -v gbpln1.prt

will generate GenBank flatfile records for the nucleotide sequences as well as GenPept flaftile records for protein sequences contained within gbpln1.aso. These two sets will appear in the files 'gbpln1.nuc' and 'gbpln1.prt', respectively. A

Table 8
Output Format Options for 'asn2all'

Value of the '-f' flag	Resulting format
G	GenBank (nucleotide) or GenPept (protein)
F	FASTA
T	Five-column feature table
S	INSD-formatted XML
Y	TinySet XML (XML version of FASTA)
A	ASN.1 of entire record
X	XML version of entire record (structured as in the ASN.1 format)

ASN.1, Abstract Syntax Notation One and INSD, international nucleotide sequence database.

remote fetching option, '-r T', allows the download of an ASN.1 record from NCBI over a network connection using an accession number or NCBI 'gi' identifier (*see* **Note 6**) as an identifier. For instance, to perform a remote fetch of the feature table within the GenBank record for the *Epifagus virginiana* chloroplast genome, with accession number M81884, use

asn2all -r T -A M81884 -f t

The output of this command for the first feature of M81884 is given in **Fig. 1**. The five-column feature table format used is identical to that required as input to generate an ASN.1 sequence file using tbl2asn, described in **Note 4**. These utilities are available for five computer platforms and may be downloaded at asn1-converters, ftp://ftp.ncbi.nih.gov/asn1-converters.

14. **Figure 10** shows an example of a GenBank sequence in FASTA format. The initial line is called the FASTA definition line and must begin with a carat ('>') that is followed by a unique sequence identifier. The sequence is represented using the standard IUB/IUPAC nucleic acid codes as given in **Table 9**. Lower-case letters are accepted and are mapped to upper-case; a single hyphen or dash can be used to represent a gap of indeterminate length.

```
>gi|19172021|gb|AF474889.1| Oryza sativa putative drought resistant protein mRNA, partial cds
GAGAGAGTCATCCATGGAGGTGGAGGAGGCGGCGTACAGGACGGGGAGGGCGGAAAACACTTATGATGTC
GGCGGGAAGGACCAAGTTCAGGGAGACGAGGCACCCGGTGTACCGCGGCGTGCGGCGGCGCGGGGGGCGG
CCGGGCGCGGCGGGGAGGTGGGTGTGCCACGTGCGGGTGCCCGGGGCGCGCGGC
```

Fig. 10. A GenBank sequence in FASTA format. The 'definition line', beginning with a '>' and containing sequence identifiers and other descriptive information, is highlighted. The sequence is given using one-letter nucleotide codes (*see* **Table 9**).

Table 9
One-Letter Nucleotide Codes
Used in GenBank Sequences

A	adenosine
B	G T C
C	cytidine
D	G A T
G	guanine
H	A C T
K	G T (keto)
M	A C (amino)
N	A G C T
R	G A
S	G C (strong)
T	thymidine
V	G C A
W	A T (weak)
Y	T C
-	gap

15. Scripting languages allow large chunks of data to be subjected to automated analysis 'pipelines' that may involve using the output of one program as the input to another. The scripting language can be used to run the requisite programs as well as to reformat the output of an upstream program to conform to the expected input of a downstream program. Many such languages are available; however, three of the most popular are Perl, Python, and Ruby (http://www.activestate.com/Products/ActivePerl/, http://www.python.org, and http://www.ruby-lang.org/en/).

16. A GenBank flatfile can be parsed using a script written in a language such as Perl. The code shown in **Fig. 11** will parse a single flatfile to create a multidimensional hash table called '$f' that allows the contents of major flatfile fields and their subordinate fields to be addressed by name. The partial output of the script when applied to the GenBank flatfile for accession 'AJ270058', the short arm of *A. thaliana* chromosome 4, is shown in **Fig. 12**. To print the value of the 'LOCUS' field, for instance, one prints $f{'LOCUS'}{'val'}{'0'}; to print the entire set of qualifiers on the 101st gene in the 'FEATURES' field, one prints $f{'FEATURES'}{'gene'}{'101'}.

17. Spreadsheets are useful for getting an overview of a set of data and for sorting and grouping based on the value of particular fields. A simple addition to the script of **Fig. 11**, shown in **Fig. 8**, allows the data in a GenBank flatfile to be

```
 1  while(<>){
 2      next if (/\/\//);
 3      /^(\s*.+?)\s\s+(.*)/;$tag=$1;$value=$2;
 4      s/\s\s+//g;
 5      if ($tag=~/^(\s*)([A-Za-z_]+)/){
 6          if($1 eq ""){
 7              $field=$2;$i_field='val';$num{$i_field}=0;
 8              $f{$field}{$i_field}{$num{$i_field}}.=$value;
 9          }else{
10              $i_field=$2;$num{$i_field}++;
11              $f{$field}{$i_field}{$num{$i_field}}.=$value;}
12      }
13      else{
14          $f{$field}{$i_field}{$num{$i_field}}.=$_;}
15  }
16
17  print "LOCUS= $f{'LOCUS'}{'val'}{'0'}\n";
18  print "DEFINITION= $f{'DEFINITION'}{'val'}{'0'}\n";
19  print "gene FEATURE 101= $f{'FEATURES'}{'gene'}{'101'}\n";
20
21  $feat_to_get='repeat_region';
22
23  for($k=1;$k<=$num{$feat_to_get};$k++){
24      print "$feat_to_get $k= $f{'FEATURES'}{$feat_to_get}{$k}\n";}
25
```

Fig. 11. A Perl script for parsing a GenBank flatfile. The block of code beginning with line 1 performs the parsing. Lines 17–19 print the 'Locus' line, 'Definition' line, and the 101st 'gene' feature. The loop beginning at line 23 prints all annotated features of type 'repeat_region'. The output is shown in **Fig. 12**.

```
LOCUS= AJ270058            3052119 bp    DNA    linear    CON 09-FEB-2004
DEFINITION= Arabidopsis thaliana DNA chromosome 4, short arm.
gene FEATURE 101= complement(join(265039..265440,265594..265756,265883..266595))
/gene="AT4g00640"

repeat_region 1= 462916..463556/note="function=unclassified similar to F9H3, GenBank
accession number AF071527 similar to T17A5, GenBank
accession number AF024504"

repeat_region 2= 467547..467805/note="function=putative_soloLTR similar to T17A5, GenBank
accession number AF024504"

repeat_region 3= 473308..473985/note="function=unclassified similar to T3F12, GenBank
accession number AC002983 similar to F15K20, GenBank
accession number AC005824 similar to F4C12, GenBank
accession number AC005275"

repeat_region 4= 524882..524945/note="function=unclassified similar to F6H11, GenBank
accession number AL021614 similar to MPA24, GenBank
accession number AB010075"

repeat_region 5= 552826..553097/note="function=putative_solo_LTR similar to T17A5,
GenBank accession number AF024504"

repeat_region 6= 554206..554545/note="function=putative_MITES similar to T32N15, GenBank
accession number AC002534"

repeat_region 7= 571005..571567/note="function=unclassified similar to MRH10, GenBank
accession number AB006703 similar to MAE1, GenBank
accession number AB015472 similar to MXE2, GenBank
accession number AB018121 similar to F9H3, GenBank
accession number AF071527 similar to F17O7, GenBank
accession number AC003671"
```

Fig. 12. Partial output of the script of **Fig. 11** when applied to the flatfile for GenBank accession 'AJ270058'. After printing the 'Locus' line, the 'Definition' line, and the 101st 'gene' feature, the script prints all 'repeat_region' features.

written to a tab-delimited file for import into a spreadsheet (*see* **Fig. 13**). The result of such an import can be seen in **Fig. 14**. Database software such as MySQL (http://www.mysql.com) readily imports tab-delimited data, so that the file produced by the modified script can be uploaded directly into a database table where it can be queried with the powerful SQL query language.

18. The program 'fastacmd' is part of the standalone BLAST package. When a BLAST database is formatted with the '-o' switch set to 'T', additional index files are created that are used by 'fastacmd' to extract sequences, portions of sequences, and sequence ids from the formatted database. The ability to selectively extract portions of the BLAST database is helpful in the analysis of a sequence following the identification of a region of interest using a BLAST search. The operations

```
1    $feat_to_get='CDS';
2
3    $qualifiers="gene,protein_id,product,note,translation";
4    for $k (split(/,/,$qualifiers)){
5         $row.="$k\t";}
6    chop $row;print "$row\n";
7
8    for($k=1;$k<=$num{$feat_to_get};$k++){
9         %item="";$row="";
10        while($f{'FEATURES'}{$feat_to_get}{$k}=~/\/([^=]+?)=\"(.+?)\"/sg){
11             $item{$1}.=$2;print STDERR "($1\t$2)\n";}
12        for $k (split(/,/,$qualifiers)){
13             $row.="$item{$k}\t";}
14        chop $row;$row=~s/\n//g;print "$row\n";
15   }
```

Fig. 13. Perl-code fragment to produce a tab-delimited file suitable for import into a spreadsheet. To use the code, replace lines 17–24 of the code in **Fig. 11** with lines 1–15 of this figure and save this as a new script. The output is shown in **Fig. 14**.

	A	B	C	
1	gene	protein_id	product	note
2	AT4g00010	CAB80759.1	hypothetical protein	
3	AT4g00020	CAB80760.1	putative BRCA2 homolog	contains similarit
4	AT4g00030	CAB80761.1	predicted protein of unknown function	coded for by A. th
5	AT4g00040	CAB80762.1	putative chalcone synthase	strong similarity t
6	AT4g00050	CAB80763.1	putative transcriptional regulator	contains similarit
7	AT4g00060	CAB80764.1	hypothetical protein	Contains Prokary
8	AT4g00070	CAB80765.1	putative RING-finger protein	contains similarit
9	AT4g00080	CAB80766.1	putative protein	contains similarit
10	AT4g00090	CAB80767.1	putative WD repeat membrane protein	contains similarit
11	AT4g00100	CAB80768.1	putative ribosomal protein S13	S15.hmm, score:
12	AT4g00110	CAB80769.1	putative nucleotide sugar epimerase	contains similarit
13	AT4g00120	CAB80770.1	hypothetical protein	contains similarit
14	AT4g00130	CAB80771.1	hypothetical protein	similarity toConta
15	AT4g00140	CAB80772.1	hypothetical protein	contains similarit

Fig. 14. Portion of spreadsheet data generated by combining the code fragment of **Fig. 13** with that of **Fig. 11** and operating on the GenBank flatfile for accession AJ270058. Viewing data in a spreadsheet format allows live sorting by columns. Tab-delimited data can also be readily imported into database tables where it can be processed using powerful query languages such as SQL.

summarized in **Table 10** can be performed on a database of plant sequences that has been created from a GenBank ASN.1 file using the formatdb command line:

formatdb -aT -b T -p F -o T -i gbpln1.aso

19. The phrase 'low-complexity sequence' refers to a stretch of nucleotide or protein sequence that is repetitive or simple in composition *(32)*. Extreme examples include runs of A's in a nucleotide sequence, such as the poly-A tails of eukaryotic mRNAs or the poly-proline tracts found in some proteins, but the runs need not be limited to repeats of a single base or amino acid. BLAST detects and filters these runs in the 'query' by default because they often lead to false starts when BLAST initiates alignments; beginning an alignment in the poly-A tail of an mRNA is not very likely to lead to a meaningful alignment between related mRNA sequences. Low-complexity filtering may be turned off from the BLAST Web pages using a checkbox or using either blastall or blastcl3 with the '-FF' option. Nucleotide sequences are filtered using a program called Dust (Tatusov and Lipman, *Dust*, Unpublished); protein sequences are filtered with SEG.

Table 10
Sequence Extractions from Local BLAST Databases Supported by fastacmd

Command Line	Result	Sample Output
fastacmd -D1 -f -d gbpln1.aso	Extracts the whole database in FASTA format	>gi|19352189|dbj|AB071376.1| Panax ginseng... AGCAGATTTTATACTACACAGAGAAAATGGGTTTGCAGGAGG... >gi|19386651|dbj|AB081324.1| Populus nigra... AATATGAAGATCACTAAATTTCTAGGGCTCTCCTTCCTTCTC... >gi|19386653|dbj|AB081325.1| Populus nigra... AATATGAAGATCACTGAAGTTCTTGGGCTCTCGTTCCTTCTC...
fastacmd -D2 -pF -d gbpln1.aso	Extracts the whole database as a list of National Center for Biotechnology Information 'gi' identifiers	14588671 19570875 19570876
fastacmd -D3 -pF -d gbpln1.aso	Extracts the whole database as a list of GenBank accession numbers	U12859.1 U12860.1 U12856.1
fastacmd -s -pF -d gbpln1.aso	Extracts several sequences from the databases using various identifiers	>gi|549980|gb|U12856.1|ATU12856 Arabidopsi... CATTTCCTCCTTCTTCTCTCTTCTATCTGTGAACAAGGCAC... >gi|549976|gb|U12859.1|ATU12859 Arabidopsi... AAAACAACAATGGAGCTACTCACAAGCTTGACCAACTGGTTA... >gi|549978|gb|U12860.1|ATU12860 Arabidopsi... ACAAGTAAAAGAAAGAGCGAGAAATCATCGAAATGGATTTCA...
fastacmd -s u12856,u12859,u12860 L100,120 -pF -d gbpln1.aso	Extracts bases 100–120 from the sequence with identifiers U12856, U12859, and U12860	>gi|549980:100-120 Arabidopsis thaliana Co... TTATAGTTGGAAATTAATTGA... >gi|549976:100-120 Arabidopsis thaliana Co... AACCGGTTCAGTTCCTCTTCA... >gi|549978:100-120 Arabidopsis thaliana Co... AGAGAAGAGGACATAAGACTG...

Table 11
Command Lines for Remote Searches of Plant Sequences in GenBank Using the 'blastcl3' Network Client

blastcl3 command line	Task
blastcl3 -i myseq -p blastn -d nr -u "gbdiv pln[prop]" -o outfile	Search only the plant division in GenBank using 'blastn'
blastcl3 -i myseq -p blastn -d est_other -u "plants[orgn]"-o outfile	Search plant expressed sequence tags in GenBank using 'blastn'
blastcl3 -i myseq -p blastp -d nr -u "gbdiv pln[prop]" -o outfile	Search only the protein translations of CDSs annotated on plant division GenBank sequences using 'blastp'

20. It is usually necessary to have a look at the data prior to analysis to determine its exact structure or to reformat the data for input into another program. For these operations, one needs an editor capable of gracefully dealing with extremely large files and one that will not introduce extraneous characters into data files when saved. Word processors are not recommended as they tend to corrupt data files by introducing formatting codes. Powerful text editors that are freely available across all platforms are VIM (http://www.vim.org/), Nedit (http://www.nedit.org/), and Emacs (http://www.gnu.org/software/emacs/).

21. Many NCBI databases, including the sequences in GenBank, are available for BLAST searching over a network connection using a command line BLAST network client application called 'blastcl3' (http://www.ncbi.nlm.nih.gov/BLAST/download.shtml). The parameters accepted by 'blastcl3' are similar to those of 'blastall', the program used for searches of local BLAST databases. Database names are those found on the NCBI Web BLAST pages, and the database searched may be limited using Entrez queries. To see the complete list of 'blastcl3' parameters, type 'blastcl3 -' at the command prompt (*see* **Note 11**). Some command lines for representative searches of GenBank using 'blastcl3' are summarized in **Table 11**.

References

1. Benson, D.A., Karsch-Mizrachi, I., Lipman, D.J., Ostell, J., and Wheeler, D.L. (2006) Genbank. *Nucleic Acids Res.*, 34,16–20.
2. Mizrachi, I. (2004) Genbank, in *The NCBI Handbook*. National Center for Biotechnology Information.
3. Cochrane, G., Aldebert, P., Althorpe, N., Andersson, M., Baker, W., Baldwin, A., Bates, K., Bhattacharyya, S., Browne, P., van den Broek, A., Castro, M., Duggan, K., Eberhardt, R., Faruque, N., Gamble, J., Kanz, C., Kulikova, T., Lee, C., Leinonen, R., Lin, Q., Lombard, V., Lopez, R., McHale, M.,

McWilliam, H., Mukherjee, G., Nardone, F., Pastor, M.P., Sobhany, S., Stoehr, P., Tzouvara, K., Vaughan, R., Wu, D., Zhu, W., and Apweiler, R. (2006) EMBL nucleotide sequence database: developments in 2005. *Nucleic Acids Res.*, 34,10–15.

4. Ohyanagi, H., Tanaka, T., Sakai, H., Shigemoto, Y., Yamaguchi, K., Habara, T., Fujii, Y., Antonio, B.A., Nagamura, Y., Imanishi, T., Ikeo, K., Itoh, T., Gojobori, T., and Sasaki, T. (2006) The rice annotation project database (rap-db): hub for *Oryza sativa* ssp. *japonica* genome information. *Nucleic Acids Res.*, 34, 741–744.

5. Wheeler, D.L., Barrett, T., Benson, D.A., Bryant, S.H., Canese, K., Chetvernin, V., Church, D.M., DiCuccio, M., Edgar, R., Federhen, S., Geer, L.Y., Helmberg, W., Kapustin, Y., Kenton, D.L., Khovayko, O., Lipman, D.J., Madden, T.L., Maglott, D.R., Ostell, J., Pruitt, K.D., Schuler, G.D., Schriml, L.M., Sequeira, E., Sherry, S.T., Sirotkin, K., Souvorov, A., Starchenko, G., Suzek, T.O., Tatusov, T., Tatusova, T.A., Wagner, L., and Yaschenko, E. (2006) Database resources of the national center for biotechnology information. *Nucleic Acids Res.*, 34, 173–180.

6. Federhen, S. (2003) The taxonomy project, in *The NCBI Handbook*. National Center for Biotechnology Information.

7. The Arabidopsis Genome Initiative. (2000) Analysis of the genome sequence of the flowering plant *Arabidopsis thaliana*. *Nature*, 408, 796–815.

8. Wang, J., Wong, G.K., Li, S., Liu, B., Deng, Y., Dai, L., Zhou, Y., Zhang, X., Yu, J., and Hu, S. (2002) A draft sequence of the rice genome (*Oryza sativa* L. ssp. *indica*). *Science*, 296, 79–92.

9. Yamamoto, K., Sakata, K., Baba, T., Katayose, Y., Wu, J., Niimura, Y., Cheng, Z., Nagamura, Y., Sasaki, T., and Matsumoto, T. (2002) The genome sequence and structure of rice chromosome 1. *Nature*, 420, 312–316.

10. Gocayne, J.D., Dubnick, M., Polymeropoulos, M.H., Xiao, H., Merril, C.R., Wu, A., Olde, B., Moreno. R.F., Adams, M.D., and Kelley, J.M. (1991) Complementary DNA sequencing: expressed sequence tags and human genome project. *Science*, 252, 1651–1656.

11. Tolstoshev, C.M., Boguski, M.S, and Lowe, T.M. (1993) dbEST–database for expressed sequence tags. *Nat. Genet.*, 4, 332–333.

12. Boguski, M.S. (1995) The turning point in genome research. *Trends Biochem. Sci.*, 20, 295–296.

13. Wagner, L., Pontius, J.U., and Schuler, G.D. (2003) Unigene: A unified view of the transcriptome in *The NCBI Handbook*. National Center for Biotechnology Information.

14. Kitts, A., and Sherry, S. (2003) The single nucleotide polymorphism database (dbSNP) of nucleotide sequence variation, in *The NCBI Handbook*. National Center for Biotechnology Information.

15. Ostell, J.M. (2003) The entrez search and retrieval system, in *The NCBI Handbook*. National Center for Biotechnology Information.

16. Anderson, J., Fedorova, N., DeWeese-Scott, C., Geer, L.Y., Hurwitz, D., Jackson, J.J., Jacobs, A., Lanczycki, C., Liebert, C., and Marchler-Bauer, A. (2005) MMdb: Entrez's 3D-structure database. *Nucleic Acids Res.*, 33, D192–D196.

17. Sayers, E., and Bryant, S. (2003) Macromolecular structure databases, in *The NCBI Handbook*. National Center for Biotechnology Information.

18. Jentsch, J., Canese, K., and Myers, C. Pubmed: the bibliographic database, in *The NCBI Handbook*. National Center for Biotechnology Information.

19. Beck, J., and Sequeira, E. (2003) Pubmed central (PMC): an archive for literature from life sciences journals, in *The NCBI Handbook*. National Center for Biotechnology Information.

20. Madden, T. (2003) The blast sequence analysis tool, in *The NCBI Handbook*. National Center for Biotechnology Information.

21. Kwan, K. Linkout: linking to external resources from entrez databases, in *The NCBI Handbook*. National Center for Biotechnology Information.

22. Jaiswal, P., Ni, J., Yap, I., Ware, D., Spooner, W., Youens-Clark, K., Ren, L., Liang, C., Zhao, W., Ratnapu, K., Faga, B., Canaran, P., Fogleman, M., Hebbard, C., Avraham, S., Schmidt, S., Casstevens, T.M., Buckler, E.S., Stein, L., and McCouch S. (2006) Gramene: a bird's eye view of cereal genomes. *Nucleic Acids Res.*, 34, D717–D723.

23. Garcia-Hernandez, M., Berardini, T.Z., Chen, G., Crist, D., Doyle, A., Huala, E., Knee, E., Lambrecht, M., Miller, N., Mueller, L.A., Mundodi, D., Reiser, L., Rhee, S.Y., Scholl, R., Tacklind, J., Weems, D.C., Wu, Y., Xu, I., Yoo, D., Yoon, J., and Zhang, P. (2002) TAIR: a resource for integrated Arabidopsis data. *Funct. Integr. Genomics*, 2, 239.

24. Gundlach, H., Lemcke, K., Rudd, S., Kolesov, G., Arnold, R., Mewes, H.W., Mayer, K.F., Schoof, H., and Zaccaria, P. (2002) MIPS *Arabidopsis thaliana* database (MAtdb): an integrated biological knowledge resource based on the first complete plant genome. *Nucleic Acids Res.*, 30, 91–93.

25. Dong, Q., Polacco, M.L., Seigfried, T.E., Lawrence, C.J., and Brendel, V. (2004) MaizeGDB, the community database for maize genetics and genomics. *Nucleic Acids Res.*, 32, D393–D397.

26. Gene Ontology Consortium. (2006) The gene ontology (GO) project in 2006. *Nucleic Acids Res.*, 34, 322–326.

27. Sayers, E., and Wheeler, D. (2004) Building customized data pipelines using the entrez programming utilities (eutils), in *NCBI Short Courses*. National Center for Biotechnology Information.

28. Pruitt, K.D., Tatusova, T., and Maglott, D.R. (2005) NCBI reference sequence (RefSeq): a curated non-redundant sequence database of genomes, transcripts and proteins. *Nucleic Acids Res.*, 33(1), D501–D504.

29. Tatusova, T., Pruitt, K.D., and Ostell, J.M.(2003) The reference sequence (refseq) project, in *The NCBI Handbook*. National Center for Biotechnology Information.

30. Madden, T.L., Schaffer, A.A., Zhang, J., Zhang, Z., Miller, W., Altschul, S.F., and Lipman, D.J. (1997) Gapped blast and psi-blast: a new generation of protein database search programs. *Nucleic Acids Res.*, 25, 3389–3402.
31. Madden, T.L., and McGinnis, S., (2004) BLAST: at the core of a powerful and diverse set of sequence analysis tools. *Nucleic Acids Res.* 32, W20–W25.
32. Wootton, J.C., and Federhen, S. (1996) Analysis of compositionally biased regions in sequence databases. *Methods Enzymol.*, 266, 554–571.
33. Zhang, Z., Schwartz, S., Wagner, L., and Miller, W. (2000) A greedy algorithm for aligning DNA sequences. *J. Comput. Biol.*, 7(1–2), 203–214.

3

A Collection of Plant-Specific Genomic Data and Resources at NCBI

Tatiana Tatusova, Brian Smith-White, and James Ostell

Summary

The National Center for Biotechnology Information (NCBI) provides a data-rich environment in support of genomic research by collecting the biological data for genomes, genes, gene expressions, gene variation, gene families, proteins, and protein domains and integrating the data with analytical, search, and retrieval resources through the NCBI Web site. Entrez, an integrated search and retrieval system, enables text searches across various diverse biological databases maintained at NCBI. Map Viewer, the genome browser developed at NCBI, displays aligned genetic, physical, and sequence maps for eukaryotic genomes including those of many plants. A specialized plant query page allows maps from all plant genomes available in the Map Viewer to be searched to produce a display of aligned maps from several species. Customized Plant Basic Local Alignment Search Tool (PlantBLAST) allows the user to perform sequence similarity searches in a special collection of mapped plant sequence data and to view the resulting alignments within a genomic context using Map Viewer. In addition, pre-computed sequence similarities, such as those for proteins offered by BLAST Link (BLink), enable fluid navigation from un-annotated to annotated sequences, quickening the pace of discovery. Plant Genome Central (PGC) is a Web portal that provides centralized access to all NCBI plant genome resources. Also, there are links to plant-specific Web resources external to NCBI such as organism-specific databases, genome-sequencing project Web pages, and homepages of genomic bioinformatics organizations.

Key Words: Bioinformatics; plant genomics; plant genetics; genetic map; physical map; comparative genomics; sequence analysis.

From: *Methods in Molecular Biology, vol. 406: Plant Bioinformatics: Methods and Protocols*
Edited by: D. Edwards © Humana Press Inc., Totowa, NJ

1. Introduction

Molecular biology is generating a deluge of data that is dramatically altering our understanding of the processes that underlie all living beings. This new knowledge is already affecting medicine, agriculture, biotechnology, and basic science in a fundamental way.

The plant genomic effort has one technical hurdle relative to other genomic efforts. The range of plant genome size is very large, extending from approximately the genome size of small animals to more than five times the size of the human genome. However, the availability of different genetic strains and high-throughput sequencing technologies has generated a substantial amount of molecular and genetic data for the analysis of plant genome structure and function.

The National Center for Biotechnology Information (NCBI) is collecting the data from various resources and integrating the data with the analysis and retrieval tools. Direct access to this information is available from Plant Genome Central (PGC) (*see* **Table 1**). Map Viewer and Basic Local Alignment Search Tool (BLAST) are two powerful tools for the analysis of plant genes, genome structure and function, and for exploring the molecular basis of biological

Table 1
Resources Accessed Through a Browser and by Download of Files

Resource	URL
Through a browser	
NCBI homepage	http://www.ncbi.nlm.nih.gov
Entrez Global Query	http://www.ncbi.nlm.nih.gov/gquery/ gquery.fcgi
Plants in Entrez Genome Project	http://www.ncbi.nlm.nih.gov/entrez/ query.fcgi?PureSearch&db=genomeprj& details_term=%22Viridiplantae%22%5B Organism%5
Map Viewer	http://www.ncbi.nlm.nih.gov/mapview
Multi-species plant genome search	http://www.ncbi.nlm.nih.gov/mapview/ map_search.cgi?chr=plants.inf
BLAST	http://www.ncbi.nlm.nih.gov/BLAST/
Customized plant genome BLAST	http://www.ncbi.nlm.nih.gov/BLAST/ Genome/PlantBLAST.shtml
Plant Genome Central	http://www.ncbi.nlm.nih.gov/genomes/ PLANTS/PlantList.html

Entrez Gene	http://www.ncbi.nlm.nih.gov/entrez/query.fcgi?db=Gene
Entrez Probe	http://www.ncbi.nlm.nih.gov/entrez/query.fcgi?db=probe
Reference Sequences	http:/www.ncbi.nlm.nih.gov/RefSeq/
By download of files	
Genome records	ftp://ftp.ncbi.nlm.nih.gov/genomes
Reference Sequences	ftp://ftp.ncbi.nlm.nih.gov/refseq
RefSeq records for all plants	ftp://ftp.ncbi.nlm.nih.gov/refseq/release/plant/
Arabidopsis thaliana data for each chromosome	ftp://ftp.ncbi.nlm.nih.gov/genomes/Arabidopsis_thaliana
Oryza sativa data for each chromosome	ftp://ftp.ncbi.nlm.nih.gov/genomes/Oryza_sativa
GenBank gis used to create customized plant genome BLAST	ftp://ftp.ncbi.nlm.nih.gov/PLANTS/BLASTDB
BAC-end sequences in FASTA format, organized by organism	ftp://ftp.ncbi.nlm.nih.gov/genomees/BACENDS

BLAST, Basic Local Alignment Search Tool; NCBI, National Center for Biotechnology Information.

phenomena such as salt tolerance, disease resistance, herbicide tolerance, viral infection, stress tolerance, and crop production.

NCBI has developed a multi-organism query page that uses a single query to search the molecular and genetic information for the available plant genomes. The completion of the *Arabidopsis thaliana (1)* and *Oryza sativa (2)* genome projects has permitted plant researchers to take advantage of these genomes for comparative plant genomics where, in most cases, all that are available are high-throughput uncharacterized sequence data and mapped markers. NCBI has a collection of databases of mapped and unmapped plant sequences for sequence similarity searching using specialized plant genomic BLAST pages.

2. Materials

2.1. Primary Sequence Data

The bulk of the primary plant genomic data available at NCBI falls into one of three categories that mirror the types of projects currently conducted by the plant research community. These are genomic assemblies, batches of expressed sequence tags (ESTs), and genetic or physical genome maps.

2.1.1. International Nucleotide Sequence Databases

The main source of the primary sequence data is nucleotide sequences submitted by individual researchers or sequencing centers from around the world. These sequences are submitted directly to GenBank *(3)*, an archival database of nucleotide sequences built and maintained by NCBI, or are replicated from one of the collaborating databases, the DNA Databank of Japan *(4)* or the European Molecular Biological Laboratory Data Library *(5)*. Protein sequence data are taken from several sources *(6–8)*, including the conceptual translations of coding regions annotated on GenBank records.

Complete genome assemblies include that of the *A. thaliana* genome produced by the Arabidopsis Genome Initiative *(1)*, the assembly of *O. sativa* ssp. *japonica* resulting from the International Rice Genome Sequencing Project *(2)*, and that of *O. sativa* L. ssp. *indica*, produced by the Chinese Academy of Sciences in Beijing, China *(9)*. It is expected that genome sequencing projects for *Populus trichocarpa* from the DOE Joint Genome Institute (http://www.jgi.doe.gov/sequencing/why/cottonwood.html) and *Medicago truncatula* from the Medicago truncatula International Consortium (http://www.medicago.org/genome/) will be complete this year and submitted to public databases.

2.1.2. Trace Archive

Most of the data generated in genome sequencing projects is produced by whole-genome shotgun sequencing, resulting in random short (600–800 nucleotides) fragments (traces). The fragments are assembled by computer programs to generate a consensus sequence.

For many years, the traces (raw sequence reads) remained out of the public domain because the scientific community has focused its attention primarily on the end product: the fully assembled final genome sequence. But in many cases, it is important to have precise information about a specific position in the sequence. To this end, it would be helpful to be able to go back to the experimental evidence that underlies the genome sequence at that position and to see if there is any ambiguity or uncertainty about the sequence. To meet these needs, NCBI and The Wellcome Trust Genome Campus in Hinxton, UK created in 2001 a repository of the raw sequence traces generated by large sequencing projects, that allows retrieval of both the sequence file and the underlying data that generated the file, including the quality scores. The Assembly Archive *(10)* created at NCBI in 2004 links the raw sequence information found in the Trace Archive with consensus genomic sequence. Currently, trace data from 42 plant sequencing projects are stored in Trace Archive at NCBI (*see* **Table 2**).

2.1.3. Expressed Sequence Tags

NCBI dbEST contains sequence data and other information developed from single-pass sequencing of cDNA inserts. Frequently, this is the only data from large-scale projects to develop genome-wide sequence data. The summary of organisms in the EST public collection is presented at http://www.ncbi.nlm.nih.gov/dbEST/dbEST_summary.html

2.1.4. Bacterial Artificial Chromosome End Sequences

Many genome projects generate a library of bacterial artificial chromosomes (BACs) each with an insert between 100 and 200 kbp. These BACs are grouped into contigs by FPC. An inexpensive way to gather some sequence data for the contigs is to determine the nucleotide sequence for 300–500 nucleotides at the ends of the BACs. These BAC-end sequences (BES) are deposited into the international nucleotide sequence repositories. On a regular basis, NCBI collects organism-specific subsets of the plant BES into FASTA format files. These files are available on the FTP server (*see* **Table 1**).

Table 2
Plant Organisms in Trace Archive Database

Acorus americanus	*Nuphar advena*	*Oryza sativa* (*indica* cultivar-group)
Brassica oleracea	*Opuntia cochenillifera*	*O. sativa* (*japonica* cultivar-group)
Ceratopteris richardii	*Oryza alta*	*Oryza* sp.
Chara braunii	*Oryza australiensis*	*Ostreococcus* sp. CCE9901
Chlamydomonas reinhardtii	*Oryza brachyantha*	*Physcomitrella patens*
Cicer arietinum	*Oryza coarctata*	*Pinus taeda*
Citrus sinensis	*Oryza glaberrima*	*Populus balsamifera*
Coleochaete orbicularis	*Oryza granulata*	*Selaginella moellendorffii*
Glycine max	*Oryza minuta*	*Solanum demissum*
Gossypium herbaceum	*Oryza nivara*	*Sorghum bicolor*
Liriodendron tulipifera	*Oryza officinalis*	*Triphysaria versicolor*
Medicago truncatula	*Oryza punctata*	*Volvox carteri*
Meiracyllium	*Oryza ridleyi*	*Volvox carteri f. nagariensis*
Mimulus guttatus	*Oryza rufipogon*	*Zea mays*

2.1.5. Probe Database

The NCBI Probe database is a public registry of nucleic acid reagents designed for a wide range of applications, together with links to information available from data repositories external to NCBI (*see* **Table 1** for URL). The current role of this database for plant resources is to bring some of the data associated with mapped loci across a wide range of organisms into a structure amenable to intensive computation and data mining. For plants, the current information in Probe is the source organism of a probe, any GenBank records generated from the effort to determine some, or all, of the nucleotide sequence of the cloned nucleic acid, the link(s) in external database(s) to the information concerning this nucleic acid, and the NCBI maps with loci identified using the nucleic acid.

2.2. Derived (Pre-Calculated) Data

NCBI has developed several methods of organizing the relationships between the elements in primary sequence data records to facilitate movement by the user along these relationships. All of these involve pre-computing the relationship and storing this information in a database.

2.2.1. UniGene Clusters

Organisms for which there are 70,000 or more EST sequences available are listed in UniGene. UniGene is an experimental system for automatically partitioning GenBank sequences into a nonredundant set of gene-oriented clusters. Each UniGene cluster contains sequences that represent a unique gene, as well as related information such as the tissue types in which the gene has been expressed, and map location. The UniGene collection currently contains clusters for over 40 organisms, including more than 200,000 clusters for 20 plants (*see* **Table 3**).

2.2.2. UniSTS

UniSTS is a database of primer pairs that have been publicly reported as useful for STS efforts. SSR is considered a subset of STS satisfying the additional condition that the amplified material contains a simple nucleotide repeat. Thus, UniSTS includes the primer pairs used to identify an SSR locus. The data in UniSTS are collected from mapping efforts or GenBank submissions. The main role of this database for plant resources, like Probe, is to bring some of the data associated with mapped loci across a wide range of organisms into a structure amenable to intensive computation and data mining. The records

Table 3
Plant Organisms in UniGene with Number of UniGene Clusters

Plant organisms	Number of UniGene clusters
Bryopsida	
Physcomitrella patens	7962
Coniferopsida	
Picea glauca	4929
Picea sitchensis	6907
Pinus taeda	14, 198
Eudicotyledons	
Arabidopsis thaliana	25, 697
Brassica napus	3452
Citrus sinensis	8786
Glycine max	15, 047
Gossypium raimondii	3120
Helianthus annuus	2653
Lactuca sativa	4879
Lotus corniculatus	8193
Lycopersicon esculentum	12, 746
Malus × domestica	15, 274
Medicago truncatula	5415
Populus balsamifera	4783
Populus tremula × Populus tremuloides	8127
Solanum tuberosum	16, 899
Vitis vinifera	14, 608
Liliopsida	
Hordeum vulgare	13, 485
Oryza sativa	44, 394
Saccharum officinarum	4787
Sorghum bicolor	8704
Triticum aestivum	35, 263
Zea mays	26, 797

in UniSTS contain the sequence of the primers, the size of amplified product if known, PCR conditions if known, and GenBank records germane to the STS and other genomes where the STS has been identified either empirically or computationally by ePCR.

2.2.3. Entrez Gene

Entrez Gene *(11)* (*see* **Table 1**) has been implemented at NCBI to organize information about genes, serving as a major node in the nexus of genomic map, sequence, expression, protein structure, function, and homology data. Gene records are established for known or predicted genes, which are defined by nucleotide sequence or map position. Not all taxa are represented, and the current scope matches that of NCBI's Reference Sequences group (*see* **Table 1**) and NIH's Mammalian Gene Collection covering over a million loci, including over 81,000 loci for plants.

2.2.4. HomoloGene

HomoloGene is a system for the automated detection of homologs among the annotated genes of several completely sequenced eukaryotic genomes. It contains clusters of homologous genes for over a dozen model organisms, including almost 16,000 clusters for plants.

2.2.5. Conserved Protein Domains

Conserved Domain Database (CDD) *(12)* is a collection of sequence alignments and profiles representing protein domains conserved during molecular evolution. Molecular evolution uses such domains as building blocks, and these may be recombined in different arrangements to make different proteins with different functions. The collections of domain alignments in the CDD are imported from two databases outside of the NCBI, named Pfam *(13)* and SMART *(14)*; from NCBI collection of COGs *(15)*; and from a database curated by the CDD staff. CDD release v2.05 contains over 10,000 alignment-based protein domain profiles, including over 250 that are specific to plants.

2.2.6. BLAST Results with Links

The BLAST Link (BLink) report represents a pre-computed list of similar proteins for all sequences in Entrez Protein database. BLink offers various display options, including the distribution of hits by taxonomic grouping, the best hit to each organism, the protein domains in the query sequence, similar sequences that have known 3D structures, and more. The "3DStructures" option on any BLink report shows the BLAST hits that have 3D structure data in MMDB, whereas the "CDD-Search" button displays the CD-Search results page for the query protein. Additional options allow you to filter out some taxa from the result list, increase or decrease the BLAST cutoff score, or filter the BLAST hits to show only those from a specific source database, such as RefSeq or Swiss-Prot.

2.2.7. Gene Expression Omnibus

The Gene Expression Omnibus (GEO) *(16)* developed at NCBI serves as a public repository for data generated from high-throughput microarray experiments. GEO has a flexible and open design that allows the submission, storage, and retrieval of many types of data sets, such as those from high-throughput gene expression, genomic hybridization, and antibody array experiments. The 2005 collection represents the data from over 20 organisms, including 38 data sets and over 600,000 expression profiles for plant tissues.

2.3. Plant-Specific Data Resources

There are three broad groups of databases that contain data supporting the plant genomic resources at NCBI: the Map Viewer database containing plant genomic and genetic maps, Probe database and UniSTS containing probes and primers used to identify the allelic state of a locus in a mapping endeavor, and the databases for PlantBLAST which consist of various subsets of GenBank accessions associated with mapped loci through the probes and primers used to identify the loci. There are two sets of plant-specific data available for download through FTP: the BES and the accessions comprising the databases underlying PlantBLAST.

2.3.1. PlantBLAST Databases

The process of locating GenBank accession(s) for a particular probe allows NCBI to build databases for BLAST. Each of these databases contains the subset of GenBank accessions that are associated with mapped loci for a particular organism. The customized databases include four data sets:

1. GenBank sequences associated with the probe used to identify mapped genetic loci in genetic maps of *Avena sativa* (oat), *Hordeum vulgare* (barley), *O. sativa* (rice), *Lycopersicon esculentum* (tomato), *Glycine max* (soybean), *Zea mays* (corn), and *Triticum aestivum* (wheat).
2. The two collections of rice contigs developed by the Chinese WGS endeavors.
3. The whole-genome material for the *A. thaliana* genome and the *O. sativa* genome. These are the resource behind PlantBLAST at http://www.ncbi.nlm.nih.gov/ BLAST/Genome/PlantBlast.shtml. Use of this resource will be explained through examples later in the chapter.
4. A collection of databases built by NCBI with organism-specific ESTs that are available for BLAST query. The two criteria for inclusion in this group of databases are that more than 50,000 ESTs have been submitted for the organism or that the organism has maps visible in Map Viewer regardless of the number of EST submitted.

2.3.2. EST/UniGene Alignments

ESTs arising from several plant species, as well as the UniGene clusters derived from them, are aligned to the well-assembled *A. thaliana* and *O. sativa* genomes. ESTs from the monocots (*Liliopsida*) are aligned to the *O. sativa* genome; these include *H. vulgare, O. sativa, Sorghum bicolor, T. aestivum*, and *Z. mays.* Those from the Eudicotyledons are aligned to the *A. thaliana* genome and include *Arabidopsis* itself, *G. max, Lactuca sativa, Lotus corniculatus, Malus* × *domestica, M. truncatula, Populus tremula* × *Populus tremuloides, Solanum tuberosum*, and *V. vinifera.* These alignments make an important link between EST data, which are relatively inexpensive to obtain, and assembled, well-annotated genomic sequence, a more expensive commodity.

2.3.3. Genetic Map Data

Genetic map data in Map Viewer are developed only from data in the public domain. These data arrive from three broad sources.

1. Organism-specific databases external to NCBI have provided data for soybean, Poaceae (grains, corn, rice, sorghum, and foxtail millet), and Solanaceae (tomato, pepper, eggplant, and potato). The maps in this group are available concurrently at NCBI and the data provider. This includes maps from MaizeGDB (*17*), Gramene (*18*), GrainGenes (2005, http://wheat.pw.usda.gov/GG2/index.shtml), SoyBase (2005, http://soybase.org), and the Solanaceae Genomics Network (*19*).
2. Individual researchers have provided map data for sorghum, barley, onion, cassava, and beet.
3. Peer-reviewed publications satisfying the criterion that the number of linkage groups not exceed 110% of the number of chromosome pairs have been the source of map data for asparagus, almond, tef, some Solanaceae, foxtail millet, lotus, alfalfa, barrel medic, cocoa, sorghum, and rubber tree. The loci in the excess linkage groups are included in the map data as "Unplaced" loci. **Table 4** summarizes the number of maps per organism currently available in Map Viewer. Map data will continue to be developed from these three sources.

Routine processing of genetic map data at NCBI includes ascertaining that locus mnemonics conform to the nomenclature rules for those organisms that have nomenclature rules, ascertaining the accuracy of the probe name, normalization of the probe names as necessary, locating the GenBank accession(s) that are the nucleotide sequence for the cloned nucleic acid of the probe, and determining the method used to identify loci. AFLP primer names use the vocabulary elaborated by Keygene regardless of the name used by the mapping project. A table of this vocabulary is available through a URL in PGC. Special organism-specific processing can include normalizing the linkage group names across

Table 4
Plant Genetic Maps Available at NCBI

Organism	Maps at NCBI		Nonredundant	
	Solely	Concurrently	GenBank accessions	Loci
Aegilops longissima		1	50	74
Aegilops taushci		2	71	356
Aegilops umbellulata		1	37	87
Allium cepa (onion)	1		186	229
Asparagus officinalis (aspargus)	2			278
Avena (oat)		3	604	1053
Beta vulgaris (beet)	2		25	319
Capsicum annuum (pepper)	1	2	101	1565
Eragrostis tef (tef)	1		77	149
Glycine max (domesticated soybean)		1	1547	2132
Heva brasiliensis (rubber tree)	1		31	711
Hordeum bulbosum		1	134	138
Hordeum vulgare (barley)		18	1079	3306
Lotus corniculatus (lotus)	1		273	164
Lolium perenne (perennial rye grass)	1		74	607
Manihot esculenta (cassava)	1			73
Medicago sativa (alfalfa)	2		0	315
Medicago truncatula	2		36	858
Oryza sativa (rice)	1			289
Prunus dulcis (almond)	2		116	126
Secale cereale (rye)		15	4404	9296

(Continued)

Table 4
(Continued)

Organism	Maps at NCBI		Nonredundant	
	Solely	Concurrently	GenBank accessions	Loci
Setaria italica (foxtail millet)		7	410	864
Solanum peruvianum	1	1	209	399
Solanum lycopersicum (tomato)	1	4	1066	2386
Solanum lycopersicoides (nightshade)	2		32	158
Solanum melongena (eggplant)		1	139	228
Solanum tuberosum (potato)		2	59	269
Sorghum bicolor (sorghum)	7	4	2480	9475
Theobroma cacao (cocoa)	2		20	435
Triticum aestivum (bread wheat)		13	11519	24371
Triticum monococcum		1	248	357
Triticum turgidum (durum wheat)		2	141	462
Zea mays (corn)		11	4067	15903

NCBI, National Center for Biotechnology Information.

the maps or normalizing the coordinate system across the chromosomes of a particular map group. Any special processing is explicitly described in the documentation for the organism. The aggregate results of this data processing are in **Table 4**: the number of nonredundant loci in the maps for an organism as well as the number of nonredundant probe-associated GenBank accessions for each organism.

3. Methods

3.1. Accessing the Data

There are many ways of accessing plant data at NCBI. The list of available online resources and sites with files available for download is provided in **Table 1**. Depending on the focus and the goal of the research project or the level of interest, the user would select a particular route for accessing the plant data and resources. These are (1) text searches either in Entrez or Map Viewer, (2) direct browsing through Map Viewer, and (3) searches by sequence similarity using BLAST. All these entries enable navigation through pre-computed links to other NCBI resources. The usage scenarios outlined below illustrate these three entry mechanisms and the opportunity to use other NCBI resources.

3.2. Entrez

Entrez is the text-based search and retrieval system used at NCBI for all the major databases, including PubMed, Nucleotide and Protein Sequences, Genomes, Genes, and many others. The actual databases from which records are retrieved and on which the Entrez indexes are based have different designs, based on the type of data, and reside on different machines. Entrez is the tool that integrates these into a uniform information model with a concomitant retrieval system. This allows a researcher to pose the question "How many tomato ESTs have an annotation for drought?" and receive an answer, or the question "How many Arabidopsis genes have 'salt' as part of the annotation?", or the question "How many PubMed articles are there concerning 'heavy metal' and 'flowering' in Arabidopsis?". At the top of PGC are links enabling the two routes to Entrez. In the black bar are links to the homepage of a selected group of Entrez databases. In the gray bar is the Entrez query tool. This tool allows a search of a single database at a time. The database to be queried is chosen from the drop-down list at the left side of the bar, and the query is typed into the adjacent text box. Results are provided in the format for the chosen database. It is not necessary to know which database holds the information being sought. The entire suite of Entrez databases may be searched using the Entrez Global Query page (*see* **Note 1**). This is available through the "Entrez" link in the black bar (*see* **Table 1**).

A network of links allows fluid navigation between these databases. Links may interconnect a sequence in Entrez Nucleotide to the abstract, in PubMed, of the paper in which it is cited. If a sequence in Entrez Nucleotide contains annotated coding regions, then the nucleotide accession will be interconnected to the sequences in Entrez Protein which are derived from the respective conceptual translations. Many Entrez links are computationally derived, such as links between protein sequences made on the basis of sequence similarity. These pre-computed protein alignments, updated daily, are available for viewing using the BLink resource. To see the various links to other databases or resources available for any record in Entrez, click on the "Links" link in the upper right-hand corner of the Entrez display.

3.2.1. Examples: Text-Based Search in Entrez

Start with Global Query page (*see* **Fig. 1**). This is a search engine that allows an investigator to perform a database query across all the Entrez databases at NCBI in a single search (*see* **Note 1**). As an example, consider a search for stress response-related genes in *A. thaliana*. One may begin with the following query: "response regulator" AND *Arabidopsis thaliana* [organism], (*see* **Note 2**). The search matches entries in the Entrez Nucleotide, Protein, Genome, Gene, UniGene, HomoloGene, Structure, CDD, 3D Domains, and GEO databases.

Fig. 1. The Entrez Global Query output for the query "response regulator" AND Arabidopsis thaliana[organism]. The entry points for two paths of searching are encircled.

Clicking on the number of the hits found in a particular Entrez databases starts the navigation through the inter-connected system of links. Consider two possible scenarios: explore gene-oriented resources including genetic map information, or follow the route via protein sequence to 3D structure and related sequences in other organisms.

Start with the Entrez Gene database—highlighted by a square rectangle in Figure 1. Global Query generates a display of summaries (data not shown) for 44 genes. The Entrez Gene report for the *ARR4* gene of *A. thaliana* (*see* **Fig. 2A**) starts with the gene name, gene ID, and short gene description. The "Link" menu in the right top corner provides connections to many NCBI resources such as UniGene, HomoloGene, and GEO Profiles, as well as links to other organism-specific sites such as The Arabidopsis Information Resource (TAIR), Munich Information Center for Protein Sequence (MIPS), and The Institute for Genomic Research

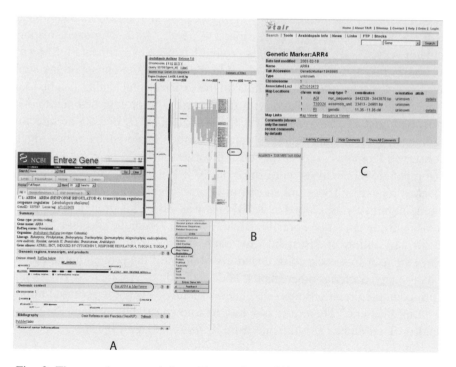

Fig. 2. The search traversal from Entrez Gene. (**A**) The Entrez Gene display for gene *ARR4*. The links to the Map View display for this gene are highlighted with a circle. (**B**) The Map Viewer display elicited from either link in the Entrez Gene display. The link to The Arabidopsis Information Resource (TAIR) marker report is highlighted with a circle. (**C**) The TAIR display for marker ARR4.

(TIGR). The Summary section includes gene type, organism information, and gene aliases. Following this is a graphical representation of a structure of the gene and genomic context with the link to see the gene in the Map Viewer. The two links to Map Viewer are highlighted in **Fig. 2A**. Following either link to Map Viewer brings up a region of the *A. thaliana* chromosome 1. Using Map Viewer navigation (Maps and Options) described in the next section, you can adjust the display to the region of interest and bring up additional tracks of UniGene and EST pre-computed alignments (*see* **Fig. 2B**). Additional links from Gene sequence maps provide link outs to Arabidopsis-specific resources like SIGnAL, Salk Institute Genomic Analysis Laboratory, TIGR, MIPS, and TAIR. Following the marker link on the genetic map highlighted in **Fig. 2B** brings up a TAIR report for the genetic marker (*see* **Fig. 2C**).

Returning to the Global Query page, it may be of interest to see if additional information can be found using a different route through the data at NCBI. Proceed to the Protein database—highlighted by a rounded rectangle—and select the SwissProt record (Accession: O82798): "Two-component response regulator ARR4" (*see* **Fig. 3A**). This protein has been experimentally shown to function as response regulator involved in the His-to-Asp phosphorelay signal transduction system *(20)*. More details on the experimental studies can be found in eight publications associated with the record. The protein belongs to ARR family and contains one response regulatory domain. The domain architecture, alignments, and domain relatives can be further explored by following the "Conserved Domains" link at the left top corner of the page.

Computational analysis performed at NCBI includes pre-computing the sequence relatives. These are shown in BLink, a graphical viewer of BLAST protein alignments developed at NCBI (*see* **Fig. 3B**). This protein is conserved across multiple taxa (there are 196 BLAST hits to 78 unique species). Selecting the display option of "Best hit" for each species from the top level menu allows the user to see the color coded distribution of the protein homologs that include many bacteria, other plants, and some fungi. Another NCBI resource of pre-computed relationships—"Related structures"—can be accessed through the "Links" menu. There are 144 hits to known structures, including many found in Bacteria but also the one from *A. thaliana* (*see* **Fig. 3C**).

3.3. Map Viewer

The Map Viewer provides a common interface to the visual display of genomic data at NCBI. This interface is an interactive tool for viewing an ensemble of aligned genetic, physical, or sequence-based maps with a user-adjusted focus ranging from an entire genome to a single gene. The "Maps and

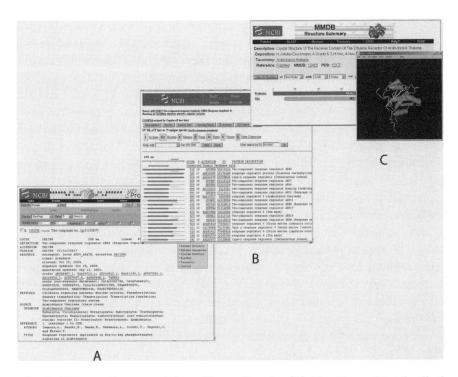

Fig. 3. The search traversal from Entrez Protein. (**A**) The Entrez Protein display for accession O82798. (**B**) The BLink display elicited from the link in **A**. (**C**) The summary and the 3D structure displayed by Cn3D for the receiver domain of the ethylene receptor of *Arabidopsis thaliana*.

Options" utility, available in the upper right corner of the display, allows the user to choose specific maps from those available as well as the order of the chosen maps across the display. Alignment between adjacent maps in a display is achieved on the basis of shared markers and is displayed as lines between the shared markers provided that the "Show connections" option is checked. Each plant that can be browsed in Map Viewer has an overview page. The organism-specific help documentation is available through this overview page. Each display has the ability to perform an Entrez query of the Map Viewer database.

3.3.1. Examples: Text-Based Search in Map Viewer

The Map Viewer provides a central interface to genomic data at NCBI, serving as an interactive tool for viewing an ensemble of aligned genetic, physical, or sequence-based maps with an adjustable focus ranging from that of a complete chromosome to that of a portion of a gene. The maps displayed in the Map Viewer may be derived from a single organism or from multiple organisms; map alignments are performed on the basis of shared markers.

Since the early 1990s, various lines of research have shown that large-scale genome structure is conserved in blocks across the grasses *(21–25)*. The result is a circular consensus grass genome *(26)*. The taxonomic groups that exhibit blockwise conservation of gene order within the group have been expanded to include the Solanaceae *(27–29)* and the legumes *(30–32)*. Locus nomenclature is organism specific and thus is unreliable as a query method between species. However, the regular nomenclature of plasmids *(33)* is not influenced by how the plasmid or insert is used. The data for the plant maps available through Map Viewer include the relationship between the organism-specific locus mnemonic and the organism-independent plasmid name. In principle, this relationship allows a cross-species text search.

Any text search term can be used as a query at the top of the Map Viewer homepage (*see* **Table 1**). These include, but are not limited to, a GenBank accession number or other sequence-based identifier, a gene symbol or alias, or the name of a genetic marker. For more complex queries, any query can be combined with one of three Boolean operator terms (AND, OR, and NOT). Wild cards, which are denoted by placing a "*" to the right of the search term, are also supported. The user can select the "all plant" group or the user can select a single organism. To demonstrate this resource, we will survey the different plant genomes for the probe RZ69.

Choose "All Plants" from the choices offered in the menu, enter the search term "RZ69" in the search box, and click "Go". The resulting page will show the number of hits to each plant genome and on which genetic map the hits are located. **Figure 4** shows that RZ69 has been identified 11 times: once in oat maps, six times in rice maps, once in bread wheat maps, and three times in corn maps. There are two directions available from this page. If the results extend across multiple organisms and only the results from a single organism are desired, then clicking on the name of the organism will display the Map Viewer results for that organism only. To exclude any map from an organism in the display, uncheck that organism. The default display chromosome for an organism is the first chromosome with a hit. To change the displayed chromosome for an organism, use the drop-down menu at the right. To exclude any map in the display, uncheck that map. To view all the selected chromosomes/maps, click on the "Display" link (*see* **Note 3**).

The dark gray line, a solid red line on the material displayed on a monitor, connects the mapped object on the different plant genetic maps and enables

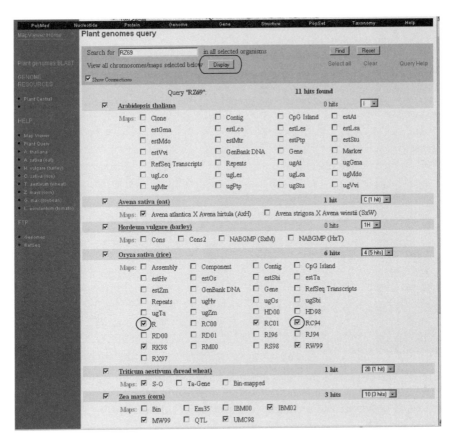

Fig. 4. The output of the RZ69 search of all plants. The display shown in **Fig. 5** is generated from this by unchecking the boxes highlighted with a black circle and clicking the button highlighted with a rounded rectangle.

one to see the real-time connection of loci on different plant chromosomes identified using an identical probe (*see* **Fig. 5**). Light gray lines, drawn in real time, show the connection of loci identified using identical probes (*see* **Note 4**).

The probe RZ69 can be considered as the name of a radius in the collection of concentric circles comprising the model of grass genomes (Figure 2 in **ref. *26***). One could draw a straight line for RZ69 and have each of the various chromosomes that the RZ69 radius marks extending above and below the straight line some amount. Alternatively, the chromosomes that the RZ69 radius marks could be arranged in a rectangular manner and the RZ69 radius drawn as a series of lines connecting the intersection points. The Map Viewer display is the latter.

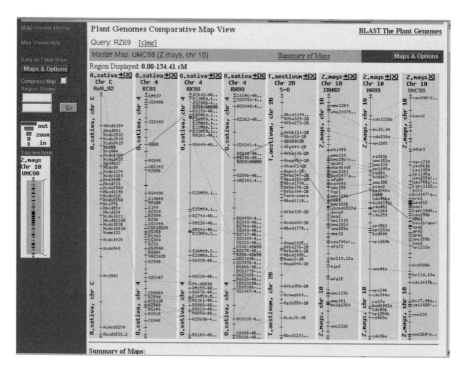

Fig. 5. The display of the maps identified through query with RZ69. The red (dark gray on the figure) lines connect the location of RZ69 in adjacent maps. The light gray lines extending between adjacent maps connect the loci identified using the same probe.

3.4. BLAST

BLAST finds regions of local similarity between sequences. The program compares nucleotide or protein sequences with members of a database and calculates the statistical significance of matches. NCBI provides the search software and the software to create a database that the search software can examine. NCBI also provides a Web services portal. For routine large searches—more than approximately 200 query sequences—or when the investigator wishes to examine nonpublic sequence data, NCBI recommends that the researcher installs BLAST locally. Otherwise, the researcher can use the Web portal (*see* **Table 1**) to search the various databases that NCBI creates and routinely updates.

3.4.1. Examples: Sequence-Based Search Using PlantGenomes BLAST

GenBank accession DT039940 is a grape EST. If this accession were one of the grape ESTs mapped to the *A. thaliana* genome sequence, performing

BLASTN against all mapped sequences would locate the Vvi_EST map, which contains the EST as a map object. However, this EST was deposited after pre-computation of the Vvi_EST map for build 5.0. Thus, the example can be considered a fairly close replica of an EST sequence produced by the sequencing machine during an EST project. The URL in either the Map Viewer portal or the Plant Central portal (encircled in **Fig. 6A**) directs the user to PlantGenomes BLAST (*see* **Fig. 6B**). A search of the *A. thaliana* proteins with BLASTX and the default parameters is done by changing the "Source" to *A. thaliana*, changing the database to proteins and entering the sequence—in this example conveniently encapsulated by a GenBank accession. Clicking on "Begin Search" yields two hits. Both have a similar score and a similar E value. Following the "Genome View" yields **Fig. 6C**. Clicking on chromosome 1 yields the default display. Use "Maps and Options" to modify this default display to create a display of, from left to right, the At_EST map, the RefSeq_mRNA map, and the Genes_seq map. Adjust the

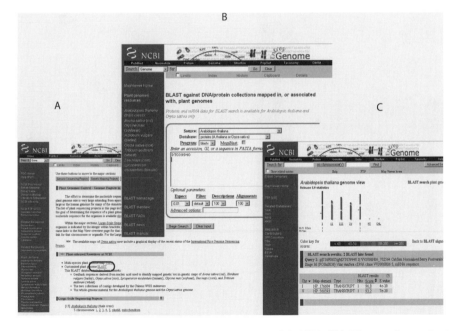

Fig. 6. The search traversal using PlantGenomes BLAST. (**A**) The starting point in Plant Genomes Central with the link to the BLAST page encircled in black. (**B**) The PlantGenomes BLAST page. (**C**) The display elicited by the "Genome View" button in the BLAST output page.

displayed region from 23,830K to 23,840K to generate **Fig. 7A**. The display shows gene *At1g64220* and that no *A. thaliana* ESTs have been recovered from this gene, though flanking genes are substantiated by EST and/or cDNA records. Clicking on the link out to SIGNaL shows in the display the absence of substantiating EST or cDNA data. However, the whole-genome chip part of the display, the MPSS part of the display, and the SAGE Target part of the display show that RNA is produced from this region of chromosome 1 and it appears that all tissues examined exhibit expression. Performing the same operations with "Maps and Options" on chromosome 5, with a display region between 16,670K to 16,710K, shows gene *At5g41685* with a couple of confirming ESTs (*see* **Fig. 7B**). It could be that this grape EST is unique—and thus not representative of a difference between *V. vinifera* and *A. thaliana*. Performing PlantEST BLAST using this EST against the *V. vinifera* ESTs shows that there are 18 other similar ESTs. Of the almost 421,000 *A. thaliana*

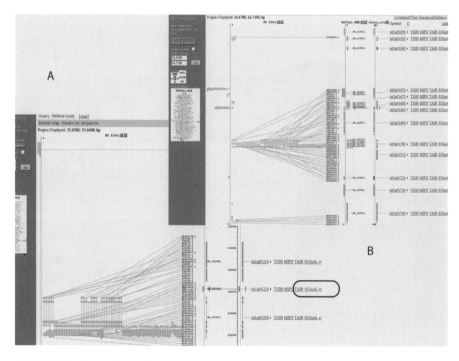

Fig. 7. The Map View display for the two regions of *Arabidopsis thaliana* identified using PlantGenomes BLAST example in **Subsection 3.4.** (**A**) The identified region on chromosome 1. (**B**) The identified region on chromosome 5.

ESTs, 16 are considered to represent gene *At5g41685* and none are considered to represent gene *At1g64220*. Of the slightly more than 190,400 *V. vinifera* ESTs, 18 are 96–100% identical to DT039940, with the differences found in regions expected to exhibit low sequence quality. Either one gene in wine grape exhibits more than twice the expression of a homolog in *A. thaliana* or possibly each of two genes in grape exhibits similar expression. Either way, one could conclude that this EST is part of the difference between *V. vinifera* and *A. thaliana*. It is possible to perform the BLAST-dependent part of the above example at other *Arabidopsis*-specific sites. However, the only way for a user

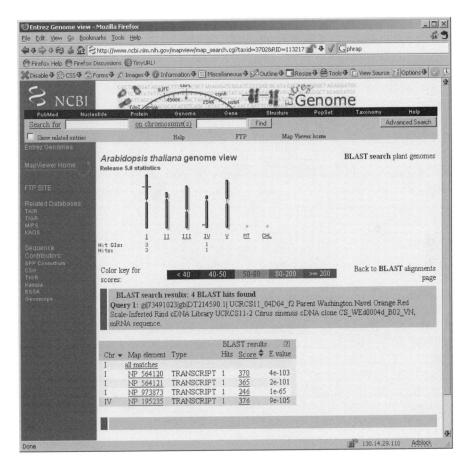

Fig. 8. The Map View display for the second PlantGenomes BLAST example in **Subsection 3.4.**

to perform the entire examination of available evidence is by manually moving text information from the page of one Web site into a query of another Web site. The previous example shows how using PlantGenomes BLAST to quickly get to a genome map enables a more user-friendly search traversal involving data external to NCBI. PlantGenomes BLAST also enables a more user-friendly search traversal involving data internal to NCBI as will be shown with the next example. GenBank accession DT214590 is a citrus EST. Searching the *A. thaliana* proteins with BLASTX and the default parameters yields four hits. The score and E value of all four are similar. Clicking on the "Genome View" yields **Fig. 8**, a display that indicates that there are three objects on chromosome 1 and one object on chromosome 4. Clicking on the URL for chromosome 1 produces a display with a gene having alternatively spliced transcripts adjacent to a gene with only a single mRNA species. Using the same GenBank accession as a regular BLASTX query of nr proteins limited by "Arabidopsis thaliana[ORGN]" yields 23 hits (*see* **Note 5**). Seventeen of the hits each have a score and an E value comparable with those instances found by PlantGenomes BLAST. In some BLAST searches, the user encounters results that can quickly be fractionated into highly interesting results and less interesting results using the score and/or E value. Not so with this example. To identify the two adjacent genes from BLAST, the user is required to sequentially examine each individual Entrez Gene entry to gain access to the Map Viewer display, and this effort still does not generate a display with both genes and the supporting evidence in a single display.

4. Notes

1. Using the Entrez Global Query, it is not necessary to know in advance in which databases the data will be found, and as such, it is an excellent starting point for an initial survey of a new research area. It is also possible to go directly to the data set of interest if you do know what you are looking for in advance.
2. The query consists of two parts, linked by a Boolean "AND", which is always given in capital letters. For the query to be successful, the quoted phrase "stress response" must be found in some part of a database record. The second portion of the query limits the search to *A. thaliana* within the Entrez organism field, given in the square brackets. Hence, for databases such as the sequence databases that classify records by organism, only records for that organism will be returned.
3. The Map Viewer display in **Fig. 5** shows the full marker names for each map track because the "Compress Map" check box, checked by default to enable the display of a large number of maps, has been unchecked. Mouseovers reveal more information about each marker. This includes links to sites internal and external to NCBI.

4. This display is both the input form for a text search of all plants and the display of identified matches to the query string. This display is interactive. The user can adjust which maps to display with the check boxes next to the map name. The user can adjust which organisms with hits to display with the check boxes by the organism name. When this page is used as the input form, the user can selectively exclude some of the organisms from the search space.

5. BLAST searches for other plant species may be carried out from any regular BLAST homepage by choosing the appropriate database and program and entering the name of the organism followed by the Entrez query field limiter [organism] in the "Limit by Entrez Query" box.

Acknowledgments

The authors thank Vyacheslav Chetvernin, Deanna Church, Peter Meric, and Sergei Resenchuk (in alphabetic order) for their expertise and diligence in the development and maintenance of many of the databases highlighted in this chapter. The authors also thank the members of other NCBI groups who enable this sea of databases through software coding.

References

1. Arabidopsis Genome Initiative (2000) Analysis of the genome sequence of the flowering plant *Arabidopsis thaliana. Nature* 408, 796–815.

2. International Rice Genome Sequencing Project (2005) The map-based sequence of the rice genome. *Nature* 436, 793–800.

3. Benson, D.A., Karsch-Mizrachi, I., Lipman, D.J., Ostell, J., and Wheeler, D.L. (2005) GenBank: update. *Nucleic Acids Res.* 33, D34–D38.

4. Tateno, Y., Saitou, N., Okubo, K., Sugawara, H., and Gojobori, T. (2005) DDBJ in collaboration with mass-sequencing teams on annotation. *Nucleic Acids Res.* 33, D25–D28.

5. Kanz, C., Aldebert, P., Althorpe, N., Baker, W., Baldwin, A., Bates, K., Browne, P., van den Broek, A., Castro, M., Cochrane, G., et al. (2005) The EMBL Nucleotide Sequence Database *Nucleic Acids Res.* 33, D29–D33.

6. Boeckmann, B., Bairoch, A., Apweiler, R., Blatter, M.C., Estreicher, A., Gasteiger, E., Martin, M.J., Michoud, K., O'Donovan, C., Phan, I., et al. (2003) The SWISS-PROT protein knowledgebase and its supplement TrEMBL in 2003. *Nucleic Acids Res.* 31, 365–370.

7. Bairoch, A., Apweiler, R., Wu, C.H., Barker, W.C., Boeckmann, B., Ferro, S., Gasteiger, E., Huang, H., Lopez, R., Magrane, M., et al. (2005) The Universal Protein Resource (UniProt). *Nucleic Acids Res.* 33, D154–D159.

8. Mulder, N.J., Apweiler, R., Attwood, T.K., Bairoch, A., Bateman, A., Binns, D., Bradley, P., Bork, P., Bucher, P., Cerutti, L., et al. (2005) InterPro, progress and status in 2005. *Nucleic Acids Res.* 33, D201–D205.

9. Yu, J., Hu, S., Wang, J., Wong, G.K., Li, S., Liu, B., Deng, Y., Dai, L., Zhou, Y., Zhang, X., et al. (2002) A draft sequence of the rice genome (*Oryza sativa* L. ssp. *indica*). *Science* 296, 79–92.

10. Salzberg, S.L., Church, D., DiCuccio, M., Yaschenko, E., and Ostell, J. (2004) The genome Assembly Archive: a new public resource. *PLoS Biol.* 2(9), E285.

11. Maglott, M., Ostell, J., Pruitt, K.D., and Tatusova, T. (2005) Entrez Gene: gene-centered information at NCBI. *Nucleic Acids Res.* 33, D54–D58.

12. Marchler-Bauer, A., Anderson, J.B., Cherukuri, P.F., Weese-Scott, C.D., Geer, L.Y., Gwadz, M., He, S., Hurwitz, D.I., Jackson, J.D., Ke, Z., et al. (2005) CDD: a Conserved Domain Database for protein classification. *Nucleic Acids Res.* 33, D192–D196.

13. Bateman, A., Coin, L., Durbin, R., Finn, R.D., Hollich, V., Griffiths-Jones, S., Khanna, A., Marshall, M., Moxon, S., Sonnhammer, E.L., et al. (2004) The Pfam protein families database. *Nucleic Acids Res.* 32, D138–D141.

14. Letunic, I., Copley, R.R., Schmidt, S., Ciccarelli, F.D., Doerks, T., Schultz, J., Ponting, C.P., and Bork, P. (2004) SMART 4.0: towards genomic data integration. *Nucleic Acids Res.* 34, D142–D144.

15. Tatusov, R.L., Fedorova, N.D., Jackson, J.D., Jacobs, A.R., Kiryutin, B., Koonin, E.V., Krylov, D.M., Mazumder, R., Mekhedov, S.L., Nikolskaya, A.N., et al. (2003) The COG database: an updated version includes eukaryotes. *BMC Bioinformatics* 4, 41.

16. Barrett, T., Suzek, T.O., Troup, D.B., Wilhite, S.E., Ngau, W.-C., Ledoux, P., Rudnev, D., Lash, A.E., Fujibuchi, W., and Edgar, R. (2005) NCBI GEO: mining millions of expression profiles—database and tools. *Nucleic Acids Res.* 33, D562–D566.

17. Lawrence, C.J., Dong, Q., Polacco, M.L., Seigfried, T.E., and Brendel, V. (2004) MaizeGDB, the community database for maize genetics and genomics. *Nucleic Acids Res.* 32, D393–D397.

18. Ware, D., Jaiswal, P., Ni, J., Pan, X., Chang, K., Clark, K., Teytelman, L., Schmidt, S., Zhao, W., Cartinhour, S., et al. (2002) Gramene: a resource for comparative grass genomics. *Nucleic Acids Res.* 30, 103–105.

19. Mueller, L.A., Solow, T.H., Taylor, N., Skwarecki, B., Buels, R., Binns, J., Lin, C., Wright, M.H., Ahrens, R., Wang, Y., et al. (2005) The SOL genomics network. A comparative resource for Solanaceae biology and beyond. *Plant Physiol.* 138, 1310–1317.

20. Brandstatter, I. and Kieber, J.J. (1998) Two genes with similarity to bacterial response regulators are rapidly and specifically induced by cytokinin in Arabidopsis. *Plant Cell* 10, 1009–1019.

21. Ahn, S.N. and Tanksley, S.D. (1993) Comparative linkage maps of the rice and maize genomes. *Proc. Natl. Acad. Sci. USA* 90, 7980–7984.

22. Devos, K.M., Chao, S., Li, Q.Y., Simonetti, M.C., and Gale, M.D. (1994) Relationship between chromosome 9 of maize and wheat homologous group 7 chromosomes. *Genetics* 138, 1287–1292.

23. Kurata, N., Moore, G., Nagamura, Y., Foote, T., Yano, M., Minobe, Y., and Gale, M.D. (1994) Conservation of genome structure between rice and wheat. *Biotechnology (NY)* 12, 276–278.

24. van Deynze, A.E., Nelson, J.C., O'Donoghue, L.S., Ahn, S.N., Siripoonwiwat, W., Harrington, S.E., Yglesias, E.S., Braga, D.P., McCouch, S.R., and Sorrells, M.E. (1995) Comparative mapping in grasses: oat relationships. *Mol. Gen. Genet.* 249, 349–356.

25. Ahn, S., Anderson, J.A., Sorrells, M.E., and Tanksley, S.D. (1993) Homologous relationships of rice, wheat and maize chromosomes. *Mol. Gen. Genet.* 241, 483–490.

26. Moore, G., Devos, K.M., Wang, Z., and Gale, M.D. (1995) Cereal genome evolution. Grasses, line up and form a circle. *Curr. Biol.* 5, 737–739.

27. Tanksley, S.D., Ganal, M.W., Prince, J.P., de Vicente, M.C., Bonierbale, M.W., Broun, P., Fulton, T.M., Giovannoni, J.J., Grandillo, S., Martin, G.B., et al. (1992) High density molecular linkage maps of the tomato and potato genomes. *Genetics* 132, 1141–1160.

28. Livingstone, K.D., Lackney, V.K., Blauth, J.R., van Wijk, R., and Jahn, M.K. (1999) Genome mapping in capsicum and the evolution of genome structure in the Solanaceae. *Genetics* 152, 1183–1202.

29. Doganlar, S., Frary, A., Daunay, M.C., Leste, R.N., and Tanksley, S.D. (2002) A comparative genetic linkage map of eggplant (Solanum melongena) and its implications for genome evolution in the Solanaceae. *Genetics* 161, 1697–1711.

30. Gepts, P., Beavis, W.E., Brummer, E.C., Shoemaker, R.C., Stalker, H.T., Weeden, N.F., and Young, N.D. (2005) Legumes as a model plant family. Genomics for food and feed report of the Cross-Legume Advances Through Genomics Conference. *Plant Physiol.* 137, 1228–1235.

31. Young, N.D., Mudge, J., and Ellis, T.H. (2004) Legume genomes: more than peas in a pod. *Curr. Opin. Plant Biol.* 6, 199–204.

32. Yan, H.H., Mudge, J., Kim, D.J., Shoemaker, R.C., Cook, D.R., and Young, N.D. (2004) Comparative physical mapping reveals features of micro synteny between *Glycine max*, *Medicago truncatula*, and *Arabidopsis thaliana*. *Genome* 47, 141–155.

33. Lederburg, E.M. (1986) Plasmid prefix designations registered by the Plasmid Reference Center 1977–1985. *Plasmid* 1, 57–92.

4

UniProtKB/Swiss-Prot
The Manually Annotated Section of the UniProt KnowledgeBase

Emmanuel Boutet, Damien Lieberherr, Michael Tognolli, Michel Schneider, and Amos Bairoch

Summary

The Swiss Institute of Bioinformatics (SIB), the European Bioinformatics Institute (EBI), and the Protein Information Resource (PIR) form the Universal Protein Resource (UniProt) consortium. Its main goal is to provide the scientific community with a central resource for protein sequences and functional information. The UniProt consortium maintains the UniProt KnowledgeBase (UniProtKB) and several supplementary databases including the UniProt Reference Clusters (UniRef) and the UniProt Archive (UniParc). (1) UniProtKB is a comprehensive protein sequence knowledgebase that consists of two sections: UniProtKB/Swiss-Prot, which contains manually annotated entries, and UniProtKB/TrEMBL, which contains computer-annotated entries. UniProtKB/Swiss-Prot entries contain information curated by biologists and provide users with cross-links to about 100 external databases and with access to additional information or tools. (2) The UniRef databases (UniRef100, UniRef90, and UniRef50) define clusters of protein sequences that share 100, 90, or 50% identity. (3) The UniParc database stores and maps all publicly available protein sequence data, including obsolete data excluded from UniProtKB. The UniProt databases can be accessed online (http://www.uniprot.org/) or downloaded in several formats (ftp://ftp.uniprot.org/pub). New releases are published every 2 weeks. The purpose of this chapter is to present a guided tour of a UniProtKB/Swiss-Prot entry, paying particular attention to the specificities of plant protein annotation. We will also present some of the tools and databases that are linked to each entry.

Key Words: Swiss-Prot; TrEMBL; UniProt; protein database; amino-acid sequence; manual annotation.

From: *Methods in Molecular Biology, vol. 406: Plant Bioinformatics: Methods and Protocols*
Edited by: D. Edwards © Humana Press Inc., Totowa, NJ

1. Introduction

In late 2002, the Swiss Institute of Bioinformatics (SIB), the European Bioinformatics Institute (EBI), and the Protein Information Resource (PIR) (*see* **Note 1**) joined forces by creating the Universal Protein Resource (UniProt) consortium *(1)*. The aim of this consortium is to provide high-quality protein databases that are freely accessible to the scientific community.

The centerpiece of UniProt is the UniProt Knowledgebase (UniProtKB, http://www.uniprot.org), a comprehensive and annotated protein sequence knowledgebase, which consists of two sections: UniProtKB/Swiss-Prot, containing manually annotated entries, and UniProtKB/TrEMBL, containing computer translation and annotation of CoDing Sequences (CDS) extracted from the European Molecular Biology Laboratory (EMBL) nucleotide sequence database *(2,3)*.

UniProtKB/Swiss-Prot is characterized by extended manual annotation (sequence properties, corresponding literature, etc.), minimal redundancy (separate entries for the same gene product in a given species are merged into a single protein entry), integration with other databases (cross-links to other life science databases including sequence-related databases as well as specialized data collections), and documentation (large number of index files and specialized documentation files) (*see* **Note 2**).

UniProtKB/TrEMBL, a computer-annotated database, mainly consists of translations of all CDS proposed by the submitters to the EMBL/GenBank/DNA Databank of Japan (DDBJ) nucleotide databases, which are not yet integrated into UniProtKB/Swiss-Prot. Some additional protein sequences are also extracted from the literature or directly submitted to UniProtKB. In addition to the preliminary information given by the submitters, UniProtKB/TrEMBL entries are processed according to automatic annotation procedures such as (1) transfer of domains and functional sites from well-characterized UniProtKB/Swiss-Prot entries belonging to protein family groups defined by InterPro *(4)*, (2) removal of redundancy by merging identical full-length sequences from the same organism, and (3) attribution of evidence to identify the source of individual data items (*see* **Note 3**).

In addition to UniProtKB, the UniProt consortium maintains several other protein databases.

1. The UniProt Archive (UniParc), which contains all publicly available protein sequences from numerous databases, including UniProtKB, RefSeq, Patent offices, and so on.

2. The UniProt Reference Clusters (UniRef), which consists of clusters of sequences sharing 100% identity for UniRef100, 90% for UniRef90, and 50% for UniRef50 (*see* **Note 4**). These databases are based on both UniProtKB and UniParc.

The Swiss-Prot group has initiated the Plant Proteome Annotation Program in 2001 *(5,6)* (http://www.expasy.org/sprot/ppap/). The current priority of this program is to annotate the proteomes of *Arabidopsis thaliana* and *Oryza sativa* but without neglecting to annotate the proteins from other plant species. Our goals are the annotation of plant-specific and plant family proteins according to the Swiss-Prot standards *(3)*. At the beginning of November 2005 (UniProt release 6.4), 14,004 plant sequence entries are present in UniProtKB/Swiss-Prot. Among them, 3665 are from *A. thaliana* and 575 from *O. sativa*. In UniProtKB/Swiss-Prot, 15 plant species are represented with more than 100 annotated proteins, whereas more than 2700 different plant species are present with at least one annotated protein.

2. Materials

Although UniProtKB is hosted by three different servers [Expert Protein Analysis System proteomic server (ExPASy), EBI, and PIR] (*see* **Note 5**), this chapter will always refer to the UniProtKB interface format used by the ExPASy server (http://www.expasy.org/uniprot/) and will focus on UniProtKB/Swiss-Prot entries. The database is updated every 2 weeks, and major releases occur approximately every 3 months. It is possible to download a local version of UniProtKB (*see* **Notes 6** and **7**).

2.1. Structure of the UniProtKB Entries

The main distribution format of UniProtKB is a custom text-based format. Entries are represented by lines beginning with a two-letter code that identifies the type of data contained in the line. Each line follows a strictly defined format, and the lines themselves are organized in such a way as to be easily legible to human users and simple to parse by computer programs (http://www.expasy.org/sprot/userman.html#entrystruc). When accessing UniProtKB entries from the ExPASy server, or one of its mirror sites, the default format is the NiceProt view, a user-friendly format when compared with the text-based format (*see* **Fig. 1**).

The general elements of an entry in the NiceProt view format are (from top to bottom) (i) ExPASy header and search tool, (ii) UniProt header, (iii) the UniProtKB entry, (iv) ExPASy tools such as BLAST *(7)* or ScanProsite *(8)*, and (v) ExPASy footer including links to mirror sites.

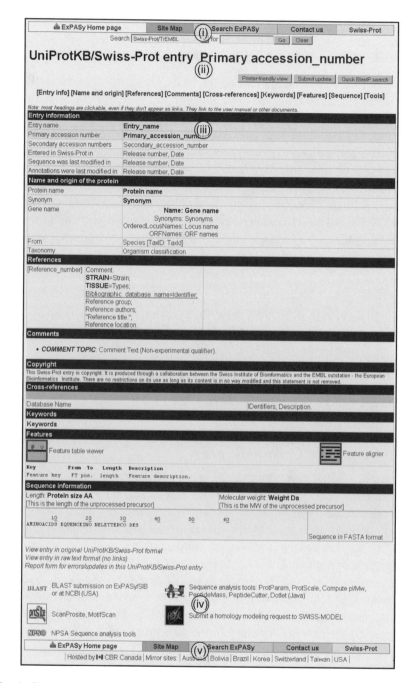

Fig. 1. Skeleton of a UniProtKB entry in the NiceProt display format on ExPASy.

Each entry corresponds to a protein sequence encoded by a single gene locus (*see* **Note 8**). References to residue positions within a sequence are made using sequential numbering starting with 1 at the N-terminal position. Displayed sequences correspond to the precursor forms of proteins, before posttranslational modifications and processing.

2.2. Tools and Databases Linked to UniProtKB

The SIB has developed ExPASy, which is one of the main entry points to UniProtKB *(9,10)*. On http://www.expasy.org/, tools are available to deal with several aspects of protein analysis, including BLAST search, proteomics, and sequence analysis, and take into account all splice variants as annotated in UniProtKB (*see* **Note 9**). Results obtained by these tools or links from other specific databases point to the corresponding UniProtKB entries.

3. Methods

3.1. Introduction

The main goal of UniProt is to provide a central resource for protein sequences and functional annotation. Together with UniProtKB/TrEMBL, UniProtKB/Swiss-Prot contains all known proteins, without species restriction. Currently, the plant protein entries represent 7% of the total content of UniProtKB/Swiss-Prot, and our main effort is focused on the annotation of plant-specific proteins from *A. thaliana* and *O. sativa*. Any new genome fully sequenced, deposited in the public nucleotide database (EMBL/GenBank/DDBJ), and for which a gene prediction has been performed will be processed automatically. The predicted set of proteins is added to the UniProtKB/TrEMBL section as soon as the data are publicly available.

One of the greatest strengths of the UniProtKB and the ExPASy proteomics Web server is the extensive integration and interconnectivity of numerous tools and external databases. The knowledgebase is cross-linked to about 100 other databases, whereas most of the tools are adapted to allow analysis of all annotated splice isoforms.

The UniProtKB is constantly evolving, and all recent modifications are detailed at http://www.expasy.org/sprot/relnotes/sp_news.html, whereas the forthcoming modifications are listed in http://www.expasy.org/sprot/relnotes/sp_soon.html.

To further improve the quality of our annotation, we encourage users to submit comments and update requests (http://www.expasy.org/cgi-bin/sp_update_forms.pl?ac=primary, accession number or the two buttons or links present in each UniProtKB entries).

Db	AC	Description	Score	E-value
☐ sp	O82804	ELF3_ARATH EARLY FLOWERING 3 protein (Nematode respons...	1417	0.0
ⓐ sp_vs	O82804-2	Splice isoform 2 of O82804 [ELF3] [Arabidopsis th...	676	0.0
☐ tr	Q6UEI3	_MESCR Early flowering 3 [ELF3] [Mesembryanthemum cryst...	348	3e-94
☐ tr	Q9SNQ6	_ORYSA Putative early flowering 3 [P0535G04.25] [Oryza ...	255	2e-66
☐ tr	Q50K78	_9ARAE ELF3 homologue [LpELF3 H1] [Lemna paucicostata]	211	5e-53
☐ tr	Q5Q0C8	_ARATH Hypothetical protein [AT3G21320] [Arabidopsis th...	122	3e-26

Fig. 2. BLAST result. Partial view of the result of the BLAST made on UniProtKB with O82804 entry as query.

3.2. Accessing UniProtKB Entries

Quick text search (for example by accession number) can be accessed directly from the UniProtKB home page hosted by ExPASy (http://www.expasy.uniprot.org) (*see* **Note 10**).

BLAST and full text search are available at http://www.expasy.uniprot.org/search/SearchTools.shtml. The BLAST output gives a list of sequences classified by level of similarity to the query and is linked to the corresponding UniProtKB entries (*see* **Fig. 2**). The "Db" column indicates to which section of the UniProtKB the protein belongs ("sp" for Swiss-Prot and "tr" for TrEMBL). All splice variants annotated in UniProtKB are considered during the BLAST ("sp_vs" for Swiss-Prot VarSplic) (*see* **Fig. 2-a**).

Database entries can be retrieved by advanced search, thanks to the sequence retrieval system (SRS) (*11*) accessible at http://www.expasy.org/srs5/ (*see* **Note 11**).

A quick search tool in different databases (e.g., UniProtKB, NEWT, PROSITE, ENZYME, and SWISS-2DPAGE) is available at http://www.expasy.org/sprot/, as well as at the top of any other page of ExPASy. Results are linked to the corresponding UniProtKB entries.

Finally, several databases and tools on ExPASy directly return lists of UniProtKB accession number or entries.

3.3. Working with a UniProtKB Entry

3.3.1. Expasy Headers

When accessing the UniProtKB through ExPASy, some elements are always present at the top of an entry page.

1. The first line (*see* **Fig. 3A-1**) indicates the source of the entry (Swiss-Prot or TrEMBL), as well as its primary accession number, which is clickable to access the original flat file format of the entry instead of the NiceProt view.

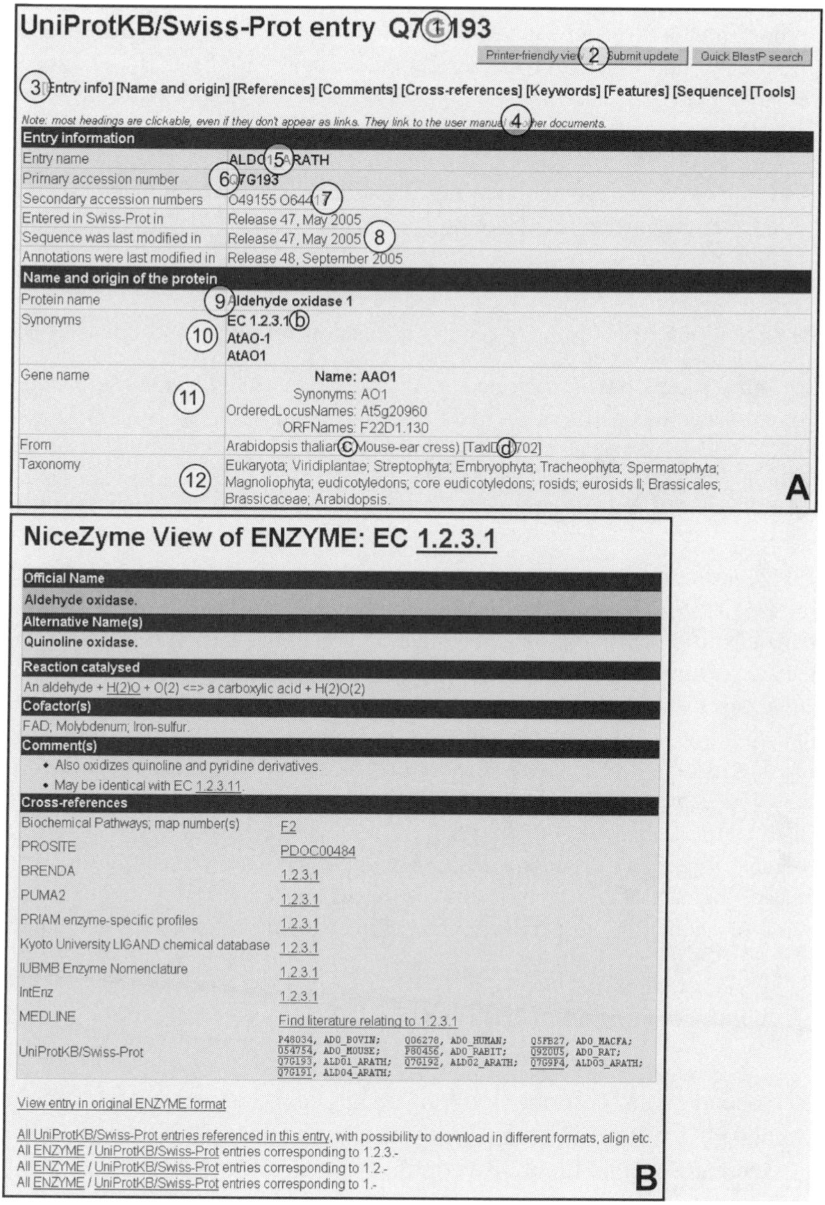

Fig. 3. Entry information and name and origin blocks of a UniProtKB entry. (**A**) Part of the NiceProt view of Q7G193 entry. (**B**) NiceZyme view of the EC 1.2.3.1 information file.

2. On the right side of the window, three clickable buttons allow the retrieval of a printer-friendly version of the entry, to submit updates or corrections of the current entry and to BLAST the current entry against UniProtKB (*see* **Fig. 3A-2**) (*see* **Note 12**).

3. The next line contains direct links to the different blocks of the entry detailed below (*see* **Fig. 3A-3**).

4. On the top of the entry, a note (*see* **Fig. 3A-4**) indicates the links to the user manual and to other documents (*see* **Note 2**).

3.3.2. Entry Information (ID, AC, and DT Lines)

The first block of each entry details accession numbers and creation dates.

1. The entry name (ID) is indicated in the first line (*see* **Fig. 3A-5**). The Swiss-Prot entry name consists of up to 11 characters and takes the general form X_Y. Both X and Y represent mnemonic codes of up to five alphanumeric characters for both the protein name (X) and the species (Y). Entry names are subject to revision and therefore do not provide a stable means of identifying individual entries (*see* **Note 13**).

2. The accession number (AC) line gives the primary accession number of the entry (*see* **Fig. 3A-6**). The primary accession number remains fixed and provides a unique stable identifier that allows unambiguous citation of the entry (*see* **Notes 14** and **15**). Because entry names are prone to change, researchers who wish to cite entries in publications should always cite the primary accession number.

3. The AC line may also contain one or more secondary accession numbers (*see* **Fig. 3A-7**), which follow the primary accession number. These are usually accession numbers of UniProtKB/TrEMBL entries that have been merged into a single UniProtKB/Swiss-Prot entry.

4. The next three lines (DT) (*see* **Fig. 3A-8**) give the date when the entry was first created, the date of last modification of the sequence, and the date of last modification of annotation, respectively. The corresponding releases are also indicated.

3.3.3. Name and Origin of the Protein (DE, GN, OS, OG, OC, and OX Lines)

The second block describes protein names, gene names, and taxonomy of the organism. UniProtKB/Swiss-Prot entries are manually checked and offer a high accuracy in this block, by contrast with UniProtKB/TrEMBL, directly depending on the diverse information given by sequence submitters and the scientific literature.

1. The recommended protein name is given in the first line (DE) (*see* **Fig. 3A-9**), followed by the alternative names used in the literature (*see* **Fig. 3A-10**). In the

case of an enzyme, the Enzyme Commission (EC) number is given as the first synonym (*see* **Fig. 3A-b**). This EC number is an active link to the enzyme database (http://www.expasy.org/enzyme/) *(12)*, which groups detailed information about enzyme activity and lists all UniProtKB/Swiss-Prot entries having the same EC number (*see* **Fig. 3B**).

2. The third row of this block (GN) (*see* **Fig. 3A-11**) describes the gene encoding the protein in the following order: gene name, synonyms, ordered locus name when applicable (*see* **Note 16**), and open reading frame (ORF) names used by the genomic sequencing projects.

3. Following the gene description, the organism name (OS) (*see* **Fig. 3A-c**) and the NCBI taxonomy identifier (OX) (*see* **Fig. 3A-d**) are actively linked to a table containing all UniProtKB entries of that organism and to the NEWT database (http://www.ebi.ac.uk/newt/), respectively (*see* **Note 17**). The summarized taxonomic hierarchy gives taxa (OC) (*see* **Fig. 3A-12**) with a link to tables listing all UniProtKB entries sharing identical taxa.

3.3.4. References (RN, RP, RC, RX, RG, RA, RT, and RL Lines)

This block lists all references used for the annotation of the protein entry. The first references are usually associated with sequence submission, followed by references providing other information concerning the function and structure of the protein (*see* **Fig. 4**).

1. Each reference starts with a number in square bracket (RN) followed by an indication of what information was extracted from the article (RP) (*see* **Fig. 4-13**). In the case of references associated with a sequence submission, the sequenced molecule type is mentioned and, if relevant, the corresponding isoform is indicated. The strain and tissues used as source for sequencing are also mentioned when available (RC).

2. Before the list of authors, cross-links to PubMed (RX) are given (*see* **Fig. 4-14**). When available, the digital object identifier (DOI) is linked to the corresponding article (*see* **Fig. 4-15**) and allows retrieval of the electronic version of the article.

3. Each author name (RA and RG) is linked to a list of all UniProtKB entries where that author is cited (*see* **Fig. 4-16**). The title of the reference (RT) follows the author names. The last line of each reference (RL) gives the conventional citation information for the reference (*see* **Note 18**).

3.3.5. Comments (CC Lines)

Most of the information in this section is extracted from the literature. Some information is also based on unproven empirical biological evidence, determined by computer prediction, or propagated from homologous members of the family (http://www.expasy.org/cgi-bin/lists?annbioch.txt). In these cases, non-experimental qualifiers are added (see below).

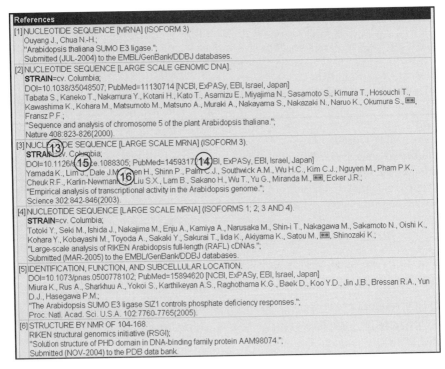

Fig. 4. Reference block of a UniProtKB entry. Part of the NiceProt view of Q680Q4 entry.

1. When experimental support for a particular assertion in the CC lines is lacking, then one of the three non-experimental qualifiers is added in parentheses at the end of the subsection (*see* **Note 19**). The qualifiers are "Potential" for computer predicted, logical, or conclusive evidence (*see* **Note 20**); "Probable" for non-direct experimental evidence (*see* **Note 21**); and "By similarity" for experimental evidence in a close member of the family. Explanations of non-experimental qualifiers can be obtained by clicking on them in the entry (*see* **Fig. 5A-17**).

2. The comment section is divided into subsections. Each subsection contains information pertinent to particular properties of the protein such as its function, subcellular location, biophysicochemical properties, and tissue specificity or induction (*see* **Fig. 5A-18**), where information is reported from literature (*see* **Note 22**).

3. The alternative product subsection describes the proteins that may be produced by alternative splicing or promoter usage and features two kinds of identifier. (1) Isoform IDs (ISO_ID) (*see* **Fig. 5A-e**) define the ID of each sequence derived from alternative splicing. They confer active links to respective sequences (*see* **Fig. 6A**),

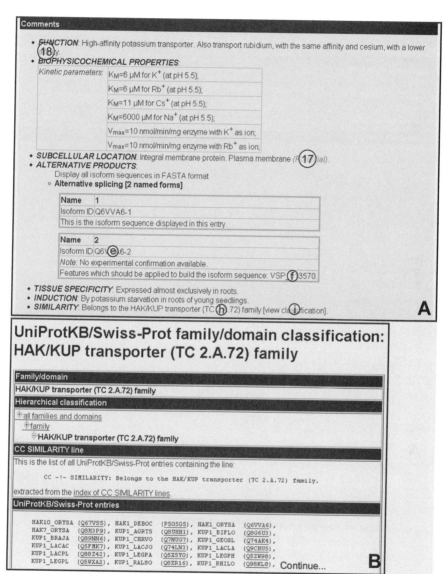

Fig. 5. Comments block of a UniProtKB entry. (**A**) Part of the NiceProt view of Q6VVA6 entry. (**B**) View of the UniProt information file about the TC 2.A.72 family.

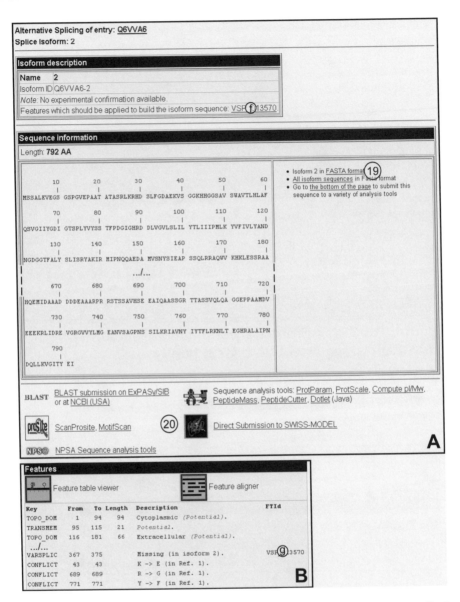

Fig. 6. Alternative splicing view and associated features table. (**A**) ExPASy display of the alternative product of the UniProtKB entry Q6VVA6. (**B**) Part of the NiceProt view of Q6VVA6 entry.

with the possibility to retrieve the FASTA format of the sequence for further analyses (*see* **Fig. 6A-19**). This view gives also links to various analysis tools to be applied on the corresponding sequence such as BLAST and ScanProsite (*see* **Fig. 6A-20**). (2) A separate feature ID (FT_ID) (*see* **Figs. 5A-f** and **6A-f**) points to each feature line of the entry necessary to describe the alternative product sequence, as a modification of the sequence shown in the entry (*see* **Fig. 6B-g**).

4. The topic "SIMILARITY" may include a comment describing to which family the protein may belong. It is actively linked to a file listing all UniProtKB entries belonging to the same family (*see* **Fig. 5B**) (*see* **Note 23**). In the case of transporter families, the transport classification (TC) number is present when available (*see* **Fig. 5A-h**). A link to the TC database (http://www.tcdb.org/tcdb/) is also present (*see* **Fig. 5A-i**).

3.3.6. Cross-References (DR Lines)

This section links the protein to several other databases that contain information relevant to that protein. Many of these cross-links are automatically added to UniProtKB/TrEMBL entries, but some are manually created in UniProtKB/Swiss-Prot entries. In the following part, a small selection of more than 100 possible database cross-references will be described in detail (*see* **Note 22**).

1. Each row of this block corresponds to a single database, the name of which is indicated in the first column (*see* **Fig. 7A-21**). A link to the relevant data in the cross-linked database is present in next columns.
2. The cross-reference section starts with EMBL lines (http://www.embl-heidelberg.de/) that are displayed in the same order as the corresponding references associated with a sequence submission. An EMBL line contains a nucleic acid sequence ID (*see* **Fig. 7A-j**) and a protein sequence ID (*see* **Fig. 7A-k**). A molecule type indicates the origin of the sequence (e.g., mRNA or Genomic_DNA) (*see* **Fig. 7A-l**) (*see* **Note 24**). Links in square brackets retrieve either the complete nucleotide sequence in EMBL/GenBank/DDBJ databases (*see* **Fig. 7A-m**) or only the CDS of the protein (*see* **Fig. 7A-n**), which is useful in the case of genomic sequences.
3. Cross-references to the protein data bank (PDB) (http://www.rcsb.org/pdb/) are present when a protein structure is available (*see* **Fig. 7A-o**). The line is actively linked to the PDB entry and contains information about the crystallographic method, the number of chains, and the range of residues present in the structure. The final line of the PDB subsection, visible only in the NiceProt view, is a link to a detailed list of all the PDB structures available for this protein (*see* **Fig. 7A-p**).
4. A link to the PROSITE database (*see* **Fig. 7A-q**) (http://www.expasy.org/prosite/) opens the corresponding documentation (*see* **Fig. 7B**). From this PROSITE page, it is possible to retrieve the list of all UniProtKB entries matching the same profile or pattern (*see* **Fig. 7B-22**) (*see* **Note 25**).

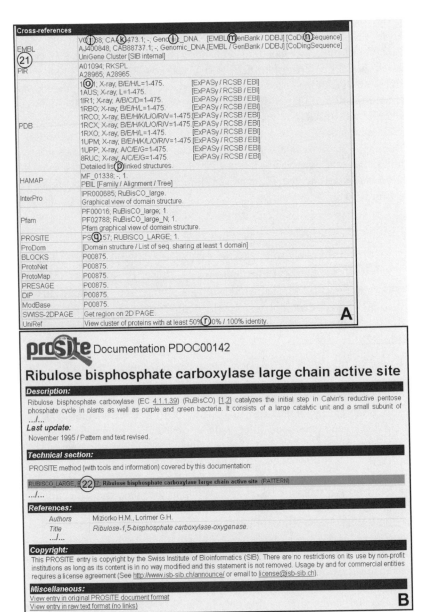

Fig. 7. Cross-references block of a UniProtKB entry and PROSITE documentation file. (**A**) Part of the NiceProt view of P00875 entry. (**B**) PROSITE documentation file of the pattern PS00157.

5. The last line of the cross-reference part contains links to UniRef sections, that is, to all entries sharing 50, 90, or 100% identity to that sequence in the UniProtKB (*see* **Fig. 7A-r**).

6. Plant-specific databases that are currently cross-linked in UniProtKB entries are summarized in **Table 1**. They have been chosen because of their content, their stability, and their frequent updates. All of them give additional information about the protein and are linked back to UniProtKB.

3.3.7. Keywords (KW Lines)

Some keywords come from automatic annotation in UniProtKB/TrEMBL entries, but most of them are added manually in UniProtKB/Swiss-Prot entries. They describe the main characteristics of the protein.

The keywords are listed before sequence features (*see* **Fig. 8A-23**). Each of them is clickable (*see* **Fig. 8A-s**) and linked to its definition and a list of all UniProtKB entries having the same keyword (*see* **Fig. 8B**).

Table 1
Plant-Specific Cross-References Present in UniProtKB

Database name and URL and goals	DR line format
GeneFarm *(13)* (http://genoplante-info.infobiogen.fr/Genefarm/). Structural and functional annotation of *Arabidopsis thaliana* gene and protein families	DR GeneFarm; GeneID; FamilyID *In UniProtKB/Swiss-Prot only*
Gramene—a comparative mapping resource for grains *(14)* (http://www.gramene.org/). Curated, open-source, Web-accessible data resource for comparative genome analysis in the grasses	DR Gramene; UniProtKB_AC; *In UniProtKB/Swiss-Prot and UniProtKB/TrEMBL*
Maize Genetics/Genomics Database (MaizeGDB) *(15)* (http://www.maizegdb.org/). Central repository for public maize information	DR MaizeDB; ProteinID; *In UniProtKB/Swiss-Prot only*
The Arabidopsis Information Resource (TAIR) *(16)* (http://www.arabidopsis.org/index.jsp). Searchable relational database on *A. thaliana*, which includes many different molecular data types and provides a comprehensive resource for the scientific community	DR TAIR; Order_locus_name; *In UniProtKB/Swiss-Prot and UniProtKB/TrEMBL*

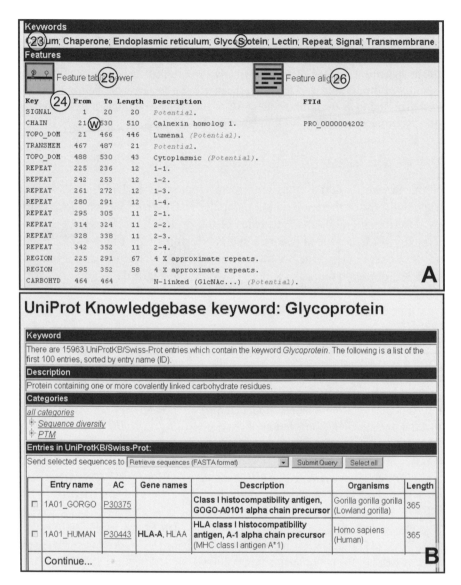

Fig. 8. Keywords and features blocks of a UniProtKB entry. (**A**) Part of the NiceProt view of P29402 entry. (**B**) Part of the UniProt keyword "Glycoprotein" definition file.

3.3.8. Features (FT Lines)

Each line of this block describes a different feature (*see* **Fig. 8A-24**) and contains a feature key such as "SIGNAL" or "TRANSMEM" (*see* **Note 22**). The limits of the feature are given according to the amino acid residue positions of the displayed sequence (*see* **Note 26**), whereas the length of the feature is also given just before the description of the feature (*see* **Note 27**). UniProtKB/Swiss-Prot entries contain extensive annotation of all features that are predicted (and compatible with the protein function), experimentally proven, or determined by resolution of the protein structure.

1. Features can be visualized with a graphical viewer by clicking on the link "Feature table viewer" (*see* **Fig. 8A-25**).
2. Individual features of a given protein can be aligned by clicking on "Feature aligner" (*see* **Fig. 8A-26**). The retrieved document gives a list of all features that can be aligned (*see* **Fig. 9A**), where features of interest are selected by checking boxes (*see* **Fig. 9A-t**). The alignment is performed by clicking on the corresponding button (*see* **Fig. 9A-u**), and the result is a ClustalW multiple sequence alignment (*see* **Fig. 9B**). Sequences of the selected features can be obtained in the FASTA format by clicking the corresponding button (*see* **Fig. 9A-v**).
3. The sequence corresponding to a feature can also be selected by clicking on its positions (*see* **Fig. 8A-w**). The full sequence of the protein, with the feature highlighted in color, is obtained in both one-letter and three-letter codes (*see* **Fig. 10-x**). The subsequence corresponding to the feature can be analyzed by several tools such as BLAST or ScanProsite (*8*) by using the links at the bottom of this page (*see* **Fig. 10-27**).

3.3.9. Sequence Information (SQ Line)

The last section in a UniProtKB entry gives information about the sequence (*see* **Fig. 11A**).

1. The header line indicates the sequence length, the molecular weight, and the CRC64 checksum (64 bit Cyclic Redundancy Check value) (*17*) (*see* **Note 28**) (*see* **Fig. 11A-28**).
2. The protein sequence is given in one-letter code with numbering every 10 amino acids. For further analysis, it is possible to retrieve this sequence in FASTA format (*see* **Fig. 11B**) by clicking on the relevant link (*see* **Fig. 11A-29**).
3. Several links are present for retrieving the current entry in other formats (*see* **Fig. 11A-30**). The link "View entry in original UniProtKB/Swiss-Prot format" displays the entry in the raw flat file format described in the **Subsection 2.1.** but with clickable links. The link "View entry in raw text format (no links)" retrieves the same format without links. The last link "Report form for errors/updates in this

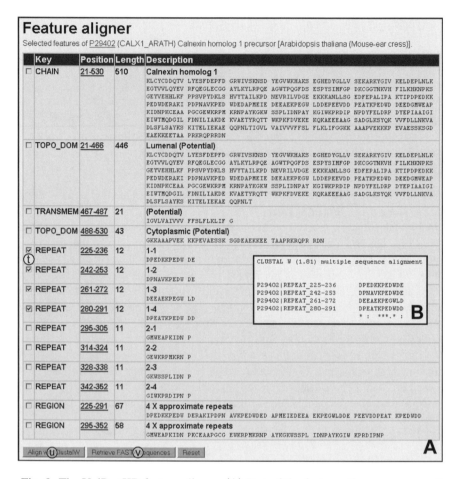

Fig. 9. The UniProtKB feature aligner. (**A**) Part of the feature aligner corresponding to P29402 entry. (**B**) Clustal alignment view of the features 1-1, 1-2, 1-3, and 1-4 of P29402 entry.

UniProtKB/Swiss-Prot entry" displays a form to submit updates and comments for the current entry (*see* **Note 12**).

3.3.10. Expasy Footers

Following any entry, some ExPASy tools are listed (*see* **Fig. 11A-31**). When a link is selected, the sequence is directly pasted in the accurate field of the tool interface, thus avoiding further manual copy/paste operations. Links to mirror sites allow switching to another server (*see* **Note 29**).

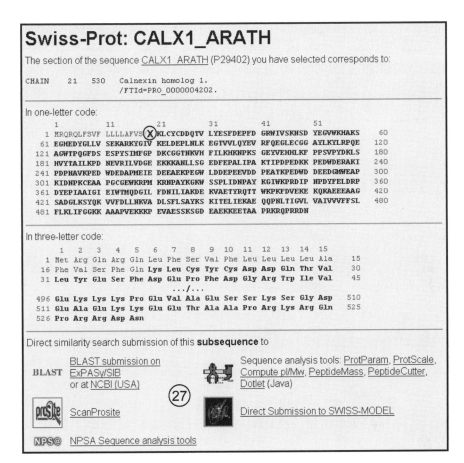

Fig. 10. Feature sequence viewer. View of the sequence of the mature chain of P29402 entry.

4. Notes

1. The SIB (Geneva, Switzerland), in collaboration with the EBI (Hinxton, United Kingdom), develops the Swiss-Prot protein database. The EBI also develops the TrEMBL protein database. PIR (Georgetown University Medical Center and National Biomedical Research Foundation, USA) used to develop the PIR-Protein Sequence Database (PSD) and is now involved in the creation of the UniRef clusters.

2. For more information, see http://www.expasy.org/sprot/userman.html and http://www.expasy.org/sprot/sp-docu.html.

3. For more information, see http://www.uniprot.org/support/docs/evidence.shtml.

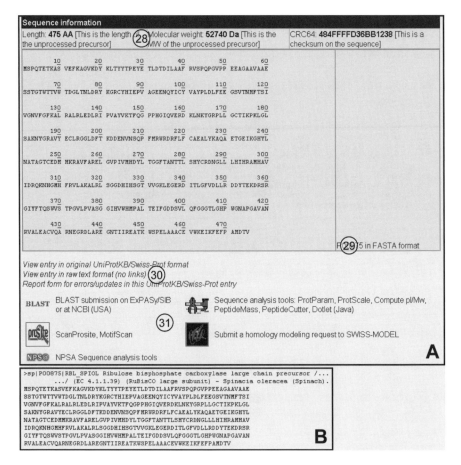

Fig. 11. Sequence block of a UniProtKB entry in NiceProt and FASTA views. (**A**) Part of the NiceProt view of P00875 entry. (**B**) FASTA format of P00875 entry sequence.

4. The UniRef reference clusters combine closely related sequences into a single record to speed sequence similarity searches. The UniRef100 database combines identical sequences and subfragments of the UniProtKB (from any species) and selected UniParc records into a single UniRef entry (ftp://ftp.expasy.org/databases/uniprot/uniref/uniref100/README). UniRef90 and UniRef50 yield a database size reduction of approximately 40 and 65%, respectively, providing for significantly faster sequence searches.

5. UniProt is currently hosted by three sites. In this chapter, we refer to the ExPASy interface (<http://www.expasy.uniprot.org/>). ExPASy is

developed by the SIB *(9,10)*. Other sites are <http://www.ebi.uniprot.org/> and <http://www.pir.uniprot.org/>. The Consortium is currently working on a unified UniProt website. <http://www.uniprat.org/>

6. Major releases usually introduce important format changes. They are distinguishable from other releases by a new primary number followed by ".0".

7. To download a local version of UniProtKB, use the web page ftp://ftp.uniprot.org/pub.

8. When different genes of a given species encode similar proteins, they are merged into a single UniProtKB entry (e.g., Ubiquitin, entry P59263). In this case, all ordered locus names are listed in the gene name section.

9. Other tools and databases developed by the EBI and PIR are available at http://www.ebi.ac.uk/services/ *(18)* and http://pir.georgetown.edu/, respectively.

10. An ExPASy navigation bar for users of the Mozilla Web browser (http://www.mozilla.org/) is available at http://expasybar.mozdev.org. It allows searches to be performed in several databases hosted by ExPASy, including UniProtKB.

11. Another version of SRS is available at EBI on http://srs.ebi.ac.uk/srsbin/cgi-bin/wgetz?-page+top+-newId.

12. Your feedback is highly important and allows us to continuously improve our knowledgebase according to your needs.

13. The IDTracker allows the retrieval of the current primary accession number for obsolete Swiss-Prot entry names (IDs) (http://www.expasy.org/cgi-bin/idtracker).

14. It can also (but rarely) happen that the primary accession number becomes a secondary accession number (e.g., when an entry is split into two entries).

15. An accession number uniquely identifies an entry. If an entry is deleted, its AC will never be attributed to another entry.

16. In the case of *A. thaliana* and *O. sativa*, we use the following nomenclature according to the standard defined for *A. thaliana*: [first letter of the genius name]-[first letter of the species name]-[chromosome number]-[g, for gene]-[locus number] (e.g., At1g15690 and Os03g16440).

17. Currently, *O. sativa* has three different taxonomy identifiers in UniProtKB/: 39947 for japonica cultivars, 39946 for indica cultivars, and 4530 for unspecified rice cultivars. In UniProtKB/Swiss-Prot, when possible, cultivars are specified for each reference related to a sequence deposition.

18. Examples for journal, book, patent, and so on references are given in the user manual (http://www.expasy.org/sprot/userman.html#Ref_line).

19. Concerns only CC and FT lines.

20. A typical example is the annotation of N-glycosylation sites in the entries of non-cytoplasmic domains or proteins.

21. A typical example is the annotation of nuclear subcellular location in the entries of active transcription factors in eukaryotic organisms.

22. Complete lists of the different comment topics, cross-referenced databases, and feature types are available in the user manual (http://expasy.org/sprot/userman.html#CC_line, http://expasy.org/sprot/userman.html#DR_line, and http://expasy.org/sprot/userman.html#FT_line, respectively).
23. The family classification is exclusively based on sequence similarities, not on functions.
24. Additional qualifiers may be present: ALT_SEQ, ALT_INIT, ALT_TERM, or ALT_FRAME. These are used in the case of discrepancies between the EMBL-derived CDS and the displayed protein sequence. These may be due to gross differences in the predicted CDS sequence (arising from the failure to correctly predict all exons for a given gene for instance), incorrect selection of the initiating methionine, termination of the sequence, or frameshifts, respectively. For more details, see the user manual (http://www.expasy.org/sprot/userman.html#DR_line).
25. Three qualifiers may be applied to a link to the PROSITE database: FALSE_NEG indicates that although the pattern or profile is not detected in the protein sequence, it is a member of that particular family or domain; PARTIAL indicates that the pattern or profile is not detected in the sequence because the sequence is not complete and lacks the region on which the pattern/profile is based; and UNKNOWN indicates doubts that the sequence is a member of the family or contains the domain described by the pattern/profile.
26. Amino acid residue numbering begins at the N-terminus of the precursor protein (the displayed sequence).
27. The description of the feature may contain a non-experimental qualifier (*see* **Subsection 3.3.5.**, **step 1**).
28. The algorithm to compute the CRC64 is described in the ISO 3309 standard *(17)*.
29. In case of problems encountered with a server, it may be helpful to switch to an alternative mirror site.

Acknowledgments

This work is partially funded by the National Institute of Health (NIH) grant 1 U01 HG02712-01, the Swiss Federal Government through the Federal Office of Education and Science, and GENOPLANTE (project Bi2001071). We thank Elisabeth Gasteiger, Eric Jain, and Alan James Bridge for critical reading of the manuscript.

References

1. Bairoch, A., Apweiler, R., Wu, C.H., Barker, W.C., Boeckmann, B., Ferro, S., Gasteiger, E., Huang, H., Lopez, R., Magrane, M., Martin, M.J., Natale, D.A., O'Donovan, C., Redaschi, N., and Yeh, L.S. (2005) The Universal Protein Resource (UniProt). *Nucleic Acids Res.* 33(Database issue), D154–D159.
2. Bairoch, A., Boeckmann, B., Ferro, S., and Gasteiger, E. (2004) Swiss-Prot: juggling between evolution and stability. *Brief. Bioinform.* 5, 39–55.

3. Boeckmann, B., Bairoch, A., Apweiler, R., Blatter, M.-C., Estreicher, A., Gasteiger, E., Martin, M.J., Michoud, K., O'Donovan, C., Phan, I., Pilbout, S., and Schneider, M. (2003) The SWISS-PROT protein knowledgebase and its supplement TrEMBL in 2003. *Nucleic Acids Res.* 31, 365–370.

4. Mulder, N.J., Apweiler, R., Attwood, T.K., Bairoch, A., Bateman, A., Binns, D., Bradley, P., Bork, P., Bucher, P., Cerutti, L., Copley, R., Courcelle, E., Das, U., Durbin, R., Fleischmann, W., Gough, J., Haft, D., Harte, N., Hulo, N., Kahn, D., Kanapin, A., Krestyaninova, M., Lonsdale, D., Lopez, R., Letunic, I., Madera, M., Maslen, J., McDowall, J., Mitchell, A., Nikolskaya, A.N., Orchard, S., Pagni, M., Ponting, C.P., Quevillon, E., Selengut, J., Sigrist, C.J., Silventoinen, V., Studholme, D.J., Vaughan, R., Wu, C.H. (2005) InterPro, progress and status in 2005. *Nucleic Acids Res.* 33(Database issue), D201–D205.

5. Schneider, M., Tognolli, M., and Bairoch, A. (2004). The Swiss-Prot protein knowledgebase and ExPASy: providing the plant community with high quality proteomic data and tools. *Plant Physiol. Biochem.* 42, 1013–1021.

6. Schneider, M., Bairoch, A., Wu, C.H., and Apweiler, R. (2005) Plant protein annotation in the UniProt Knowledgebase. *Plant Physiol.* 138, 59–66.

7. Altschul, S.F., Gish, W., Miller, W., Myers, E.W., and Lipman, D.J. (1990) Basic local alignment search tool. *J. Mol. Biol.* 215, 403–410.

8. Gattiker, A., Gasteiger, E., and Bairoch, A. (2002) ScanProsite: a reference implementation of a PROSITE scanning tool. *Applied Bioinformatics* 1, 107–108.

9. Gasteiger, E., Gattiker, A., Hoogland, C., Ivanyi, I., Appel, R.D., and Bairoch, A. (2003) ExPASy: the proteomics server for in-depth protein knowledge and analysis. *Nucleic Acids Res.* 31, 3784–3788.

10. Gasteiger, E., Hoogland, C., Gattiker, A., Duvaud, S., Wilkins, M.R., Appel, R.D., and Bairoch, A. (2005) Protein identification and analysis tools on the ExPASy server, in *The Proteomics Protocols Handbook* (Walker, J.M., ed.). Humana, Totowa, NJ, pp. 571–607.

11. Etzold, T. and Argos, P. (1993) SRS – an indexing and retrieval tool for flat file data libraries. *Comput. Appl. Biosci.* 9, 49–57.

12. Bairoch, A. (2000) The ENZYME database in 2000. *Nucleic Acids Res.* 28, 304–305.

13. Aubourg, S., Brunaud, V., Bruyere, C., Cock, M., Cooke, R., Cottet, A., Couloux, A., Dehais, P., Deleage, G., Duclert, A., Echeverria, M., Eschbach, A., Falconet, D., Filippi, G., Gaspin, C., Geourjon, C., Grienenberger, J.-M., Houlne, G., Jamet, E., Lechauve, F., Leleu, O., Leroy, P., Mache, R., Meyer, C., Nedjari, H., Negrutiu, I., Orsini, V., Peyretaillade, E., Pommier, C., Raes, J., Risler, J.-L., Riviere, S., Rombauts, S., Rouze, P., Schneider, M., Schwob, P., Small, I., Soumayet-Kampetenga, G., Stankovski, D., Toffano, C., Tognolli, M., Caboche, M., and Lecharny, A. (2005) GeneFarm, structural and functional annotation of Arabidopsis gene and protein families by a network of experts. *Nucleic Acids Res.* 33, D641–D646.

14. Ware, D.H., Jaiswal, P., Ni, J., Yap, I.V., Pan, X., Clark, K.Y., Teytelman, L., Schmidt, S.C., Zhao, W., Chang, K., Cartinhour, S, Stein, L.D., and McCouch, S.R. (2002) Gramene, a tool for grass genomics. *Plant Physiol.* 130, 1606–1613.

15. Lawrence, C.J., Dong, Q., Polacco, M.L., Seigfried, T.E., and Brendel, V. (2004) MaizeGDB, the community database for maize genetics and genomics. *Nucleic Acids Res.* 32(Database issue), D393–D397.

16. Rhee, S.Y., Beavis, W., Berardini, T.Z., Chen, G., Dixon, D., Doyle, A., Garcia-Hernandez, M., Huala, E., Lander, G., Montoya, M., Miller, N., Mueller, L.A., Mundodi, S., Reiser, L., Tacklind, J., Weems, D.C., Wu, Y., Xu, I., Yoo, D., Yoon, J., and Zhang, P. (2003) The Arabidopsis Information Resource (TAIR): a model organism database providing a centralized, curated gateway to Arabidopsis biology, research materials and community. *Nucleic Acids Res.* 31, 224–228.

17. Press, W.H., Flannery, B.P., Teukolsky, S.A., and Vetterling, W.T. (1993) *Numerical Recipes in C*, 2nd edition. Cambridge University Press, Cambridge, pp. 896–902.

18. Harte, N., Silventoinen, V., Quevillon, E., Robinson, S., Kallio, K., Fustero, X., Patel, P., Jokinen, P., and Lopez, R. (2004) European Bioinformatics Institute. Public web-based services from the European Bioinformatics Institute. *Nucleic Acids Res.* 32(Web Server issue), W3–W9.

5

Plant Database Resources at The Institute for Genomic Research

Agnes P. Chan, Pablo D. Rabinowicz, John Quackenbush,
C. Robin Buell, and Chris D. Town

Summary

With the completion of the genome sequences of the model plants *Arabidopsis* and rice, and the continuing sequencing efforts of other economically important crop plants, an unprecedented amount of genome sequence data is now available for large-scale genomics studies and analyses, such as the identification and discovery of novel genes, comparative genomics, and functional genomics. Efficient utilization of these large data sets is critically dependent on the ease of access and organization of the data. The plant databases at The Institute for Genomic Research (TIGR) have been set up to maintain various data types including genomic sequence, annotation and analyses, expressed transcript assemblies and analyses, and gene expression profiles from microarray studies. We present here an overview of the TIGR database resources for plant genomics and describe methods to access the data.

Key Words: Sequence annotation; Gene index; Transcript assembly; Microarray.

1. Introduction

The Institute for Genomic Research (TIGR) was established in 1992 in Rockville, MD, USA. TIGR is a not-for-profit organization dedicated to analyzing genomes and studying the underlying biological information by high-throughput sequencing as well as the development of databases and computational tools. The sequence data, analyses and tools are all freely available to the scientific community and the public.

From: *Methods in Molecular Biology, vol. 406: Plant Bioinformatics: Methods and Protocols*
Edited by: D. Edwards © Humana Press Inc., Totowa, NJ

After the completion of the first genome sequence of a free living organism, *Haemophilus influenzae (1)*, and many other microbial genomes (http://pathema.tigr.org/tigr-scripts/CMR/CmrHomePage.cgi) *(2)*, TIGR expanded its research interests and has participated as a key member of international collaborative efforts for high-throughput sequencing and analysis of a wide array of genomes including plants, animals, fungi, and protozoans *(3–6)*. The analyses include genomic and cDNA sequencing, microarray studies, and development of software for data management and display, sequence assembly, gene finding, genome annotation, and microarray data processing (http://www.tigr.org/software). By the mid–late 1990s, TIGR became involved in plant genomics as a member of international consortia focused on sequencing the genomes of the model plant *Arabidopsis thaliana (7)* and that of the economically important cereal crop plant, *Oryza sativa (8)*.

Typically, a genome is sequenced by a whole-genome shotgun sequencing (WGS) approach or a bacterial artificial chromosome (BAC)-based approach. In the WGS approach, the genomic DNA is sheared and cloned into libraries with various insert sizes (e.g., 2–3, 6–8, and 10–12 kbp). Paired-end sequence reads from the libraries are assembled into contigs that are linked to form scaffolds, according to shared sequence overlaps and the mate-pair information, respectively *(9,10)*. In the BAC-based approach, partially digested genomic DNA is cloned into a low copy number vector to produce a large-insert size (100–150 kbp) BAC library. BAC clones are positioned along the length of the chromosomes in a physical map, which is constructed based on the sharing of similar restriction patterns by overlapping BAC clones *(11)*. BAC clones with minimal but reliable overlaps form a "minimal tiling path" (MTP) and are selected for the construction of individual BAC shotgun libraries. BAC end sequences are also used to select additional BAC clones to extend existing sequence contigs. Paired-end reads from the BAC shotgun library are generated and assembled to reconstruct the consensus BAC sequence. If gaps exist in a BAC assembly, targeted sequencing reactions are carried out to fill-in the missing sequences. The BAC assemblies are subsequently merged with neighboring BAC clones as defined in the MTP to construct contiguous sequences that are referred to as virtual contigs or as pseudomolecules if spanning entire chromosomes.

The biological features in the resulting assembled genome sequence are identified through a process referred to as annotation. Several sources of information are used to identify the biological features of a genome. *Ab initio* gene finders predict putative coding regions based on nucleotide sequence context and splice site signals using computational algorithms. Other supporting evidences such as expressed transcripts and protein sequences provide concrete evidence for gene expression and the refinement of gene structures.

Evidence of gene expression can be collected by sequencing random clones from cDNA libraries derived from different tissue types, developmental stages, or treatment conditions to generate expressed sequence tags (ESTs). ESTs provide experimental evidence of gene expression and are also used in the annotation process to define the exon–intron boundaries of genes. The abundance of EST sequences present in different cDNA libraries obtained from a given species provides information of the relative levels of gene expression and insights into the temporal and spatial regulation of gene expression. To create a catalog of expressed gene sequences for a target species, the EST sequences in the expressed transcript population should be reduced to a non-redundant set. This can be achieved by clustering all available expressed sequences from a target species, including ESTs and expressed transcript sequences, based on shared sequence identity. The TIGR Gene Index databases *(12)* currently maintain assemblies of gene transcripts from 33 different plant species. A new project focusing on creating a comprehensive collection of plant-specific expressed transcripts, the TIGR Plant Transcript Assembly (TA) project, is underway to complement and expand the gene index project and will include approximately 233 plant species with publicly available expressed transcript sequences.

Microarray analysis is a high-throughput approach to capture the gene expression profiles of a biological sample by hybridizing the expressed transcripts from an RNA sample to tens of thousands of gene sequences immobilized on a microscope slide. Microarray expression profiling provides an efficient approach to compare gene expression levels from different experimental conditions, treatments, developmental stages, or mutants. The potato cDNA microarray as well as the maize and rice oligonucleotide microarray databases at TIGR create and maintain the probe sequences representing the gene sequences, the associated gene annotation, and the gene expression profiles generated using the microarray platforms.

The Plant Genomics databases at TIGR contain sequence, annotation, and expression information generated in-house as well as by third parties. The TIGR databases are implemented using a Sybase relational database system as the back-end component. In this review, we describe each type of plant genomic database at TIGR, including the sequence and annotation databases, the gene index databases, and the microarray databases (*see* **Table 1**). In each database section, an introduction of the data type used to create the database is presented,

Table 1
The Institute for Genomic Research (TIGR) Plant Genomics Databases
(http://www.tigr.org/plantProjects.shtml)

Database	URL
Sequence and annotation databases	
Arabidopsis	http://www.tigr.org/tdb/e2k1/ath1
Rice	http://rice.tigr.org/
Maize	http://maize.tigr.org/
Medicago	http://www.tigr.org/tdb/e2k1/mta1
Wheat	http://www.tigr.org/tdb/e2k1/tae1
Dana Faber Cancer Institute	http://compbio.dfci.harvard.edu/tgi/plant.html
Plant transcript assembly database	http://plantta.tigr.org
Microarray databases	
Potato	http://www.tigr.org/tdb/potato
Maize	http://www.maizearray.org
Rice	http://www.ricearray.org
Pine	http://www.tigr.org/tdb/e2k1/pine
Other databases	
Plant repeat database	http://www.tigr.org/tdb/e2k1/plant.repeats
Transposon database	ftp://ftp.tigr.org/pub/data/TransposableElements/transposon_db.pep

followed by the methods and informatics tools available for accessing the sequence data and analyses.

2. Sequence and Annotation Databases

2.1. Background

2.1.1. Major Sequence and Annotation Databases at TIGR

TIGR has participated in genome sequencing and annotation projects of a number of model plants including *A. thaliana*, the cereal species *O. sativa* subsp. *japonica* cv. Nipponbare and *Zea mays*, and the model legume species *Medicago truncatula*.

2.1.1.1. *A. THALIANA*

TIGR was part of an international consortium (*Arabidopsis* Genome Initiative) to sequence the complete genome of the model plant *Arabidopsis* and has contributed about one-third of the sequences generated. TIGR first completed and published the sequence of chromosome 2 (19.6 Mbp) in

1999 *(13)*, and the complete genome was published in December 2000 *(7)*. TIGR also completed a genome reannotation project in 2004 to provide uniform high-quality annotation across the *Arabidopsis* genome *(14)*. In conjunction with Cold Spring Harbor Laboratory and Washington University, TIGR also performed limited WGS on *Brassica oleracea* ($\sim$0.5$\times$) to exploit the sequence conservation between the two species, with the goals to improve existing *Arabidopsis* gene annotation and to discover potential novel gene sequences.

2.1.1.2. RICE

Rice is one of the major food sources worldwide and also serves as a model for cereal species such as maize, wheat, barley, sorghum, sugarcane, and millet. The rice genome is the smallest among the cereals ($\sim$400 Mbp) and is highly collinear with other cereal species. TIGR participated in the international public rice genome sequencing project and has generated extensive sequence and bioinformatics resources for a thorough analysis of this cereal model plant at the genomic level *(15–17)*. TIGR is currently funded to annotate the rice genome, and the results of the annotation project have been made publicly available *(18)*. In collaboration with an international consortium, TIGR is also participating in a rice resequencing effort to sequence an additional 19 rice lines (http://rice.tigr.org).

2.1.1.3. MAIZE

Maize is a classical genetic model from which the first mobile element was discovered and is also an economically important crop in the USA. The size (est. 2500–2700 Mbp) and repetitiveness of the maize genome have been a major obstacle for whole-genome sequencing. Current estimates indicate that repetitive DNA sequences, particularly retrotransposons, constitute approximately 80% of the maize genome. As a member of the Consortium for Maize Genomics, TIGR evaluated two gene-enrichment approaches including methylation filtration (MF) and high Cot (HC) selection to selectively sequence the maize gene space. Results showed that these gene-enrichment approaches can provide a rapid and cost-effective alternative for sequencing maize or other cereal genomes that are also often relatively large and highly repetitive. Approximately 300 Mbp of maize gene-enriched genomic assemblies (assembled *Z. mays* or AZMs) have been generated from the MF and HC libraries for gene discovery and analysis. In addition, draft assemblies of approximately 300 gene-containing maize BAC clones have been generated to provide an overview of the organization of the maize genome *(19,20)* (*see* **Note 1**) (http://maize.tigr.org).

2.1.1.4. MEDICAGO

Legumes are unique in their ability to fix atmospheric nitrogen through a symbiotic relationship with *Rhizobia*, gram-negative bacteria, and have remarkably high levels of protein. Nearly 33% of all human nutritional requirements for nitrogen come from legumes. In developing countries, legumes serve as the single most important source of protein. Legumes also play a central role in nearly all crop rotation systems. Secondary compounds synthesized in legumes, such as isoflavonoids and triterpene saponins, have also been shown to possess anti-cancer and other health-promoting effects. *M. truncatula* (barrel medic) is a model legume with a genome size of approximately 500 Mbp, closely related to alfalfa (*Medicago sativa*), which is the number one forage crop in the USA. As part of an international consortium to sequence the euchromatic gene-rich portion of the *M. truncatula* genome (estimated at 250–300 Mbp), TIGR is sequencing chromosomes 2 and 7 using a BAC-based approach. Sequence annotation is carried out by the International Medicago Annotation Group (IMGAG) that involves several informatics centers including TIGR (*see* **Note 2**) (http://www.tigr.org/tdb/e2k1/mta1).

2.1.2. Sequence Annotation

The TIGR sequence and annotation databases store and track the genomic sequence and annotation information of a target species. A series of informatic processes are used to aid in the identification of gene models from the genomic sequences (structural annotation), assignment of biological functions to the gene models (functional annotation), and whole-genome analysis.

An annotation pipeline for eukaryotic genomes has been developed at TIGR (*21*). The first step of the process is structural annotation which is to identify genes from the genomic sequences by defining the exon and intron boundaries. Several *ab initio* gene finder tools such as FGENESH (*22*), GeneMark.hmm (*23*), Genscan, Genscan+ (*24*), and GlimmerM (*25*) are used to predict genes in a target genome, the suite used being tailored to the species under investigation. Additional evidence from transcript and protein sequence databases are aligned to the genomic sequence by taking into account the splice site consensus (*26*). To generate automated annotation, the gene model is produced from a selected gene finder that shows the best performance for the target species. During manual annotation, gene models are individually curated by human inspection to combine evidence from gene predictions and transcript and protein alignments. Programs are now being deployed to automatically combine both experimental and *ab initio* predictions in gene model construction (*27,28*). The functional annotation process involves assigning biological functions to the gene models based on high-scoring matches from protein or domain database searches (*see* **Note 3**).

In the following section, the TIGR rice database is used as an example to describe methods available for accessing the sequence data and genome annotation. Similar approaches apply to other sequence and annotation databases at TIGR and are also described where applicable. **Table 2** summarizes the data types and analyses available from the major TIGR sequence and annotation databases.

Table 2
An Overview of the Major Plant Sequence and Annotation Databases at The Institute for Genomic Research (TIGR)

	Arabidopsis	Maize	*Medicago*	Rice	Wheat
Data type					
Complete genome	+			+	
BAC assembly (finished)	+		+	+	+
BAC assembly (draft)		+		+	
Pseudomolecule	+			+	
Gene-enriched assembly		+			
Annotation and analysis					
Ab initio gene prediction	+	+	+	+	+
EST/TC alignment	+	+	+	+	+
Protein alignment	+	+	+	+	+
Full-length cDNA evidence	+			+	
Protein domain	+			+	
Transposon-related gene models	+			+	
Gene ontology	+			+	
Paralogous protein family	+			+	
Repeat element	+			+	
Genetic marker	+	+		+	
Insertional mutant	+			+	
Oligonucleotide microarray probe	+			+	
Data access					
MANATEE	+	+	+	+	+
GBrowse	+			+	
DAS	+			+	
FTP flat file	+	+	+	+	+
Customized flat file				+	
BLAST server	+	+	+	+	+

BAC, bacterial artificial chromosome; EST, expressed sequence tag; TC, tentative consensus.

2.2. Methods for Accessing Sequences, Annotation, and Analyses

2.2.1. The TIGR Rice Genome Annotation Project

2.2.1.1. SEQUENCE ASSEMBLIES

The rice genome was sequenced using a BAC-based approach *(8)*. A BAC clone of interest can be identified and retrieved based on sequence similarity using the BLAST server or using the clone name identifier or GenBank accession. A BAC report shows the gene models with putative biological functions and distribution within the BAC clone. A genome browser, the GBrowse viewer *(29)*, is used to display the order of the BAC clones along the pseudomolecules. The nucleotide sequences of the BAC clones and pseudomolecules are available through FTP download.

2.2.1.2. GENE MODEL ANNOTATION

The gene models on the BAC assemblies or pseudomolecules can be accessed by several approaches including sequence similarity, functional annotation, or locus name. A BLAST server is available for sequence comparison to identify a gene model of interest. Alternatively, gene models can be retrieved based on a putative biological function (text search), a protein domain such as Pfam (http://pfam.janelia.org) or Prosite (http://www.expasy.org/prosite) entries, or Gene Ontology (GOSlim) assignments (http://www.geneontology.org/GO.slims.shtml). Gene models can also be accessed using a gene model identifier such as the locus name or a TIGR internal gene model identifier. A gene model report with functional and structural annotation is available as part of the MANATEE Web page display (http://manatee.sourceforge.net). The gene model can also be viewed in the context of the pseudomolecule using the GBrowse viewer (http://www.gmod.org). The TIGR Rice GBrowse has been set up to display a large range of genome features such as the predicted gene models from different gene finders, transcript and protein alignments, genetic markers, and so on (*see* **Fig. 1**). To retrieve information for a set of gene models with similar properties, a data extraction tool is available for batch retrieval. This tool allows the selection of a nucleotide range from a given chromosome and the selection of one or more of the associated data such as genomic sequence, exon, intron, untranslated region, upstream sequence from translational start codon, intergenic sequence, transcript sequence, coding region, protein sequence, functional annotation, gene models with a Pfam domain match, and so on.

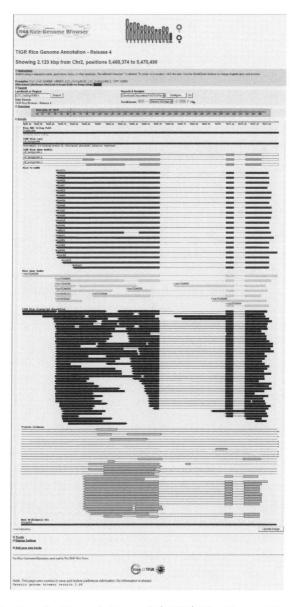

Fig. 1. The Institute for Genomic Research (TIGR) Rice Genome Browser. The TIGR Rice Genome Browser is setup based on the GBrowse genome viewer for displaying multiple genome features along individual BAC clones or pseudomolecules. Examples of data tracks include gene models predicted by different gene finders, transcript and protein alignments, genetic markers, transposon insertion flanking sequence tags, and so on.

2.2.1.3. GENOME-BASED TRANSCRIPT ALIGNMENTS

ESTs and full-length cDNAs are valuable resources for improving annotation quality based on experimental evidence. The Program to Assemble Spliced Alignments (PASA) is a tool developed at TIGR, which performs genome-based EST and full-length cDNA alignments and clustering *(30)*. This tool can be used to aid in the discovery of new gene models not originally predicted by the gene finders, to update gene structures such as the exon–intron boundaries and untranslated regions, and to detect putative alternative splice variants. A list of alternatively spliced variants has been created for the individual rice chromosomes.

2.2.1.4. PARALOGOUS FAMILIES

The identification of paralogous protein families based on protein domains allows not only the study of gene duplication events but also the annotation of gene models that originally had no known function. To generate the paralogous families, a novel set of protein domains generated from an all-against-all BLASTP search of the rice gene models is created. The rice gene models are organized into paralogous protein families based on Pfam domain matches and the new BLASTP-based domains, according to domain composition as defined by the type and the number of domains present in the gene models *(14)*. A list of the paralogous protein families with associated gene models is available through FTP download.

2.2.1.5. GENETIC MARKERS

A set of sequenced rice genetic markers obtained from the Japanese Rice Genome Program (http://rgp.dna.affrc.go.jp) and Gramene (http://www.gramene.org/markers) have been mapped in silico to the rice BAC/PAC clones that were used to construct the pseudomolecules. The marker mapping results can be retrieved according to a specific range of cM units along the pseudomolecules, the name or GenBank accession of the markers, or the name of the BAC/PAC clones. For comparative genomics studies, both the maize and the wheat genetic markers obtained from the MaizeGDB database (http://www.maizegdb.org) and the GrainGenes Wheat database (http://wheat.pw.usda.gov/cgi-bin/graingenes/browse.cgi?class=marker), respectively, have been anchored to the rice pseudomolecules through *in silico* mapping to the BAC/PAC clones. The syntenic relationship can be viewed through an alignment of conserved chromosomal regions from the selected species.

2.2.1.6. INSERTIONAL MUTANTS

Flanking sequence tags (FSTs) are sequences next to insertion sites of mobile elements such as Tos17, T-DNA, or Ac/Ds in the rice genome *(31)*. FST sequences were collected from GenBank and aligned to the rice genome. To search for possible insertional mutations in the vicinity of a gene of interest, FSTs can be located based on the functional description of the gene model or the sequence coordinates of a chromosomal region or a BAC/PAC clone.

2.2.1.7. SIMPLE SEQUENCE REPEATS

Simple sequence repeats (SSRs) have been identified in the rice pseudo-molecules using the SSR Identification Tool (SSRIT) *(32)*. The results can be retrieved according to the SSR type (e.g., mononucleotide, dinucleotide) or a given sequence motif (e.g., actg).

2.2.2. Other Annotation Features for Plant Genomes

2.2.2.1. GENOME-BASED TRANSCRIPT ALIGNMENTS

In the *Arabidopsis* project, the alternatively spliced gene models have been further classified into subcategories such as alternative donor/acceptor, alternate terminal exons, exon skipping, unspliced introns, and others. A splicing variation report shows an alignment of the alternatively spliced gene models with the EST and full-length cDNA sequences and also highlights the matching exon–intron boundaries in the sequence alignments.

2.2.2.2. SEQUENCE DUPLICATIONS

In the *Arabidopsis* genome reannotation project, segmentally duplicated chromosomal regions have been identified by a pair-wise comparison of gene models along all chromosome pairs *(14)*. Different combinations of chromosome pairs can be selected for segmental duplication analysis. In addition, clusters of tandemly duplicated genes have been identified by searching the protein products of the gene models using BLASTP. Different cluster sizes can be selected for the tandem duplication analysis.

2.2.2.3. PLANT REPEAT DATABASES

The TIGR Plant Repeat Databases represent a collection of high copy number sequences of four plant families including the Brassicaceae, Fabaceae, Gramineae, and the Solanaceae *(33)*. The repeat sequences include annotated repeats collected from GenBank based on a keyword search. In addition, the *Oryza* Repeat Database is augmented with iterative searches to include new repeat sequences identified from the GenBank genome survey sequences (GSS) and BAC sequences based on sequence similarity to the annotated repeats collected from GenBank. The repeat databases that are available at the TIGR BLAST server can be obtained through FTP downloads (http://www.tigr.org/tdb/e2k1/plant.repeats).

2.2.2.4. Transposon Database

TIGR also maintains a FASTA file of annotated transposon-encoded proteins that is intended to be representative of all known prokaryotic and eukaryotic transposons. The database is used internally for transposon annotation and is available on the FTP site for download (ftp://ftp.tigr.org/pub/data/TransposableElements/transposon_db.pep).

3. Plant Gene Index Databases
3.1. Background

Despite the rapid advances of genome sequencing, EST sequencing and analysis remains an important tool for gene discovery and genome annotation. The TIGR Gene Index databases contain 88 organisms including 31 animal, 33 plant, 15 protist, and 9 fungal species. Each database represents a catalog of expressed gene sequences for a target species or, in some cases, genus. The gene catalogs are created by the clustering and assembly of EST and cDNA sequences to generate a set of consensus sequences. The databases are updated periodically based on the availability of newly submitted EST or cDNA sequences.

The 33 plant gene index databases (http://compbio.dfci.harvard.edu/tgi/plant.html) include model plants such as *Arabidopsis*, rice, and *Medicago*, crop plants such as maize, wheat, barley, rye, soybean, potato, tomato and an assortment of other species. For each target species, publicly available EST sequences are downloaded from the GenBank dbEST division. Expressed gene transcripts representing cloned cDNAs or gene sequences are parsed from the GenBank nucleotide section. The sequences are trimmed to remove contaminants including vector and, in some species, ribosomal sequences. To create the tentative consensus (TC), virtual transcript assemblies, the EST, and transcript sequences are clustered based on shared sequence similarity from an all-against-all search using the TGICL tool developed at TIGR *(34)*. On the basis of the sequence overlaps, a multiple sequence alignment layout is generated in each cluster, and the sequences are assembled into a TC sequence using a CAP3-based assembly program. The resulting assemblies represent a non-redundant catalog of expressed gene sequences in the target species.

For their functional annotation, the TCs are searched against a non-redundant protein database, and putative biological function is assigned based on high-scoring protein hits. Annotations for protein domains and Gene Ontology (GO) assignments are generated from searches against the Pfam domain database and a protein-GO association database excluding inferred by electronic annotation (IEA) evidence, respectively. The TCs are associated with metabolic pathway components by searching against characterized pathway components with assigned Enzyme Commission (EC) numbers obtained from Kyoto Encyclopedia of Genes and Genomes (KEGG) (http://www.genome.ad.jp/kegg).

3.2. Methods for Accessing Sequences, Annotation, and Analyses

3.2.1. TC Sequence and Annotation

The TC sequence for a gene of interest can be retrieved according to functional annotation, sequence similarity using a BLAST search, or identifiers of the TCs or the component reads. A list of TC identifiers can also be provided for batch retrieval. The TC report displays the nucleotide sequence and annotation of the TC consensus from assembled ESTs and cDNA sequences. The potential open-reading frames are predicted using ESTScan *(35)* and framefinder *(36)*. The orientation of the TC is determined by a voting scheme that combines orientations of the EST and cDNA sequences, predicted genes, high-scoring protein hits, and the position of the poly-A or poly-T tract. The component EST and cDNA sequences are shown to indicate the relative positions of the component reads with reference to the TC consensus. A list of the component reads with the source library information, GenBank accessions, and the coordinates is provided. The TCs can also be subjected to a text search based on their functional annotation.

3.2.2. Expression Summaries

A user-defined percentage cutoff can be used to select for transcripts that are predominantly expressed in a given tissue. For example, a cutoff of 50% in a given tissue will return a list of TCs each containing at least 50% of component ESTs derived from libraries related to the selected tissue. Differentially expressed transcripts can be identified through the use of a statistical test to analyze the EST expression data derived from multiple library sources *(37)*. A list of TCs exclusively expressed in one of the libraries can also be provided given two user-selected library sources. To obtain the EST expression data of a TC sequence, an expression report showing the frequency of distribution of the component ESTs among different cDNA libraries is provided. The report represents a digital Northern showing the expression levels of the TC in different tissue types or libraries.

3.2.3. Single-Nucleotide Polymorphism Report

Putative single-nucleotide polymorphisms (SNPs) are collected from the multiple sequence alignments generated during the assembly stage of TC construction. An SNP is scored when the polymorphism is found in sequences originating from more than one independent library source. The report shows the position, the flanking sequences, and the putative SNP identified.

3.2.4. Oligomer Prediction

For each target species, a set of 70-mer oligonucleotides that are unique among the TC sequences are generated using OligoPicker *(38)*. The unique oligonucleotides are selected based on criteria that include cross-hybridization potential, melting temperature, and secondary structure. These oligonucleotides can be used for designing microarrays for gene expression profiling studies. An oligomer report is provided for each TC sequence and shows its nucleotide sequence, length, position, and melting temperature of the oligonucleotide.

3.2.5. Genome Mapping

The plant TC sequences are mapped to the complete *Arabidopsis* genome, and the rice genomes, to provide a comprehensive overview of alignments among all available plant TCs to the model plant genomes. The results can be displayed through two different genome viewers: the TGI viewer and the GBrowse.

3.2.6. Orthologous Groups

The Eukaryotic Gene Orthologs (EGO) database was developed to create tentative orthologous families based on the nucleotide sequence similarity among the TC sequences *(39)*. A pair-wise comparison across the TC sequences available from all species including plant, animal, protest, and fungi is used to search for the reciprocal best matches to construct tentative orthologous groups. In the report, orthologous TC sequences are listed and presented in a graphical display to indicate the relationship of the TCs among different species. The search results and multiple alignments of the TCs in the orthologous cluster are also provided.

Fig. 2. The Institute for Genomic Research (TIGR) Plant Transcript Assembly (TA). The TIGR Plant Transcript Assembly database generates clustering and assembly of public sources of cDNAs and ESTs from GenBank into a non-redundant set of transcript assemblies, for all plant species (~233) with available EST and cDNA sequences. More than one species, such as a particular order, can be selected for sequence search.

3.3. The TIGR Plant TA Database

With the migration of the TIGR Gene Index project to the Dana Farber Cancer Center and the need to have access to transcript assemblies for multiple plant species, a Plant TA database has been developed at TIGR *(40)*. The Plant TA database is set up to create clustering and assembly of expressed transcript sequences into sets of non-redundant transcript assemblies, for all plant species with publicly available EST and cDNA sequences (currently $\sim$233 species) (*see* **Fig. 2**). While similar to the previous TIGR Gene Index project, the TA assembly includes expressed sequences such as ESTs and cDNAs. However, the TA assembly selectively excludes but not predicted transcript sequences derived from genome sequencing projects. The TA assembly report includes the assembled consensus sequence, a graphical display, and the positional information of the component reads. Additional functions such as putative annotation and orientation are also included in the TA assembly report. The TA database is searchable by sequence similarity (BLAST) or by its annotation (text search). More than one species can be selected for sequence search (e.g., a particular phylogenetic order or family). The TA assemblies are also available through FTP download.

4. Microarray Databases

4.1. Background

TIGR has participated in several collaborative efforts to provide efficient and economical microarray platforms and databases for performing gene expression profiling analysis in the crop plants potato, rice, and maize as well as pine (*see* **Note 4**). These projects provide low-cost microarrays and hybridization services through the design and development of microarrays, the distribution of microarrays to the research community, and the establishment of a centralized public repository of gene expression data in a standard data format to allow easy public sharing and analysis.

To design the maize oligonucleotide-based microarray, gene sequences were collected from ESTs, cDNAs, genomic sequences, and predicted gene models. The orientation and accuracy of the target sequences were validated by several approaches including comparison to gene models identified in related species, alignment to expressed transcripts, protein sequences, and the presence of poly-A or poly-T tracts. A small subset of repetitive sequences, mitochondrial, and chloroplast gene sequences, are also included in the microarray design. The maize oligonucleotide microarray contains more than 57,000 oligonucleotides derived from gene sequences obtained from the AZMs and the maize TCs (http://www.maizearray.org). The 70-mer oligonucleotides were selected based on a series of criteria including cross-hybridization potential, melting temperature, and secondary structure. The potato microarray project includes the development of a cDNA microarray with approximately 12,000 clones, and the establishment of an expression profiling service (http://www.tigr.org/tdb/potato). For the rice oligonucleotide microarray, the oligonucleotides were designed from Release 2 of the TIGR Rice Genome Annotation using only models based on finished sequence (http://www.ricearray.org). In the first release of the rice oligonucleotide array, approximately 20,000 oligonucleotides were designed and printed. In a pending release, approximately 45,000 oligonucleotides will be available on a single array. Both the microarrays and a hybridization service are available to the potato, maize, and rice research communities on a cost recovery basis.

4.2. Methods for Data Access and Analyses

As the potato, rice, and maize array projects use a similar schema and platform, the bioinformatics that support the arrays and use of the expression data is highly similar among the projects. Thus, a generic discussion of the tools and analyses provided from these array projects is presented below.

4.2.1. Data Submission

The TIGR expression databases are MIAME compliant *(41)*. MIAME describes the minimum information about a microarray experiment, which are the experimental parameters needed for the interpretation of the experimental results and replication of the experiment. The microarray expression data to be deposited to one of the TIGR expression databases (potato, rice, and maize) is subject to a five-step processing and analysis pipeline. First, users perform array hybridizations and scan the arrays to generate raw tiff images. Second, the images are quantified using data extraction software, such as GenePix *(42)*. The image quantification can either be performed by the user, or the raw images can be forwarded to TIGR for extractions. Third, users submit descriptions of the experimental parameters through the MIAME Submission Tool, which is a series of Web pages specifically designed to capture MIAME information. Fourth, the printtip lowess method of normalization is applied to the array images using the Limma package of Bioconductor *(43)*. Finally, both the normalized and the raw expression data are loaded into the expression database and remain private for a defined period (2 months for potato, 6 months for maize, and 3 months for rice). Users are able to view the data through a password-protected Web site prior to public release. TIGR also provides a service to submit the microarray data on the user's behalf to the NCBI Gene Expression Omnibus *(44)* (http://www.ncbi.nlm.nih.gov/geo). The TIGR expression databases and processing pipeline can be adapted to accommodate additional microarray platforms, such as Affymetrix or Agilent as they become publicly available.

4.2.2. Gene Expression Data

To identify a probe sequence that represents a gene of interest, a BLAST server for searching against the original gene sequences used for oligonucleotide design is provided. Probe sequences can also be identified based on other criteria, including putative function of the gene product, gene model identifier, GenBank accession, oligonucleotide identifier, GOSlim term, and EC number. Large-scale gene expression data are accessible by two approaches. First, a study browser provides an overview of all the studies stored in the database and permits queries on conditions such as study variables, treatment, and tissue type. Second, a gene-level expression search capable of querying the expression levels of a gene of interest across one or more hybridizations under various experimental conditions. Furthermore, the project Web site provides a service for bulk FTP download of project data, enabling user-defined analyses.

4.2.3. GenePix Result Metrics Tool

The GenePix result (GPR) metrics tool can be used to perform a quality assessment on scanned arrays by generating array metrics including total spots, percentage missing spots, percentage bad spots, average foreground, and average background intensities across different channels. To submit GPR files for assessment, users first login to ftp://metrics:metrics@ftp.tigr.org (username:metrics and password:metrics), set the transfer mode to binary (by default), and upload the GPR files for analysis.

4.2.4. Pre-Computed Statistical Analyses

A list of pre-computed analysis based on different combinations of selected expression data sets is provided. These analyses include hierarchal clustering, k-means clustering, and identification of differentially expressed genes. The pre-computed analyses allow for a preliminary "first pass" examination of the expression data for an entire study by providing an overview of the data quality and an initial assessment of the biological significance of the data.

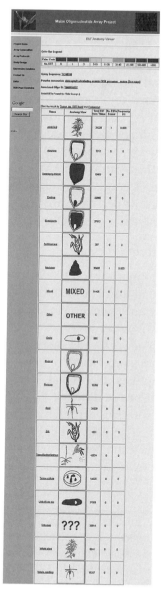

Fig. 3. The Oligo and EST Anatomy viewer (OEAV) for gene expression in microarray analysis. An anatomy viewer is provided for each of the probe sequence on the microarray, which shows the expression levels of the corresponding gene with reference to the anatomical structures of plants.

4.2.5. Custom-Defined Statistical Analyses

An array analysis server has been developed (http://plantarray.tigr.org) for performing custom analysis on the expression data. The Gohan server is capable of analyzing expression data and allows users to select a set of hybridizations from one or more studies and choose a method of analysis with defined parameters. The results are delivered to an email address upon completion and are available on the server for 48 h prior to deletion. The types of analyses available are based on open-source array analysis packages implemented with the R-statistical language (http://www. r-project.org). The analyses currently include Significance Analysis for Microarrays (SAM) with a Two-Class Unpaired analysis *(45)* (http://cran.r-project.org/src/contrib/Descriptions/samr.html). This service is currently only available for maize expression studies but will be expanded for use with potato and rice expression studies in the near future.

4.2.6. Oligo and EST Anatomy Viewer

The Oligo and EST Anatomy viewer (OEAV) tool is used for the analysis of gene expression levels among different tissues or cDNA libraries. It provides a quick reference for users to compare gene expression data based on EST frequency (digital Northern). A set of genes related to a specific biological function can be analyzed. For each gene sequence, the search results include a digital Northern and an anatomy viewer of gene expression (*see* **Fig. 3**). On the basis of EST frequency data, the digital Northern indicates the levels of expression of the gene sequence in different tissue types. The anatomy viewer shows the levels of gene expression with reference to the plant anatomy. The OEAV tool uses an approach similar to that of the NCBI SAGE Genie and the TIGR gene index project and will be expanded to incorporate the gene expression levels using the microarray expression data.

5. Notes

1. The TIGR maize database contains 300 Mbp of gene-enriched genomic assemblies (AZMs) from the maize gene space and close to 300 gene-rich BAC assemblies. Gene model predictions and alignments to transcripts and protein sequence databases are available through the MANATEE Web display.
2. The TIGR *Medicago* database project has developed an Automated *Medicago* Sequence Contig Pipeline to generate the most up-to-date *Medicago* BAC contig assemblies on a daily basis. The pipeline performs automated download and assembly of publicly available *Medicago* BAC clones from GenBank. The assembled BAC contigs are annotated for predicted gene content using the gene finder FGENESH. The results are displayed through GBrowse along with IMGAG

gene model annotation, the *Medicago* gene index, ESTs, BAC ends, and genetic marker data. Toward the end of the project, as the BAC overlaps and resulting sequence-based contigs stabilize, it is anticipated that this pipeline will be replaced by a representation of the BAC tiling path for each chromosome similar to those provided for the *Arabidopsis* and rice genomes.

3. During functional annotation, the biological function of the gene model is described by a "gene name," which is determined based on results obtained from searching the gene model against transcript and protein sequence databases. The gene name is assigned by one of the following approaches: direct transfer from the best protein or domain hit for perfect database matches, the addition of the qualifier "putative" for non-perfect but significant database matches, the use of "expressed protein" if the gene model is supported only by EST evidence, or the use of "hypothetical protein" if the gene model is supported only by gene predictions.

4. The goal of the pine functional genomics project is to study gene expression during various stages of somatic and zygotic embryogenesis to understand the development of the pine embryo (http://www.tigr.org/tdb/e2k1/pine). cDNA libraries have been constructed from pine embryos and were end sequenced to generate EST sequences. A set of non-redundant clones have been selected for the design of a pine cDNA microarray to measure gene expression profiles during embryogenesis. These arrays are currently available to the community on a cost recovery basis.

5. In October 2006, The Institute for Genomic Research (TIGR) and several affiliated organizations merged to become the new J. Craig Venter Institute (JCVI). However, the domain name of "tigr.org" in the URLs of the plant databases remains active.

References

1. Fleischmann, R.D., Adams, M.D., White, O., Clayton, R.A., Kirkness, E.F., Kerlavage, A.R., Bult, C.J., Tomb, J.F., Dougherty, B.A., Merrick, J.M., et al. (1995) Whole-genome random sequencing and assembly of *Haemophilus influenzae* Rd. *Science* 269, 496–512.

2. Peterson, J.D., Umayam, L.A., Dickinson, T., Hickey, E.K., and White, O. (2001) The comprehensive microbial resource. *Nucleic Acids Res* 29, 123–125.

3. Berriman, M., Ghedin, E., Hertz-Fowler, C., Blandin, G., Renauld, H., Bartholomeu, D.C., Lennard, N.J., Caler, E., Hamlin, N.E., Haas, B., et al. (2005) The genome of the African trypanosome *Trypanosoma brucei*. *Science* 309, 416–422.

4. Carlton, J.M., Angiuoli, S.V., Suh, B.B., Kooij, T.W., Pertea, M., Silva, J.C., Ermolaeva, M.D., Allen, J.E., Selengut, J.D., Koo, H.L., et al. (2002) Genome sequence and comparative analysis of the model rodent malaria parasite *Plasmodium yoelii yoelii*. *Nature* 419, 512–519.

5. El-Sayed, N.M., Myler, P.J., Bartholomeu, D.C., Nilsson, D., Aggarwal, G., Tran, A.N., Ghedin, E., Worthey, E.A., Delcher, A.L., Blandin, G., et al. (2005)

The genome sequence of *Trypanosoma cruzi*, etiologic agent of Chagas disease. *Science* 309, 409–415.

6. Gardner, M.J., Hall, N., Fung, E., White, O., Berriman, M., Hyman, R.W., Carlton, J.M., Pain, A., Nelson, K.E., Bowman, S., et al. (2002) Genome sequence of the human malaria parasite *Plasmodium falciparum*. *Nature* 419, 498–511.

7. The Arabidopsis Genome Initiative (2000) Analysis of the genome sequence of the flowering plant *Arabidopsis thaliana*. *Nature* 408, 796–815.

8. International Rice Genome Sequencing Project (2005) The map-based sequence of the rice genome. *Nature* 436, 793–800.

9. Pop, M., Kosack, D.S., and Salzberg, S.L. (2004) Hierarchical scaffolding with Bambus. *Genome Res* 14, 149–159.

10. Sutton, G., White, O., Adams, M., and Kerlavage, A.R. (1995) TIGR Assembler: a new tool for assembling large shotgun sequencing projects. *Genome Sci Technol* 1, 9–19.

11. Pampanwar, V., Engler, F., Hatfield, J., Blundy, S., Gupta, G., and Soderlund, C. (2005) FPC web tools for rice, maize, and distribution. *Plant Physiol* 138, 116–126.

12. Lee, Y., Tsai, J., Sunkara, S., Karamycheva, S., Pertea, G., Sultana, R., Antonescu, V., Chan, A., Cheung, F., and Quackenbush, J. (2005) The TIGR gene indices: clustering and assembling EST and known genes and integration with eukaryotic genomes. *Nucleic Acids Res* 33, D71–D74.

13. Lin, X., Kaul, S., Rounsley, S., Shea, T.P., Benito, M.I., Town, C.D., Fujii, C.Y., Mason, T., Bowman, C.L., Barnstead, M., et al. (1999) Sequence and analysis of chromosome 2 of the plant *Arabidopsis thaliana*. *Nature* 402, 761–768.

14. Haas, B.J., Wortman, J.R., Ronning, C.M., Hannick, L.I., Smith, R.K., Jr., Maiti, R., Chan, A.P., Yu, C., Farzad, M., Wu, D., et al. (2005) Complete reannotation of the Arabidopsis genome: methods, tools, protocols and the final release. *BMC Biol* 3, 7.

15. Buell, C.R., Yuan, Q., Ouyang, S., Liu, J., Zhu, W., Wang, A., Maiti, R., Haas, B., Wortman, J., Pertea, M., et al. (2005) Sequence, annotation, and analysis of synteny between rice chromosome 3 and diverged grass species. *Genome Res* 15, 1284–1291.

16. Rice Chromosome 10 Sequencing Consortium (2003) In-depth view of structure, activity, and evolution of rice chromosome 10. *Science* 300, 1566–1569.

17. Rice Chromosomes 11 and 12 Sequencing Consortia (2005) The sequence of rice chromosomes 11 and 12, rich in disease resistance genes and recent gene duplications. *BMC Biol* 3, 20.

18. Yuan, Q., Ouyang, S., Wang, A., Zhu, W., Maiti, R., Lin, H., Hamilton, J., Haas, B., Sultana, R., Cheung, F., et al. (2005) The institute for genomic research Osa1 rice genome annotation database. *Plant Physiol* 138, 18–26.

19. Chan, A.P., Pertea, G., Cheung, F., Lee, D., Zheng, L., Pontaroli, A.C., SanMiguel, P., Yuan, Y., Bennetzen, J.L., Barbazuk, W.B., et al. (2006) The TIGR maize database. *Nucleic Acids Res* 34, D771–776.

20. Whitelaw, C.A., Barbazuk, W.B., Pertea, G., Chan, A.P., Cheung, F., Lee, Y., Zheng, L., van Heeringen, S., Karamycheva, S., Bennetzen, J.L., et al. (2003) Enrichment of gene-coding sequences in maize by genome filtration. *Science* 302, 2118–2120.

21. Wortman, J.R., Haas, B.J., Hannick, L.I., Smith, R.K., Jr., Maiti, R., Ronning, C.M., Chan, A.P., Yu, C., Ayele, M., Whitelaw, C.A., et al. (2003) Annotation of the Arabidopsis genome. *Plant Physiol* 132, 461–468.

22. Salamov, A.A., and Solovyev, V.V. (2000) Ab initio gene finding in Drosophila genomic DNA. *Genome Res* 10, 516–522.

23. Lukashin, A.V., and Borodovsky, M. (1998) GeneMark.hmm: new solutions for gene finding. *Nucleic Acids Res* 26, 1107–1115.

24. Burge, C., and Karlin, S. (1997) Prediction of complete gene structures in human genomic DNA. *J Mol Biol* 268, 78–94.

25. Pertea, M., and Salzberg, S.L. (2002) Computational gene finding in plants. *Plant Mol Biol* 48, 39–48.

26. Huang, X., Adams, M.D., Zhou, H., and Kerlavage, A.R. (1997) A tool for analyzing and annotating genomic sequences. *Genomics* 46, 37–45.

27. Allen, J.E., Pertea, M., and Salzberg, S.L. (2004) Computational gene prediction using multiple sources of evidence. *Genome Res* 14, 142–148.

28. Allen, J.E., and Salzberg, S.L. (2005) JIGSAW: integration of multiple sources of evidence for gene prediction. *Bioinformatics* 21, 3596–3603.

29. Stein, L.D., Mungall, C., Shu, S., Caudy, M., Mangone, M., Day, A., Nickerson, E., Stajich, J.E., Harris, T.W., Arva, A., et al. (2002) The generic genome browser: a building block for a model organism system database. *Genome Res* 12, 1599–1610.

30. Haas, B.J., Delcher, A.L., Mount, S.M., Wortman, J.R., Smith, R.K., Jr., Hannick, L.I., Maiti, R., Ronning, C.M., Rusch, D.B., Town, C.D., et al. (2003) Improving the Arabidopsis genome annotation using maximal transcript alignment assemblies. *Nucleic Acids Res* 31, 5654–5666.

31. Miyao, A., Tanaka, K., Murata, K., Sawaki, H., Takeda, S., Abe, K., Shinozuka, Y., Onosato, K., and Hirochika, H. (2003) Target site specificity of the Tos17 retrotransposon shows a preference for insertion within genes and against insertion in retrotransposon-rich regions of the genome. *Plant Cell* 15, 1771–1780.

32. Temnykh, S., DeClerck, G., Lukashova, A., Lipovich, L., Cartinhour, S., and McCouch, S. (2001) Computational and experimental analysis of microsatellites in rice (*Oryza sativa* L.): frequency, length variation, transposon associations, and genetic marker potential. *Genome Res* 11, 1441–1452.

33. Ouyang, S., and Buell, C.R. (2004) The TIGR Plant Repeat Databases: a collective resource for the identification of repetitive sequences in plants. *Nucleic Acids Res* 32, D360–D363.

34. Pertea, G., Huang, X., Liang, F., Antonescu, V., Sultana, R., Karamycheva, S., Lee, Y., White, J., Cheung, F., Parvizi, B., et al. (2003) TIGR Gene Indices

clustering tools (TGICL): a software system for fast clustering of large EST datasets. *Bioinformatics* 19, 651–652.

35. Iseli, C., Jongeneel, C.V., and Bucher, P. (1999) ESTScan: a program for detecting, evaluating, and reconstructing potential coding regions in EST sequences. *Proc Int Conf Intell Syst Mol Biol* 138–148.

36. Slater, G. (2000), Algorithms for analysis of ESTs, PhD thesis, University of Cambridge, UK.

37. Stekel, D.J., Git, Y., and Falciani, F. (2000) The comparison of gene expression from multiple cDNA libraries. *Genome Res* 10, 2055–2061.

38. Wang, X., and Seed, B. (2003) Selection of oligonucleotide probes for protein coding sequences. *Bioinformatics* 19, 796–802.

39. Lee, Y., Sultana, R., Pertea, G., Cho, J., Karamycheva, S., Tsai, J., Parvizi, B., Cheung, F., Antonescu, V., White, J., et al. (2002) Cross-referencing eukaryotic genomes: TIGR Orthologous Gene Alignments (TOGA). *Genome Res* 12, 493–502.

40. Childs, K.L., Hamilton, J.P., Zhu, W., Ly, E., Cheung, F., Wu, H., Rabinowicz, P.D., Town, C.D., Buell, C.R., and Chan, A.P. (2007) The TIGR plant Transcript Assemblies database. *Nucleic Acids Res* 35, D846–851.

41. Brazma, A., Hingamp, P., Quackenbush, J., Sherlock, G., Spellman, P., Stoeckert, C., Aach, J., Ansorge, W., Ball, C.A., Causton, H.C., et al. (2001) Minimum information about a microarray experiment (MIAME)-toward standards for microarray data. *Nat Genet* 29, 365–371.

42. Fielden, M.R., Halgren, R.G., Dere, E., and Zacharewski, T.R. (2002) GP3: GenePix post-processing program for automated analysis of raw microarray data. *Bioinformatics* 18, 771–773.

43. Smyth, G.K., and Speed, T. (2003) Normalization of cDNA microarray data. *Methods* 31, 265–273.

44. Barrett, T., Suzek, T.O., Troup, D.B., Wilhite, S.E., Ngau, W.C., Ledoux, P., Rudnev, D., Lash, A.E., Fujibuchi, W., and Edgar, R. (2005) NCBI GEO: mining millions of expression profiles–database and tools. *Nucleic Acids Res* 33, D562–D566.

45. Tusher, V.G., Tibshirani, R., and Chu, G. (2001) Significance analysis of microarrays applied to the ionizing radiation response. *Proc Natl Acad Sci USA* 98, 5116–5121.

6

MIPS Plant Genome Information Resources

Manuel Spannagl, Georg Haberer, Rebecca Ernst, Heiko Schoof, and Klaus F. X. Mayer

Summary

The Munich Institute for Protein Sequences (MIPS) has been involved in maintaining plant genome databases since the *Arabidopsis thaliana* genome project. Genome databases and analysis resources have focused on individual genomes and aim to provide flexible and maintainable data sets for model plant genomes as a backbone against which experimental data, for example from high-throughput functional genomics, can be organized and evaluated. In addition, model genomes also form a scaffold for comparative genomics, and much can be learned from genome-wide evolutionary studies.

Key Words: *Arabidopsis*; *Medicago*; *Lotus*; maize; tomato; genome annotation.

1. Introduction

Currently, the Munich Institute for Protein Sequences (MIPS) maintains genome databases for the plant model organisms *Arabidopsis thaliana* [MIPS *A. thaliana* database (MatDB)], *Oryza sativa* [rice, MIPS *O. sativa* database (MOsDB)], *Medicago truncatula* [European *Medicago* and Legume Database (UrMeLDB)], *Lotus japonicus*, *Zea mays* and *Solanum lycopersicum* (tomato). Besides information on individual genetic elements, their physical and functional properties and information about the underlying nucleotide or amino acid sequence, more and more combinatorial and comparative queries become important to address complex scientific questions.

From: *Methods in Molecular Biology, vol. 406: Plant Bioinformatics: Methods and Protocols*
Edited by: D. Edwards © Humana Press Inc., Totowa, NJ

In the first section of this chapter, we aim to introduce content, technical setup and architecture of the plant genome resources available at MIPS. In additional sections, we introduce analytical and bioinformatics analytical features that have been developed within and around PlantsDB. The features selected and introduced broaden the capabilities of PlantsDB to undertake complex queries. The third section is devoted to the introduction and usage of Web services to use different, distributed databases and to make use of them to query for complex questions.

2. Materials

The MIPS PlantsDB system has been designed as an information resource for plant genomes. Its aim is to structure and communicate plant genomic data and communicate knowledge and data on information-enriched plant genomic backbones. Besides classical genetic elements, for example, protein-coding genes, additional genomic features such as repeat and transposable elements, *cis*-elements, non-coding RNAs, miRNAs etc., gain more and more attention and importance in genome research and analysis. PlantsDB aims to meet these demands in plant genome research and accommodates and communicates data on different genetic element features.

MIPS PlantsDB database architecture has been designed as a generic architecture, which is applied for all plant genome databases hosted at MIPS. The generic outline of PlantsDB enables users to easily apply comparative and analytical modules on one or several different databases and to easily carry out comparative and integrative analysis over several databases. However, frequently different analytical preferences and emphasis for different plant genomes and their respective communities exist. Besides the advantages generated by the generic design and outline, it is of course possible to add specific commodities to individual species databases that are of special interest for particular user communities.

Genome analysis often leads to inherently heterogeneous analytical data. Contemporary genome projects are implemented as community efforts that are designed to deliver transparent and stable data sets that have been accomplished by transparent and documented analytical pipelines and methods. However, owing to often distributed and non-centralized genome projects and consequently the distributed analysis and annotation, frequently differing annotation procedures (e.g., curated vs. automated annotation) and analytical results exist.

The MIPS PlantsDB system aims to consider this in storing and communicating parallel annotations. Parallel annotations are not restricted to annotations from different groups or resources on the same underlying nucleotide sequence but can also be handled for sequences based on differing underlying sequences, a scenario often happening during ongoing and not yet finalized genome projects, frequently resulting from discrepancies in the assembly and conflicting sequences in clone overlaps. Parallel annotations and data sets are separated both within the physical data scheme and within the Web interface.

MIPS PlantsDB consists of different "species pillars" that are introduced in the following.

2.1. Arabidopsis thaliana

MAtDB contains both the original MIPS assembly of the *Arabidopsis* Genome Initiative genome sequence and the TIGR version 5 assembly and annotation *(1,2)*. Manual curation of gene models and functional assignments, which partially continued at MIPS in parallel to the whole-genome reannotation performed by TIGR, has been mapped to TIGR version 5 and resulted in a merged annotation set. Both gene prediction sets can be browsed independently. In addition, data from many European *Arabidopsis* resources are integrated through the use of BioMoby Web services within the frame of the EU PlaNet project.

2.2. Medicago truncatula

The genome of the model legume *M. truncatula* is currently being sequenced within an international collaboration. As part of PlantsDB, the UrMeLDB integrates data generated within the international *Medicago* sequencing project as well as the European *Medicago* project. All publicly available *Medicago* genomic clone and sequence data, completed as well as unfinished, are integrated into UrMeLDB on a regular basis. To prevent duplication of work and to assure uniform annotation standards applied on all *Medicago* genomic sequences and chromosomes, gene prediction and protein annotation is performed within an international collaborative framework, the International *Medicago* Genome Annotation Group (IMGAG).

Automated annotation of *Medicago* sequences, including gene prediction, is also performed at other sites. Given the current accuracy of gene prediction, multiple alternative predictions can be valuable for individual genes, allowing possible alternative models to be considered (e.g., in planning experiments or designing primers).

2.3. Lotus

Lotus is an evolutionary close relative to *Medicago* and represents a complimentary legume model genome to *Medicago*. Publically available sequence data are integrated, and gene prediction on these sequences has been performed by Eugene (INRA Toulouse) using an automated pipeline adopted for legumes.

2.4. Solanum (Tomato)

The tomato genome project is in an already advanced state. Large-scale genome sequencing has already been initiated. Similar to *Medicago*, the project is an international effort. The MIPS tomato genome database aims to integrate all sequence data from the international collaboration as well as associated analytical data.

2.5. Grass Genomes: Rice and Maize

Both *Arabidopsis* and *Medicago* are dicotyledonous model plants. Albeit rice and maize have a long tradition as model plants as well, they are at the same time crop plants of major importance for the world food supply. They are representatives of the monocotyledonous plants that diverged from the dicotyledonous plants approximately 150–250 million years ago. Thus, all different levels of comparative analysis between representatives of the monocotyledonous and the dicotyledonous plants are of special interest. A consistent analysis and representation of these genomes is indispensable to perform such analysis. However, in some respects, the data types and research focus differ between the species. For example, the analysis of repeat elements and transposons played only a subordinated role in *Arabidopsis* but is a major focus for the analysis of maize. Thus, an important requirement for a generic data model is flexibility to encompass species-specific data without affecting comparability.

For rice, the MIPS data resource MOsDB includes the publicly available *O. sativa* ssp. japonica cv. Nipponbare assembly and annotation as provided by TIGR. When the International Rice Genome Sequencing Project (IRGSP) genome sequence and annotation becomes public, this will be included in parallel.

The maize database includes approximately 250 bacterial artificial chromosome (BAC) sequences available to date (~30 Mb) with manually curated gene predictions. In addition, a detailed analysis and classification of repeat elements is available. These data allow a first insight into the structure and composition of the maize genome and provide a basis for comparative and combinatorial analysis.

2.6. PlantsDB Analysis Tools, Web Interface and Data Retrieval

An important tool for comparative genomics is the prediction of orthologous genes (genes in different species that evolved from a common ancestral gene by speciation) between genomes. We apply a tool designed for the detection of conserved orthologous sequence (COS) markers to extract ortholog from the Similarity Matrix of Proteins (SIMAP) database. SIMAP is a database containing a precomputed similarity matrix covering the similarity space formed by >4 million amino acid sequences from public databases and completely sequenced genomes of all-against-all protein similarity searches.

SIMAP is a database for the precomputed homologies of protein sequences computed using best bidirectional FASTA matches *(3)*. Within a gene report, putative orthologous sequences from more than 200 fully sequenced genomes (October 2005) can be derived for the gene's protein sequence using the build-in SIMAP retrieval.

2.6.1. PlantsDB Functionalities

The consistent application interface to all MIPS plant databases enables the implementation of advanced query tools that can be applied on all individual databases of the PlantsDB family. For example, we developed a sequence export tool (Genetic Element Retrieval System; GenERSys) (http://mips.gsf.de/proj/plant/jsf/general/genersysIndex.jsp) that allows users to download specific sequence data sets, such as all the first introns of all protein-coding genes on a selected contig, or on a selected number of basepairs upstream of all start codons in a genome.

The aim of the Web interface to the MIPS plant genome resources is to provide access to all included genomes in a common format, as well as to tools for cross-species comparisons.

To browse data, the user can navigate in a genome-oriented way. Assuming one would start from the chromosome list, all contigs anchored to each chromosome can be retrieved. A contig report contains detailed information on the entry as well as links to sequence, European Molecular Biology Laboratory (EMBL) database records, a list of annotated genetic elements, or a graphical viewer. The genetic element list links to reports on the protein genes or other features. Sequences can be viewed and downloaded as HTML, XML or FASTA format. For protein-coding genes, unspliced, spliced (transcript) and coding DNA sequences as well as protein sequences are available. Moreover, cross-references in the reports allow easy access to entries in external databases associated with the entry (e.g., relevant literature in PubMed).

Alternatively, complete lists of all sequenced contigs, all genetic elements, or all elements of a selected type are available for browsing. To visualize and browse genetic elements on a specified contig, a graphical interface, Gbrowse *(4)*, has been integrated.

Search options include search by name, free text, or sequence. The free text search option allows inspection of the content of all text fields, and it is available for individual genomes or across all databases. BLAST is used as a homology search engine. The target databases for similarity searches include clones (completed and unfinished), contigs, and genetic elements (e.g., coding sequences). Besides data sets from the plant projects at MIPS (*Arabidopsis, Medicago*, maize and rice), Swissprot/SWALL and plant-specific data sets selected from the EMBL nucleotide database [e.g., all *Arabidopsis* expressed sequence tags (ESTs)] are searchable.

Finally, the download section provides FTP access to various data downloads. This includes FASTA-formatted sequence files for all clones/contigs and protein-coding genes. Beside this, the download section contains functionality to create and download a GAMEXML file for a specified contig and coordinate range. The GAMEXML format is used by the Apollo Genome Browser, which has become a widespread and well-accepted genome annotation viewer and editor (*see* **Fig. 1**).

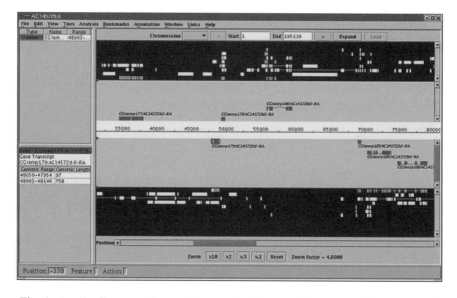

Fig. 1. Apollo Genome Viewer. The Apollo Genome Viewer provides opportunities beyond browser-based genome browsers. By downloading genomic regions or genomes of interest in GAME XML, Apollo can be used to visualize and inspect the regions of interest. Edits and changes can be undertaken, and the changes undertaken can be saved.

Apollo provides a detailed graphical viewer for genome data with more flexible interaction possibilities than a browser-based display. In addition, it enables interactive curation of the genome annotation to save the results locally. Thus, downloading a GAMEXML file of the region of interest enables the user to inspect (and modify) all gene annotation data, thus building an own local annotation data set. This also provides an infrastructure for community-based distributed manual annotation.

2.7. PlantsDB System Architecture and Design

The core data are stored in an ORACLE relational database management system. A modular approach has been chosen for the implementation of MIPS plant genome resources. To manage these data modules in a component-oriented manner, a multi-tier architecture (*see* **Fig. 2**) following the J2EE standard has

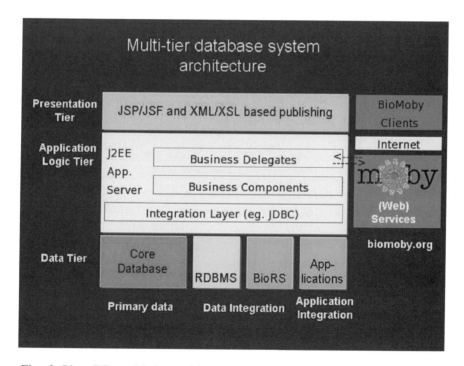

Fig. 2. PlantsDB multi-tier architecture. PlantsDB separates data tier and core databases from the application logic and the presentation tier. Suitable middleware methods allow any JAVA application to access a cluster of databases through JDBC. Publishing and presentation are accomplished by JAVA server page technology as a simple and fast way to create dynamic Web content.

been implemented. For integration with remote databases, the Web services-based interoperability solutions of the BioMOBY initiative are implemented *(5)* (see also below).

The core of the system is constituted by a flexible and generic data model for the representation of genome sequence and annotation. Three basic entities are defined: *Clone, Contig* and *GeneticElement. Clones* store sequence and attached information that relates to a physical clone. To assemble a representation of a genome sequence, clone sequences are processed to remove overlaps and redundancy, ambiguous sequence or vector contamination and then stored as *Contigs.* The *Contig* data module also stores information on how to assemble individual clones into longer contigs and pseudomolecules representing whole chromosomes.

The third data module, *GeneticElement,* stores all genetic elements anchored on the genome sequence: protein-coding genes, non-coding RNAs, repeats, sequenced markers, transposons and so on. *GeneticElements* can consist of subelements, for example, exons, introns, untranslated regions (UTRs), or domains, that constitute a particular *GeneticElement.* Such *GeneticElements* associated to a gene (promoter, transcript, alternative transcripts, regulatory elements, cDNA matches, etc.) can be identified through a single group entry.

For every plant species, a separate physical instance of all three generic data modules is created to ensure scalability and separation of name spaces.

3. Methods

A common task in biology is the identification of genes that are orthologous to a gene of interest (GOI) for evolutionary comparisons. Knowledge about the biological or molecular functions can be transferred from other organisms to the organism/biological process under study. Often, experimentalists are not familiar or simply have no time to extract and reorganize such information from raw program outputs such as BLAST, as this task frequently requires programming skills to manage large amounts of result data. Assistance in problems referring to comparative genomics is a reoccurring theme in the specification of modern genome databases.

PlantsDB contains several unique features facilitating the determination of orthologous sequences from a wide range of species. Orthology/homology can currently be addressed by three integrated tools that are based on the principles of similarity (SIMAP) *(3)*, bidirectional best hit strategy (CosMarkers) *(6,7)* and synteny (SynBrowse) *(8)*. We use example cases to illustrate how within the PlantsDB environment these tools aid support to detect and retrieve homologs, paralogs and orthologs from several plant species.

SIMAP is a database containing the similarity space formed by amino acid sequences from public databases and completely sequenced genomes *(3,9)*. Sequence similarity calculations are updated on a regular basis, and SIMAP sequences represent all publicly available sequences, including sequences from genome projects in PlantsDB. Gene entries are linked to their precomputed FASTA alignments within SIMAP. A list of homologous sequences derived from both protein and EST data can be either accessed for all proteins present within SIMAP or separated for the three major taxonomic groups: higher plants (Viridiplantae), fungi and mammalia. This grouping allows preselection for taxonomic groups of interest and a restriction to a relevant set of sequences.

3.1. Selection of Orthologous Genes Within PlantsDB

In the following, we use example cases to illustrate how different tools can be used to address typical problems arising during the search for orthologs. Consider a researcher who is interested in the sulfolipid-biosynthesis protein SQD1 in *A. thaliana* and aims to study SQD1 presence and function in a broad range of plant species including monocotyledonous plants. In particular, the researcher may be interested to generate specific hybridization probes or polymerase chain reaction (PCR) primers to map and isolate the respective genes. The enzyme contains a glucosyl-transferase domain that is a prominent domain in many species including plants. Using the PlantsDB services and infrastructure, the researcher will need to undertake the following actions:

1. In the start page of MAtDB2 (http://mips.gsf.de/proj/plant/jsf/athal/index.jsp), follow the link to the search form.
2. Find the respective database identifier for the GOI, by specifying either the AGIcode 'At4g33030' or the description 'SQD1' in the free text search field of the section for genetic elements.
3. The search will result in two identifiers linking to either coding sequence (At4g33030) or transcript (At4g33030.1) associated to the entry.
4. To open the element report page, the user follows one of the links, for example, click on the coding sequence link 'At4g33030'. The element report contains various details about sizes and coordinates of the gene and its exons. At the bottom line of the page, CDS, spliced, unspliced and protein sequences can be retrieved for the respective element.
5. If the user is mainly interested in plant orthologs, a direct access to the highest-ranking SIMAP entries for plant species using the 'Viridiplantae' link of SIMAP is provided. Homologous genes and EST clusters are shown in a table describing hit identifier, species, a short description line, hit similarity, identity and alignment length. In the last column, a link to the alignment between query and hit sequences is given. These data enable the user to judge the quality of the results and hits detected.

For extended searches, similarly best-scoring alignments and database identifiers can be obtained for entries of mammalian and/or fungal origin within SIMAP using the respective links on the element report page. For SQD1, besides the self-hit, as best hit two proteins from the *Medicago* database UrMeLDB and the rice plants DB, as well as EST matches from various species such as sorghum, sugar cane and maize are found. Amino acid identities between *Arabidopsis*, rice and *Medicago* are about 78%, and similarities are about 90% indicating high homology between these proteins. Interestingly, searches in the fungal and mammalian protein SIMAP sets reveal only modest identities (~25%) and similarities (~50%) to GAL10-like UDP-glucose epimerases. Therefore, obvious orthologs for SQD1 seem to be restricted to plants.

6. For queries without taxonomic restrictions, the full SIMAP similarity space for SQD1 can be searched. The SIMAP icon on the element report page links to a general input form of SIMAP in which the protein sequence of SQD1 can be copy-pasted. By activating a check box in the SIMAP entry form, alignments can optionally be retrieved for all significant SIMAP entries to address the quality of individual hits. The results in the 'SIMAP SeqFinder results' page confirm specificity of SQD1 to plants and, in addition, to bacterial species. For each SIMAP entry, links to their UNIPROT entry pages at http://www.expasy.org/uniprot/<EntryAccession> are given to provide fast access to a comprehensive functional description of the respective entry. Detailed listing for various scores such as bitscore or identity for homologous proteins can be obtained by following the link '>Show homologs' in the SIMAP sequence ID bar. Several bacterial sulfolipid biosynthesis protein (SqdB) genes show a high homology to the *Arabidopsis* protein with an identity range from approximately 42% up to approximately 78%, which is significantly higher than the identities obtained for the alignments of SQD1 against fungal or mammalian homologs.

In summary, the database search provides biologists with numerous homologous proteins. Literature and knowledge-based databases (such as UNIPROT) can be queried using these accessions to retrieve possible functions and interaction partners. In particular, many homologous/orthologous genes were identified in plants, and it seems likely that in higher eucaryotes, SQD1-like proteins are restricted to plants.

3.2. Selection of COS Markers from PlantsDB

In the example given above, the user has been enabled to identify plant homologs for SQD1. In the next example, we aim to illustrate how to develop molecular markers and/or hybridization probes to screen genomic or cDNA libraries. A frequent complication for this kind of experiments is the presence of highly similar, paralogous sequences within the genome(s) of interest. PlantsDB provides the 'COSMarker' (conserved ortholog sets) tool that retrieves orthologous cliques formed from bidirectional best SIMAP hits.

1. The COSMarker page can be accessed in the navigation bar at the menu item 'comparative genomics' and the submenu 'COSMarker' (http://mips.gsf.de/cgi-bin/proj/planet/cos/seedSelect.pl).
2. First, the user has to specify a seed species that would be, in the case of SQD1, the '*Arabidopsis thaliana* FL' genome within the listbox. Genomes with full-length proteins are indicated by the abbreviation 'FL'. This separates them from EST databases that are used for many plant genomes for which no or only insufficient genome data are available. In our example, we are solely interested in SQD1 orthologs. Thus, we fill in the AGIcode 'At4g33030' in the text field and press submit.
3. On the next page, the species to be queried as well as the stringency parameter settings for the analysis can be specified. In our example, we are interested in putative orthologs in *Brassica, Medicago*, rice and maize and activate the respective checkboxes 'brassica_napus', 'Medicago_truncatula_UrMeLDB', 'rice_plantsDB' and 'zea_mays'. The *Brassica* and maize data sets are derived from EST sets.
4. The stringency of the analysis can be defined by two main parameters. 'Minscore' determines the minimal Smith–Waterman alignment score for pairwise alignments to be considered, whereas the parameter 'maxnext' affects how strictly paralogs will be suppressed during the analysis. In the default setting, COSMarkers requires the presence of orthologs for all non-seed organisms to result in a reported COS cluster. However, this restriction can be disabled, and incomplete COS marker sets can be retrieved by activating the respective checkbox. In our example, we will allow for 'incomplete sets' and use the default parameters. In case of a failure in detecting COS clusters, less stringent parameters can be applied. In this case, the report sheet lists in detail how many putative orthologs were unreported because of their exclusion by the parameters.
5. Using default values, the user will find one cluster reported for the *Arabidopsis* SQD1 and detailed information about alignments following the link 'cluster details'.

Besides the determination of orthologs to single user-specified genes, the COSMarker tool allows genome-wide investigations for orthologs between whole genomes. In this use case, the steps are similar to the above-listed ones; however, the user specifies only a seed species and omits a specific gene identifier in step 2.

The COSMarker tool already extends some of the common tasks to a whole-genome scale; however, it still relies on precomputed result data. In **Subsection 3.3**, however, data from PlantsDB are integrated and analyzed in powerful workflows that enable users to perform highly specific, complex, yet freely adjustable high-throughput comparative genome analysis.

3.3. Web Services as a Means to Use Distributed Data Resources

Over the past few years, plant genomic research, especially in the model system *A. thaliana*, has generated unprecedented amounts of data of broad scientific interest. Exploiting the full potential of these data requires accessibility and integration. Although all data are readily being made public, the mechanism used is mostly project-specific Web sites. This leads to collections of databases that cater excellently to specialized fields of interest. However, it also leads to a situation where it is very tedious to find all these sites and query each of them to compile all data available for a single *Arabidopsis* gene. On the contrary, more and more researchers are using comprehensive, genome-wide, integrative approaches and are requesting easy access to comprehensive integrated data sets and to the facilities to conceive and execute custom analysis pipelines tailored to their needs, as opposed to being limited to the analyses set up elsewhere.

There are two approaches to data integration applicable in this context: data warehousing and service-based networks. Although data warehousing enables rich integration, quality control and curation, the effort required for collecting all data, transforming it into a unified schema, extending the schema to encompass new data types and maintaining a current data set is large. This approach is frequently employed by computational biology groups who maintain local databases and analysis systems and tailor them to their needs. But it is not available to many experimentalists without sufficient programming expertise and compute resources.

For a number of years, a service-based approach for distributed computing has been proposed as an alternative and established in several collaborative projects. This does not require integration of all data into central resources but relies on developing mechanisms to enable data sharing and integration from diverse sources using common protocols and data formats. This can be implemented on the basis of Web services, an increasingly used Internet standard built around XML. Client interfaces can assimilate and display current data from any number of remote providers. Analyses based on this system are always based on the most up-to-date data, as it is queried directly from the primary source.

This approach was employed successfully in the European PlaNet project *(5)* (http://www.eu-plant-genome.net) aimed at improving availability and integration of European plant genomic data. Several tutorials, applications and examples are available on the PlaNet Web page. The use cases described in **Subsections 3.3.3–3.3.5** are designed as illustrative examples.

3.3.1. Custom Workflows Using PlantsDB Web Services and Taverna

To demonstrate what kind of custom analysis can be performed using Web services, we will present three examples here.

1. Use sequence similarity and keywords to retrieve *Arabidopsis* proteins and compare the lists.
2. Perform some automatic annotation for all proteins on a *Medicago* contig.
3. Do a genome-wide evaluation of the function of *Arabidopsis*-specific proteins when compared to proteins conserved between *Arabidopsis*, *Medicago* and rice, based on Gene Ontology (GO) terms.

Each of these implements a 'workflow' or an analysis pipeline, where a series of database lookup, data transformation or analysis tasks are combined by joining outputs of one step to the inputs of the next step. The complete procedure can then be executed as one.

We will use a standalone application provided by the MyGrid project, Taverna, to execute these workflows *(10)*. The same services can also be accessed using other clients, for example, the Web-based REMORA, available at http://bioinfo.genopole-toulouse.prd.fr/remora. We introduce Taverna here as it is a very powerful and comprehensive tool and is well worth the initial learning curve. A list of client applications that can be used to access PlantsDB services is available at http://www.eu-plant-genome.net.

The workflows presented here can be used as is with own inputs. Outputs can then be saved as spreadsheet files for further evaluation. However, workflows can also be modified, and individual analysis steps can be added or exchanged to create customized workflows from scratch.

3.3.2. Custom Workflows Using PlantsDB Web Services and Taverna Custom Workflows

The selected examples use the PlantsDB application programming interface (API) in addition to and services provided by other projects and databases.

3.3.2.1. THE TAVERNA WORKBENCH

The Taverna workbench is an application developed by the MyGrid project (http://www.mygrid.org.uk) *(10)*. It can be downloaded from http://taverna.sf.net and runs on Windows and Unix/Linux-like systems, including MacOSX. Thus, any standard personal computer connected to the Internet should be able to run the examples. It is one of the advantages of the service-based approach that most calculations are done remotely on computer systems dedicated to the task. However, Taverna can be memory consumptive for complex workflows. Generally, the user experience will be more pleasant with a minimum of 512 MB of main memory.

The workflow files for the examples discussed here, as well as many other useful workflows, can be downloaded from the PlaNet Web page (http://www.eu-plant-genome.net). Also, example input data and images of the workflow are provided there and can be found by going to 'Tools' and 'Taverna workflows'.

Most of Web services introduced and used in our example cases are Biomoby compliant. Biomoby is a project to establish a standard for providing and discovering biological data and analysis services *(5,11)*. The project Web page http://www.biomoby.org has documentation, tutorials and code. Very few key facts are necessary to understand Biomoby in the context of these examples (*see* **Notes 1–3**).

3.3.3. Example 1: Compare Lists of Arabidopsis Proteins Selected by Keyword and by Sequence Similarity

1. This workflow consists of two subworkflows plus the additional comparison of the results (*see* **Fig. 3**). The first workflow undertakes a keyword query of PlantsDB using the service 'getAGILocusCodes'. To this end, a base Biomoby object is created (*see* **Note 1**), and 'Global_Keyword' is given as namespace (*see* **Notes 2 and 3**). The ID input is used for the search word that will be entered when running the workflow. The result from 'getAGILocusCodes' is XML formatted, containing the XML representation of Biomoby objects. To render it more human readable,

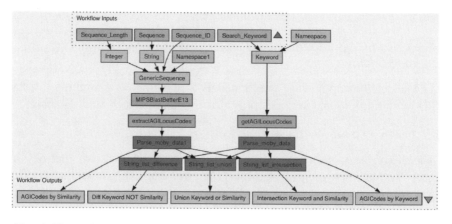

Fig. 3. Example 'Compare lists of *Arabidopsis* proteins selected by keyword and sequence similarity'. Graphical workflow presentation to retrieve *Arabidopsis* proteins using similarity and keyword-based search. For details, *see* **Subsection 3.3.3**.

the 'Parse_moby_data' service is used. This extracts namespace and ID, and the 'moby-data' output of 'getAGILocusCodes' is connected to its input. Actually in this case, we only use the IDs that are all AGI locus codes. The list of these IDs would be the output of the first subworkflow and is fed into Taverna output 'AGIcodes by keyword'.

2. The second part starts with a sequence, which is needed to enter when running the workflow. This is input into a Biomoby 'AminoAcidSequence' object. In Taverna, you can see that a String 'Sequence' and an Integer 'Length' go into the 'AminoAcidSequence' object, and the sequence is input into the 'Value' input of the String 'Sequence'. The output 'mobyData' of 'AminoAcidSequence' is then connected to 'MIPSBlastBetterE13'. This service performs a protein BLAST search against the *Arabidopsis* protein database and reports all matches with an E-value of less than 10^{-13}. To create a list with only the IDs of the matched proteins from the BLAST output, the output is fed into the 'extractAGILocusCodes' service and the resulting XML output fed into 'Parse_moby_data' to get a plain text list of IDs. This list is the output of the sequence similarity part, 'AGIcodes by similarity'.

3. The two lists of IDs can then be compared using standard string list functions also available in Taverna. Union will give a non-redundant list of all IDs found by keyword or by sequence similarity, intersecting those IDs that are found by both. Difference gives IDs in one list but not the other. The results are fed into Taverna outputs, as well as the original lists.

4. To execute this workflow, Taverna has to be started and 'Load from web' has to be activated (*see* **Fig. 4**). Subsequently, the following URL has to be pasted: http://mips.gsf.de/proj/planet/Taverna_workflows/compareKeywordAnd-Similarity.xml (*see* **Fig. 4A**). After selecting 'Run workflow' from the menu, the required inputs, keyword and sequence, are need to be entered. Select them by clicking, then click 'New input', and a field to paste the respective data will be available (*see* **Fig. 4B** and **C**). Example input data are provided on the PlaNet Web page. For testing, there is also a version available that has built-in example data and will run without the need for giving input data.

5. Once the workflow has been completed, individual panes for each of the outputs can be selected, and a list of AGI codes will be available (*see* **Fig. 4D**). The output can be saved as an Excel file.

3.3.4. Example 2: Automatic Annotation of Proteins on a Medicago contig

1. Load the workflow (*see* **Fig. 5**) from http://mips.gsf.de/proj/planet/Taverna_workflows/characterizeGeneticElement.xml into Taverna.

2. This workflow uses the methods of PlantsDB to (1) look up a contig by its name, (2) get all annotations on that contig, (3) get the protein sequences, and (4) use several analysis methods on those protein sequences.

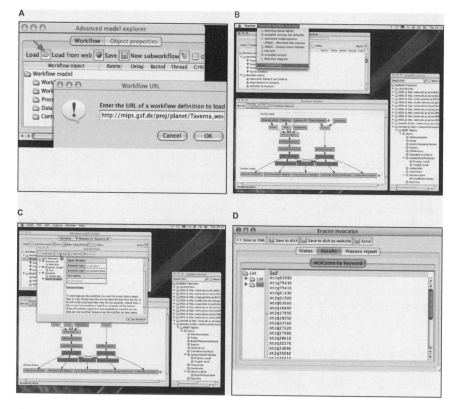

Fig. 4. Workflow execution within Taverna. Screenshots depicting individual steps while using Taverna for executing sequential query commands. (**A**) Load pre-existing workflow by pasting the respective URL. (**B**) Selecting 'Run workflow' will cause opening of a dialogue box. (**C**) The relevant data are pasted into the dialogue box. (**D**) Depicts the outcome of the workflow executed within Taverna as a list of AGI locus codes.

3. A contig in PlantsDB can correspond to a BAC clone and thus can be retrieved not only by the BAC name but also by the EMBL accession of the submitted BAC sequence. The initial look up of the contig using a 'Global_Keyword' as in the previous workflow circumnavigates problems with wrong capitalization of the contig name or of using a name that is not actually the primary ID in PlantsDB.

4. 'GetElementsForContig' retrieves all annotation, and 'ExtractElementNames' filters out only the IDs that are then used by "GetProteinSequence" to individually look up the sequence for each ID. Taverna automatically iterates over any such list of results. However, this creates 'lists of lists' when there are multiple inputs, and to maintain readability, 'FlattenList' is included at every step this happens.

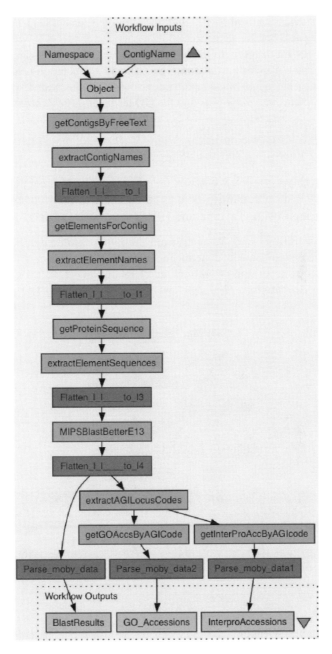

Fig. 5. Example 'Automatic Annotation of proteins on a *Medicago* contig'. Graphical workflow presentation to undertake automatic annotation of proteins on a selected *Medicago* contig. For details, *see* **Subsection 3.3.4**.

5. PlantsDB services return XML, and the 'ExtractElementSequence' service is provided to parse out the sequences.
6. The sequences are then sent to Interpro and to a BLAST search against *Arabidopsis* proteins. For the matching *Arabidopsis* proteins, GO annotation is looked up.
7. As a result, a list of *Medicago* proteins is output, along with Interpro domains, matching proteins in *Arabidopsis* and the GO terms of those *Arabidopsis* matches.

3.3.5. Example 3: Genome-Wide Analysis of the Functions of Conserved Versus Arabidopsis-Specific Proteins

This is a highly complex workflow (*see* **Fig. 6**), so analyzing it step by step is not feasible. It should mainly serve as an example of the richness of analyses accessible through using Taverna and (Biomoby) Web services.

1. Load the workflow into Taverna from http://mips.gsf.de/proj/planet/Taverna_workflows/cos_araricemedi_GOtermcount.xml.
2. As the execution of this workflow involves retrieving tens of thousands of genes from *Arabidopsis*, *Medicago* and rice, plus retrieving GO annotation for

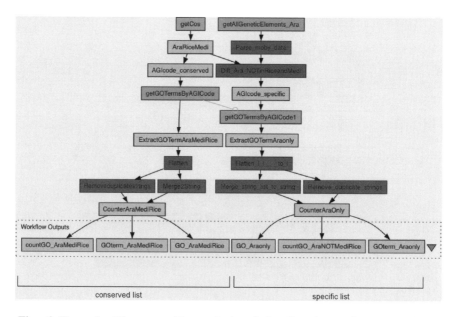

Fig. 6. Example 'Genome-wide analysis of the functions of conserved versus *Arabidopsis*-specific proteins'. Workflow depiction for the selection of function from conserved versus *Arabidopsis*-specific proteins. Detailed description to be found within text (*see* **Subsection 3.3.5.**).

each *Arabidopsis* gene, it is quite compute intensive. Even with a fast network connection and with a low load on the service providers, it will take several hours. Storing the results additionally causes the local machine (where Taverna is running) to require a large amount of memory. It has been successfully run on a Linux workstation with 1 GB of memory. However, for a demo, we suggest downloading the example results from http://mips.gsf.de/proj/planet/Taverna_workflows/Cos_AraMediRiceCount.xls as a spreadsheet.

3. The procedure to generate sets of putative orthologues is described above (*see* **Subsection 3.1.**). This is used in this workflow to retrieve the orthologues between *Arabidopsis*, *Medicago* and rice. All orthologues detected in *Arabidopsis* and *Medicago* are retrieved (AraMedi) and all orthologues detected in *Arabidopsis* and rice (AraRice). The lists of IDs are output for each species separately as Ara_inMedi and Medi_inAra, respectively, Ara_inRice and Rice_inAra.

4. Using 'GetAllGeneticElements', all proteins annotated in PlantsDB for *Arabidopsis* and rice are retrieved. These lists are necessary to retrieve all proteins that were not detected as ortholog pairs. To do so, the difference between these full lists and the ortholog lists is generated, similar to the list evaluations performed in workflow two.

5. The intersection of *Arabidopsis* proteins in AraMedi and AraRice (Ara_inMediRice), that is proteins that have an ortholog both in *Medicago* and in rice, is taken as the 'conserved' protein list. Their number is counted and reported as 'Ara_inMediRice_count'. Using 'getGOAccsByAGICode', a service provided by the Nottingham *Arabidopsis* Stock Centre (http://arabidopsis.info), GO terms for each of these proteins are retrieved.

6. Similarly, all *Arabidopsis* proteins not contained in either of the ortholog lists are used as the 'specific' list (Ara_NOTinMediNOTinRice). Also for each of these, GO terms are retrieved.

7. Further analysis of these data can reveal which GO terms are overrepresented or underrepresented in conserved proteins, with respect to the set of specific proteins (*see* **Table 1**).

4. Notes

1. Biomoby objects are a key component of the Biomoby standard. They define the data that are passed from service to service, that is, the inputs and outputs of Biomoby services. An object ontology defines the relationships between objects, and thus what data they contain and how they are represented in XML. All objects are based on 'Object', which is characterized by a namespace and ID (*see* **Notes 2** and **3**). Thus, all objects have namespace and ID. More complex objects, for example, CommentedAASequence, contain additional data such as sequence, sequence length and description. Taverna handles these details for you. When you add a Biomoby object to your workflow, you will see inputs that allow you to fill in all the data required for that object.

Table 1
Example Output of the Workflow 'Genome-Wide Analysis of the Functions of Conserved Versus *Arabidopsis*-Specific Proteins'

CountGO_AraMediRice	GOterm_AraMediRice	GO_AraMediRice	GO_Araonly	CountGO_Ara NOTMediRice	Goterm_Araonly
215	Cellular_component, endomembrane system	GO:0012505	GO:0000004	7034	Biological_process, biological_process unknown
2	Biological_process, cell wall modification	GO:0042545	GO:0008372	6452	Cellular_component, cellular_component unknown
2	Cellular_component, cell wall	GO:0005618	GO:0012505	4082	Cellular_component, endomembrane system
4	Molecular_function, pectinesterase activity	GO:0030599	GO:0003824	793	Molecular_function, catalytic activity
408	Cellular_component, cellular_component unknown	GO:0008372	GO:0004185	52	Molecular_function, serine carboxypeptidase activity
1	Biological_process, protein kinase C activation	GO:0007205	GO:0006508	598	Biological_process, proteolysis and peptidolysis
5	Biological_process, defense response	GO:0006952	GO:0016757	298	Molecular_function, transferase activity, transferring glycosyl groups

1	Molecular_function, diacylglycerol kinase activity	GO:0004143	GO:0000148	10	Cellular_component, 1,3-beta-glucan synthase complex
1	Molecular_function, metalloendopeptidase inhibitor activity	GO:0008191	GO:0016020	1930	Cellular_component, membrane
2	Biological_process, ribosome biogenesis	GO:0007046	GO:0006075	9	Biological_process, beta-1,3 glucan biosynthesis
34	Biological_process, carbohydrate metabolism	GO:0005975	GO:0003843	12	Molecular_function, 1,3-beta-glucan synthase activity
282	Cellular_component, chloroplast	GO:0009507	GO:0008168	68	Molecular_function, methyltransferase activity
20	Molecular_function, molecular_function unknown	GO:0005554	GO:0006633	66	Biological_process, fatty acid biosynthesis

The workflow results in various numbers for genes conserved among *Arabidopsis*, *Medicago* and *Rice* and the respective GO terms and GO identifier are given (columns 1–3). In analogy, *Arabidopsis* proteins without counterparts in *Medicago* and *Rice* and their respective GO term and GO identifier are listed in columns 4–6.

2. Identifiers should unambiguously refer to a single piece of data, for example, a sequence. For this purpose, many databases maintain lists of identifiers and ensure each identifier is only used once. However, beside the identifier itself, it is also necessary to know the database it stems from. Sometimes, but not always, this can be guessed because the identifier has a special format such as IPR000023. But explicitly stating the namespace, for example INTERPRO, is much more reliable. This is required for Biomoby, and Biomoby maintains a list of namespaces that is also used in other projects. Whenever you create a Biomoby object in Taverna, you will find an input for the namespace. If you need to look up a valid namespace, you can use a search form at http://mobycentral.icapture.ubc.ca/cgi-bin/GO_XrefAbbs_Maint.cgi, but there is also a list in the 'Tools' section of http://www.eu-plant-genome.net.

3. A somewhat special namespace in Biomoby is 'Global_Keyword', which does not point to any database but was introduced to make it easier to enter the system with free text. So use this if you intend to enter free text, much like you would into a search box on a Web page.

References

1. The Arabidopsis Genome Initiative (2000) Analysis of the genome sequence of the flowering plant *Arabidopsis thaliana*. *Nature* 408, 796–815.
2. Haas, B.J., Wortman, J.R., Ronning, C.M., Hannick, L.I., Smith, R.K., Jr., et al. (2005) Complete reannotation of the *Arabidopsis* genome: methods, tools, protocols and the final release. *BMC Biol* 3, 7.
3. Arnold, R., Rattei, T., Tischler, P., Truong, M.D., Stumpflen, V., et al. (2005) SIMAP–the similarity matrix of proteins. *Bioinformatics* 21 (Suppl 2), ii42–ii46.
4. Stein, L.D., Mungall, C., Shu, S., Caudy, M., Mangone, M., et al. (2002) The generic genome browser: a building block for a model organism system database. *Genome Res* 12, 1599–1610.
5. Wilkinson, M., Schoof, H., Ernst, R., Haase, D. (2005) BioMOBY successfully integrates distributed heterogeneous bioinformatics web services. The PlaNet exemplar case.*Plant Physiol* 138, 5–17.
6. Fulton, T.M., Van der Hoeven, R., Eannetta, N.T., Tanksley, S.D. (2002) Identification, analysis, and utilization of conserved ortholog set markers for comparative genomics in higher plants. *Plant Cell* 14, 1457–1467.
7. Gupta, P.K., Rustgi, S. (2004) Molecular markers from the transcribed/expressed region of the genome in higher plants. *Funct Integr Genomics* 4, 139–162.
8. Pan, X., Stein, L., Brendel, V. (2005) SynBrowse: a synteny browser for comparative sequence analysis. *Bioinformatics* 21, 3461–3468.
9. Rattei, T., Arnold, R., Tischler, P., Lindner, D., Stumpflen, V., et al. (2006) SIMAP: the similarity matrix of proteins. *Nucleic Acids Res* 34, D252–D256.

10. Oinn, T., Addis, M., Ferris, J., Marvin, D., Senger, M., et al. (2004) Taverna: a tool for the composition and enactment of bioinformatics workflows. *Bioinformatics* 20, 3045–3054.

11. Wilkinson, M.D., Links, M. (2002) BioMOBY: an open source biological web services proposal. *Brief Bioinform* 3, 331–341.

7

HarvEST
An EST Database and Viewing Software

Timothy J. Close, Steve Wanamaker, Mikeal L. Roose, and Matthew Lyon

Summary

Key Words: Barley; Brachypodium; Citrus; Coffea; Cowpea; rice; soybean; wheat; EST; Affymetrix.

1. Introduction

HarvEST is principally an expressed sequence tag (EST) database and viewing software that emphasizes gene function and is oriented to comparative genomics and the design of oligonucleotides. HarvEST supports activities such as microarray content design, function annotation, and interfacing with physical and genetic maps. The software is freely downloadable from the HarvEST Web site (http://harvest.ucr.edu) for installation in Windows. Unless otherwise stated, EST sequences in HarvEST have been trimmed to regions of high phred score *(1)*, cleaned of vector, and assembled using CAP3 *(2)* with quality values. Further details on the assembly method were given in **refs** *3* and *4*. An exception is the assemblies created by Affymetrix (http://www.affymetrix.com) to design genome arrays for several organisms including wheat, rice, and soybean; for these assemblies, the sequences do not have associated quality values. HarvEST

From: *Methods in Molecular Biology, vol. 406: Plant Bioinformatics: Methods and Protocols*
Edited by: D. Edwards © Humana Press Inc., Totowa, NJ

includes an ACE file viewer that allows the user to examine the sequence alignment and readily determine where individual sequences reliably deviate from a consensus sequence, for example, single-nucleotide polymorphisms (SNPs).

Key features of HarvEST that are standard to all versions require no Internet connection. Key features include a choice of assemblies, sequence alignment viewing, archived BLAST hit information, unigene export, Boolean searching with user-defined quantitative settings, retrieval of annotations for genome array probe sets, outputs of searches displayed as orthologous loci on rice or *Arabidopsis* genomes, and various other searching, reporting, and export functions. HarvEST also contains hyperlinks to other databases and facilitates connection to NCBI and a server at UC Riverside for live BLAST searches. HarvEST:Barley and HarvEST:Citrus contain additional functions related to genetic linkage maps and genome resources that are specific to these organisms.

The primary goal of HarvEST is to maximize the efficiency of the user by taking advantage of fast processors and the portability of personal computers and through a user interface that requires no memorization of command codes. The functions contained in HarvEST have evolved gradually since HarvEST was initiated in 2001 and continue to evolve. Every function has been the outcome of simplifying repetitive tasks in our research projects at UC Riverside or in response to specific requests by users. All graphical layouts and query operations are the original work of the authors.

2. Materials

2.1. System Recommendations and Requirements

HarvEST is a relational database with a graphical user interface, built using Microsoft Visual FoxPro 9.0 (http://msdn.microsoft.com/vfoxpro/). A Windows installer for each HarvEST version is created using Inno Setup (http://www.jrsoftware.org/isinfo.php) and then posted on a server at http://harvest.ucr.edu. The installation files may be freely downloaded, installed, copied, and further distributed (*see* **Note 1**). Windows is required for installation (Windows 95, 98, Me, NT4, 2000, or XP), and video resolution of 1024×768 is also required. A 1-GHz or higher processor and a minimum of 512-MB RAM are recommended. Free hard disk space requirement varies between species, for example, 800 MB is required for HarvEST:Barley. A fast Internet connection (to use hyperlinks to other databases from HarvEST) is optional but recommended.

2.2. The HarvEST Database System Is Currently Available for Barley, Brachypodium, Citrus, Coffea, Cowpea, Rice, Soybean, and Wheat (see Note 2)

2.2.1. HarvEST:Barley

Version 1.50 of HarvEST:Barley (436 MB download and 690 MB installed) contains four assemblies (*see* **Note 3**). Two of these (assemblies #21 and #25) are very closely related to the "Barley1" microarray produced by Affymetrix (http://www.affymetrix.com/products/arrays/specific/barley.affx) as part of the USDA-IFAFS project, "An Integrated Physical and Expression Map of Barley for Triticeae Improvement". Assembly #31 was the central point of organization for the design of overgo probes utilized in the National Science Foundation project, "Coupling Expressed Sequences and Bacterial Artificial Chromosome Resources to Access the Barley Genome" *(5)* (http://oligospawn.ucr.edu). HarvEST:Barley links out to the "Barley Genome" physical map database developed in that project, available at http://phymap.ucdavis.edu:8080/barley. Assembly #32 is a more "relaxed" assembly created using CAP3 parameter settings that tended to join alleles into single contigs rather than separate them. This assembly has been more suitable for SNP discovery than the other, more stringent, assemblies and is the main reference point for a high-density SNP-based genetic linkage map that is utilized in HarvEST:Barley for barley genetic map and synteny displays.

Version 1.50 is enabled with functions for viewing the Affymetrix "Barley1" microarray *(3)* content, including probe set location, probe sequences, and enhanced probe set annotations. Version 1.50 contains barley EST data sets of more than 30,000 ESTs each from projects in the USA, Germany/Australia, Japan, Scotland and Finland, as well as smaller (less than 3,000) data sets of barley ESTs, whole cDNAs, and genomic sequences from several other contributors (reviewed in **ref. 5**). HarvEST:Barley contains best BLASTX matches to the UniProt database (http://www.pir2.uniprot.org/), the annotated rice genome [The Institute for Genomic Research (TIGR), version 4; January 2006; http://www.tigr.org/tdb/e2k1/osa1/], and the annotated *Arabidopsis* genome (TAIR, version 6; November 2005; http://www.arabidopsis.org/). Further development of HarvEST:Barley is funded by the USDA Plant Genome program and the NSF Plant Genome Research Program DBI-0321756.

2.2.2. HarvEST:Brachypodium

Version 0.50 of HarvEST:Brachypodium (21 MB download and 30 MB installed) contains two assemblies of 20,137 sequences from six libraries from *Brachypodium distachyon* (*see* **Note 3**). All sequences were downloaded from the GenBank dbEST database by Steve Wanamaker at UC Riverside. HarvEST:Brachypodium, like HarvEST:Barley, contains best BLASTX matches to UniProt and the annotated rice and *Arabidopsis* genomes. This initial version of HarvEST:Brachypodium was funded by the USDA Plant Genome program.

2.2.3. HarvEST:Citrus

Version 1.09 of HarvEST:Citrus (246 MB download and 425 MB installed) displays 173,630 sequences from 77 libraries from *Citrus* and *Poncirus*. ESTs from 16 libraries produced at University of California Riverside (98,638 ESTs), 10 at University of California Davis (Abhaya Dandekar; 17,469 ESTs), five at USDA/ARS US Horticultural Research Lab in Ft. Pierce, Florida (Robert Shatters, Michael Bausher, Jose Chaparro, and Greg McCollum; 16,399 ESTs), and two from Volcani Center, Israel (Avi Sadka; 1,764 ESTs) have been derived from chromatograms using the full HarvEST pipeline (*see* **Note 3**). These 134,270 ESTs retain their phred quality values and can be viewed more extensively than other sequences in HarvEST:Citrus. All other sequences were downloaded from the GenBank dbEST database by Matthew Lyon and Steve Wanamaker at UC Riverside. The latter include about 10,000 additional ESTs from USDA/ARS US Horticultural Research Lab in Ft. Pierce; about 21,000 ESTs from Universidad Politecnica de Valencia, Valencia, Spain; about 3100 ESTs from Cecilia McIntosh at East Tennessee State University; about 2500 ESTs from the Laboratory of Biotechnology & Citrus Genome Analysis Team (CGAT), Shizuoka, Japan; about 250 cDNA or genomic sequences from the GenBank nr database; and contributions from various others.

Version 1.09 contains five assemblies and is enabled with functions for viewing the Affymetrix Citrus genome array (http://www.affymetrix.com/products/arrays/specific/citrus.affx) content, including probe set location, probe sequences, and enhanced probe set annotations. Assembly C37 was used as the basis of the expression portion of the Citrus genome array, whereas assembly C38 was used to identify SNPs for the genotyping portion of the Citrus genome array. HarvEST:Citrus, like HarvEST:Barley, contains best BLASTX matches to UniProt and the annotated *Arabidopsis* genome. Further development of HarvEST:Citrus is funded by the USDA Plant Genome Program, the California Citrus Research Board, and the University of California Discovery Grants Program.

2.2.4. HarvEST:Coffea

Version 0.10 of HarvEST:Coffea (55 MB download and 75 MB installed) contains two assemblies of 46,701 sequences from 12 libraries from *Coffea* (*see* **Note 3**). All sequences were downloaded as chromatograms and matching sequence and quality value files from Sol Genomics Network(http://www.sgn. cornell.edu) by Steve Wanamaker at UC Riverside with permission of Chenwei Lin, Cornell University, Ithaca, NY (Steve Tanksley Research Group). HarvEST:Coffea, like HarvEST:Citrus, contains best BLASTX matches to UniProt and the annotated *Arabidopsis* genome. The initial development of HarvEST:Coffea was funded by the USDA Plant Genome program.

2.2.5. HarvEST:Cowpea

Version 0.04 of HarvEST:Cowpea (11 MB download and 5 MB installed) contains one assembly of 334 sequences from five libraries from *Vigna unguiculata* (*see* **Note 3**). ESTs from one library *(6)* produced at University of California Riverside (259 ESTs) were derived from chromatograms using the full HarvEST pipeline (*see* **Note 3**). These ESTs therefore retain their phred quality values and can be viewed more extensively than other sequences in HarvEST:Cowpea. The other sequences were downloaded from the GenBank dbEST database by Steve Wanamaker at UC Riverside. HarvEST:Cowpea, like HarvEST:Citrus, contains best BLASTX matches to UniProt and the annotated *Arabidopsis* genome. Development of this prototype version of HarvEST:Cowpea was funded by the USDA Plant Genome program.

2.2.6. HarvEST:Rice

HarvEST:Affymetrix Rice version 1.01 (144 MB download and 320 MB installed; not an Affymetrix product) utilizes the same probe set display and annotation functions as HarvEST:Barley but includes only the assembly that was produced by Affymetrix for rice genome array content (http://www. affymetrix.com/products/arrays/specific/rice.affx). This assembly includes 268,014 sequences, mainly ESTs, displayed as just a single library. Development of this prototype version of HarvEST:Affymetrix Rice was funded by the USDA Plant Genome program.

2.2.7. HarvEST:Soybean

HarvEST:Affymetrix Soybean version 0.96 (125 MB download and 344 MB installed; not an Affymetrix product) utilizes the same probe set display and annotation functions as HarvEST:Barley but includes only the assembly that was produced by Affymetrix for soybean genome array content (http://www. affymetrix.com/products/arrays/specific/soybean.affx). This assembly includes 318,858 sequences, mainly ESTs, displayed as 136 libraries. This includes 11 libraries from the soybean cyst nematode *Heterodera glycines* and one library of sequences from the water mold *Phytophthora sojae*. Development of this version of HarvEST:Affymetrix Soybean was funded by the USDA Plant Genome program.

2.2.8. HarvEST:Wheat

The standard HarvEST:Wheat version 1.13 (182 MB download and 330 MB installed) displays about 101,000 wheat and other Triticeae ESTs produced mainly by a NSF-sponsored wheat project *(7)*. HarvEST:Wheat contains four different assemblies and has most of the same functions as HarvEST:Barley, including viewing functions for the Affymetrix wheat genome array (http://www.affymetrix.com/products/arrays/specific/wheat.affx). A separate version of HarvEST:Wheat called "HarvEST:Affymetrix Wheat" (161 MB download and 414 MB installed; not an Affymetrix product) utilizes the same probe set display and annotation functions as HarvEST:Wheat but includes only the assembly that was produced by Affymetrix for wheat genome array content. This assembly includes 389,065 sequences, mainly ESTs, from 201 libraries. Development of HarvEST:Wheat and HarvEST:Affymetrix Wheat was funded by the USDA Plant Genome program.

2.2.9. HarvEST:Other Species *(see **Note 4**)*

3. Using HarvEST

3.1. Download and Installation

Open http://harvest.ucr.edu, scroll to the version of HarvEST that is desired, and download to a local hard drive. To install, double click the H*.exe file to launch the Windows installer (*see* **Note 5**). The installation directory defaults to a sub-directory of C:\HarvEST. You may select an alternate location if you wish. The different HarvEST programs can be installed without conflicting with each other. The Visual FoxPro runtime libraries will be installed, and the executable and data files will be decompressed to the installation directory.

To upgrade to a new version, first uninstall the previous version through Windows "Add or Remove Programs", then install the new version. Go to Control Panel, Add or Remove Programs, click on the HarvEST version that is to be removed and then click Remove. A separate uninstallation through Windows must be performed for each HarvEST software.

3.3. Searching the HarvEST Database (see also Note 6)

You may find information in any version, searching by (1) expression pattern; (2) GenBank ID, EST name, or Unigene ID (*see* **Note 7**); or (3) best BLAST hit keyword. You may also search the Affymetrix genome array for barley, citrus, rice, soybean, or wheat.

3.3.1. Search by Expression Pattern

Select the libraries in which you wish to find ESTs by clicking the checkboxes in the "Include" column (*see* **Fig. 1**). Also set the minimum frequency of a unigene (composed of one or more ESTs) in the selected libraries by entering a value in the "Include if Min %" field above the "Include" column. If you wish to include all the contents of a library, then set the minimum unigene frequency to 0.000, otherwise choose a larger number such as 0.2% (1 per 500).

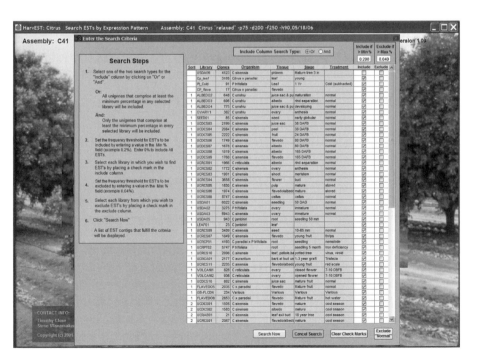

Fig. 1. Searching the database by expression pattern.

If more than one library is selected for inclusion then also decide, using the OR/AND control in the "Include Column Search Type" area near the top of the screen, whether a unigene must exist at the selected frequency in all of the selected libraries (select AND) or in at least one of the selected libraries (select OR). The default setting is OR as this is the most commonly applicable choice.

If desired, select libraries from which you wish *not* to find ESTs, by clicking the checkbox in the "Exclude" column. Set the maximum tolerable frequency of a unigene in the libraries selected for exclusion by entering a value in the "Exclude if Max %" field above the "Exclude" column. If you wish to exclude all the contents of a library, then set the maximum frequency to 0.000, otherwise choose a larger number such as 0.04% (1 per 2500).

Note that the descriptor columns are sortable by clicking on the field name Library, Clones, Organism, Tissue, Stage, and Treatment. It is often helpful to sort by these descriptors to speed up the selection of multiple libraries that are related to a particular biological theme. The column labeled "Sort" on the far left enables the user to have complete control over the sorting, by typing in numbers and then sorting, by the entries in this column.

The searches are Boolean operations for which a user sets qualitative and quantitative parameters according to their interest. For example, the choice could be genes associated with a particular stress (e.g., low temperature, heat, drought, aluminum, and nitrogen deprivation), developmental stage (e.g., malted seed, pre-anthesis spike, and mature fruit), or tissue (e.g., endosperm, pericarp, embryo sac, rachis, root tip, and rind). The user decides what percentage representation is relevant. Once the choices have been made, the search is executed locally and an interactive output is displayed. The output can be viewed on-screen or exported as tab-delimited summary tables or FASTA sequence files.

3.3.2. Search by GenBank ID, EST Name, or Unigene ID

You may search the database with GenBank accession numbers, EST names (the names used elsewhere in HarvEST), or unigene number (*see* **Fig. 2**). For unigene number searches, it is important to be aware of which assembly is currently being viewed. Different assemblies can be chosen from the Main Menu using the "Select a Different Assembly" screen. The currently viewed assembly number is shown on the upper left of many screens as a reminder. Lists may be used to initiate queries by clicking on "Find a List" and then navigating to a tab-delimited input list of the selected type (e.g., GenBank ID numbers).

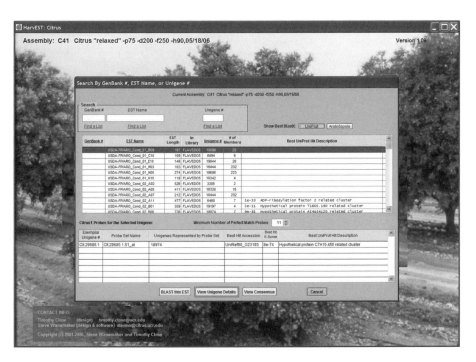

Fig. 2. Searching the database with GenBank accession numbers, expressed sequence tag (EST) names, or unigene numbers.

3.3.3. Search by Best BLAST Hit Keyword

You may also type in one or more keywords to search the BLAST annotations for each sequence to create an output list.

3.3.4. Search the Barley, Citrus, Rice, Soybean, or Wheat Chip

HarvEST contains functions that facilitate the interpretation of data from Affymetrix genome arrays by providing flexible annotation tools and visual displays detailing the position of each probe relative to sequences used in their design. You may input single probe set names or browse to a list of probe set names (*see* **Fig. 3**). You may output tab-delimited annotation files stating the best BLAST hits and the genome sequence location of putative orthologs within fully sequenced *Arabidopsis* and rice reference genomes. To generate annotations, you may decide how many probes in a probe set must match an annotated unigene to absorb its annotation. The highest BLASTX score from

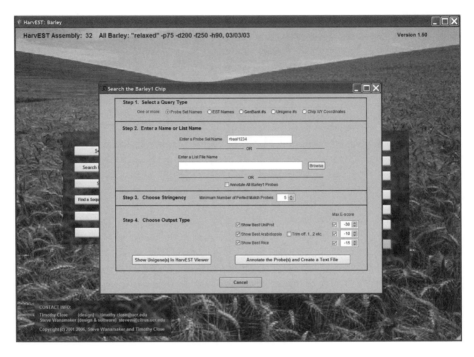

Fig. 3. Searching the barley chip.

any unigene touched by the probe set is then reported. You may export the sequences of 25-mer probes covering any selected gene.

3.4. Visualization of Search Results

The search output is displayed on a standard "Output of EST Search" screen for all search operations (*see* **Fig. 4**). This output screen is displayed after clicking on "Search Now" from the "Search by Expression Pattern" or "Search by Best BLAST Hit Keyword" screens, "View Unigene Details" from the "Search by GenBank #, and EST Name or Unigene #" screen or "Show Unigenes in HarvEST Viewer" from the "Search the Chip" screen. The left-hand side of this screen provides a sortable list of unigenes that fulfilled the search criteria, with the number of clones and number of sequences comprising each unigene also listed. Often a given clone was sequenced on both ends; so, the number of sequences in these cases is larger than the number of clones. Unigene frequency is counted as the number of clones in a library that are members of a unigene divided by the total number of clones in that same library. Complete details on the source of all members of a unigene, when

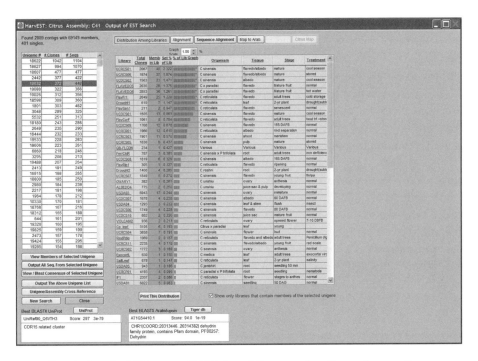

Fig. 4. Standard "Output of EST Search" screen for all search operations.

highlighted on the left, are shown on the right-hand side of the screen, and the best BLAST hits are shown on the bottom of the screen with hyperlinks to external databases (e.g., UniProt). An Internet connection is required for these hyperlinks to be utilized. As you scroll down the list on the left, the composition information and BLAST annotations change to show the pertinent information for each unigene. Several export operations and alternative views of the unigene composition can be accessed from this screen.

The menu bar on the top right provides access to an EST source list for the selected unigene ("Distribution Among Libraries"), a linear alignment view of the CAP3 assembly ("Align"), a sequence alignment ("Sequence Alignment"), or a display of the unigenes on genetic linkage maps. Information present in the best BLAST hit annotations is used to direct a map view of putative orthologs on rice or *Arabidopsis* reference genomes.

The "Sequence Alignment" and "Genetic Map" views, the latter of which was available only in HarvEST:Barley at the time of this paper, provide several additional layers of information. The "Sequence Alignment" view highlights nucleotide positions with red color if they deviate from the consensus sequence and have a phred quality value above a user-adjustable threshold, and gold if the deviation from the consensus sequence is at or below this threshold value (*see* **Fig. 5**). Where trace files were not available, all nucleotides were assigned dummy quality values not greater than 17 (lower on the ends and adjacent to bases called N). The alignment is sortable by EST name, start position in the ACE file, start position relative to the consensus sequence, or genotype. Sorting by genotype is especially useful when inspecting the evidence for SNPs and is accompanied by a companion filter that can reduce the number of genotypes displayed.

The "Barley Map" view shares the same display functions as the more comprehensive "Barley Genetic Map" display available from the HarvEST:Barley main menu. This display function was inactive in other versions of HarvEST at the time when this paper was written. In the display shown in **Fig. 6**, one can see the conservation of gene order between barley chromosome 5H and rice chromosome 12 (5HS distal region), rice chromosome 9 (5HS proximal, centromere, and 5HL proximal), and rice chromosome 3 long arm (5HL distal region).

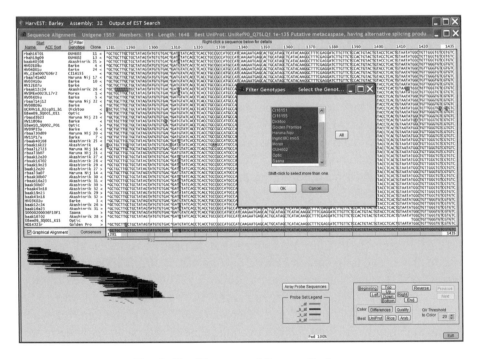

Fig. 5. The "Sequence Alignment" view.

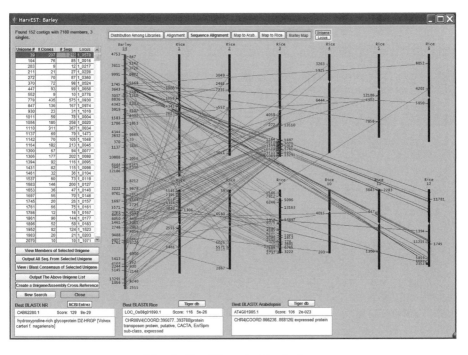

Fig. 6. The "Genetic Map" view.

From the standard Search Output screen, some additional layers of supporting information can be displayed and/or exported including the following examples.

1. Click on a library name to view a detailed description sheet for a library.
2. Click the "View Members of Selected Unigene" button for a browsable list of the ESTs in the highlighted unigene.
3. Click the "Output all Sequences from Selected Unigene" button to export the highlighted unigenes to a FASTA-formatted text file.
4. Click the "View/Blast Consensus of Selected Unigene" button to perform an NCBI BLASTX-nr search (this requires an Internet connection). For some versions of HarvEST, this screen also has the option to BLAST against HarvEST assemblies that are archived on a BLAST server at UC Riverside (this also requires an Internet connection).
5. Click the "Output the above Unigene List" button to create a FASTA or tab-delimited text file. This operation opens the same type of export screen that can be accessed from "Output a Unigene Set" from the HarvEST main menu. The exported unigenes are limited to those included in the Search Output screen when the unigene export function is evoked from the Search Output screen, whereas an entire assembly can be exported when starting from "Output a Unigene Set" on

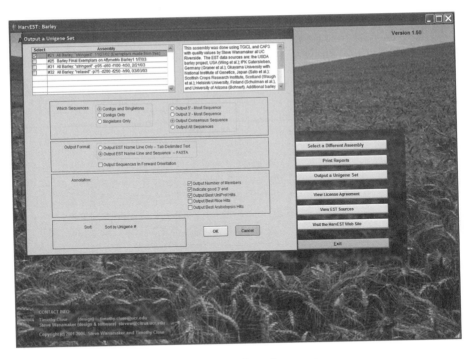

Fig. 7. Output of a unigene set.

the main menu (*see* **Fig. 7**). Search options include a choice of contigs and/or singletons; forward orientation of consensus sequences; a choice of only the unigene name, the unigene name, and consensus sequence or all sequences contained in the unigene; number of members in the unigene; and BLAST annotations. This export operation is particularly well suited to users who wish to import large batches of data from HarvEST into their own data structures. Users who do so are requested to cite the HarvEST version as the source of their information.

6. Click the "Create a Unigene/Assembly Cross-Reference" button to generate a tab-delimited text file. This operation provides a mechanism to relate unigene numbers between different assemblies.

3.5. Select a Different Assembly

Most versions of HarvEST contain more than one assembly. For example, HarvEST:Barley contains four assemblies as described in **Subsection 2.2.1**. Note that all the HarvEST operations are in the context of a specific assembly, and the active assembly can be changed by opening the "Select a Different Assembly" screen from the main menu. As stated elsewhere, unigene names are specific to each assembly; so, it is important that a user be aware of which assembly is active in the HarvEST browser, and if using unigene numbers, then also which assembly the unigene numbers came from.

3.6. Print Reports

The "Library Summary" shows the libraries comprising each assembly and how many clones, contigs, contigs unique to the library, and singletons are present in each library. The "Orientation Calls by Library Report" shows the number of forward and reverse orientation reads from each library. Orientations are determined by a combination of sequencing primer information, presence of polyT or polyA, and best BLASTX orientation. The "Print a Summary by Source" lists the number of ESTs from each person, laboratory, or group. Operations in the Print Reports category can be displayed on screen, sent to a printer, or exported to tab-delimited text files, the latter of which are particularly useful if you wish to create or modify tables for other purposes.

Notes

1. HarvEST is distributed under the HarvEST License Agreement. To receive and use this program you must agree to the following conditions:

    1. This software is available free of charge for academic purposes. Contact Steve Wanamaker or Timothy Close for a price quote if you do not make your data publicly available.

       You may re-distribute HarvEST software as long as

       a. You do not charge for it.
       b. You do not alter it in any way.
       c. You do not include it as part of software that you charge for.

       You agree to give us feedback concerning program performance, ease of use, the utility of various features, and any errors found.
2. Check this site periodically for new versions. Subsequent versions willrespond to user feedback and contain updated information and new functions, including Web operability. If you would like to be automatically notified of new versions, send a message asking that you are placed on a notification list.
3. The assemblies in HarvEST:Barley, HarvEST:Brachypodium, HarvEST:Citrus, HarvEST:Coffea, HarvEST:Cowpea, and HarvEST:Wheat were made by Steve Wanamaker using CAP3. The sequence clusters in these assemblies are not identical to clusters created by other programs or from other input sets of sequences. The unigene numbers are different for each assembly and do not correspond to unigene numbers in any other assembly or database.

The assemblies in HarvEST:Affymetrix Rice, HarvEST:Affymetrix Soybean, and HarvEST:Affymetrix Wheat were produced by Affymetrix and have been included in the HarvEST software with permission of Affymetrix.

4. HarvEST is copyrighted. The authors would be happy to apply HarvEST to any organism. See http://harvest.ucr.edu/requirements.htm for filerequirements.

5. If you received the error message, "Error 2355 during installation: The setup file HarvESTxxx.exe is damaged". You will need to re-download the file.

If you receive the following error message,

PROGRAM: FORM_HARVEST.LOAD

LINE.........: 54

ERROR # ..: 1961 A subdirectory or file c:\temp\already exists

then you will need to delete the file or folder c:\temp. HarvEST will re-create it properly. Open My Computer, C-drive, find the temp file or folder. Right click on it. Click delete.

6. The following features are currently under development and may be available: Larger assemblies.

Global SNP finder.

Links to BAC clones and contigs.

Links to OligoSpawn overgo probe designer and database.

Web-based operation of HarvEST.

7. HarvEST unigene numbers change between versions and differ between assemblies.

References

1. Ewing, B., Green, P. (1998) Base-calling of automated sequencer traces using Phred II. Error probabilities. *Genome Res* 8, 186–194.

2. Huang, X., Madan, A. (1999) CAP3: a DNA sequence assembly program. *Genome Res* 9, 868–877.

3. Close, T.J., Wanamaker, S., Caldo, R., Turner, S.M., Ashlock, D.A., Dickerson, J.A., Wing, R.A., Muehllbauer, G.J., Kleinhofs, A., Wise, R.P. (2004) Global expression profiling in cereals: 22K barley GeneChip comes of age. *Plant Physiol* 134, 960–968.

4. Close, T.J. (2005) The barley microarray: a community vision and application to abiotic stress. *Czech J Genet Plant Breed* 41,144–152.

5. Zheng, J., Svensson, J.T., Close, T.J., Jiang, T., Lonardi, A. (2006) OligoSpawn: a software tool for the design of overgo probes from large unigene datasets. *BioMed Central Bioinformatics* 7, 7.

6. Ismail, A.M., Hall, A.E., Close, T.J. (1999) Allelic variation of a dehydrin gene co-segregates with chilling tolerance during seedling emergence. *Proc Natl Acad Sci USA* 96, 13566–13570.

7. Zhang, D., Choi, D.W., Wanamaker, S., Fenton, R.D., Chin, A., Malatrasi, M., Turuspekov, Y., Walia, H., Akhunov, E.D., Kianian, P., Otto, C., Simons, K., Deal, K.R., Echenique, V., Stamova, B., Ross, K., Butler, G.E., Strader, L., Verhey, S.D., Johnson, R., Altenbach, S., Kothari, K., Tanaka, C., Shah, M.M., Laudencia-Chingcuanco, D., Han, P., Miller, R.E., Crossman, C.C., Chao, S., Lazo, G.R., Klueva, N., Gustafson, J.P., Kianian, S.F., Dubcovsky, J., Walker-Simmons, M.K., Gill, K.S., DvoÃák, J., Anderson, O.D., Sorrells, M.E., McGuire, P.E., Qualset, C.O., Nguyen, H.T., Close, T.J. (2004) Construction and evaluation of cDNA libraries for large-scale EST sequencing in wheat (*Triticum aestivum* L.) *Genetics* 168, 595–608.

8

The TAIR Database

Rebecca L. Poole

Summary

The *Arabidopsis* Information Resource (TAIR) is a highly sophisticated, extensive, user friendly, Web-based resource for researchers working on the model plant *Arabidopsis thaliana*. The main gateway to this resource is through TAIR's homepage http://www.arabidopsis.org. It is a repository of large amounts of data including gene, mapping, protein, expression and community data in the form of a relational database. These data can be searched, downloaded and analysed using the tools provided. Here, the simple search (for retrieval of information), SeqViewer (for the visualization of the five *Arabidopsis* chromosomes and associated annotations) and AraCyc (database of *Arabidopsis* biochemical pathways, with a graphical overview onto which large data sets, such as gene expression data, can be overlaid) tools are described with examples.

Key Words: *Arabidopsis*; genome; genetic map; transcript; polymorphism; transposon; biochemical pathways; omics; database; ontology.

1. Introduction

1.1. Resource overview

The *Arabidopsis* Information Resource (TAIR) is an extensive Web-based resource for researchers working with the model plant *Arabidopsis thaliana* (*1,2*). It is a collaborative project between the Department of Plant Biology, Carnegie Institution of Washington, Stanford, California, and the National Centre for Genome Resources (NCGR), Sante Fe, New Mexico.

The aim of TAIR is to provide an integrated, up-to-date view of *Arabidopsis* biology from genome to phenome by collating information from various

From: *Methods in Molecular Biology, vol. 406: Plant Bioinformatics: Methods and Protocols*
Edited by: D. Edwards © Humana Press Inc., Totowa, NJ

sources, from personal communications to published literature *(1–4)*. Users are invited to contribute to the data with up-dates and corrections. As a result, TAIR is constantly evolving; so, the data and tools will change over time. The examples described in this chapter use the data and tools available in January 2006.

Data collated by TAIR include information about genes, proteins, pheno-types, clones, biochemical pathways, experimental protocols, labs and researchers working with *Arabidopsis*, microarray data, publications and ontologies. As a member of the Gene Ontology Consortium, TAIR contributes to the development of three ontologies (process, molecular function and component). TAIR's Web site also provides ontologies for plant structures and *Arabidopsis* developmental stages. An *Arabidopsis* phenotype ontology is under development in association with Gramene and the Plant Ontology Consortium. It is possible to search and order seed and DNA stocks from TAIR through their collaboration with the *Arabidopsis* Biological Resource Centre (ABRC).

In addition to the wealth of data stored in the TAIR database, the Web site provides a number of tools to search, view and analyse this information *(1,2,5)*.

The first set of tools are used for data retrieval and include a simple text-based query that returns all objects containing the search term, and more complex searches (advanced searches) that allow the user to greatly refine the query. Bulk data retrieval tools are also available to provide access to large data sets. One example is the gene hunter tool, which searches for all available gene information from a list of selectable databases and Web sites. The TAIR data sets can be downloaded through the file-transfer protocol (FTP) site ftp://ftp.arabidopsis.org/home/tair/.

The second group of tools are graphical display tools. They include SeqViewer, a tool to visualize the *Arabidopsis* genome, MapViewer: which allows alignment of the sequence information with genetic and physical maps, the chromosome map tool: which draws maps of the *Arabidopsis* genome from user-defined lists of genes, and AraCyc: a searchable tool, which allows the user to view *Arabidopsis* biochemical pathways *(6,7)*. A third group constitutes the data analysis tools, which include the two microarray analysis tools, VxInsight® (Sandia National Laboratories, which can be used for free to analyse TAIR data sets) and Java TreeView (an open source tool for use on all array platforms). The Sequence data analysis tools include the BLAST, FASTA, WU-BLAST and Patmatch tools that are used for sequence similarity searches against a number of protein and sequence data sets. TAIR suggests Patmatch as the best choice for searching with short sequences (less than 30 nucleotides) and amino acid motifs. The motif finder, another sequence analysis tool, searches the sequences upstream of genes of interest for the frequency of 6-mers to identify potential cis-regulatory elements. Finally, it is possible to search for restriction sites within a sequence of interest using the restriction analysis tool.

To expand upon information gained from TAIR, links are also provided to external databases and tools that may be of interest. Completing this comprehensive resource is the news section, which provides the community with announcements from TAIR, events and job listings and a link to the *Arabidopsis* newsgroup.

The extent of this resource is far beyond the scope of this chapter; so, the examples will be limited to just two of the featured tools (SeqView and AraCyc) and will commence with navigating the Web site and information retrieval. This may be the first action of a researcher starting work on a gene and wishing to obtain all data currently available for that gene.

1.2. Protocols in this Chapter

1.2.1. Web Site Navigation and Information Retrieval

The main access point to the TAIR database is from the homepage http://www.arabidopsis.org. From here, it is possible to access all data set tools and links to external sites. This can be done directly by following the appropriate links on the homepage or by using the search facilities.

1.2.2. SeqViewer

SeqView provides a graphical representation of the *Arabidopsis* genome from a whole-genome view down to a 5-kb nucleotide view. It is also extensively annotated with features, i.e. gene models, markers, polymorphisms, transcripts [expressed sequence tags (ESTs) and full-length cDNAs], T-DNA/transposon insertions and annotation units, which are useful not only for visualization of the genome but also for procedures such as positional cloning.

1.2.3. AraCyc

AraCyc is a tool for the visualization of *Arabidopsis* metabolic pathways. AraCyc is derived from the MetaCyc reference database and has been modified to remove non-*Arabidopsis* pathways and to include additional pathways *(7–9)*. The pathways are not yet complete and may contain errors. Users are encouraged to share their knowledge to build on this resource. As AraCyc, like the rest of TAIR's data, is constantly updated, outputs may change over time as more accurate data are incorporated. AraCyc can be used to gain information about pathways, metabolites, enzymes, reactions and the associated genes. The database is searchable in a manner of ways, described below. There is also a large overview map displaying all of the pathways and reactions contained in the database onto which large data sets, such as transcriptomic, proteomic and metabolomic data, can be overlaid.

2. Materials

2.1. Hardware and Software Requirements

A computer with Internet access; the recommended Internet browsers are Internet Explorer 6 and above, Netscape 6 and above, Firefox (PC) and Internet Explorer 5 and above, Firefox and Safari (Mac) (*see* **Note 1**).

2.2. TAIR System

The TAIR system comprises distinct layers to facilitate system management and development: the database tier (a relational database), middle tier (comprising servers to handle transactions and abstraction layers to interface object-oriented applications with relational databases) and a client tier (including Web browser-based clients, Java applets and other applications developed by TAIR or third parties). Communications between system components are accomplished using standard technologies such as hyper-text transfer protocol (HTTP), Java DataBase Connectivity (JDBC) and common object request broker architecture (CORBA) as well as by proprietary protocols providing interfaces to external databases. In addition to dynamic access to the database, a bulk download facility is provided using FTP.

TAIR's underlying database is based on an object-oriented schema, in unified modelling language (UML) form, implemented using an relational database management system (RDBMS), in this case Sybase Adaptive Server RDBMS. It provides controlled access to the twenty-seven individual data sets currently available. These data sets are updated from various sources including user contributions and major repositories of data. Five formal update paths are identified: (1) computer-assisted entry by curators from journals or other publications, (2) import from other databases, (3) synthesis of new information such as by sequence analysis programs, (4) annotation of information by curators, and (5) user submissions. Quality of data is maintained with reference to external standards governing the ontological basis (i.e. structure) and controlled vocabularies (i.e. content). Mechanisms to ensure quality control over all aspects of data capture, storage and dissemination are also in place. The database resides on a dedicated domain of a Sun Enterprise 10000 server equipped with twelve 333-MHz central processor units (CPUs) and 12-GB random access memory

(RAM), 2 GB of which is dedicated for exclusive use by the Sybase application. The system has funding for its development and maintenance to the end of the current TAIR project, with plans for handover to a suitable repository for long-term curation.

Full system documentation and specific details about these aspects of the project are available at the TAIR Web site.

2.3. TAIR Data

The TAIR database comprises 27 data sets (http://www.arabidopsis.org/about/datasources.jsp) that can be grouped into 12 major data types summarized in **Table 1**.

3. Methods
3.1. Web Site Navigation and Data Retrieval

3.1.1. Navigation

The TAIR homepage (http://www.arabidopsis.org) is the main access point to the database (*see* **Fig. 1**). At the top of the homepage is a tool bar that appears at the top of every page, which allows quick and easy navigation of the Web site. The 'About TAIR' link contains a plethora of information about the resource including database documentation, publications and how to cite the Web page. The 'sitemap' link is sorted by subject and contains links to every page on the Web site. The 'contact' link provides information on how to find and contact the Web site's curators. The 'help' link is an extremely useful part of this Web site as following this link will open help files relevant to the current page. Finally are the 'order' and 'login' links, here you can register for the Web site, which allows the user to order DNA and seed stocks. The homepage itself is split into seven sections (advanced search, tools, *Arabidopsis* information, stocks, external links, FTP downloads, and news), which form the links in a second navigation tool bar that appears underneath the first on all other pages. There is also an additional 'breaking news' banner on the right of the homepage.

3.1.2. Text-Based Search

From the TAIR homepage, a search can be performed by using the search box at the top of the page, and refined (e.g. by gene, protein, marker or keyword) by using the drop-down menu at the top of the page (*see* **Fig. 1**). This search facility looks for any entries that contain the search term (e.g. 'MRP' will return AtMRP1, AtMRP5 and DMRPA1, to name but a few). Wildcards (*)

Table 1
The *Arabidopsis* Information Resource (TAIR) Database Contains 27 Data Sets
that Can be Grouped into 12 Major Data Types, Summarized from
http://www.arabidopsis.org/help/quickstart.jsp#data

Data type	Description
Genes and loci	Includes experimentally determined and predicted genes
Sequences	Full genome sequences of the Columbia ecotype from the *Arabidopsis* genome initiative (AGI), mRNA sequences, expressed sequence tags (ESTs) and genome survey sequences (GSS). Available to registered users from academic or non-profit organizations is 95 Mb of shotgun sequence which form the Landsberg *erecta* ecotype. Also included are protein sequences downloaded from the AGI and SwissProt
Maps	Sequence, genetic and physical maps
DNA	Primarily made up of *Arabidopsis* Biological Resource Centre (ARBC) stock, including clones, vectors, filters, libraries and pooled DNA for insertion screening
Polymorphisms	Includes single-nucleotide polymorphisms (SNPs), insertions/deletions (INDELs) and T-DNA/transposon insertion lines
Keywords	Controlled vocabularies (ontologies) to describe gene function, biological process, subcellular localization, developmental stage and anatomy. Also used to describe genes, loci, microarray experiments, community and publications
Microarray data	Includes both raw and normalized publicly available data, mostly obtained from the *Arabidopsis* functional genomics consortium (AGC)
Genetic markers	Including CAPS, SSLPs and QTL. Primer sequences are also available for the markers
Germplasms	Mainly from the ABRC and include mapping populations and mutant strains
Proteins	Information includes molecular weight, isoelectric point, domains and motifs
Community	Lists of researchers and groups working primarily on *Arabidopsis*
References	Publications, meeting abstracts and personal communications

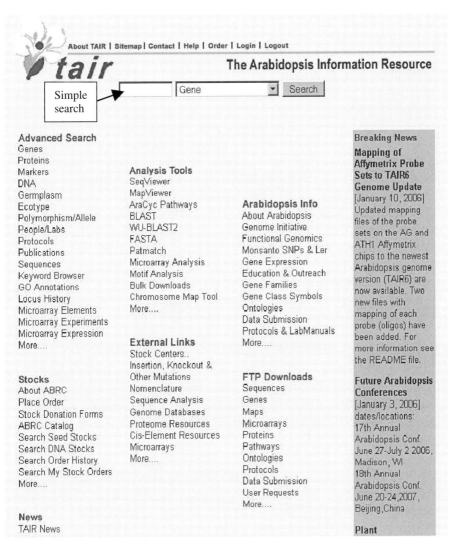

Fig. 1. TAIR's homepage, the gateway to the database and website. The page is separated into 7 categories, search, tools, Arabidopsis informations, stocks, external links, FTP downloads and news. The navigation tool bar at the top of the page provides links to information about the website including a site map. The simple search at the top of the page can be refined using the drop-down menu next to the search box.

are accepted in this search facility, though are not always necessary. For this example, type vrn* into the search box and refine the search to 'genes' to find the vernalization genes *vrn1* and *vrn2* (*see* **Note 2**). The result page contains a summary of all matches with associated links to locus and gene model information (*see* **Fig. 2**) (*see* **Note 3**). Clicking on the question mark button, located next to table headers opens a pop-up window containing the header definition (*see* **Fig. 3**). A summary of the search results can be downloaded, in tab-delimited text format, by checking the selection box next to each match (alternatively use the 'check all' to select all results) and clicking on the download button. Comprehensive information about a gene of interest can be obtained by following the locus link. The locus page describes, in detail, all of the features of the particular locus (*see* **Fig. 4**). The data includes a history of the locus (*see* **Note 4**), the 'representative gene model' (the gene model from which other data are derived, e.g. gene features), a list of alternative gene models, annotation (including how this annotation was derived), sequence information (genomic and transcripts), expression data, links to maps and external information, polymorphisms, and a list of publications.

If the search returns no results, try searching the entire Web site (*see* **Note 5**) or one of the advanced search tools (*see* **Note 6**).

3.2. Browsing the Arabidopsis Chromosomes Using SeqViewer

The SeqViewer tool can be reached from TAIR's homepage or directly from http://www.arabidopsis.org/servlets/sv (*see* **Note 7**). The five chromosomes can be searched by name to find genes, polymorphisms, annotation unit, marker, transcript or T-DNA/transposon insertions or by sequence match.

3.2.1. SeqViewer Name Search

Up to 250 names can be searched simultaneously. These can be entered manually on separate lines or uploaded from a list of names.

To search for the three related disease resistance genes *RPP1*, *RPP5* and *RPP13*, which confer resistance to downy mildew, enter 'RPP1', 'RPP5' and 'RPP13' into the search box as shown in **Fig. 5A** (*see* **Note 8**). Here, the default result output settings are applied. The results are returned with search matches marked on the whole-genome view in red (each chromosome is represented by a green line, with landmark markers in black, and centromeres in blue, colours refer to those as seen on the Web site) (*see* **Fig. 5B**). A summary of the matches, listing the locus name, gene description and sequence coordinates is opened on a separate page by clicking on the link at the top of the results page (*see* **Fig. 6**). To zoom-in on a chromosome region, click on the green

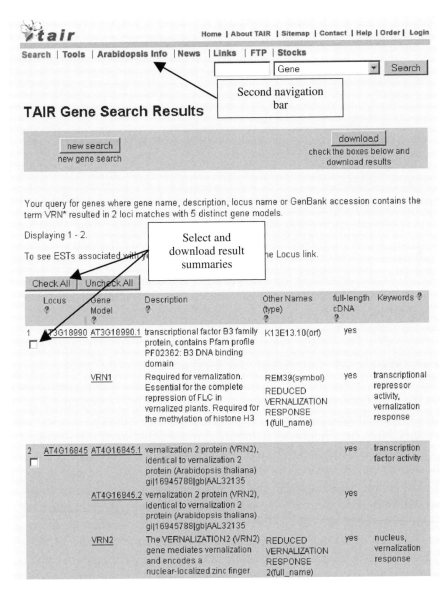

Fig. 2. Simple search. Results are grouped by locus (represented as alternating blocks) and include gene model and corresponding descriptions, alternative names, whether a full-length cDNA is available and keywords. From here it is possible to download a text summary of selected results. A second navigation tool bar appears beneath the first on all pages except the homepage, which provides links to the 7 categories as set out on the homepage.

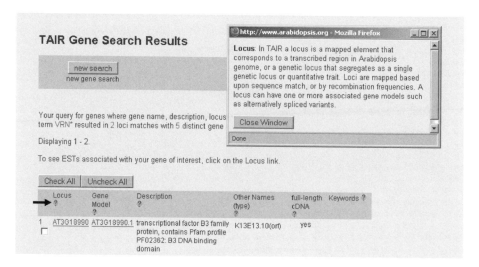

Fig. 3. An information pop-up box, containing a definition of the associated term (in this case locus), can be opened by clicking on the question mark button (indicated by the arrow).

chromosome bar (*see* **Note 9**). The close-up view (showing the selected segment of the chosen chromosome) is then displayed beneath the whole-genome view, with the search result(s) highlighted in yellow (*see* **Fig. 7**). Multiple close-up views can be opened; each one is highlighted on the whole-genome view by a numbered black box. Information displayed in the close-up view is separated by type in colour-coded bands. The top band contains the markers and is magenta, followed by polymorphisms in red, T-DNA insertions/transposons in purple, genes in orange, transcripts in grey, cDNAs in blue, and annotation units (*see* **Note 10**) in pink (colours are as seen on the Web page). A description box is opened by holding the cursor over a particular object, clicking on the object opens the full detail page. In the case of genes, this is the locus information page described earlier. At the bottom of each band is a coloured bar with tick marks that represent every object in that band. Not all of the objects are displayed in the default view, those present are denoted with a red flag and those not shown with a black flag. All objects can be displayed by checking 'show all data' in the bar between the genome view and the close-up view. It is possible to display only selected bands; depending on information required, bands are selected using the appropriate tick boxes. Users may re-centre the view (a yellow bar, denoting the region to be centred on, will appear above

tair

[   ] | Gene   ▼ | Search

Locus: AT3G18990

Update History ?	AT3G18990 replaces VRN1 on 2004-02-23
Date last modified	2003-05-02
TAIR Accession	Locus:2085849
Representative Gene Model ?	AT3G18990.1
Other names:	K13E13.10, REDUCED VERNALIZATION RESPONSE 1, REM39, VRN1

Other Gene Models ?

Name	Description	Source	Date
VRN1	Required for vernalization. Essential for the complete repression of FLC in vernalized plants. Required for the methylation of histone H3	GenBank	2002-07-12
		GenBank	2002-07-12

Annotations ?

Category	Relationship Type ?	Keyword ?
GO Biological Process	involved in	vernalization response
GO Molecular Function	has	transcriptional repressor activity

Annotation Detail

RNA Data

One-channel Arrays

array element name ?	avg. signal intensity (std. error)	avg. signal percentile (std. error)
256944_AT	232.788 (2.986)	72.717 (0.232)

Associated Transcripts ?

type	number associated
EST	(13)
cDNA	(4)

Description ?	transcriptional factor B3 family protein, contains Pfam profile PF02362: B3 DNA binding domain
Chromosome	3
Nucleotide Sequence ?	full length CDS full length cDNA full length genomic

Protein Data ?

name	Length(aa)	molecular weight	isoelectric point	domains(# of domains)
AT3G18990.1	342	39275.0	9.5175	Transcriptional factor B3:IPR003340(2)

Map Locations ?

chrom	map	map type ?	coordinates	orientation	attrib
3	AGI	nuc_sequence	6548875 - 6551847 bp	reverse	details
3	K13E13	assembly_unit	27266 - 30238 bp	reverse	

Map Links ?   Map Viewer   Sequence Viewer

Gene Feature ?

type	coordinates	annotation source	date
ORF	274-2765		
5' utr	1-273		
coding_region	274-333		

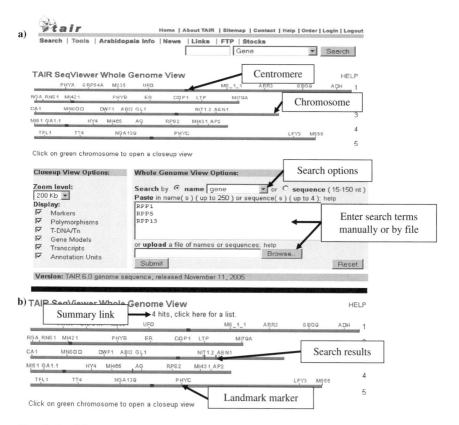

Fig. 5. SeqViewer name search. Multiple search terms can be entered manually into the search box (or uploaded as a file) on the SeqViewer frontpage (**a**), clicking the submit button starts the query. (**b**) The results are displayed on the whole-genome view as red flags on the green chromosome bars. Above the chromosome graphic is a link to the results summary. In addition to the centromeres (blue), some markers are displayed on the whole genome view as landmarks to aid orientation of the chromosomes. (Colours are as seen on the web page).

Fig. 4. Locus information page. The information associated with the selected locus is displayed in alternating colour bands representing different data types. This page links to further information pages.

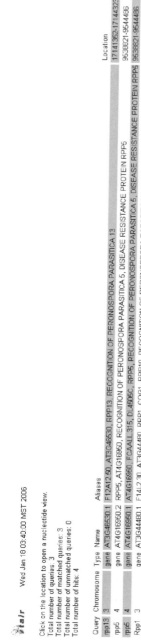

Fig. 6. The SeqViewer results text summary. The search results can be viewed in a table summarizing the query matches including chromosome location coordinates, which serve as links to the appropriate nucleotide sequence.

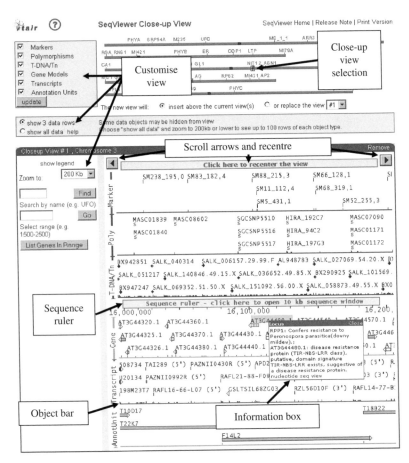

Fig. 7. SeqViewer close-up view. The 200kb display of the chromosome region is opened by clicking on the chromosome bar and is denoted in the whole genome view (top of page) by a numbered box. The search result is highlighted in the gene band and the information box opened by holding the cursor over the object. The view can be customised to change the zoom level, show only selected bands to include more lines of data per band and to change the view area by scrolling through the chromosome or by using the recentre bar.

the re-centre area when holding the cursor in this area, clicking executes the function) or scroll along the chromosomes using the scroll arrows.

The default zoom level of the close-up view is to display a 200-kb chromosome segment. This can be changed using the drop-down menu. Views include 1× (whole chromosome), 5 Mb, 1 Mb, 200 kb, 80 kb, 40 kb, 20 kb and

10 kb (*see* **Note 11**). Between the T-DNA and gene bands is the chromosome ruler, which displays the chromosome coordinates, and is clickable to display the nucleotide view.

3.2.2. Search by Sequence

The SeqViewer tool (http://www.arabidopsis.org/servlets/sv) allows searches of up to four short sequences (*see* **Note 12**). This tool can be used for testing the uniqueness of polymerase chain reaction (PCR) primer sequences (*see* **Note 13**). The example used here is for two short sequences representing potential PCR primers. Again, sequences can be entered manually or uploaded from a file. For a sequence search, ensure that 'sequence' and not 'name' is selected by checking the appropriate radio button. Enter the two sequences as follows:

>Primer F

agaagagctgaagacggag

>Primer R

tctcttcataatcaagaag

For this example, the 80-kb close-up view with only gene models and transcripts will be displayed (*see* **Fig. 8A**). As with the name search, the matches are displayed as a red flag on the whole-genome view, with a link to the nucleotide view at the top of the page (*see* **Fig. 8B**). Clicking on the red flag opens the close-up view, which shows the sequence matches in the top band and the selected bands below (*see* **Fig. 9**). Following the link at the top of the result page opens the nucleotide view in a new window, with the sequence matches highlighted in a red box (*see* **Fig. 10**). The default nucleotide view is to display the gene annotations. This view is annotated with start/stop codons (in blue boxes), introns and exons (in light purple and orange, respectively), and with gene models aligned to the right of the sequence (blue). The sequence can be copied and pasted for use in other applications, and some of the annotation will be maintained (i.e. upper case letters representing the coding sequence). The view can be further customized to show either the forward or the reverse strand, and the annotation from the selected or both strands. As with the close-up view, it is possible to scroll through the sequence (moving 5 kb at a time) using the arrow keys. The other objects that can be highlighted on the nucleotide view are those that form the bands from the close-up view, e.g. T-DNA insertions (*see* **Fig. 11**). At the top of each view is a legend describing how each feature is highlighted in the sequence. In the case of the T-DNA insertion view, the sequence matches are still highlighted in a red box, the T-DNA inserts/transposons are denoted by dotted lines (double for a match and

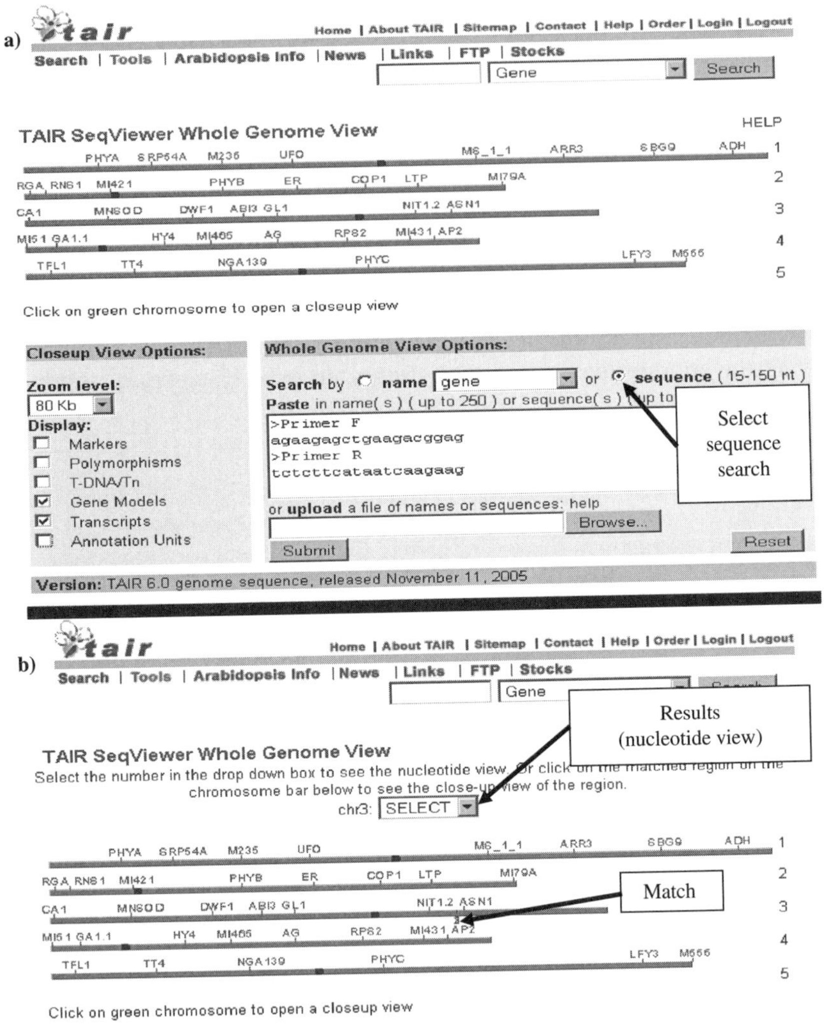

a)

b)

Fig. 8. SeqViewer sequence search. The search is performed by entering the sequence of interest into the search box (or by uploading a file) and clicking 'submit'. The close-up view can be customised, here the 80kb zoom level has been selected to display only the gene models and transcript bands (**a**). The results are returned positioned on the whole chromosome view (linking to the close-up view) and as a drop-down above the whole chromosome view, which links to the nucleotide view (**b**).

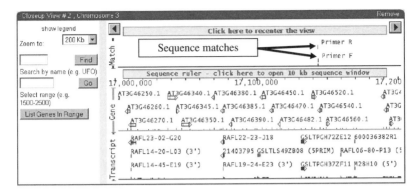

Fig. 9. SeqViewer sequence search close-up view. The sequences are displayed in the top band aligned to the appropriate place on the map.

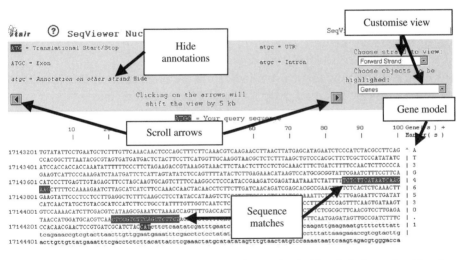

Fig. 10. The SeqViewer nucleotide view (gene view). The default nucleotide view is the gene view here gene models are aligned to the sequence with, introns, exons, UTRs and stop/start codons annotated. The results of the sequence search results are also highlighted. The view can be customised, using the drop-down menus and buttons at the top of the page, to view the alternative strand, display other features and to hide the annotations from the complementary strand. Above the sequence is a sequence ruler, demarking 10bp intervals.

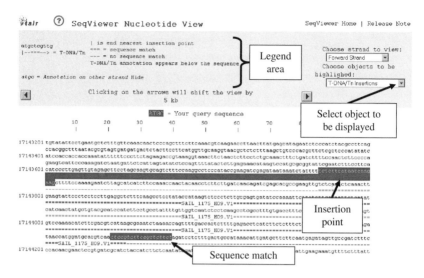

Fig. 11. SeqViewer nucleotide view (T-DNA view). An alternative nucleotide view displays T-DNA/transposon insertions aligned to the selected *Arabidopsis* sequence, with insertion points indicated by vertical lines and sequence matches by a double dashed vertical lines (single lines indicate no match). As with all view options the legend appears at the top of each page.

single for no match) with an arrow indicating the orientation of the insert, and the end nearest the insertion point is denoted by a vertical line.

3.3. Visualising Metabolic Pathways Using AraCyc

The AraCyc tool can be accessed from TAIR's homepage or directly at http://www.arabidopsis.org/tools/aracyc/. From the AraCyc frontpage, there are three options to access and use the AraCyc tool. The first is a search facility (main query page), the second provides a graphical overview of all of the metabolic pathways (metabolic map), and the third allows the user to overlay experimental data on the metabolic map (omics viewer) (*see* **Fig. 12**).

3.3.1. AraCyc: Main Query Page

The main query page can be searched in several ways, the first is to search by name or Enzyme Commission (EC) number (the EC number associated with either the enzyme or the reaction). This search can be performed on all data or refined by choosing an object from the drop-down menu (*see* **Fig. 13**). Searching all data with the term 'vitamin' returns two pathways (Vitamins E

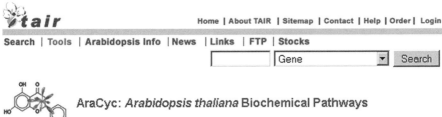

AraCyc: *Arabidopsis thaliana* Biochemical Pathways

SeqViewer | MapViewer | AraCyc | BLAST | WU-BLAST2 | FASTA | Pattern Matching | Restriction Analysis | Gene Hunter | Motif Analysis | Bulk Downloads

AraCyc is a tool for visualizing biochemical pathways of *Arabidopsis thaliana*. It is supported by the Pathway Tools software developed by Peter Karp's group at SRI. AraCyc was computationally predicted for the sequenced Arabidopsis genome using the MetaCyc reference database. The predicted pathways were then manually validated and non-Arabidopsis pathways were removed. The manual curation of the database, including correcting pathways and adding missing pathways, is on-going. Many pathways still contain errors and incomplete reactions and enzyme/gene assignments.

We encourage AraCyc users to share their pathway knowledge with the research community. You may use this Pathway Data Submission Form to send us data for updating existing pathways or adding new pathways. Please send your completed form to: curator@arabidopsis.org

Main Query Page.
Search or browse Pathways, Compounds, Enzymes and more. See a QuickTime movie demo on searching AraCyc.

Metabolic Map UPDATE
A 'bird's eye' view of Arabidopsis metabolism. (Note: This page may take a moment to load). Click here for a QuickTime demo.

Omics Viewer UPDATE
Overlay large scale data from experiments such as microarray expression, proteomics, and metabolic profiling onto the Metabolic Map. Click here for a QuickTime demo.

Fig. 12. AraCyc: metabolic pathways tool front page. There are three ways to access and use the AraCyc database; via the main query page, using the metabolic map or the omics Viewer. From the AraCyc front page, tutorials and movie demonstrations can also be accessed (this portion of the page is not shown).

and C biosynthesis), several compounds and one reaction (*see* **Fig. 14**). Each of these results links to a page of further detailed information relating to that result. The pathway detail page provides an overview of the genes, compounds, enzymes and reactions that make up the pathway, in this case in the biosynthesis of vitamin E. Clicking on the 'more detail' button reveals molecular diagrams of each compound (*see* **Fig. 15**). Holding the cursor over each feature opens a

tair

Pathway Tools Query Page

This form provides several different mechanisms for querying Pathway/Genome Databases.

Select a dataset: [Arabidopsis thaliana COL ▾]

- **Query** [All (by name or EC#) ▾] [Vitamin] [Submit] [?]

To retrieve objects by name, first select the type of object you wish to retrieve, then enter the name of the object and click Submit. All objects containing that name as a substring will be returned. You may also enter multiple names or EC numbers, separating them with commas.

- **Browse Ontology:** [Pathways ▾] [Submit] [?]

Each dataset contains classification hierarchies for pathways, for reactions (the enzyme nomenclature system), for compounds, and for genes. Select a classification system to browse.

- **Choose from a list of all** [Pathways ▾] [Submit]

- **Links to summary information about the selected organism:**
  - Summary page for dataset
  - **Cellular Overview Diagram/Omics Viewer**
  - **History of updates to this dataset**
  - **PathoLogic Pathway Analysis** (not available for *E. coli* or MetaCyc)

- **Comparative Analysis** NEW

Generate summary tables that compare various properties across one or more selected organisms.

[Help] [Advanced Query Form] [Pathway Tools Home] [Feedback]

Fig. 13. AraCyc main query page. The AraCyc dataset can be searched by name, or browsed by ontology or object list, using the appropriate section. In addition to the search functions are links to summary pages and a comparative analysis tool.

tair

Query Results

The query **Vitamin** matched the following objects:

Pathways   Pathway pages contain: Depiction of metabolic pathway, of chromosomal locations of pathway genes, and of regulation of pathway genes.

- **vitamin** E biosynthesis
- ascorbate biosynthesis (**vitamin** C biosynthesis)

Compounds   Compound pages contain: compound structural information, and links to all reactions and pathways in which the compound participates.

- anthranilate (**vitamin** L1)
- biotin (**vitamin** H)
- L-ascorbate (**vitamin** C)
- L-carnitine (**vitamin** B T)
- menaquinol (**vitamin** K_2)
- menaquinone-8 (**vitamin** K_2)
- nicotinamide (**vitamin** PP)
- nicotinate (**vitamin** B_3)
- pantothenate (**vitamin** B_5)
- pyridoxine (**vitamin** B_6)
- riboflavin (**vitamin** B_2)
- thiamine (**vitamin** B_1)

Reactions   Reaction pages contain: reaction equation with chemical structures, links to all enzymes that catalyze the reaction, and all pathways in which the reaction participates.

- ATP + pyridoxal = ADP + pyridoxal 5'-phosphate (**Vitamin** B6 kinase)

Fig. 14. AraCyc, name search results. The tool returns all objects containing the search term, in this case 'vitamin'. The results are grouped by type with each result serving as a link to a page containing further details.

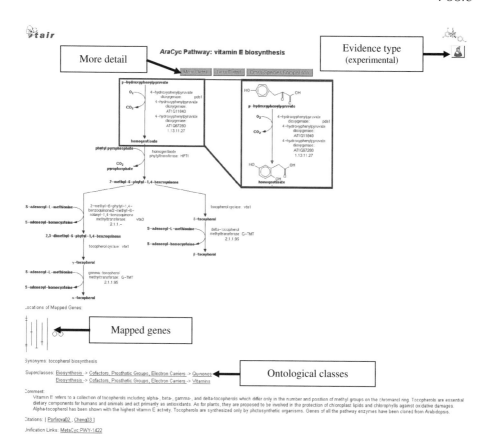

Fig. 15. Vitamin E biosynthesis metabolic pathway details page. The metabolic pathway is shown diagrammatically, with compounds linked by reactions (lines) enzymes involved are aligned to the appropriate reaction. The genes assigned to each pathway are also displayed on the relevant parts of the pathway diagram, but also as points on a schematic genome map beneath the pathway schematic. The pathway can also be displayed with increased detail (see inset). Small images in the top right hand corner of the page indicate the evidence type for the information.

brief information box, and clicking the feature opens the full information page specific to that gene or compound. At the top right corner of the page is a symbol that denotes the evidence from which the pathway was derived, a flask represents experimental evidence (as seen in this example), a book indicates that the evidence comes from literature, and a computer indicates pathway that has been derived computationally (*see* **Note 14**).

It is also possible to browse the database using ontologies. Here, the data are organized into classification hierarchies. Start by selecting 'pathways' from the drop-down menu; from here, it is possible to expand each nested 'child'

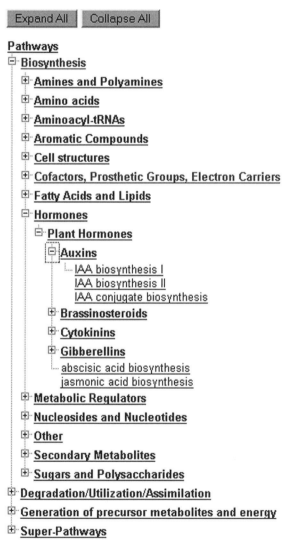

Fig. 16. AraCyc, ontology browser. This tool allows all AraCyc data to be searched using the ontology displayed as a tree view. The tree can be expanded or collapsed and has been expanded here to reveal the auxin biosynthetic pathways.

class going down through the hierarchy until the lowest possible class, e.g. biosynthesis, hormones, plant hormones and auxins (*see* **Fig. 16**). Clicking on one of these three options (IAA biosynthesis I, II and IAA conjugate biosynthesis) opens the relevant pathway information page (as described above). Each class above the individual pathway level is linked to an information page that details the associated parent and child classes.

The final method of searching the AraCyc database from the main query page is to search from a list of objects. The lists can be chosen from the drop-down menu.

A recent addition to AraCyc is the 'comparative analysis' tool that generates comparison tables for the metabolic network between organisms. It is important to note that such comparisons are only as good as the data currently available for the selected organisms. At the time of writing, only two organisms could be compared: *A. thaliana* COL and *Escherichia coli* K-12.

3.3.2. AraCyc: Metabolic Map

The metabolic map provides a 'bird's eye' view of the metabolic pathways in the AraCyc database, with metabolites represented as nodes (node shape indicates metabolite class), and reactions as lines. Holding a cursor over a node

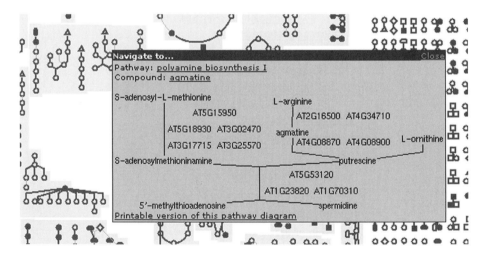

Fig. 17. AraCyc, metabolic map overview. From the metabolic map overview individual pathways can be seen in more detail from the navigation box, opened by clicking on a compound. From here it is possible to navigate to the associated pathway, compound and reaction information pages.

opens a pop-up window with a description of the metabolite and pathway. Clicking on the node expands this box to include a diagram of the entire pathway and links to the relevant detail pages (*see* **Fig. 17**).

3.3.3. AraCyc: Omics Viewer

Finally, the omics viewer allows overlaying of data, such as expression, proteomic and metabolomic data, onto the metabolic map (*see* **Note 15**). Data for use with the omics viewer tool should be saved as a tab-delimited text file and should have the gene, protein or enzyme name in the first column (called column 0), with the remaining data in the following columns (*see* **Note 16**). **Figure 18** shows a portion of a gene expression data file for an experiment with six time points, ready for use with the omics viewer. Only one data point can be input for each gene within a treatment or time point. Therefore, replicates need to be combined and the average of these used (*see* **Note 17**).

For this example, gene expression data will be used (*see* **Note 18**). An example gene expression data set can be downloaded from the FTP site,

<table>
<tr><td colspan="2">Identifier
(Column zero)</td><td colspan="2">Data
(columns 1-6)</td></tr>
</table>

	A	B	C	D	E	F	G
1	# Arabidopsis locus id	time_point_1	time_point_2	time_point_3	time_point_4	time_point_5	time_point_6
2	At1g77760	1.15	2.3	3.2	2.15	1.53	1.75
3	At2g13360	0.7	-0.53	0	-0.73	0.03	-0.72
4	At3g10230	-1.1	-0.05	1.05	1.15	1.25	0.05
5	At3g10230	-0.65	-0.58	1.13	1.23	0.67	-0.12
6	At3g01120	-1.08	-0.15	-1.2	-1.15	-1.15	-0.58
7	At3g01500	0.07	-0.72	-0.68	-1.4	-1.93	-9.23
8	At3g02470	0.03	-0.53	0.58	1.28	0.55	1.4
9	At3g02470	0.55	-0.12	0.62	0.65	-0.05	1.22
10	At3g02580	0.6	-0.55	0.08	0.55	-2.2	-1.65
11	At3g02580	1.15	0.7	0.03	-0.6	-2.4	-1.65
12	At3g02780	-1.15	0.05	0.1	-0.08	-0.57	-0.28
13	At3g04120	-0.15	-1.55	0.12	-0.3	0.23	1.77
14	At3g04120	-0.15	-1.5	0.05	-0.32	0.25	1.7
15	At3g04120	-0.07	-0.85	0.1	-0.75	0.2	1.55
16	At3g04870	1.05	-1.08	-0.05	-1.1	0.05	-1.33

Fig. 18. Sample gene expression file. To overlay expression data on the metabolic map, data must be saved as tab-delimited text files and the first column (column zero) should contain the gene name (AGI locus identifier), with numeric data in the following columns (extra columns such as descriptions can be kept in the file as data columns to be analysed are defined later).

Pathway Tools Omics Viewer

The Pathway Tools [...] (formerly the Pathw[...] Viewer) paints data [...] user's high-through[...] experiments onto t[...] Overview diagram

Dataset and data upload

The Omics Viewer can be used for:

- **Microarray Expression Data:** Reaction lines (and protein icons, where present) are color-coded according [...] absolute e[...] gene that c[...] that cataly[...]

Describe data

The Omics Viewer allows a scientist to interpret the results of gene-expression experiments in a pathway context.
- **Proteomics Data:** Reaction lines (and protein icons, where present) are color-coded according to the concentration of the enzyme that catalyzes that reaction step.
- **Metabolomics Data:** Compound icons are color-coded according to the concentration of the compound.
- **Reaction Flux Data:** Reaction lines are color-coded according to reaction flux values.
- **Other Experimental Data:** Any experiment, high-throughput or otherwise, in which data values are assigned to genes, proteins, reactions or metabolites can be viewed in a pathway context using the Omics Viewer.

More information about the Omics Viewer, including sample datafiles and displays.

The Omics Viewer takes as input a tab-delimited data file that is stored on your local computer. The file contains relative or absolute data for a single gene, protein, reaction, or metabolite, or some combination. Each row of the file contains data for a single gene, protein, reaction, or metabolite, and begins with the object name, ID or EC number. You may choose to display either data from a single column, the ratio of two columns, or a time series animation of multiple columns. Data columns other than those specified are ignored.

For an example data file, see here.

In addition to the color-coded Metabolic Overview diagram, the resulting display includes a histogram and basic statistics computed from the supplied data.

Download sample data file here

Select a dataset [...] file containing [...] [...]mental data [...] a URL):

Arabidopsis thaliana COL ▼

Do you want to display absolute or relative data values?

C:\Documents and Setti[...] Browse...

Relative ▼

If displaying relativ[...] data values, us[...]

○ a single data column
○ the ratio of two data columns
◉ 0-centered scale (e.g. log scale)
○ 1-centered scale (negative values will be discarded)

Data values are a [...] [...]e items in the first [...]zeroth) column of your datafile are

Describe data

Choose data type e.g. genes

Genes ▼

Note: By selecting *Any of the ab[...]*, you can combine, for example, gene expression and met[...] display. There are some dangers in[...] however. Some names may be amb[...] they refer to genes, proteins or meta[...] values from different kinds of experi[...] comparable, so the resulting diagra[...] some important ways.

Single Experiment Time Step or Animated Time Series

To display a single experiment time step, enter a single column number in one or both of the column number fields below.

To display an animated time series, enter a list of column numbers (with each column number corresponding to a single timepoint), one per line, in the first column number field below. If you wish to include a denominator column for a ratio calculation, you can enter either a single column number (in which case the same data column will be used as the denominator for all timepoints), or one column number for each numerator column number. Note that zoomed views of individual pathways are not available with animations.

Data column (numerator or ratios)

2
3
4
5
6

If using tw[...] date c[...]

Enter columns to be displayed

Note: For column numbering purpose[...] contains the gene name, whose column nu[...] data column is column number 1.

Color Scheme

Data values are divided into color bins, with the highest value bins displayed in red, the lowest value bins displayed in green or yellow, and the middle bins displayed in blue. Three color scheme options are available.

- By default, the color bins range over the entire spectrum, and the cutoff values for the color bins are derived from the data itself. This means that different experiments could be displayed using different color schemes, making it difficult to directly compare them.
- Alternatively, you may specify a value for the maximum value cutoff bin. All displays that use the same maximum value cutoff will use the same color scheme (assuming other settings, such as relative vs absolute, or log format vs [...])

Define colour scheme and display

[...] therefore directly [...] should be a [...]es greater than the [...] in red. [...] color bins: red for [...]old, yellow for data [...]eshhold, and blue for values in between. The threshold value should be a number, e.g. 3 or 10.

Choose a color scheme:
☑ Full color spectrum, computed from data provided (default)
○ Full color spectrum with a maximum cutoff:
◉ Three color display with specified threshhold: `1`

Display Type

By default, data values are painted on the cellular overview chart. However an alternative display is to generate a table containing all individual pathways which have one or more data values that exceed some threshhold (or are less than the inverse of that threshhold). To select this alternative display, choose the corresponding option below and specify the threshhold.
◉ Paint data on overview chart (default)
○ Generate a table of individual pathways exceeding threshhold: [...]
○ Combine both displays (not yet implemented for animations)

Submit data

Submit Note that this request will take several minutes to complete (possibly longer for large datasets).

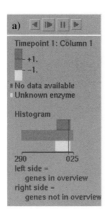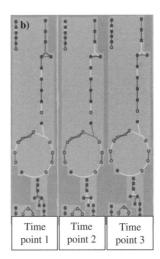

Fig. 20. Omics viewer output. Data is overlaid on the metabolic map. A legend describing the colour scheme and data statistics (histogram) appears to the left of the map (**a**). A cropped image displaying the TCA cycle alone at times points one, two and three is shown in **b**, red lines indicate up-regulated genes, yellow down regulated and blue no change, using a two fold change as the cut-off (colours are as seen on the Web site).

ftp://ftp.arabidopsis.org/home/tair/tmp/ExpressionSample.txt, or more simply by following the link from the omics viewer front page (*see* **Fig. 19**). This file should be saved to a local drive (*see* **Note 19**). From the omics viewer front page, select the *Arabidopsis* data set and upload the example data using the browse function. Next, select 'relative' from the drop-down menu and 'a single data column' and 'zero centred' by checking the appropriate radio button (*see* **Note 20**). The next drop-down menu allows the selection of the data in the first column (column 0), as this is gene expression data select 'genes' (*see* **Note 21**). In the next section of the Web page, single experiment time step or animated

◄───

Fig. 19. AraCyc, omics viewer. The Omics viewer frontpage provides links to information on how to use the tool, example data files and data input options. The data file used for the worked example can be obtained by following the relevant link on this page. The input options shown here are to display log ratio values for a 6 point time course experiment overlaying the data onto the metabolic map using three colour bins, with a cut-off of 1.

Total number of data rows (not including comment lines): 487
Number of rows for which the gene could not be found: 18
Number of rows for which the gene name was ambiguous: 0
Number of rows for which the gene is valid, but for which a data value was missing or malformed: 0

The following table shows statistics for the selected column/column ratio for all genes and for those genes whose product appears in the Overview.

Data Statistics	All Genes	Overview Genes
Number of values:	469	434
Minimum value:	-2.42	-2.42
Maximum value:	4.95	4.95
Median:	-0.03	-0.03
Mean:	-3.999976680e-2	-2.82255920e-2
Standard deviation:	0.92669874	0.93623257

Objects that could not be found: At4g13610 At4g19020 At4g20380 At4g21610 At4g28120 At4g34880 At4g39280 At5g26140 At1g03040 At1g14090 At1g15110 At2g40660 At2g46860

Objects with missing or malformed data:

Instructions for saving this diagram to your local disk

Fig. 21. Omics viewer, results summary. These data is displayed underneath the overview map and summarises the data in the output view and lists features with no match on the map or that have missing or malformed data. At the very bottom is a link with instruction to save the diagram.

time series, enter 1–6 in the left-hand side box to display all six time points (*see* **Note 22**). Next select the colour scheme for the output (*see* **Note 23**), for this example, select the three-colour option with a cut-off value of 1, assuming the data are log ratios to a base of 2, anything with a ratio of >1 or <−1 represents a fold change of 2 or more. Finally, use the default display option (display data on the metabolic map overview) to be the final output of the data. To run the tool, click the submit button, this may take a few minutes for larger data sets.

As this example uses a time course experiment, a separate image is generated for each time point, and an animation will run displaying one image at a time (*see* **Note 24**). A legend is opened on the left, with the animation controls at the top (*see* **Fig. 20A**). The data are displayed on the right, aligned to the appropriate reactions on the entire metabolic map (*see* **Fig. 20B**). When single experiments are displayed on the omics viewer, a summary of the data, including a list of genes not present on the map, and features with missing or malformed data, is generated and displayed under the map. This feature is not incorporated with multiple experiments, such as this time course. Repeat the submission of these sample data, but display only column 1 by entering '1' into the left-hand side data selection box. At the bottom is a summary of the data (*see* **Fig. 21**). To save the images, use the 'save page' as function in the Web browser. For most Web browsers, this will save the overview, but not the individual pathways. To save the individual pathways, follow the link at the bottom of the omics viewer page and save all HTML files.

From the output, it is easy to see how gene expression changes over time within pathways. This tool can be used to aid interpretation of omics data providing an easy way to interpret visual representation of the biochemical processes that may be regulated temporally, spatially or by a chosen treatment.

4. Notes

1. As many parts of the Web site use Javascript, this should be enabled in the Web browser, as should cookies if you wish to register. Some features may not work well if pop-ups are blocked. In addition, TAIR have noted that some users have had problems with TAIR BLAST when using Netscape, old versions of Mac OS and Safari on OSX. With Safari, the problem is with the BLAST result output. This can be overcome by choosing the 'e-mail results' option.
2. Repeat the gene search without the wildcard, i.e. 'vrn'. This will return the same matches as with the wildcard. It is worth getting into the habit of using the wildcards as these are necessary for searching other tools, e.g. SeqViewer.
3. TAIR defines a locus as an element that maps to a transcribed region of the *Arabidopsis* genome or a genetic locus that segregates as a single genetic locus or quantitative trait. As such, a locus can have multiple gene models such as splice

variants. Gene model is the term used for any description of a gene product, e.g. mRNA sequence or computationally predicted sequence.

4. The locus history provides information about changes to the locus identifier. As sequence information becomes more complete, some loci have been split, others merged and some have become obsolete. This may be important when using 'old' information.

5. The Web site search (select 'Google TAIR website' from the drop-down menu at the top of each Web page) uses the Google search engine, which allows searching for phrases. To search for a phrase, it must appear in speech marks, e.g. 'drought stress'. Without the marks, the search will return anything with the terms drought or stress.

6. It is possible to perform highly refined name searches for specific information using the advanced searches. In the case of the advanced gene search, it is possible to refine the search by keyword (including the type of evidence supporting the annotation, e.g. experimental), by feature (e.g. full-length cDNA) and by map location. It is also possible from here to specify the number of search results per page. This is important if there are more records identified than those displayed per page. The download button will only retrieve information for the results on the current page; so, for a large number of results, it will be necessary to download each page in turn.

7. Before using SeqViewer, it is advisable to look at the help page, as this contains a link to the release update notes providing users with up-to-date information about noted problems and changes to the tool.

8. In SeqViewer, wildcards (*) can be used in the name search. The search term 'RPP*' would return RPP1, RPP8 and RPP13, as shown in the example above, but would also include, amongst others, RPP4 and RPP8. The search term 'RPP' alone would return nothing.

9. In the SeqViewer tool, clicking the chromosome bar next to the red flag denoting the search result may not open the close-up view centred on the object of interest (often it is not even in the view); so, the user may be required to use the scroll arrows to centre on the gene of interest. It is possible to decide which way to scroll by using the sequence ruler and the coordinates of the object as described in the results summary.

10. TAIR's definition of an annotation unit is a clone that has been artificially extended by the addition of neighbouring clones to improve the annotation of genes that fall in the overlapping region. The name is derived from the clone that contributed the most sequence. Information pages about the clones that make up the annotation units are linked to from the annotation unit detail page. Owing to artificial extension of the sequences, the gene of interest may not be present in every clone. Therefore, before purchasing, it is advised that a BLAST search using the gene of interest be performed against the GenBank BAC sequences data set.

11. From the whole-genome view, it is possible to choose the zoom level of the close-up view from the drop-down list (whole chromosome, 5 Mb, 1 MB, 200 kb and 80 kb), the option to zoom-in further on a chromosome segment (40, 20 or 10 kb) can only be selected once the close-up view has been opened.

12. The sequences must be between 15 and 150 bp in length. Single sequences can be entered in raw or FASTA format Multiple sequences must be entered in FASTA format.

    Raw

      agaagagctgaagacggag

    FASTA (single)

      >Seq 1

      agaagagctgaagacggag

    FASTA (multiple)

      >Seq 1

      agaagagctgaagacggag

      >Seq 2

      tctcttcataatcaagaag

13. The SeqViewer sequence search finds exact matches only. If testing for the specificity of primers, it is important to remember this, as sequence mismatches between closely related genes or repeat regions within a gene will not be identified. For longer sequences that do not return an exact match, it is suggested that BLAST (>30 bp) or Patmatch (<30 bp) be used to identify the best match. This sequence can then be entered into the sequence search.

14. For more information about the controlled vocabulary of evidence codes used for AraCyc (and all databases in the MetaCyc family), visit http://bioinformatics.ai.sri.com/evidence-ontology/.

15. The data entered into AraCyc are not restricted to expression, proteomic and metabolomic data. Any form of numerical data can be displayed, e.g. sequence similarity scores.

16. In data files for use with the omics viewer, lines that start with # or; are ignored by the program, allowing the addition of notes and column headers.

17. It is important to note that the AraCyc tool does not perform statistical analysis to verify any changes in gene expression, and so will report all differences irrespective of statistical significance. It may therefore be worth using this tool for a set of data that have been statistically verified by other means (e.g. for change in gene expression).

18. For gene expression data, it is recommended that the *Arabidopsis* genome initiative (AGI) locus number is used as the gene identifier. The AGI locus number can be obtained for a supplied list of array elements (from

AFGC, CATMA and Affymetrix arrays only) using the array element advanced search tool, accessed through TAIR's homepage or directly from http://www.arabidopsis.org/tools/bulk/microarray/index.jsp. The full data set can be downloaded from ftp://ftp.arabidopsis.org/home/tair/Microarrays/. Some array elements will map to multiple loci because of the detection of paralogous sequences. The user must decide to display one or all loci depending on the aim of their experiment. In addition, this tool cannot be used to add the AGI locus identifiers to the data file. Therefore, the user will have to do this manually or by writing their own script to combine the two tables (table of array elements mapped to AGI loci and the data table).

19. Briefly, these data are derived from cDNA experiments over a time course and are expressed as log ratio normalized values. More details can be found at http://www.arabidopsis.org/info/expression/microarrayFunctionalV2.jsp and follow the link for low temperature regulatory circuits and gene regulons in higher plants.

20. Typically, for Affymetrix expression data, choose absolute and relative for two-channel array data. Also select 'relative' if wishing to display data as ratios of a control sample (must be included in the uploaded data file). Select 'one data column' for Affymetrix expression intensity scores or where ratios have already been calculated. Select 'the ratio of two data columns' where you wish AraCyc to calculate and display the ratio of two or more samples.

21. It is also possible to select proteins, compounds, reactions or all data depending on the data to be viewed. The 'any of the above' option allows the user to combine all data types. However, if this is chosen, there is no differentiation between each data type on the resulting metabolic map.

22. The left-hand side (numerator) box must always have an input value; 1 represents the first potential data column (the column next to the gene identifier). If there is one set of data (a single experiment) in the form of either a list of ratios from a comparison or the expression intensities from one Affymetrix array, simply enter the number of the column containing these data into the left-hand side box (the data do not have to be the first column). The right-hand side box is required to define the denominator for the calculation of ratios. For example, in an experiment with four treatments (treatment 1 = control), each treatment can be displayed as a ratio of the control by entering 2, 3 and 4 in the left-hand side box and 1 in the right-hand side box (assuming the data are entered in the first four potential data columns).

23. The default colour setting is to assign colour bins over the entire spectrum, with cut-off values derived from the data set itself. This means different experiments may be displayed with different colour schemes. To avoid this (assuming all other selections are the same, e.g. ratio, log scale), use one of the other two options. The first is to select a maximum cut-off. This will display everything above this cut-off in red, leaving the rest of the spectrum for those below. This can aid detection of

smaller changes. The third option relies on a cut-off and just three colour bins. Anything greater than the cut-off will be displayed in red, those less than the inverse of the cut-off in yellow and everything else in blue.

24. The animation may not start in the overview map, even if the legend is rolling through the animation. This can be overcome by using the animation control scroll buttons and waiting for each page to load. Generally, once each time point has been loaded, the animation can be restarted.

References

1. Garcia-Hernandez, M., Berardini, T.Z., Chen, G., Crist, D., Doyle, A., Huala, E., Knee, E., Lambrecht, M., Miller, N., Mueller, L.A., Mundodi, S., Reiser, L., Rhee, S.Y., Scholl, R., Tacklind, J., Weems, D.C., Wu, Y., Xu, I., Yoo, D., Yoon, J., and Zhang, P. (2002). TAIR: a resource for integrated *Arabidopsis* data. *Funct Integr Genomics* 2, 239–253.

2. Huala, E., Dickerman, A., Garcia-Hernandez, M., Weems, D., Reiser, L., LaFond, F., Hanley, D., Kiphart, D., Zhuang, J., Huang, W., Mueller, L., Bhattacharyya, D., Bhaya, D., Sobral, B., Beavis, B., Somerville, C., and Rhee, S.Y. (2001) The *Arabidopsis* Information Resource (TAIR): a comprehensive database and web-based information retrieval, analysis, and visualization system for a model plant. *Nucleic Acids Res* 29, 102–105.

3. Weems, D., Miller, N., Garcia-Hernandez, M., Huala, E., and Rhee, S.Y. (2004). Design, implementation, and maintenance of a model organism database for *Arabidopsis thaliana*. *Comp Funct Genomics* 5, 362–369.

4. Rhee, S.Y., Beavis, W., Berardini, T.Z., Chen, G., Dixon, D., Doyle, A., Garcia-Hernandez, M., Huala, E., Lander, G., Montoya, M., Miller, N., Mueller, L.A., Mundodi, S., Reiser, L., Tacklind, J., Weems, D.C., Wu, Y., Xu, I., Yoo, D., Yoon, J., and Zhang, P. (2003). The *Arabidopsis* information resource (TAIR): a model organism database providing a centralized, curated gateway to *Arabidopsis* biology, research materials and community. *Nucleic Acids Res* 31, 224–228.

5. Reiser, L. and Rhee, S.Y. (2005) Using the *Arabidopsis* information resource (TAIR) to find information about *Arabidopsis* genes, in *Current Protocols in Bioinformatics* (Baxevanis, A.D. et al., eds.), John Wiley and Sons, NY, pp. 1.11.1–1.11.45.

6. Zhang, P., Foerster, H., Tissier, C., Mueller, L., Paley, S., Karp, P., and Rhee, S.Y. (2005). MetaCyc and AraCyc. Metabolic pathway databases for plant research. *Plant Physiol* 138, 27–37.

7. Mueller, L.A., Zhang, P., and Rhee, S.Y. (2003) AraCyc. A biochemical pathway database for *Arabidopsis*. *Plant Physiol* 132, 453–460.

8. Thimm, O., Bläsing, Y.G., Nagel, A., Meyer, S., Kruger, P., Selbig, J., Müller, L., Rhee, S.Y., and Stitt, M. (2004) MapMan: a user-driven tool to display genomics data sets onto diagrams of metabolic pathways and other biological processes. *Plant J* 37, 914–939.

9. Krieger, C.J., Zhang, P., Mueller, L., Wang, A., Paley, S., Arnaud, M., Pick, J., Rhee, S.Y., and Karp, P. (2004) MetaCyc: recent enhancements to a database of metabolic pathways and enzymes in microorganisms and plants. *Nucleic Acids Res* 32, D438–D442.

9

AtEnsEMBL
A Post-Genomic Resource Browser for Arabidopsis

Nick James, Neil Graham, Debbie Clements, Beatrice Schildknecht, and Sean May

Summary

A comprehensive Arabidopsis genomic resource has been developed at the Nottingham Arabidopsis Stock Centre (NASC) to support the international plant community. This browser, termed AtEnsembl, provides a detailed and user-friendly interface for accessing a wide range of Arabidopsis-based genomic information and post-genomic resources using the Ensembl browser. The resource aims to provide the broadest possible range of Ensembl features, including pointers to germplasm as well as representations of gene and protein information, links to Affymetrix gene expression data, and extensive data download capabilities.

Key Words: Arabidopsis; Ensembl; genome; Affymetrix; gene expression; gene annotation; *Brassica*.

1. Introduction

The original draft of the Arabidopsis genome was published in late 2000 *(1)* and was the first plant genome to be fully sequenced. This reflects the role of Arabidopsis as a model for all plant species and particularly for dicotyledonous plants. The focus by the plant science community on this species has led to an abundance of resources being available for this plant, including gene expression information, gene knockouts, mutant descriptions, expressed gene sequences, comparative genomic, phenotypic, and importantly, the complete genome sequence. We have assembled this information and integrated it within

From: *Methods in Molecular Biology, vol. 406: Plant Bioinformatics: Methods and Protocols*
Edited by: D. Edwards © Humana Press Inc., Totowa, NJ

a public Ensembl database (*see* **Note 1**) for access by the whole plant science community. Below, we describe the current content of this database and provide an introduction on how to access this information.

2. Materials

2.1. Arabidopsis Genomic Sequence

The annotation for this release was derived from many international sources in a rapid but partially uncoordinated manner that resulted in some inconsistencies and conflicting gene models. The Institute for Genomic Research (TIGR) (http://www.tigr.org) in the USA was therefore funded to provide a high-quality, consistent, re-annotation for the completed and assembled sequence. This work was initially carried out in cooperation with the bioinformaticians at the Munich Information Centre for Protein Sequence (MIPS) (http://mips.gsf.de/) in Germany and provided a first-pass annotation. However, these two groups have subsequently continued their annotation work with somewhat divergent approaches to *in silico* gene prediction, whereby both have released further independent annotations *(2)*. TIGR released their final Arabidopsis annotation (TIGR5) in January 2004, and responsibility for gene annotation has subsequently been passed onto The Arabidopsis Information Resource (TAIR). The first release of the TAIR annotation (TAIR6) was in December 2005. In the future, we will almost certainly see the two annotations converging to a single validated analysis, but for the present time, there are significant conflicts between the two approaches, which will need to be resolved (*see* **Note 2**). Rather than creating yet another Arabidopsis annotation, AtEnsembl objectively displays annotation from both TAIR6 (previously TIGR) and MIPS in parallel. We believe that this gives the experimental user an important opportunity to choose between putative gene structures derived by two expert groups where there may be some ambiguity that could impact on or inform their experimental work.

The first release of the TAIR annotation, TAIR6, has a sequence length of 119 Mbp, incorporating sequences from 1613 bacterial artificial chromosomes (BACs), yeast artificial chromosomes (YACs), cosmids, and polymerase chain reaction products. The MIPS annotation contains all data from the Arabidopsis Genome Initiative (AGI) project *(1)* and has a sequence length of 118 Mbp (chromosomes 1–5), derived from 1558 contigs. Both TAIR6 and MIPS (but not the earlier TIGR5) also include chloroplast and mitochondria (C + M) contigs that are considered separately. The TAIR6 annotation describes 26,751 genes and 3818 pseudogenes. With the further inclusion of alternatively spliced transcripts in this release, these genes now yield 30,695 distinct proteins. In contrast, the MIPS annotation contains 26,719 genes and 800 pseudogenes, with no alternative splicing. Both groups use a mixture of automatic and manual curation. MIPS estimates that approximately 50% of genes in their annotation have experimental evidence. AtEnsembl displays both TIGR and MIPS annotations independently, and in addition, we have mapped the MIPS annotation onto the TAIR chromosome assembly (*see* **Note 3**).

The TAIR annotation also contains information about the InterPro domains of Arabidopsis proteins. This information is displayed in the GeneView and ProteinViews of Ensembl. The DomainView page allows users to view all the proteins in Arabidopsis that share a particular InterPro domain. All predicted and annotated protein features from the TIGR annotation are stored, including Pfam, Prosite, transmembrane regions, Prints and Prodom domains. With TIGR5, we additionally supplemented the protein references supplied, with more of the information available from UniProt, but this has not been necessary with TAIR6.

The Gene Ontology (GO) project is an attempt to produce standardized vocabularies for the description of the molecular function, biological process and cellular component of gene products, to facilitate cross-database queries. The GO information included in AtEnsembl comes from TAIR6 annotation. Ensembl contains an inbuilt GO browser that enables users to identify genes, displayed by chromosomal location, with the same ontology as their gene of interest.

2.2. Expressed Sequence Tag and cDNA Data

AtEnsembl contains expressed sequence tag (EST) and cDNA data mapped onto the Arabidopsis genome using the Ensembl pipeline process. All Arabidopsis ESTs from the European Molecular Biology Laboratory (EMBL) database and Arabidopsis cDNAs from EMBL, RIKEN, Ceres and Genoscope have been used. The current AtEnsembl release (v35) contains a subset of 95,336 cDNA and 539,937 Arabidopsis EST sequences (*see* **Note 2**) considered to be the most appropriate best-match alignments from the possible 102,599 original source sequences.

2.3. Protein Alignments

AtEnsembl displays all matching Arabidopsis proteins, other plant proteins, and non-plant proteins from UniProt. We also align Arabidopsis proteins from TrEMBL in a similar pipeline process to that used for ESTs. Each protein displayed in AtEnsembl is cross-linked to any available external records that we can access (e.g. UniProt record). This currently includes (from UniProt) 3957 Arabidopsis proteins, 5476 other plant proteins, 15,865 proteins from other species, and 20,735 TrEMBL-predicted proteins.

2.4. Germplasm Information

The Nottingham Arabidopsis Stock Centre (NASC) stocks around 350,000 Arabidopsis lines that contain either a T-DNA or transposon-based insert. These have been donated into the public domain mostly as large populations created by a number of different groups, where the largest single donation from one group is a T-DNA insertion population generated by the SALK Institute (SALK lines) *(3)*. In addition, we hold insert populations containing large numbers (tens of thousands each) of genetic and genomic tools such as enhancer-trapped, gene-trapped, activation-tagged, and transactivation lines. Further information on all these, and other insert types, can be found on our website (http://www.arabidopsis.info). Genome sequences that flank the majority of these inserts have been retrieved and characterized by various donors. Many of these sequences are now widely available in public databases such as EMBL, as well as the local websites of the individual researchers. Most of the sequence coordinates displayed in AtEnsembl were provided by Sean Walsh at the John Innes Centre (ATIDB). SALK coordinates were obtained from the SIGNAL database at the SALK Institute. Other coordinates were generated locally in-house.

The inserts are displayed in AtEnsembl through the ContigView and GeneView pages, allowing users to see the number and location of inserts in a region or gene. In the ContigView display, different insert types are represented by different colored triangles to allow easier browsing for a desired type of insert in a particular gene. Inserts are then linked through these triangles as buttons to the NASC germplasm catalogue to allow ordering of the insert line, in a display that presents further insertion-associated information such as growing requirements, selection requirements and phenotype information.

2.5. Affymetrix Data

AtEnsembl currently contains data from two sets of Affymetrix GeneChip arrays: the AG and 'whole genome' (ATH1) Arabidopsis chips. In the near future, we will also be including data from the 2006 release full-genome Arabidopsis tiling chip. At NASC, we have separately mapped the AG and ATH1 probes onto the TAIR6 sequence to allow a clear display of the physical positions of the individual probes for every represented gene on both GeneChips. Individual probes (normally 11 per gene in the ATH1 GeneChip) were aligned using Mummer *(4)* before being grouped into probe sets and loaded into AtEnsembl. Because the individual probes in a probe set are stored in AtEnsembl, users can view these in ContigView and see exactly how probe sets align with the TAIR and MIPS genes displayed. Both the AG and ATH1 probe sets link through to the NASC microarray database, NASCArrays. Links are provided to the Spot History and Gene Swinger tools that allow users to analyze changes in expression of the gene viewed in AtEnsembl, currently over a range of more than 3000 GeneChip hybridizations.

2.6. Complete Arabidopsis Transcriptome MicroArray (CATMA) Data

Formed in 2000, the Complete Arabidopsis Transcriptome MicroArray (CATMA) project designed primers and generated short amplicons (150–350 bp) specific to the vast majority of all genes in the Arabidopsis genome. These gene sequence tags (GSTs) were then used in the construction and distribution of an Arabidopsis-spotted microarray. We have taken the original primer sequences from the CATMA project and aligned them individually onto the TAIR6 annotation to re-create virtual representations of the GST amplicons in their appropriate positions on the genome. To ensure specificity, we currently display in AtEnsembl only those primers that have been selected to uniquely match at 100% identity along their full length for both of the pair of primers that were used to create the GSTs.

The CATMA GST resource was further developed by the EU-funded AGRIKOLA project (http://www.agrikola.org) as a source of specific markers for individual gene manipulation. AGRIKOLA uses RNA interference (RNAi) techniques to constitutively disrupt or knockdown the expression of Arabidopsis genes, particularly where insertion lines are unavailable or inadequate. The gene specificity for these RNAi knockdowns comes from incorporation of the appropriate CATMA GST amplicon into transformation plasmids that have been designed to express the GST gene fragments as small RNAs within the plant. The project has generated transformable clones for most of the CATMA GSTs and has given these to NASC for redistribution to the plant community along with a small number of demonstration knockdown plant lines representing particular lines of interest to the AGRIKOLA consortium (5). Information from this project, such as the CATMA probe sequences, primer sequence, plate number, and location, are displayed through AtEnsembl on our CATMA View page. This detailed view page can be accessed through a standard lane in the AtEnsembl ContigView showing the exact dimensions and position of the CATMA amplicon on the appropriate chromosome. For convenience and completeness, the CATMA view page is also available from the specific GeneView pages. Users can go from either of these pages to order CATMA plates or AGRIKOLA plasmid and seed stocks from the NASC germplasm catalogue.

2.7. Brassica ESTs

Expressed *Brassica* sequences from public and proprietary sources are mapped using the Basic Local Alignment Search Tool (BLAST) to identify candidate homologous genes on the Arabidopsis genome. Complete annotation for these expressed sequences can be viewed and searched using the online, linked *Brassica* BASC database at http://bioinformatics.pbcbasc.latrobe.edu.au (*see* Chapter 10).

2.8. Brassica rapa Genomic Sequence

As the first stage of the Multinational Brassica Genome Sequencing Project (http://www.Brassica.info), end sequences are being obtained for 110,000 *B. rapa* genomic DNA fragments maintained as BACs. These genomic sequences have been mapped onto candidate syntenic regions of the Arabidopsis genome based on paired end sequences sharing identity with Arabidopsis sequence within 500 kbp, and in complementary orientation to each other. Intervening BAC sequence may be assumed to share a high degree of synteny with the corresponding Arabidopsis region. Annotation for *B. rapa* BAC end sequences is provided through links to the *Brassica* BASC database at http://bioinformatics.pbcbasc.latrobe.edu.au (*see* Chapter 10).

2.9. Brassica oleracea Genomic Sequence

In a collaboration between Cold Spring Harbor and TIGR, 0.5× coverage whole genome shotgun sequences have been produced for *B. oleracea*. These sequences have been mapped onto the Arabidopsis genome by their most significant BLAST match. Complete annotation for these expressed sequences can also be viewed and searched using the linked *Brassica* BASC database at http://bioinformatics.pbcbasc.latrobe.edu.au (*see* Chapter 10).

3. Methods

3.1. The Entry Page – A Summary and a Gateway

The entry page of AtEnsembl provides information about the latest version of the browser and the Ensembl project. It allows the user to choose to view either the MIPS assembly or the TIGR/TAIR and MIPS combined assembly. Choosing an assembly takes the user to the initial search screen (*see* **Fig. 1**). This provides information on the sources of the data for the assembly and details about the current release of AtEnsembl. From here, there are a number of ways of browsing the information in AtEnsembl, including browsing individual chromosomes, searches based on gene name, accession number or through a BLAST search.

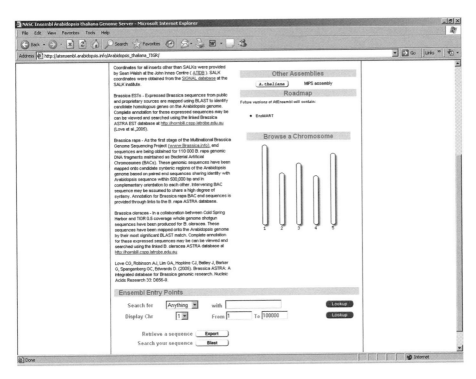

Fig. 1. Entry pages to AtEnsembl.

3.2. Browsing Chromosomes

Information about each individual chromosome can be browsed by clicking on the graphical representation of the appropriate chromosome. This will link to a screen (*see* **Fig. 2**) that provides information about the number of known genes and pseudogenes and the GC content along the chromosome. In the AtEnsembl chromosome view, the centromeric regions can easily be distinguished through simple visual inspection of the number of genes in this vertical bar chart. Clicking on a region of the chromosome will link to the more detailed ContigView (*see* **Fig. 3**) of that region.

3.3. ContigView

This provides a range of more focused and detailed information on ESTs, genes, proteins, Affymetrix data, knockout lines and other functional structures. Almost all of the features and diagrams within the ContigView can be seen as

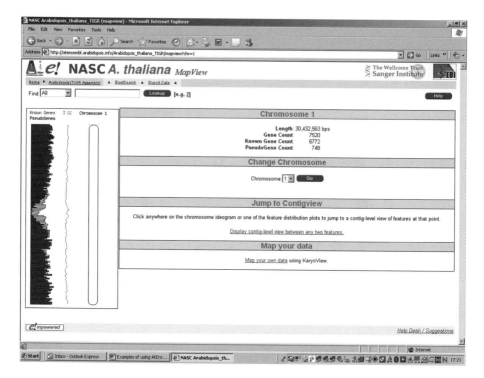

Fig. 2. Information summary for chromosome 1.

buttons linking to further material. If in doubt about the identity of an image, try clicking on it to provide yourself with more detailed information.

The ContigView page is split into four vertical sections bounded by red boxes, with progressively more detailed information about the selected chromosome region in each section as you move down the page. Each bounding red box for a section is represented on the section above by a smaller red box showing the area covered. In other words, the small red box on the top chromosome view shows you where the region overview below it sits on the chromosome, and so on down the page like Russian dolls (colors is as seen on the web page).

The first section shows the map position on the chromosome that is being viewed. Below this is an overview of that region, showing the position of genes as determined by the assemblies chosen (MIPS and/or TIGR/TAIR). Below this is a more detailed view, described below, and then the very detailed base-pair view. This latter view displays the DNA sequence and translations of that sequence in all three frames.

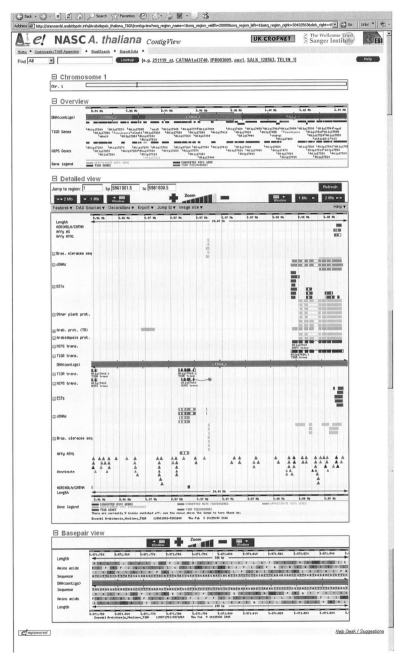

Fig. 3. ContigView page.

The detailed view provides information about both strands of the chromosome (shown above and below the horizontal central band representing the contig information) and contains the following positional information:

1. AGRIKOLA/CATMA – Position of the CATMA GST products. These are linked to information on that GST and enable ordering of the GST or AGRIKOLA clones and seed sets from the NASC germplasm catalogue.
2. Affy AG – Position of the Affymetrix probe sets on the older Arabidopsis AG chip. Clicking on the probe set gives you alternative choices to go to either the Spot History or the Gene Swinger tools within NASCArrays.
3. Affy ATH1 – Position of the Affymetrix probe sets on the Arabidopsis ATH1 chip. Clicking on these probe sets also link out to the appropriate Spot History and Gene Swinger tools of NASCArrays.
4. Bras oleracea seq – Position of *B. oleracea* sequences mapped onto the Arabidopsis genome.
5. cDNAs – Position of cDNAs mapped onto the genome.
6. ESTs – Position of Arabidopsis ESTs.
7. Arabidopsis prot – Position of Arabidopsis proteins linked to the UniProt database.
8. MIPS trans – Position of genes as determined by MIPS, and links to the gene view page.
9. TIGR trans – Position of genes as determined by TIGR, and links to the gene view page.
10. Knockouts – Position of inserts (T-DNA, transposon, enhancer trap, activation tag, etc.) from different populations (SALK, John Innes, FLAG, etc.). Many of these are linked to the NASC germplasm catalogue, so that the lines can be ordered. This is done by clicking on the triangle representing an insert and choosing order stock.

It is extremely important to realize that all of these features are simply gateway icons to more detailed information. All they represent in this particular view are the relative positions of the features on the chromosome. Everything is a button and can be pressed to either open up more information through new pages or drop-down menus, or in the case of the various tools, they can be pressed to customize the view for the user. In every case, changes can easily be reversed or you can always use your back button on the browser to return to an earlier view.

3.4. Gene and PeptideView Page

The GeneView page provides detailed information on the gene (*see* **Fig. 4**), including genome location, intron/exon structure, GO and InterPro domain annotations. From this view, there is also an option to export both DNA and peptide sequences in a number of formats (flat file, FASTA, etc.) and to include a range of annotations and flanking DNA sequence. Similar information is displayed on the PeptideView page (*see* **Fig. 5**). Here, information on the

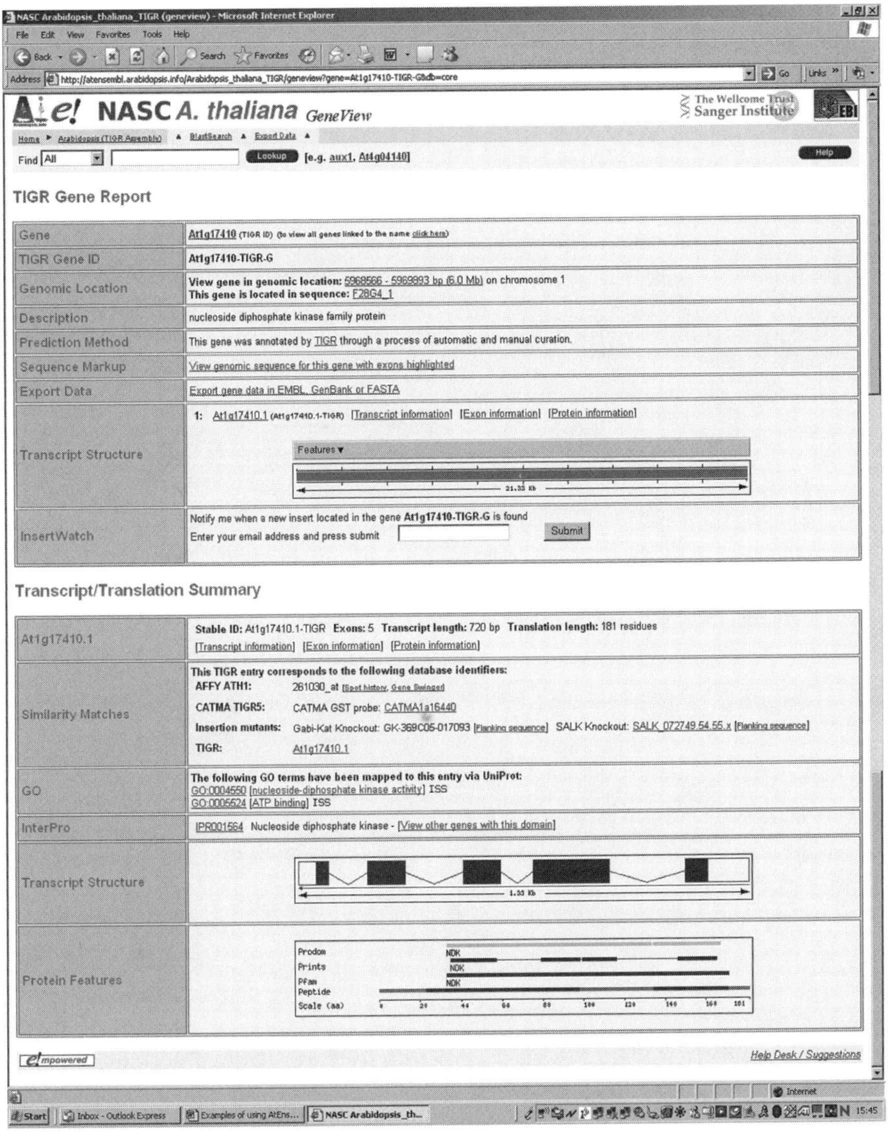

Fig. 4. Gene information page.

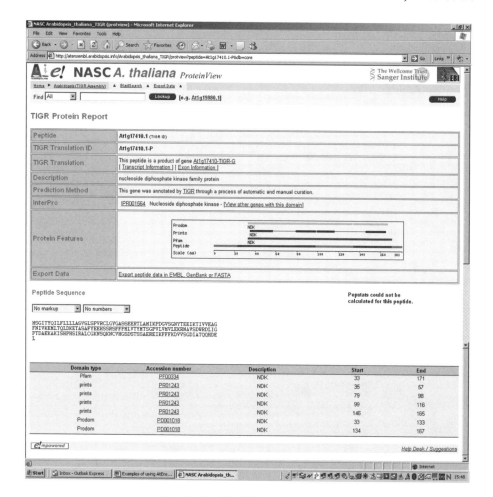

Fig. 5. Peptide information page.

peptide sequence, protein features and domains is displayed. This includes InterPro annotations of the peptide domains and links to the InterPro database at EMBL. There is also a diagram displaying protein domains within the peptide, as found by InterPro member databases, including Pfam and Prosite.

3.5. Searching Using Accession Code or Gene Name

AtEnsembl can be searched using a keyword (e.g. auxin and cyclin) or accession number [AGI code (often called the 'AT code'), Affymetrix code, CATMA code, etc.]. This can be performed on any page within the browser.

The search result page provides information about the fields in which the search term was found (e.g. gene, peptide, and transcript), and these will then link to more detailed information, for example, the GeneView page.

3.6. BLAST Search

An alternative method of searching AtEnsembl is to use BLAST. A full range of BLAST searches is available (BLASTX, BLASTN, and TBLASTX) using both DNA and peptide sequences. The result page (*see* **Fig. 6**) displays the search results in three different views, with the alignments colour coded dependent on the %ID of the alignment. The 'alignment versus karyotype' displays the location of the alignment on the chromosome, with the most significant alignment highlighted in a red box. The 'alignment versus query' displays the alignments against the query sequence, and the alignment summary displays all the information about the alignments including E-value, location of the alignment,

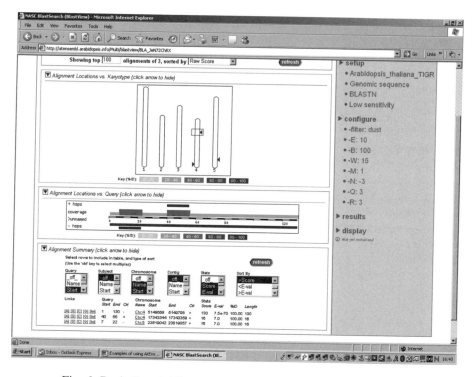

Fig. 6. Basic Local Alignment Search Tool (BLAST) result page.

and %ID. There are links from each alignment to view the query sequence, the alignment, the ContigView of the alignment and the genomic sequence.

4. Notes

1. Ensembl is a joint project between the EBI (European Bioinformatics Institute) and the Sanger Institute that provides access to genome sequence and automatically generated annotation for a large number of animal and protist species, including human, mouse, dog, *Drosophila* and yeast. The Ensembl website (http://www.ensembl.org) provides access to these data in a number of ways: through the website, which is the most commonly used method; through a Web-based data mining interface, EnsMart; and by direct access to the databases themselves through download or MySQL login. The whole Ensembl project, including all software and data is open source, and a number of businesses and other research institutions have set up their own local installations *(6)*. The genomic information for two of the species hosted at Ensembl, *C. elegans* and mosquito, is based on annotation imported from other non-local sources, demonstrating a cooperative approach that we have followed at NASC to develop our plant version of the browser.

2. The information maintained within this system is subject to change, and the authors suggest visiting the Website on a regular basis to view updates and new information.

3. The current AtEnsembl uses TAIR build 6. The database will be updated with new builds of the Arabidopsis genome as they become available.

References

1. The Arabidopsis Genome Initiative. (2000) Analysis of the genome sequence of the flowering plant *Arabidopsis thaliana*. *Nature* 408, 796–815.
2. Wortman, J.R., Haas, B.J., Hannick, L.I., Smith, R.K., Maiti, R., Ronning, C.M., Chan, A.P., Yu, C.H., Ayele, M., Whitelaw, C.A., White, O.R., Town, C.D. (2003). Annotation of the Arabidopsis genome. *Plant Physiol.* 132 (2), 461–468.
3. Alonso, J.M., Stepanova, A.N., Leisse, T.J., Kim, C.J., Chen, H.M., Shinn, P., Stevenson, D.K., Zimmerman, J., Barajas, P., Cheuk, R., Gadrinab, C., Heller, C., Jeske, A., Koesema, E., Meyers, C.C., Parker, H., Prednis, L., Ansari, Y., Choy, N., Deen, H., Geralt, M., Hazari, N., Hom, E., Karnes, M., Mulholland, C., Ndubaku, R., Schmidt, I., Guzman, P., Aguilar-Henonin, L., Schmid, M., Weigel, D., Carter, D.E., Marchand, T., Risseeuw, E., Brogden, D., Zeko, A., Crosby, W.L., Berry, C.C., Ecker, J.R. (2003) Genome-wide insertional mutagenesis of *Arabidopsis thaliana*. *Science* 301 (5633), 653–657.
4. Kurtz, S., Phillippy, A., Delcher, A.L., Smoot, M., Shumway, M., Antonescu, C., Salzberg, S.L. (2004) Versatile and open software for comparing large genomes. *Genome Biol.* 5 (2), R12.

5. Hilson, P., Small, I., Kuiper, M.T.R. (2003). European consortia building integrated resources for Arabidopsis functional genomics. *Curr. Opin. Plant Biol.* 6 (5), 426–429.

6. Hubbard, T., Andrews, D., Caccamo, M., Cameron, G., Chen, Y., Clamp, M., Clarke, L., Coates, G., Cox, T., Cunningham, F., Curwen, V., Cutts, T., Down, T., Durbin, R., Fernandez-Suarez, X.M., Gilbert, J., Hammond, M., Herrero, J., Hotz, H., Howe, K., Iyer, V., Jekosch, K., Kahari, A., Kasprzyk, A., Keefe, D., Keenan, S., Kokocinsci, F., London, D., Longden, I., McVicker, G., Melsopp, C., Meidl, P., Potter, S., Proctor, G., Rae, M., Rios, D., Schuster, M., Searle, S., Severin, J., Slater, G., Smedley, D., Smith, J., Spooner, W., Stabenau, A., Stalker, J., Storey, R., Trevanion, S., Ureta-Vidal, A., Vogel, J., White, S., Woodwark, C., Birney, E. (2005). Ensembl 2005. *Nucleic Acids Res.* 33, D447–D453.

10

Accessing Integrated *Brassica* Genetic and Genomic Data Using the BASC Server

Christopher G. Love and David Edwards

Summary

The BASC system provides tools for integrated mining and browsing of genetic, genomic and phenotypic data. The BASC demonstration server provides access to raw and analysed information for *Brassica* species and comparative information with *Arabidopsis*. We can use the tools within the *Brassica* BASC server to identify candidate genes for traits, conduct genome comparisons with *Arabidopsis*, identify syntenic regions and view gene expression profiles. The integration of BASC modules allows researchers to gain a comprehensive view of diverse *Brassica* genetic, genomic and phenotypic information. This chapter demonstrates the application of this resource, through the identification of candidate genes for an observed seed oil quality trait in *Brassica napus*.

Key Words: Comparative genomics; Data integration; Brassica; Arabidopsis; Genetic; Genomic; Microarray; Database.

1. Introduction

The increased complexity and abundance of biological information generated through current plant research has made it difficult for researchers to gain a comprehensive view of the data available. To improve accessibility to the information, numerous computational tools and databases have been established *(1)*. However, their independent development has resulted in many diverse types that lack interoperability leaving it up to the researchers to identify trends and linkages between the data contained. As we attempt to understand

From: *Methods in Molecular Biology, vol. 406: Plant Bioinformatics: Methods and Protocols*
Edited by: D. Edwards © Humana Press Inc., Totowa, NJ

the relationship between trait and gene, it is essential to maintain associations between the data to utilize information from genetics and genomics technologies. Through the integration of these data, researchers have the capability to ask complex biological questions. The Bioinformatics Advanced Scientific Computing (BASC) system provides an integrated resource for querying across a wide range of genetic, genomic and phenotypic data for a range of commercially important crop plants. The public demonstration server provides access to genetic mapping, functional annotation, comparative genomics and gene expression data for public *Brassica* data, with comparison to *Arabidopsis*.

The *Brassicas* consist of many extensively researched crop species. The species are characterized by a wide range of adaptations that have been domesticated into crops, including oilseed rape/Canola and swede (*Brassica napus*); cabbage, cauliflower, broccoli and Brussels sprout (*Brassica oleracea*); chinese cabbage, pak choi and turnip (*Brassica rapa*); and mustards (*Brassica nigra, Brassica juncea* and *Brassica carinata*), with many commercial and industrial applications *(2,3)*. Widely cultivated throughout the world, *Brassicas* play an important role in global horticulture and agriculture. *Brassica*s also share extensive synteny with the fully sequenced model plant, *Arabidopsis thaliana (4–7)*. This close relationship can assist characterization of the *Brassica* genome through comparative mapping, and utilization of the wealth of information available for *Arabidopsis (8)*.

There is a wide range of genetic and genomic data available within the public domain relating to *Brassicas*, as well as the increasing comparative data derived from *Arabidopsis*. These include reference linkage maps, a wide range of quantitative trait loci (QTL) relevant to basic processes, phenotypic data and sequence-based information. In addition, there is an increasing availability of comparative information, with extensive genetic and physical sequence maps, with common markers available to assist in extrapolating information from *Arabidopsis* to *Brassica*. Further genomic *Brassica* information will also emerge as the goal to sequence the *B. rapa* genome progresses, with an estimated completion by the end of 2007 *(9)*.

Researchers can identify QTL using CMap to compare genetic maps with common markers, and map the *Brassica* region to a syntenic location in *Arabidopsis (10)*. Users can then browse the *Arabidopsis* physical sequence and view *Brassica* genes represented within the syntenic region, as well as the associated functional annotation and expression profile of the genes.

2. Materials

The BASC resource is based on five distinct modules, with three of these providing access to expressed sequence tag (EST), MarkerQTL and microarray gene expression data. Two further modules include an *Arabidopsis* Ensembl genome viewer, and the CMap comparative genetic map viewer (*see* **Fig. 1**). The information provided in this example is based on release two of the public *Brassica* BASC (November 2005) and is subsequent to change during regular updates every three months. The site is optimized for use with the Mozilla Firefox browser (http://www.mozilla.com/).

2.1. BASC MarkerQTL

The MarkerQTL module is designed based on structuring genotypic data with an emphasis on molecular map integration, phenotypic data, experimental design and QTL analysis. MarkerQTL contains all data associated with pedigrees of individual plants, markers used for genotyping, resulting genotypes and alleles as well as QTL resulting from the analysis of this data. The relational database structure allows integration of genotypic data based on molecular marker analysis, and phenotypic data from agronomic, physiological and biochemical analysis. This data includes DNA based, or physiological markers, and any associated genetic linkage maps, combined with the details of the mapping or diversity population. The web interface is structured so researchers can easily browse between the genotypic, phenotypic and genomic information.

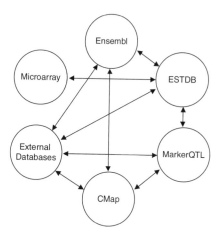

Fig. 1. Integration between the different modules of BASC.

The flexible cross-reference system allows linking to and from external databases. Markers derived from EST sequences have associated entries within Expressed Sequence Tag DataBase (ESTDB) and are hyperlinked to any associated information within the comparative mapping program, CMap. Further information regarding the origin of the marker data set is provided through a link to the original website or relevant publication.

2.1.1. The Views

MarkerView relates information from a marker assay and allows representation of Simple Sequence Repeat (SSR), Single Nucleotide Polymorphism (SNP), restriction fragment length polymorphism (RFLP) and Amplified Fragment Length Polymorphism (AFLP) marker data – the information corresponding to each marker including primer and any associated map or locus information. Where markers have been mapped, their respective map and loci are obtained through LocusView. Tools enable the selection of markers within a defined region. The frequency of alleles at a locus can also be identified through LocusView.

IndividualView describes information relating to individual plants. An individual is defined as one DNA sample originating from a single plant or fungal isolate. Data includes the association of the individual within any population, their relationship with other individuals within a collection and where the individual was derived from.

GenotypeView provides a profile of all alleles determined for an individual. The genotype for a marker is predominantly by allele size, or nucleotide base in the case of SNPs.

QTLView contains information on quantifiable traits for each individual and population. Individual traits are the calculated single value of a trait for that individual, whereas genetic and QTL mapping data are associated with groups of individuals. The basic information includes trait name and description as well as the measurement unit and the analysis method. Much of the data are analyzed to identify the association of markers to linkage groups and the calculation of Linkage Disequilibrium (LD) scores prior to insertion into the database.

MapView displays the ordering of markers represented on a map. It allows a marker to have multiple locations in a single map as well as locations on multiple maps.

2.1.2. The Data

Publicly available *Brassica* molecular genetic marker data has been incorporated within the MarkerQTL database. This includes data collated from the *B.*

napus IMSORB project (http://Brassica.bbsrc.ac.uk/IMSORB/), Tom Osborn's laboratory at the University of Wisconsin (http://osbornlab.agronomy.wisc.edu/research/maps.html), the Biotechnology and Biological Sciences Research Council (http://www.Brassica.info/ssr/SSRinfo.htm), the Celera Ag Gen *Brassica* Consortium *(11)*, IAPB (University of Goettingen), UK cropnet and GenBank. Markers include 253 RFLPs, 755 genomic SSRs, 103EST SSRs sequence repeats and 166 additional markers. Over 150 QTL for various traits have also been included.

2.2. BASC CMap

CMap is a tool for visualizing and comparing genetic and physical maps, originally developed for the Gramene project (http://www.gramene.org/). Within the public BASC system, CMap has been adopted for *Brassica* data. The BASC version of CMap contains QTL and genetic maps from *Brassica* species including eight *B. oleracea*, nine *B. napus*, three *B. rapa*, six *B. juncea*, four *Arabidopsis* sequence and two *Arabidopsis* genetic maps. Maps can be compared where there is correspondence between markers.

2.3. BASC ESTDB

The ESTDB module provides functional annotation for EST collections. An annotation pipeline that is capable of handling large-scale EST collections controls the data analysis (*see* **Fig. 2**). Information is available at multiple levels including EST, clone and contig level. The pipeline uses Phred *(12)* if the tracefiles are available, and cross_match for vector trimming *(13)* prior to assembly and annotation. All raw information regarding parameters for the relevant analysis program is recorded and presented within AnalysisView. At the EST level, the trimmed sequence is available as well as links to the associated clone and contig. The ESTs are clustered using d2_cluster *(14)* and assembled using Phrap *(15)*. The assembly builds can be viewed through ClusterView. Associated library information is available through CloneView with links to the annotated contig.

Functional annotation is displayed through ContigView. Annotation is derived from sequence similarity to entries within UniProt *(16)* and GenBank nt *(17)*, using BLAST *(18)*. The top BLAST match to UniProt is used to provide gene descriptions where possible. Details of BLAST results (such as score and e-value) are stored in the database to allow querying and filtering through the web interfaces. Gene Ontology (GO) annotation *(19)* is provided where available through intermediate mapping to UniProt. Assembled contigs are mapped onto the *Arabidopsis* genome using WU-BLAST *(20)*, and their

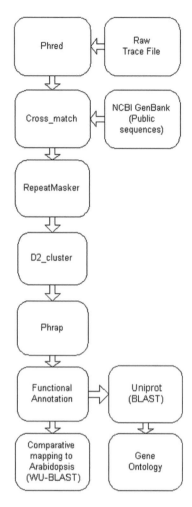

Fig. 2. ESTDB processing and annotation pipeline.

genomic location recorded within the database. This provides the basis for linking to the *Arabidopsis* Ensembl genome viewer.

Further links are available within the database to associate with other components of BASC. Where a marker is derived from an EST, there is a link to any associated information within the MarkerQTL database. Similarly, if the cDNA has been used in a microarray experiment, a link is provided to view the resulting profile of gene expression.

2.3.1. The Data

Publicly available *Brassica* sequences are downloaded from GenBank every three months, combined with proprietary sequences and processed through the annotation pipeline. Total sequence numbers within the database are 1,954 (*B. oleracea*), 3,267 (*B. nigra*), 8,576 (*B. juncea*), 13,573 (*B. rapa*) and 82,900 (*B. napus*).

2.4. Arabidopsis Ensembl

Arabidopsis Ensembl is an open source tool developed by NASC (http://arabidopsis.info/see Chapter 9). The genome browser allows biological information to be captured based on the anchoring of features to the complete genome sequence of *Arabidopsis*. The genome browser provides a comprehensive view of the complete annotated genome allowing ease of navigation between data sets *(21)*.

2.5. BASC Microarray

The microarray module provides a platform for storing and visualizing microarray gene expression data. The system provides data mining of microarray experiments, samples, spots and images. Both raw and normalized data are stored as single channel data to allow comparison across multiple slides and experiments. Information relating to how a feature performed over multiple treatment types is provided as median normalized and Z-score normalized values. Additional information relating to RNA samples and relevant scan data is provided.

2.5.1. The Data

The data contained within this public module include results from *B. napus* root and leaf tissue hybridizations of a *Brassica* 7K cDNA microarray chip.

3. Methods

To provide a representative view of a workflow using the BASC resource, three different examples are described, demonstrating the links between the different modules. The first two examples utilize MarkerQTL to identify markers associated with an oil trait, and comparison of *Brassica* genetic maps with correspondence to the *Arabidopsis* genome sequence using CMap. The third example shows how the researcher can mine the *Arabidopsis* genome annotation to identify candidate *Brassica* genes annotated with oil biosynthesis and view an associated gene expression profile.

3.1. BASC MarkerQTL

First, search for markers that may be associated with the QTL for oilseed quality using the MarkerQTL module.

1. Open http://bioinformatics.pbcbasc.latrobe.edu.au and click on BASC Project, then on 'BASC Full Text Search'. Enter the Marker ID 'pW188'. This will search all databases for this keyword (*see* **Note 1**). Click on the RFLP marker that is from IMSORB. This provides MarkerView information and links to the source of the original information, in this example IMSORB (*see* **Note 2**).
2. Clicking on the Locus will take you to LocusView. This view relates the marker to a location on a map. In this case, the marker is only found on one map, on chromosome NO3 of the IMSORB TN map.
3. To identify any QTL around this locus, click on 'Show QTLs' (*see* **Note 3**). Two QTL are found within 10 cM of this marker.
4. Click on QTL '15' and this will take you to the QTLView providing information relating to the QTL, in this case seed oil content (*see* **Note 4**).
5. Follow the 'link to CMap' feature search. This will identify all linkage groups in *B. napus* associated with this QTL. Click on the 'View on Map' for map N03 to display the trait on linkage group three.
6. Selecting the 'Show Additional Options Menu' and Show Labels 'All', followed by 'Redraw Map', will display all markers and QTL mapped onto this linkage group.
7. This genetic map shows a single marker that is underlying two QTL for seed oil content identified in different populations (*see* **Fig. 3**). Further identification of markers within this region may help fine map the trait to a smaller region.

3.2. BASC CMap

CMap provides the capability of displaying genetic maps and can identify syntenic regions by comparing maps where there is correspondence. This is of particular use as many markers have been compared between *Brassica* and *Arabidopsis*. This example utilizing CMap demonstrates its use in comparing between *Brassica* species and *Arabidopsis* to identify syntenic regions.

1. From the BASC homepage click on the link to CMap. This takes you to the BASC CMap home page.
2. From the home page click on 'Maps'. This will provide a basis for selecting the map required.
3. Select reference species *B. oleracea* from the drop-down menu. Click on 'Change Species'. Select 'Oleracea consensus' and click 'Show selected map sets'.
4. From the 'Ref. Map' list, select 'O5' and click 'Draw Maps'. This will display linkage group 5.

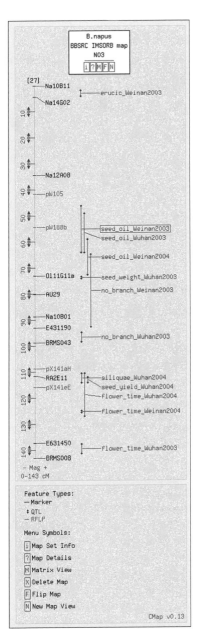

Fig. 3. CMap representation of *Brassica napus* IMSORB quantitative trait loci (QTL) map with associated markers.

5. To compare the reference *B. oleracea* linkage group 5 with the corresponding linkage group in *B. napus*, open the 'Show Comparison Menu' and select '*B. napus –* Parkin 2005' map from the Comparative Maps (right) menu. Select 'N15' and click 'Redraw Map'. This will display the extensive although imperfect collinearity between *B. oleracea* linkage group 5 and *B. napus* linkage group 15.

6. To compare the *B. napus* map with *Arabidopsis*, again, in the 'Show Comparison Menu', select '*Arabidopsis* – Lukens 2003' and chromosome '1' from the Comparative Maps (right) menu and click 'Redraw Map'. This compares *B. napus* to a sequence map of *Arabidopsis*, displaying correspondence with *B. napus* markers and *Arabidopsis*, BAC clones. The view highlights a region of collinearity.

7. Clicking on any of the BAC features (e.g. F13K23) on the *Arabidopsis* map showing correspondence will take you to the relevant feature page where users can then link to a physical location on the *Arabidopsis* Ensembl viewer and browse the gene annotations for that region.

8. To view the physical *Arabidopsis* region, click on the cross-reference to '*Arabidopsis* Ensembl'.

3.3. Arabidopsis Ensembl

The *Arabidopsis* Ensembl genome viewer allows browsing of the genome and identification of annotated genes within a specified region (also, see Chapter 9). Using this tool, we can search the *Arabidopsis* annotation for genes associated with oil biosynthesis and identify whether they have any sequence similarity to *Brassica* genes (*see* **Fig. 4**).

1. From the BASC home page, click on the BASC 'Full Text Search'. This will allow searching of all modules of BASC including the *Arabidopsis* Ensembl gene annotations by key words.

2. Select 'athaliana_gene' and insert into the text box '+acyl-carrier protein' and click 'Lookup'.

3. From the results returned, click on the second match for Ensembl Gene/Peptide: At1g08510-TIGR-G. This will link to the Ensembl GeneView entry.

4. GeneView provides information regarding annotation of the gene, transcript information and protein features (*see* **Note 5**). External links to other databases provide further information relating to the gene.

5. To view the gene with aligning *Brassica* ESTs, under the Genomic Location header, you can click to the location of the gene at chromosome 1: 2,691,083–2,694,380.

6. This takes you to ContigView. Within this view, there are mappings of both *Arabidopsis* and *Brassica* ESTs to the gene region. *Brassica* ESTs are represented as dark green bars (*see* **Note 6**).

7. Click on the *Brassica* EST. This will link to the BASC ESTDB annotation for this *Brassica* gene.

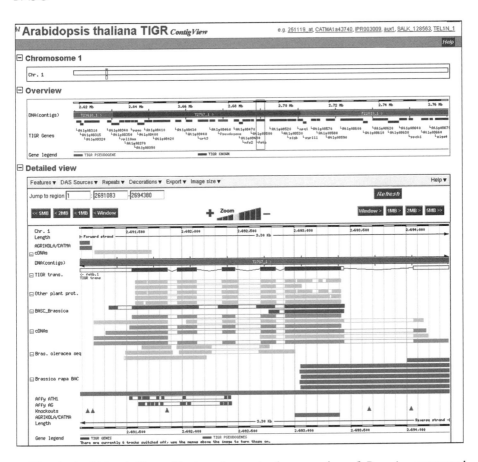

Fig. 4. *Arabidopsis* Ensembl view demonstrating mapping of *Brassica* expressed sequence tags (ESTs) to *Arabidopsis* genome with links to BASC ESTDB.

3.4. BASC ESTDB

Using ESTDB, we can further characterize the *Brassica* EST identified in **Subsection 3.3.** (*see* **Fig. 5**).

1. The ContigView of the BASC ESTDB displays the related GO and Uniprot (Uniref90) annotation for the gene. Note how the annotation is the same as for the *Arabidopsis* gene viewed earlier within *Arabidopsis* Ensembl. External links to the host databases for the relevant annotations provides further information, and at the bottom of the page, there is a diagrammatical representation of the BLAST matches in relation to the gene sequence (*see* **Note 7**).

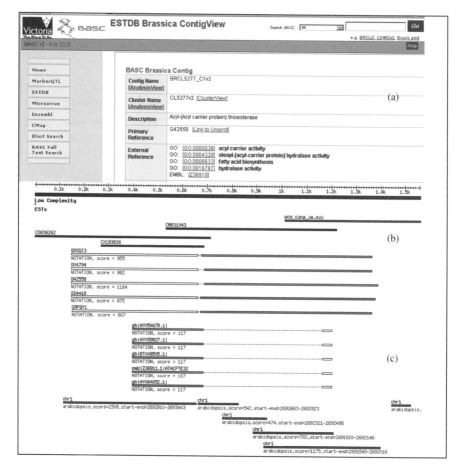

Fig. 5. BASC ESTDB screenshot demonstrating (**A**) the annotation at ContigView level with (**B**) graphical representation of members of the contig, BLAST hits aligned with the contig and (**C**) comparative mapping to the *Arabidopsis genome.*

2. Click on ClusterView to display ESTs that have been assembled into the contig.
3. Click on 'b09_03na' to take you through to ESTView.
4. Click again on clone name b09_03na to go to CloneView.
5. From CloneView, library information can be gained to identify the tissue the sample was extracted from.
6. This clone has been applied in a microarray experiment, and the results can be obtained through the 'Link to Microarray'.

3.5. BASC Microarray

From the EST, we can link to Microarray DB and look at the gene expression under different conditions. In the context of this example, it allows us to look at the expression profile for the previously identified 'Acyl carrier protein' gene.

1. From the SpotView table, the user can identify all the treatment types and tissues used in the microarray experiment (*see* **Note 8**).
2. Clicking on the tissue column will provide further details about the treatment of the RNA sample. We can see that that the highest expression of the gene is from Surpass 400 leaf sample 9 weeks following planting.

4. Notes

1. The full text search utilizes a Boolean-based search protocol. Specifying the relevant search parameters can assist in refining the search. See the help pages for more information.
2. The MarkerQTL database can contain sequence-based markers as well as physiological markers.
3. The markers displayed can be limited to particular markers and at varying intervals (e.g. 5–10 cM) using the filter option at the top of the page.
4. Direct access to this and other oil-related QTL may be gained by entering the term 'oil' in the full text search.
5. Within ContigView, the user can gain further annotation including GO and detailed information relating to the gene and exon/intron boundaries as well as annotation method by following the links on this page.
6. Note mappings of syntenic *B. rapa* BACs as part of the *B. rapa* genome sequencing, these mappings are 'in silico' predictions of syntenic regions between *B. rapa* and *Arabidopsis* based on mapping of *B. rapa* BAC ends. There is also a track for *B. oleracea* 0.5× genome shotgun sequences by TIGR that have been mapped to *Arabidopsis*.
7. The data consist of proprietary and public data, where the sequence is proprietary, the nucleotide sequence may not be viewed. Public release of proprietary data is expected during 2007.
8. Two normalized values are provided forming the basis for comparing expression under different treatment types. The Z-score and median-normalized values are different normalization methods. Normalized values that are high relate to high levels of the transcript within the tissue.

References

1. Stein, L. D. (2003) Integrating biological databases. *Nat Rev Genet* 4, 337–45.
2. McGregor, D. I., and Kimber, D. S. (1995) *Brassica Oilseeds Production and Utilization*, CAB International, Wallingford.

3. Rakow, G. (2004) Economic Importance of *Brassica* species, in *Brassica* (Pua, E. C., and Douglas, C. J., Eds.), Vol. 54, pp. 6–11, Springer, New York.

4. Parkin, I. A., Gulden, S. M., Sharpe, A. G., Lukens, L., Trick, M., Osborn, T. C., and Lydiate, D. J. (2005) Segmental structure of the *Brassica napus* genome based on comparative analysis with *Arabidopsis thaliana*. *Genetics* 171, 765–81.

5. The Arabidopsis Genome Initiative. (2000) Analysis of the genome sequence of the flowering plant *Arabidopsis thaliana*. *Nature* 408, 796–815.

6. Lukens, L., Zou, F., Lydiate, D., Parkin, I., and Osborn, T. (2003) Comparison of a *Brassica oleracea* genetic map with the genome of *Arabidopsis thaliana*. *Genetics* 164, 359–72.

7. Lagercrantz, U. (1998) Comparative mapping between *Arabidopsis thaliana* and *Brassica nigra* indicates that *Brassica* genomes have evolved through extensive genome replication accompanied by chromosome fusions and frequent rearrangements. *Genetics* 150, 1217–28.

8. Hall, A. E., Fiebig, A., and Preuss, D. (2002) Beyond the Arabidopsis genome: opportunities for comparative genomics. *Plant Physiol* 129, 1439–47.

9. Bancroft, I., and Lim, Y. P. (2003) *Concept Note for the Brassica Genome Sequencing Project*. Available at http://www.brassica.bbsrc.ac.uk/concept_note_for_brassica_genome_seq.doc.

10. Jaiswal, P., Ni, J., Yap, I., Ware, D., Spooner, W., Youens-Clark, K., Ren, L., Liang, C., Zhao, W., Ratnapu, K., Faga, B., Canaran, P., Fogleman, M., Hebbard, C., Avraham, S., Schmidt, S., Casstevens, T. M., Buckler, E. S., Stein, L., and McCouch, S. (2006) Gramene: a bird's eye view of cereal genomes. *Nucleic Acids Res* 34, D717–23.

11. Piquemal, J., Cinquin, E., Couton, F., Rondeau, C., Seignoret, E., Doucet, I., Perret, D., Villeger, M.-J., Vincourt, P., and Blanchard, P. (2005) Construction of an oilseed rape (*Brassica napus L.*) genetic map with SSR markers. *TAG* 111, 1514–23.

12. Ewing, B., Hillier, L., Wendl, M. C., and Green, P. (1998) Base-calling of automated sequencer traces using phred. I. Accuracy assessment. *Genome Res* 8, 175–85.

13. Green, P. (1996) *Cross_match*, University of Washington, Seattle, USA. Available at http://www.phrap.org/phredphrap/phrap.html.

14. Burke, J., Davison, D., and Hide, W. (1999) d2_cluster: a validated method for clustering EST and full-length cDNAsequences. *Genome Res* 9, 1135–42.

15. Green, P. (1996) *Phrap Sequence Assembly Program*, University of Washington, Seattle, USA. Available at http://www.phrap.org/phredphrap/phrap.html.

16. Bairoch, A., Apweiler, R., Wu, C. H., Barker, W. C., Boeckmann, B., Ferro, S., Gasteiger, E., Huang, H., Lopez, R., Magrane, M., Martin, M. J., Natale, D. A., O'Donovan, C., Redaschi, N., and Yeh, L. S. (2005) The universal protein resource (UniProt). *Nucleic Acids Res* 33, D154–9.

17. Benson, D. A., Karsch-Mizrachi, I., Lipman, D. J., Ostell, J., and Wheeler, D. L. (2005) GenBank. *Nucleic Acids Res* 33, D34–8.
18. Altschul, S. F., Gish, W., Miller, W., Myers, E. W., and Lipman, D. J. (1990) Basic local alignment search tool. *J Mol Biol* 215, 403–10.
19. Harris, M. A., Clark, J., Ireland, A., Lomax, J., Ashburner, M., Foulger, R., Eilbeck, K., Lewis, S., Marshall, B., Mungall, C., Richter, J., Rubin, G. M., Blake, J. A., Bult, C., Dolan, M., Drabkin, H., Eppig, J. T., Hill, D. P., Ni, L., Ringwald, M., Balakrishnan, R., Cherry, J. M., Christie, K. R., Costanzo, M. C., Dwight, S. S., Engel, S., Fisk, D. G., Hirschman, J. E., Hong, E. L., Nash, R. S., Sethuraman, A., Theesfeld, C. L., Botstein, D., Dolinski, K., Feierbach, B., Berardini, T., Mundodi, S., Rhee, S. Y., Apweiler, R., Barrell, D., Camon, E., Dimmer, E., Lee, V., Chisholm, R., Gaudet, P., Kibbe, W., Kishore, R., Schwarz, E. M., Sternberg, P., Gwinn, M., Hannick, L., Wortman, J., Berriman, M., Wood, V., de la Cruz, N., Tonellato, P., Jaiswal, P., Seigfried, T., and White, R. (2004) The gene ontology (GO) database and informatics resource. *Nucleic Acids Res* 32, D258–61.
20. Gish, W. (2006) *WU-BLAST 2.0*, University of Washington. Available at http://blast.wustl.edu.
21. Birney, E., Andrews, T. D., Bevan, P., Caccamo, M., Chen, Y., Clarke, L., Coates, G., Cuff, J., Curwen, V., Cutts, T., Down, T., Eyras, E., Fernandez-Suarez, X. M., Gane, P., Gibbins, B., Gilbert, J., Hammond, M., Hotz, H. R., Iyer, V., Jekosch, K., Kahari, A., Kasprzyk, A., Keefe, D., Keenan, S., Lehvaslaiho, H., McVicker, G., Melsopp, C., Meidl, P., Mongin, E., Pettett, R., Potter, S., Proctor, G., Rae, M., Searle, S., Slater, G., Smedley, D., Smith, J., Spooner, W., Stabenau, A., Stalker, J., Storey, R., Ureta-Vidal, A., Woodwark, K. C., Cameron, G., Durbin, R., Cox, A., Hubbard, T., and Clamp, M. (2004) An overview of Ensembl. *Genome Res* 14, 925–8.

11

Leveraging Model Legume Information to Find Candidate Genes for Soybean Sudden Death Syndrome Using the Legume Information System

Michael D. Gonzales, Kamal Gajendran, Andrew D. Farmer, Eric Archuleta, and William D. Beavis

Summary

Comparative genomics is an emerging and powerful approach to achieve crop improvement. Using comparative genomics, information from model plant species can accelerate the discovery of genes responsible for disease and pest resistance, tolerance to plant stresses such as drought, and enhanced nutritional value including production of anti-oxidants and anti-cancer compounds. We demonstrate here how to use the Legume Information System for a comparative genomics study, leveraging genomic information from *Medicago truncatula* (barrel medic), the model legume, to find candidate genes involved with sudden death syndrome (SDS) in *Glycine max* (soybean). Specifically, genetic maps, physical maps, and annotated tentative consensus and expressed sequence tag (EST) sequences from *G. max* and *M. truncatula* can be compared. In addition, the recently published *M. truncatula* genomic sequences can be used to identify *M. truncatula* candidate genes in a genomic region syntenic to a quantitative trait loci region for SDS in soybean. Genomic sequences of candidate genes from *M. truncatula* can then be used to identify ESTs with sequence similarities from soybean for primer design and cloning of potential soybean disease causing alleles.

Key Words: Legume Information System; Comparative genomics; Sudden death syndrome; Quantitative trait loci; Genetic maps; Physical maps; Linkage maps; Model legumes; Synteny; Candidate genes.

From: *Methods in Molecular Biology, vol. 406: Plant Bioinformatics: Methods and Protocols*
Edited by: D. Edwards © Humana Press Inc., Totowa, NJ

1. Introduction

A fully sequenced and annotated genome accelerates the identification of candidate genetic loci underlying phenotypes of interest. But, what can be done if the genome from a species of interest is not sequenced, nor is it likely to be sequenced in the near future? Because sequence and function of genes are largely conserved among related species, comparative genomics can leverage information and knowledge gained from a sequenced model, or reference species, to make hypotheses about the relationship between genotype and phenotype for related species.

Legumes (soybeans, dry beans, peas, alfalfa, peanuts, etc.) are important sources of proteins, oils, anti-oxidants, and anti-cancer compounds, and provide organic sources of nitrogen fertilizer. Unfortunately, most crop legumes have not been sequenced because their large and complex polyploid genomes make genome sequencing cost prohibitive. Fortunately, several legume species including *M. truncatula* (barrel medic), *Lotus japonicus* (Japanese lotus), and *Phaseolus vulgaris* (dry beans) have relatively small and tractable genomes. A genome sequencing project (http://www.medicago.org), supported by the Noble Foundation and NSF-PGRP, has produced annotated genomic sequence for most of *M. truncatula* gene space. Thus, in regions of the genome where syntenic relationships exist between barrel medic and a crop legume, the annotated genomic sequence from barrel medic can be leveraged to nominate or identify candidate genes of interest in other legumes.

As an example, we will use the Legume Information System (LIS; http://www.comparative-legumes.org), a publicly accessible, clade information resource that integrates genetic and molecular data from multiple legume species to conduct cross-species genomic and transcript comparisons and identify candidate genes *(1)*. Our goal for this tutorial is to find candidate genes for sudden death syndrome (SDS) in soybean. SDS, caused by *Fusarium solani f. sp. glycines*, creates toxins in the roots resulting in root rot and leaf scorch, severely reducing soybean production each year. SDS is a major concern and has become the focus for breeders and scientists interested in producing a more resistant soybean plant. Quantitative trait loci (QTL) for SDS in soybean have been previously identified and mapped *(2)*. The goal of QTL studies is to identify genomic regions that are statistically associated with variation in complex quantitative traits such as SDS resistance. Once QTL regions have been located, the actual genetic elements responsible for the phenotype can perhaps be identified.

To describe our approach for finding candidate genes, we will begin by using the CMap module of LIS to query and display soybean SDS QTLs on genetic maps. Next, the soybean linkage maps are compared with *M. truncatula* maps to identify syntenic regions containing SDS QTLs. Once genomic markers in *M. truncatula* have been identified as syntenic to the SDS QTL region, we utilize the *M. truncatula* physical maps to identify the sequenced genomic clones in the same regions. These genomic sequences within the physical region are then analyzed for candidate genes using annotations displayed in the LIS Comparative Functional Genomics Browser (CFGB). Finally, consensus sequences aligned to genomic sequence can be analyzed using the existing annotations to isolate soybean expressed sequence tag (EST) sequences, follow on primer design, or further analysis.

2. Materials

Clade-oriented Web-based information resources, such as LIS, offer both data and applications. Database content can consist of raw experimental data but also often consists of information resulting from preliminary analyses, such as computationally generated sequence annotation and results of QTL analyses. Available applications are generally tools for further analysis and visualization of data and information. All procedures described here will use the LIS Web site available at http://www.comparative-legumes.org. A high-speed connection is recommended. The site has been optimized to work using Netscape 7.x and Internet Explorer 5.x for Windows as well as Netscape 7.x for Macintosh. This tutorial was based on analysis done in October 2005. Annotated data available in LIS was last updated in fall of 2005 using the Genome Initiative for species X (XGI) pipeline (v. 2.0). As LIS is updated with new data, results displayed as part of this tutorial may change.

2.1. LIS Overview

LIS integrates map, genomic, and transcript data from a number of databases and allows researchers to access and compare data through a single, but multifaceted, Web interface. The LIS database content and applications that we will use includes the XGI transcript and genomic databases, CMap and SoyBase. All publicly available transcript and genomic data from *M. truncatula*, *L. japonicus*, *Glycine max*, and *Arabidopsis thaliana* have been analyzed by a variety of computational annotation algorithms (described in detail below) and stored using NCGR's XGI system (http://www.ncgr.org/xgi). The XGI genomic pipeline (XGI-g) analyzes genomic sequence data for each species, and the results can be used in cross-species comparisons. Cross-legume analyses include alignment of gene sequences to genomic contigs to validate *ab initio* gene predictions. Transcript sequences of ESTs from all available legumes are analyzed, annotated, and stored using the XGI transcript pipeline (XGI-t). CMap, developed as part of the Generic Model Organism Database project (http://www.gmod.org/cmap/index.shtml), has been incorporated into LIS to provide access to genetic and physical maps from all legume species. SoyBase (http://www.soybase.org) provides soybean map and biochemical pathway data.

2.2. Genetic and Physical Maps

2.2.1. Map Data

All curated linkage maps from SoyBase have been incorporated into the CMap module of LIS. CMap provides researchers access to curated genetic and physical maps for *Glycine max* and *M. truncatula*, as well as genetic maps for *Medicago sativa* (alfalfa), *P. vulgaris* (dry beans), and *Arachis hypogaea* (peanut). LIS will soon incorporate genetic and physical maps for peas, lentils, and other legume species as they become available. At this time, correspondences between marker loci on different maps are based on curated name matches (provided by Dr. David Grant, USDA-ARS, Ames, Iowa), taking into account the possibility of the same marker being mapped onto multiple loci within a map set for polyploid species. Marker comparisons are possible when legume genetic researchers adhere to nomenclature standards across species for mapped loci, thus further enabling identification of syntenic regions and comparative genomics.

2.2.2. Capabilities of the CMap Researcher Interface

CMap features positioned on the maps are displayed using different symbols and colors to represent the various feature types, for example, the QTL data is color coded according to classification of QTL traits. When using the tool for comparative work, the researcher should choose a map from the database to be used as a reference map for the comparison. Comparative maps can then be added to the viewer so that alignments relative to the reference map can be investigated. The compared maps are selected from a list of all maps in the database and allow the researcher to specify the number of correspondences to the reference map. Alternatively, a comparison matrix can be used to display the number of correspondences between different maps and map sets in the database, and can be used to locate maps with high levels of synteny. These features allow researchers to compare linkage groups within species, a useful capability for these polyploid species, and among species, an essential capability for identification of syntenic regions. Lines indicating relationships between features, for example, loci or QTL, are drawn between maps in a comparison. Feature details and map set information can be accessed from the "map view". Map set details include species, map type, map units, curator remarks, and the listing of the maps in the set. Feature details include the feature name, feature type, aliases or synonyms, map position, cross-references to other databases, as well as correspondence details from other maps associated with the feature. It is also possible to change the size of map images and save them for later research sessions.

2.3. Genomic Sequence Data

LIS genomic data consist of genomic sequences that have been analyzed, annotated, and stored using XGI-g. Public sequence information is gathered from the NCBI's high-throughput genomic (HTG) division *(3)* for species of interest (see Chapters 2 and 3). HTG sequences from large-scale genome sequencing centers are submitted as in-process assemblies in various stages of completeness, often containing two or more contigs. LIS does not assemble the genomic data but takes data from NCBI HTG as submitted by the sequencing centers. Each genomic sequence is then separated into its constituent contigs and analyzed using a sliding window (when appropriate to the given analysis) of length 10,000 bp with an overlap of 3000 bp. These nucleotide segments are automatically annotated using a computational pipeline that runs a series of sequence or motif similarity computational algorithms. The XGI pipeline compares nucleotide segments using BLASTX (v. 2.1.3) *(3–5)* against NCBI's nr database *(3)* and with BLASTN (v. 2.1.3) *(4,5)* and tBLASTX (v. 2.1.3) *(4,5)* against the consensus sequences produced in the LIS transcript database. BlimpSearcher (v. 3.5) analysis *(6)* against the Blocks + database *(7,8)* is used to identify protein motifs and families. InterProScan (v. 3.1) *(9)* integrates results from a variety of protein motif analysis tools using the InterPro database *(10)*. Consensus or equivalent analysis results between overlapping pieces are merged before being stored in the LIS database. GenScan (v. 1.0) *(11)* performs *ab initio* gene prediction on the genomic sequences and is used to evaluate complete contig sequences and to define and compare the genes and exon–intron organization of the sequences. The results of the genomic pipeline are stored in the LIS genomic database and are updated on a regular basis.

2.3.1. Capabilities of the Comparative Functional Genomics Browser (CFGB)

The CFGB is an application that allows the researcher to visualize genomic sequence annotations. It also enables visualization of comparative transcript data aligned to genomic contigs for purposes of validating gene predictions.

The CFGB gives researchers the freedom to add multiple contigs to the viewer allowing for comparative analysis between sequences. By using description matches of the different analysis types, contigs can be compared for regions with similar annotation. Each genomic sequence has been annotated using the XGI-g and represents the annotation types as colored blocks. The location and direction of transcript in the colored blocks are represented in relation to the genomic sequence. The size of the images in the CFGB can be manipulated with zooming, panning, and sorting functions. The CFGB also supports the ability to change the aspect ratios for all sequences at the same time.

2.4. Transcript Data

The LIS transcript database currently consists of EST and consensus sequences for *M. truncatula*, *G. max*, *L. japonicus*, and *A. thaliana*. Using the XGI-t, raw public EST data are gathered from NCBI (http://www.ncbi.nlm.nih.gov/dbEST/) and annotated automatically. Where available, quality scores for EST sequences are incorporated into the database for use in subsequent analyses. Detailed metadata concerning sequence origin, such as submitting organization, organism, clonal library, and methodology, are captured in the database and are viewable with the LIS interface. Libraries are also categorized by a manual curation process for more accurate querying through the interface.

The XGI-t process begins by screening raw EST data for quality. Screening operations include removal of most common vector sequences, poly (A/T) trimming, N trimming, adapter/linker removal, length trimming, and poor quality read trimming. Vector screening and adapter/linker screening removes sequence contamination of the insert that typically arises as part of the cloning process. In addition, the fidelity of a sequence read typically degenerates toward the end of the sequence, resulting in errors in base calling, which are trimmed out as part of this process. Finally, low-complexity sequences represented by polyadenylated regions are removed because such sequences can produce many false-positive matches in subsequent analyses. The end result of the quality screens is a high-quality "approved" sequence that is then deposited in the database. An EST that has failed the XGI vector screen analysis for one or more reasons is not included in subsequent analyses, but may still be inspected through the interface as a failed EST. Approved EST data are clustered using Phrap *(12)*, which performs clustering and contig assembly to produce "consensus" sequences. These LIS consensus sequences are used in aggregation of the high-quality sequence information of member ESTs and are used in all subsequent analyses. Consensus sequences are analyzed using

NCBI's BLASTX (v. 2.1.3) algorithm *(3–5)* to search for potential homologs against NCBI's nr database *(3)*. Blimp-Searcher (v. 3.2) *(6)* and InterProScan (v. 3.1) *(9)* are used as previously described in **Subsection 2.3**. Each of these analyses is followed by the association of gene ontology (GO) terms *(13)* with the computationally generated sequence annotation to further annotate the consensus sequences with potentially useful information. Pexfinder (v. 1.0) *(14)*, co-developed by NCGR and OSU-OARDC, based on Signal P (v. 3.0) *(15)*, has also been incorporated into the transcript analysis pipeline. Pexfinder (Protein excreted) predicts proteins excreted through the plasma membrane, based on signal peptides. The results of the pipeline analyses are stored in the LIS transcript database and are updated periodically depending on the number of publicly available sequences for analysis.

2.5. Capabilities of the Features and Annotations Viewer

The Features and Annotation (F&A) viewer displays all the meta-data and annotations for a given LIS consensus sequence. Links to sequence details, multiple sequence alignments, EST membership data, as well as library and organism metadata are provided as well. The F&A module also gives a graphical presentation of the annotations linked directly to GO where appropriate. In addition, output from the annotation details can be viewed by following the appropriate links.

3. Methods

3.1. Search and Display QTL for SDS Using LIS CMap

To compare maps, we must first find and display a reference map. To begin, search for the SDS QTL feature in the database. Using the CMap module, features are any elements that can be placed on a map, either as a point or as an interval (*see* **Note 1**). Once the QTL has been found, its corresponding genetic map can be displayed.

1. Open http://www.comparative-legumes.org in a Web browser and select "MAPS" from the navigation menu.
2. On the resulting page, select "Search".
3. Enter SDS* in the "Feature Names" text box. The asterisk (*) and percent (%) signs can be used as wildcard characters. Using wildcards, we ensure that all QTLs with the name SDS are found, for example, SDS 1-1, SDS 8-2, and so on (*see* **Note 2**).
4. Next, restrict species to "Soybean" (*see* **Note 3**).
5. Select the Submit button. The query should retrieve a number of results.
6. For this example, we will examine SDS 8-2 located on the 2003 Composite Genetic Map, linkage map C2. In CMap, a map is represented as a linear arrangement

of interconnected features. This is usually a single linkage group in the case of a genetic map. Related maps are grouped into map sets. Generally, these are the result of a particular study, such as the set of linkage groups produced by a genetic mapping study. The 2003 Composite Genetic Map consists of 20 linkage groups constructed by Cregan et al. *(16)* using JoinMap *(17)* and data from segregating progeny of *Glycine max* A81-356022 × *Glycine soja* PI468916, *Minsoy* × *Noir1*, *Minsoy* × *Archer*, *Noir1* × *Archer*, and *Clark* × *Harosoy*. This is the most recent soybean composite map available (*see* **Note 4**). By default, the results are sorted by QTL name. To find SDS 8-2, you can either page through the results by selecting "Next", or to find the QTL most easily, sort the results by linkage map by selecting "Map Name".

7. Once you have found SDS 8-2, select "Feature Details". The Feature Details page provides helpful information about the QTL such as aliases, start and stop positions, accession ID, map information, correspondence details, as well as a cross-reference to the curated data at Soybase. By selecting the "View QTL data at Soybase", you will find more attribute information such as heritability, references to curated papers, parents, sample size, and type of segregating progeny.

8. After identifying and selecting QTL, you will want to view the QTL on a linkage map with the "View on Map" tool. The map view for C2 shows that SDS 8-2 lies in a feature-rich region between marker loci K418 and Satt460 (*see* **Fig. 1**). Different feature types are represented by different shapes [such as horizontal tick marks (for points), line intervals, boxes, arrows, etc.] or different colors. For more information on any feature, click on it to view the corresponding feature detail page (*see* **Note 5**).

3.2. Comparing Maps

Now that a reference map has been selected and displayed, it serves as the basis for comparisons with maps from other species. Comparative maps may be added as vertically represented maps to both the left and the right of the reference map. The researcher may keep adding additional maps as long as curated comparisons are available. We will compare the C2 linkage map of soybean to genetic maps in *M. truncatula*. We are looking for regions of *M. truncatula* that are syntenic to the SDS 8-2 regions in the reference soybean map.

1. The C2 map will be the reference map. In the "Show Comparison Menu", choose the drop-down list for Comparative Maps (right): This will display the new map to the right of the reference map.

2. From the drop-down list, select the genetic: barrel medic—Young (U. Minn) 2004 map [7]. The number in brackets represents the number of correspondences to the reference map. Correspondences are the number of feature matches between the reference map and the comparative map.

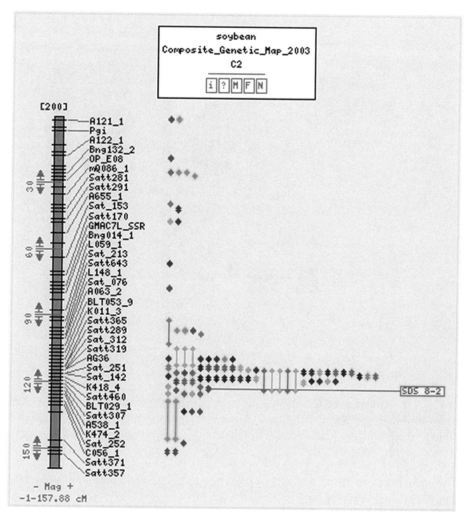

Fig. 1. The soybean linkage map C2 from the 2003 soybean composite map is visualized using CMap. The quantitative trait loci for SDS 8-2 is found in a feature-rich region at about 120 cM.

3. From the new list, select 4 [2] (*see* **Note 6**).
4. Select the "Redraw Map" button.

The comparison between the soybean C2 linkage map and *M. truncatula* linkage map 4 shows a syntenic region between K365 and A538 (*see* **Fig. 2**).

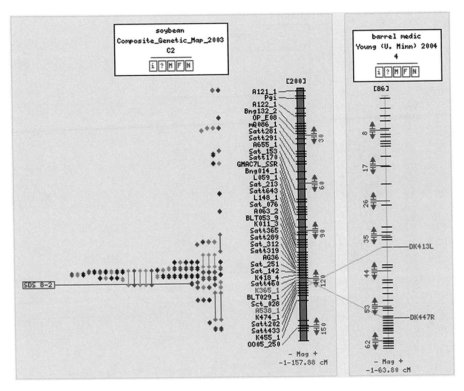

Fig. 2. Comparing the C2 linkage map from soybean against linkage map 4 from barrel medic shows syntenic regions between markers DK413L and DK447R. In this region, where syntenic relationships exist, potential candidate genes for SDS 8-2 exist.

The area between these markers represents a region with potential annotated candidate genes (*see* **Note 7**).

3.3. Position Information to Genomic Sequence

We will now use *M. truncatula* physical maps to relate the position of the genetic markers to the genomic sequence.

1. To make the map easier to read, flip the *M. truncatula* linkage map by selecting the "F" located under the map label (*see* **Note 5**).
2. Select "Show Comparison Menu" and choose the drop down-list for Comparative Maps (right).
3. From the drop-down list, select physical: barrel medic—Cook/Kim MtGenome v3 (UC Davis) [70].
4. Select 1090 [2].

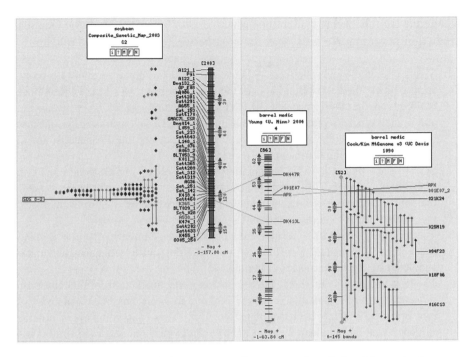

Fig. 3. In the regions of the genome where syntenic relationships exist between barrel medic and soybean, the annotated genomic sequence from barrel medic can be leveraged to identify candidate genes of interest.

5. Select the "Redraw Map" button. The *M. truncatula* genetic map and the *M. truncatula* physical map show correspondence between the ascorbate peroxidase (APX) feature in the region we determined above (*see* **Fig. 3**). APX is actually a rather interesting candidate gene, when we consider APX's function in plants. Ascorbate is essential to maintaining the anti-oxidant system that protects plants from oxidative damage due to biotic and abiotic stresses *(17)*.

6. Looking at the physical map in this region, we see that there is a genomic clone that is in phase 3 of sequencing.

7. Click on the 021K24 clone (in blue on the Web page) to view the Feature Details for this phase 3 clone.

8. Select "View annotated clone at LIS" to view the annotated genomic sequence.

3.4. Finding Candidate Genes Within Annotated Genomic Sequences

The genomic sequence has been annotated using the LIS genomic pipeline and will provide assistance in selecting candidate genes. The CFGB tool is used to visualize the genomic annotations. The regions of interest are areas

where hits from BLAST results line up with exon predictions from the GenScan
prediction program. We will view the area between 92,000 and 95,000 bp.

1. Scroll over the blocks between 92,000 and 95,000. Move your mouse over a
   feature block, the description/name of the feature pops up (*see* **Fig. 4**). Red blocks
   (colored on the Web site) are BLASTX hits against NCBI nr, whereas pink
   and orange blocks represent tBLASTX and tBLASTn hits against LIS consensus
   sequence libraries. The various shades of green represent InterPro results, and the
   brown blocks are the GenScan hits (*see* **Note 8**).
2. Scroll over the tBLASTX annotations (pink boxes) until you find Gm_
   014_240664_Apr04. This set of boxes represents a soybean consensus sequence
   that has been annotated using sequence comparisons as a peroxisomal APX
   (*see* **Note 9**). Click on the pink box for Gm_ 014_240664_Apr04.
3. The resulting tBLASTX results list the sequence similarity to the genomic
   sequence. Click on the hyperlinks to view the consensus sequence information
   annotated by the LIS transcript pipeline.
4. The resulting page is the F&A Page. Looking at the annotation for this consensus
   sequence confirms that this sequence is from soybean, has been isolated from
   root libraries, and has been annotated as APX. Thus, this comparison provides
   evidence that our candidate gene in *M. truncatula* has high sequence similarity to
   an annotated consensus sequence, expressed in soybean.

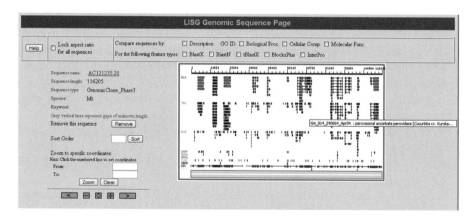

Fig. 4. The Legume Information System Comparative Functional Genomics Browser
allows the visualization of genomic analysis results, including comparative transcript
data, with the gene sequences aligned to genomic contigs to validate gene predictions.
The regions of interest are areas where hits from BLAST results line up with exon
predictions from the GenScan prediction program.

5. Now that a candidate gene has been selected, we want to find the EST sequence for the gene. To do so, select "Show Membership" to get the list of EST members used to create this consensus sequence.
6. The resulting page lists all ESTs used in the assembly of the consensus sequence. Select "sak50g04.y1" under Sequence Name to view the EST details. The sequence detail views capture all information relevant at the individual sequence level, and other information can be accessed through this display. This includes quality trimming, base composition, as well as sequence metadata and clustering information.
7. To save the EST sequence, select "FASTA" from the drop-down list.
8. Next, select the "Download" Button. Warning! The sequence is not saved automatically.
9. To save the sequence, you must select File > Save Page As.
10. Select the "Save" button. Congratulations! You now have the EST sequence that should provide a template to create primers for use in identifying polymorphic alleles. As a result, you can screen soybean cultivars for this candidate gene or use the primers for use in molecular mechanism studies.

4. Notes

1. For a complete set of features in CMap, please see http://www.comparative-legumes.org/cgi-bin/cmap/feature_type_info.
2. If you are searching for multiple names, separate them with commas or white space. To find features with spaces in the name, surround the name in double quotes, for example, "abc 123".
3. We did not select "Restrict Feature Type" by default, ALL feature types will be searched. As there are a number of possible QTL feature types, it is recommended that you query against ALL data types unless you are certain, to improve your chances of finding the Feature Name.
4. As was done for the 1999 composite maps (*17*), the SoyBase staff interpolated markers from all other published mapping studies onto each linkage group using a proportional relationship between anchor loci (i.e., loci in common between the Cregan maps and the other population). Although this method allows the inclusion of loci from many different mapping studies for which the original segregation data are not available, it results in maps where the exact order of closely spaced loci may be incorrect. For this reason, order and genetic distances between closely spaced loci should be considered approximate rather than exact.
5. CMap allows researchers a number of functions to improve map layouts. For general help using the CMap interface, please see http://www.comparative-legumes.org/cgi-bin/cmap/help?section=map_viewer.
6. As our goal is to find syntenic regions between maps, it is best to select maps with more than one correspondence.

7. You may notice the corresponding features (highlighted in red and connected by a light blue line on the Web page) from soybean do not match those of *M. truncatula*. This is because these features have more than one alias.

8. For general help and options using the CFGB, please see http://www.comparative-legumes.org/lis/lis_help.html.

9. tBLASTX compares the six-frame translations of the genomic sequence against the six-frame translations of the LIS consensus sequence data set.

Acknowledgments

We thank David Grant for his careful curation of map data. This work is supported by USDA-ARS Specific Cooperative Agreement no. 3625-21000-038-01.

References

1. Gonzales, M.D., Archuleta, E., Farmer, A., Gajendran, K., Grant, D., Shoemaker, R., Beavis, W.D., and Waugh, M.E. (2005) The Legume Information System (LIS): an integrated information resource for comparative legume biology. *Nucleic Acids Res* 33, D660–D665.

2. Njiti, V.N., Meksem, K., Iqbal, M.J., Johnson, J.E., Kassem, M.A., Zobrist, K.F., Kilo, V.Y., and Lightfoot, D.A. (2002) Common loci underlie field resistance to soybean sudden death syndrome in Forrest, Pyramid, Essex, and Douglas. *Theor Appl Genet* 104, 294–300.

3. Benson, D.A., Karsch-Mizrachi, I., Lipman, D.J., Ostell, J., and Wheeler, D.L. (2004) GenBank: update. *Nucleic Acids Res* 32, D23–D26.

4. Altschul, S.F., Gish, W., Miller, W., Myers, E.W., and Lipman, D.J. (1990) Basic local alignment search tool. *J Mol Biol* 215, 403–410.

5. Altschul, S.F., Madden, T.L., Schaffer, A.A., Zhang, J., Miller, W., and Lipman, D.J. (1997) Gapped BLAST and PSI-BLAST: a new generation of protein database search programs. *Nucleic Acids Res* 25, 3389–3402.

6. Henikoff, S. and Henikoff, J.G. (1994) Protein family classification based on searching a database of blocks. *Genomics* 19, 97–107.

7. Henikoff, J.G., Greene, E.A., Pietrokovski, S., and Henikoff, S. (2000) Increased coverage of protein families with the blocks database servers. *Nucleic Acids Res* 28, 228–230.

8. Henikoff, S., Henikoff, J.G., and Pietrokovski, S. (1999) Blocks+: a non-redundant database of protein alignment blocks derived from multiple compilations. *Bioinformatics* 15, 471–479.

9. Zdobnov, E.M. and Apweiler, R. (2001) InterProScan–an integration platform for the signature-recognition methods in InterPro. *Bioinformatics* 17, 847–848.

10. Mulder, N.J., Apweiler, R., Attwood, T.K., Bairoch, A., Barrell, D., Bateman, A., Binns, D., Biswas, M., Bradley, P., Bork, P. et al. (2003) The InterPro database, 2003 brings increased coverage and new features. *Nucleic Acids Res* 31, 315–318.

11. Burge, C. and Karlin, S. (1997) Prediction of complete gene structures in human genomic DNA. *J Mol Biol* 268, 78–94.
12. Green, P. (1993) Laboratory of Phil Green, University of Washington. Available at http://www.phrap.org.
13. Ashburner, M., Ball, C.A., Blake, J.A., Botstein, D., Butler, H., Cherry, J.M., Davis, A.P., Dolinski, K., Dwight, S.S., Eppig, J.T. et al. (2000) Gene ontology: tool for the unification of biology. The Gene Ontology Consortium. *Nat Genet* 25, 25–29.
14. Torto, T.A., Li, S., Styer, A., Huitema, E., Testa, A., Gow, N.A., van West, P., and Kamoun, S. (2003) EST mining and functional expression assays identify extracellular effector proteins from the plant pathogen *Phytophthora*. *Genome Res* 13, 1675–1685.
15. Bendtsen, J.D., Nielsen, H., von Heijne, G. and Brunak, S. (2004) Improved prediction of signal peptides: SignalP 3.0. *J Mol Biol* 340, 783–795.
16. Song, Q.J., Marek, L.F., Shoemaker, R.C., Lark, K.G., Concibido, V.C., Delannay, X., Specht, J.E., and Cregan, P.B. (2004) A new integrated genetic linkage map of the soybean. *Theor Appl Genet* 109, 122–128.
17. Stam, P. (1993) Construction of integrated genetic linkage maps by means of a new computer package: JoinMap. *Plant J* 3, 739–744.

12

Legume Resources: MtDB and Medicago.Org

Ernest F. Retzel, James E. Johnson, John A. Crow, Anne F. Lamblin, and Charles E. Paule

Summary

To identify the genes and gene functions that underlie key aspects of legume biology, researchers have selected the cool season legume *Medicago truncatula* as a model system for legume research. The mission of the *M. truncatula* Consortium is to promote unrestricted sharing of data and information that are provided by Medicago research groups worldwide. Through integration of a variety of data and tools, the medicago.org site intends to facilitate progress in the fields of structural, comparative, and functional genomics. To this goal, and as a consortium partner, the Center for Computational Genomics and Bioinformatics (CCGB) at the University of Minnesota has developed MtDB2.0, the *M. truncatula* database version 2.0. The MtDB2.0 database is the first step toward the global integration of *M. truncatula* genomic, genetic, and biological information. MtDB2.0 is a relational database that integrates *M. truncatula* transcriptome data and provides a wide range of user-defined data mining options. The database is interrogated through a series of interfaces, with 58 options grouped into two filters. Sequence identifiers from all public *M. truncatula* sites [e.g., IDs from GenBank, CCGB, The Institute for Genomic Research (TIGR), National Center for Genome Resources (NCGR), and l'Institut National de la Recherche Agronomique (INRA)] are fully cross-referenced to facilitate comparisons between different sites, and hypertext links to the appropriate database records are provided for all queries' results. MtDB's goal is to provide researchers with the means to quickly and independently identify sequences that match specific research interests based on user-defined criteria. MtDB2.0 offers unrestricted access to advanced and powerful querying tools unmatched by any other public databases. Structurad Query Language (SQL)-encoded queries with a Java-based Web user interface, incorporate different filtering that allow sophisticated data mining of the expressed sequence tag sequencing project results, including the CCGB *M. truncatula* Unigene set generated with the Phrap assembler. The underlying database and query software

From: *Methods in Molecular Biology, vol. 406: Plant Bioinformatics: Methods and Protocols*
Edited by: D. Edwards © Humana Press Inc., Totowa, NJ

have been designed for ease of updates and portability to other model organisms. Public access to the database is at http://www.medicago.org/MtDB.

Key Words: *Medicago truncatula*; MtDB; legume; database; genome analysis.

1. Introduction

The legumes are a very large family of plants. Included in the thousands of species of legumes are agriculturally critical crop plants, specifically soybeans, common beans, alfalfa, peanuts, peas, cowpeas, pigeonpeas, lentils, as well as the model organisms of *Medicago truncatula* and *Lotus japonicus*. Legumes as agricultural commodities have dramatic value, both as cash crops ($18 billion annually, US market), and in their contribution to atmospheric nitrogen fixation ($8 billion annually in the United States). Nitrogen fixation has importance both in human and animal nutrition, in crop rotation, and in environmental issues. As a whole, legumes make a critical contribution to the human food supply (through soybeans, common beans, peas, chickpeas, lentils, and peanuts), to edible oils (through soybean and peanut), and to the animal food supply (through clover and alfalfa). Crop rotation with legumes and cereals decreases the need for the use of nitrogen-containing fertilizers. Internationally, the nitrogen fixation capabilities enable farming in areas where nitrogen-based fertilizers would be prohibitively expensive. In terms of developing a sustainable agricultural environment, one of the key factors is the protection of the environment, and legumes thus contribute to this effort. The interests of the legume community range from molecular biology through evolutionary development to breeding. This broad focus, and its significance in the public and private sector, is reflected in the size of the community, third only in size to the *Arabidopsis* and cereal communities.

The success of the plant genome projects in terms of delivering massive amounts of raw sequence is evident from the representation of plants in the national data resources. Full genome sequencing has been accomplished for *Arabidopsis (1)* and rice. Most important to legume research, the sequencing of the 200 megabases (Mb) "genespace" of *M. truncatula* has been undertaken in an international effort. In addition, the sequencing of a related model organism, *L. japonicus*, is underway in Asia, and has produced a comparable amount of genomic sequence data *(2)*. Within the plant gene-discovery [expressed sequence tag (EST)] projects, the legumes are particularly heavily represented (>700,000 ESTs) because of significant investment in EST and functional genomics projects in both the model organisms (*Medicago* and *Lotus*) as well as crop plants.

Despite the success of the genomics efforts in producing data, less attention has been paid to developing the bioinformatics and biocuration issues involved in data analysis, annotation, enhancement, integration, and distribution of this information. There are notable exceptions to this: The *Arabidopsis* Information Resource (TAIR, *see* chapter 8) *(3)* is a well-funded and robust user-managed effort addressing the issues of the *Arabidopsis* community and its full genome sequence. An additional resource that was developed by TAIR is AraCyc *(4,5)*, a pathway resource developed under the BioCyc model. Gramene *(6,7)* (*see* chapter 15) is the evolution of the RiceGenes database and is a forward-looking example of co-developing both a user resource and the strategic deployment of database inter-operability, making extensive use of XML, the Distributed Annotation System (DAS) *(8)*, and BioMOBY (http://biomoby.org) *(9)* efforts. The Legume Information System (LIS, *see* chapter 11) *(10)* has been an extension of the former SoyBase database, and it is in collaboration with this group that we will develop our efforts in annotation and resource integration. Our own projects at University of Minnesota [MtDB2.0, the *M. truncatula* database (http://medicago.org/MtDB)] *(11)* is another example of plant genome resources. MtDB2.0 and MtDB3.0 are Oracle-based systems primarily focused on information that can be derived from EST and Bacterial Artifical Chromosome (BAC)-end sequencing projects. The genome sequence project has begun to integrate other databases, using the MySQL-based Ensembl *(12,13)* as the primary automated annotation tool for genomic sequence data, and "decorating" this information by integrating it with the rich, user-oriented and problem-oriented, user interface we have developed for MtDB2.0 and other locally developed databases. This philosophy of adding value to information by further analysis and organization is key to this development.

1.1. Long-Term Goals

The long-term goal of this effort is to contribute substantial information resources to the Legume Information Network (LIN) in collaboration with the National Center for Genome Resources (NCGR) LIS. At this point in time, archiving data from the genome projects and providing genome browser access to the data is no longer sufficient. As has been demonstrated in other species sequencing projects (TAIR, Gramene), genomic data must be both automatically annotated and curated by an individual, group, or community. These annotations must be sufficiently fine-grained to provide a means of developing biological inferences, and most importantly, be accessible to both human users and other machine-based bioinformatic resources through Web services protocols (rather than Web interfaces).

2. Materials

2.1. The Database and the Nimbus Web Application

The design of the Nimbus Web application attempts to alleviate a couple of issues incurred in previous designs: frequent requests for changes in Web page appearance, and requests for variations on queries. It employs a tiered architecture and dynamically generated Web pages to address these concerns.

The server architecture separates database access and Web page presentation into separate software tiers. This allows a separation in development work between the database access aspects of the software from that of the page generation, freeing the Web page designer to tailor the presentation for specific clients while maintaining common database access code.

Web pages are generated dynamically with a template language. One of the templates developed for the Web application generates a query menu from a list of predefined queries. Another template constructs an HTML form for the selected query to allow the user to supply query input parameters. As the menu and the form are generated dynamically, it is easier to add new queries to the system.

2.2. Implementation

The Nimbus Web application uses several open source packages. It uses Apache Struts as the server architecture. It generates the Web pages using Apache's Velocity. Velocity template language was chosen because of its simple, easy to learn programming design. This permits someone with limited programming knowledge to be able to write and edit templates.

Database access is handled by the Hibernate object-relational mapping package. Hibernate turns the results of relational database queries into lists of software objects that capture the structure of the information. These resulting objects are passed along to the Velocity page generator where they are referenced by the page templates. For example, if the query returned a list of unigene objects, the template could specify what information about each unigene object to display: sequence, ESTs, BLAST hits, etc. that are accessible from the object.

Hibernate, Struts, and Velocity are all Java language packages. The Nimbus Web application software is built using Apache Ant, a commonly used build tool for Java. Ant reads build instructions from a file named build.xml. Two packages are generated for the Nimbus Web application: first, the database access package using Hibernate and second, the Web application that uses Struts and Velocity.

The database access code is generated by Ant using the Hibernate Tools reverse engineering tasks. These tools access the Nimbus database and generate Java classes that match the schema structure of the database. The Nimbus build utilizes special features of the Hibernate Tools that allow for custom naming of Java classes and their fields and methods, as well as some customized templates that add extra methods to the generated classes. The resulting classes are packaged as a jar archive that will be used by the Web application.

The Web application is a combination of velocity templates for generating Web pages and Java classes that implement the Struts actions for the Web server. The Ant build compiles the Java classes, which it packages along with

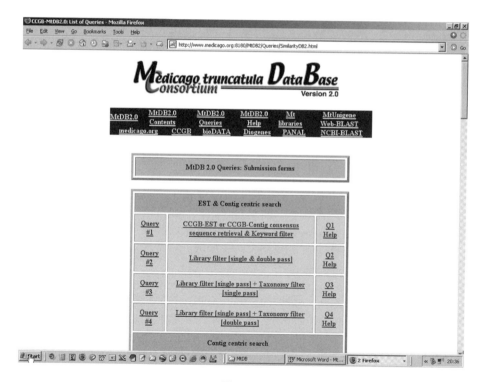

Fig. 1.

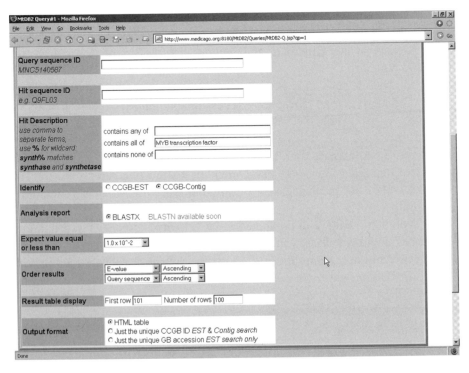

Fig. 2.

the templates and other supporting packages and configuration files into a Web archive file, namely nimbus.war. To deploy the Web service, the war file can be simply copied into the webapps directory of an Apache Tomcat Web server.

As the Nimbus Web application is currently deployed, much of the presentation can be changed without modifying the Java code but rather by editing text files: the style sheet, the Velocity templates, the Velocity macros file, and the ApplicationResources properties messages file.

The ApplicationResources.properties file contains a list of key-value properties. The templates have access to this message file through the $text variable and can request the message value for any defined key. As much displayed text as possible should be placed in that file and referenced by the templates. This allows for alternate languages, as multiple versions of the ApplicationResources.properties file can be included for other languages.

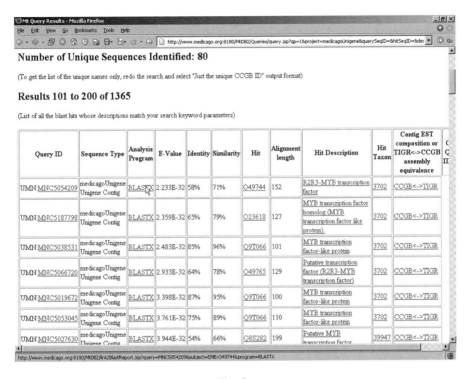

Fig. 3.

The templates have common display elements placed in the macros file, so that changes can be made in a single location. For example, the query menu might appear on every page, so the menu generation routine can be written as a macro to be placed in the macros file.

2.3. Distributed Databases: DAS and Semantic BioMoby

We have included the development of DAS, of Web services, and most recently MOBY services (*see* **Note 1**) and semantic MOBY (*see* **Note 2**). These three efforts represent differing methods of accomplishing inter-operability. If we had only established information (i.e., we were archiving data), we would likely not utilize all three of these methods. These represent three levels of complexity, the least difficult being DAS and the most difficult being semantic MOBY.

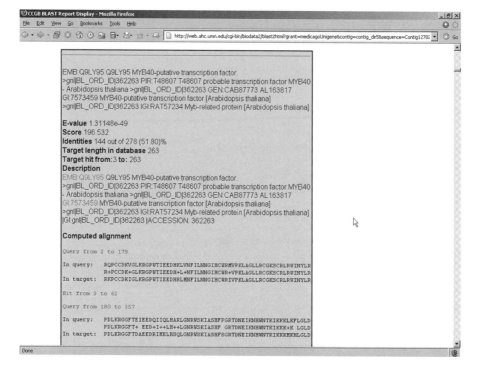

Fig. 4.

2.4. Limitations

The current design does not provide any support for paging lengthy tables of results. It was hoped that the Criteria interface in Hibernate might be used for that purpose, but the current Hibernate version does not support recursive subqueries.

Users can construct very complicated queries that can take a long time to retrieve. Ideally, the server log would record the queries, and the frequently used ones could be accommodated by defining a new query that would encompass that combination. In some cases, a materialized view should be constructed in the database schema to aggregate information such that it can be queried in a reasonable length of time.

3. Methods

Biological researchers expect to be able to access data from databases through a Web browser. Although many biological data are stored in relational databases, most of the people who wish to access the information do not have

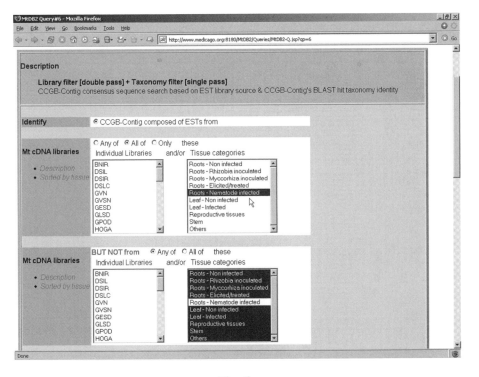

Fig. 5.

the expertise in database query languages to access the data directly. Thus, many database resources provide Web applications that allow the researchers to selectively access and present the database information through a Web browser.

The Nimbus Web application is designed to provide Web page access to the Nimbus database. It provides Web pages that display information about each of the Nimbus primary data types: Unigenes, ESTs, Clones, and Libraries (called by the more general term "CloneSet" in Nimbus). It also provides the means to specify which of the data to retrieve from the database for display.

The query menu presents the user with a list of predefined queries. The user can select one of these queries, and the Web application will build an input form to allow the user to enter input parameters for the query. The form also allows the user to combine the selected query with other prede-fined queries. As these queries are not built into the Web application, new queries may be added as needed without changing the Web application.

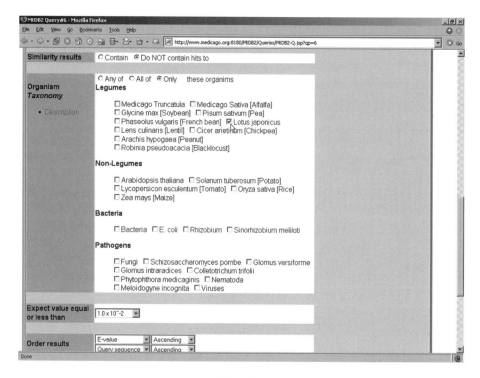

Fig. 6.

An administrator, knowledgeable in the Nimbus schema and database query languages, can define a new query and add it to the list that the template accesses.

The predefined queries are categorized by the type of data they return. For example, a number of queries may return the IDs of unigenes. The results of these queries can be combined using set operations: and, or, difference. Furthermore, the input parameters to the queries are similarly categorized so that the output of a query can be used as input to the parameter of another query.

3.1. Identification of Genes that have Specific Annotation

In this case, we will search for MYB transcription factors.

Starting from the http://www.medicago.org/MtDB/, click the link to access Mt2.0. This opens a new page with a list of search options (*see* **Fig. 1**). In this case, we want to search using a keyword filter (Query 1), using the terms MYB transcription factor (*see* **Fig. 2** and **Note 3**).

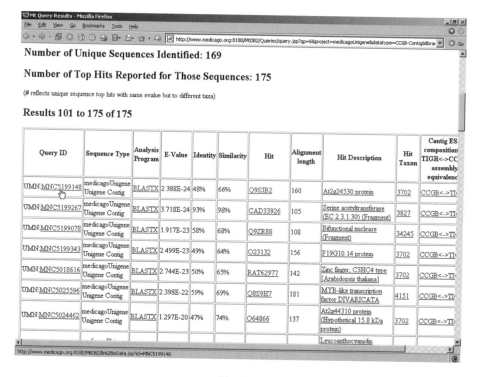

Fig. 7.

A total of 80 unigenes were identified (*see* **Fig. 3**). To view the annotation for these unigenes and compare the BLAST alignment used for the annotation, click on the BLASTX link. The resulting page (*see* **Fig. 4**) displays a long list of BLAST annotation for the identified gene.

3.2. Identification of Genes by Presence or Absence in Specific cDNA Libraries and are Likely to be Specific to Defined Species

In this example, we identify genes that are specific to the nematode root-infected cDNA library and that have not been identified in *L. japonicus*. We will use the CCGB-Contig consensus sequence search based on EST library source and CCGB-Contig's BLAST hit taxonomy identity, Query 6, "Library filter [double pass] + Taxonomy filter [single pass]". **Figure 5** shows selection for the nematode-infected root library and exclusion of the remaining libraries, whereas **Fig. 6** shows the exclusion of sequences matching *L. japonicus* sequences. A total of 169 unique sequences were identified using this method (*see* **Fig. 7**).

These sequences may be downloaded for further analysis or their annotation browsed over the Web.

4. Notes

1. DAS *(8)* is a relatively simple mechanism by which annotations can be shared across Web-based resources. It is a well-established, if presently somewhat limited mechanism for accomplishing this. We utilize this standard in part because this is well-supported under Ensembl. To provide DAS-enabled data, it is only necessary to conform to DAS XML standards, to register your server with DAS Central, and to provide a viewer of this data. The advantage of this for the biological community is that they can place their data in the context of a larger genome project and visualize the results. The *Medicago* Ensembl instance is registered as a DAS reference server and is presently utilized by several external laboratories in the public and private sector to give context to their projects. Although the browser provides an excellent view of the genomic data, gene calls, and similarity results, we believe that in the future, the DAS server (as well as the database API previously discussed) will be of considerable value to the community. What the DAS server allows a user to do is take a simple result set (e.g., a small EST project), developing the similarity results to our target *Medicago* database and marking up the results using the DAS XML standards. Users can not only get their results, but see the visual results in the context of the project without having to develop those views locally.

2. Semantic MOBY is a higher level of complexity from DAS. Semantic MOBY is built on the semantic Web design elements defined by the World Wide Web Consortium (W3C) and provides a rich mechanism for registering both object hierarchies and service hierarchies through MOBY Central. Semantic MOBY utilizes a combination of Web services, W3C-standard Web Ontology Language (OWL), OWL-DL (Description Logic), and Resource Description Framework (RDF). These standards allow the development of RDF graphs that are completely self-describing in their object content and their services. We are presently developing a project with the group at LIS to provide the first implementation of semantic MOBY for the legumes. This project, known as LIN, will first deliver semantic MOBY implementations of legume orthologs. The discovery server essential for this project will be maintained at NCGR and will be incorporated into the NSF-funded Virtual Plant Information Network. The semantic MOBY and LIN projects are described in somewhat more detail at http://services.ccgb.umn.edu.

3. Additional options allow you to refine the search, select against the presence of words, choose whether to search against annotated ESTs or assembled contigs, and order or format the results.

Acknowledgments

This database has been supported by the USDA-CSREES under the National Research Initiative, by several grants from the National Science Foundation under the Plant Genome Research Program and the Biological Databases and Informatics program.

References

1. The *Arabidopsis* Genome Initiative. (2000) Analysis of the genome sequence of the flowering plant *Arabidopsis thaliana*. *Nature* 408, 796–815.
2. Nakamura, Y., Asamizu, E., Kaneko, T., Kato, T., Sato, S., Tabata, S. (2006) A large scale genome analysis of a model legume, *Lotus japonicus*. *Plant Cell Physiol* 47, S84–S84.
3. Rhee, S.Y., Beavis, W., Berardini, T.Z., Chen, G., Dixon, D., Doyle, A., Garcia-Hernandez, M., Huala, E., Lander, G., Montoya, M., Miller, N., Mueller, L.A., Mundodi, S., Reiser, L., Tacklind, J., Weems, D.C., Wu, Y., Xu, I., Yoo, D., Yoon, J., Zhang, P. (2003). The *Arabidopsis* Information Resource (TAIR): a model organism database providing a centralized, curated gateway to *Arabidopsis* biology, research materials and community. *Nucleic Acids Res* 31, 224–228.
4. Zhang, P., Foerster, H., Tissier, C., Mueller, L., Paley, S., Karp, P., Rhee, S.Y. (2005). MetaCyc and AraCyc. Metabolic Pathway Databases for Plant Research. *Plant Physiol* 138, 27–37.
5. Mueller, L.A., Zhang, P., Rhee, S.Y. (2003) AraCyc. A Biochemical Pathway Database for *Arabidopsis*. *Plant Physiol* 132, 453–460.
6. Jaiswal, P., Ni, J., Yap, I., Ware, D., Spooner, W., Youens-Clark, K., Ren, L., Liang, C., Zhao, W., Ratnapu, K., Faga, B., Canaran, P., Fogleman, M., Hebbard, C., Avraham, S., Schmidt, S., Casstevens, T.M., Buckler, E.S., Stein, L., McCouch S. (2006) Gramene: a bird's eye view of cereal genomes. *Nucleic Acids Res* 34, D717–D723.
7. Ware, D., Jaiswal, P., Ni, J., Pan, X., Chang, K., Clark, K., Teytelman, L., Schmidt, S., Zhao, W., Cartinhour, S., et al. (2002) Gramene: a resource for comparative grass genomics. *Nucleic Acids Res* 30(1), 103–105.
8. Olason, P.I. (2005) Integrating protein annotation resources through the Distributed Annotation System. *Nucleic Acids Res* 33, W468–W470.
9. Wilkinson, M., Schoof, H., Ernst, R., Haase, D. (2005) BioMOBY successfully integrates distributed heterogeneous bioinformatics Web services. The PlaNet exemplar case. *Plant Physiol* 138(1), 4–16.
10. Gonzales, M.D., Archuleta, E., Farmer, A., Gajendran, K., Grant, D., Shoemaker, R., Beavis, W.D., and Waugh, M.E. (2005) The Legume Information System (LIS): an integrated information resource for comparative legume biology. *Nucleic Acids Res* 33, D660–D665.

11. Lamblin, A.F.J., Crow, J.A., Johnson, J.E., Silverstein, K.A.T., Kunau, T.M., Kilian, A., Benz, D., Stromvik, M., Endre, G., VandenBosch, K.A., Cook, D.R., Young, N.D., Retzel, E.F. (2003) MtDB: a database for personalized data mining of the model legume *Medicago truncatula* transcriptome. *Nucleic Acids Res* 31(1), 196–201.

12. Hubbard, T., Andrews, D., Caccamo, M., Cameron, G., Chen, Y., Clamp, M., Clarke, L., Coates, G., Cox, T., Cunningham, F., Curwen, V., Cutts, T., Down, T., Durbin, R., Fernandez-Suarez, X.M., Gilbert, J., Hammond, M., Herrero, J., Hotz, H., Howe, K., Iyer, V., Jekosch, K., Kahari, A., Kasprzyk, A., Keefe, D., Keenan, S., Kokocinsci, F., London, D., Longden, I., McVicker, G., Melsopp, C., Meidl, P., Potter, S., Proctor, G., Rae, M., Rios, D., Schuster, M., Searle, S., Severin, J., Slater, G., Smedley, D., Smith, J., Spooner, W., Stabenau, A., Stalker, J., Storey, R., Trevanion, S., Ureta-Vidal, A., Vogel, J., White, S., Woodwark, C., Birney, E. (2005). Ensembl 2005. *Nucleic Acids Res* 33, D447–D453.

13. Birney, E., Andrews, T.D., Bevan, P., Caccamo, M., Chen, Y., Clarke, L., Coates, G., Cuff, J., Curwen, V., Cutts, T., Down, T., Eyras, E., Fernandez-Suarez, X. M., Gane, P., Gibbins, B., Gilbert, J., Hammond, M., Hotz, H.R., Iyer, V., Jekosch, K., Kahari, A., Kasprzyk, A., Keefe, D., Keenan, S., Lehvaslaiho, H., McVicker, G., Melsopp, C., Meidl, P., Mongin, E., Pettett, R., Potter, S., Proctor, G., Rae, M., Searle, S., Slater, G., Smedley, D., Smith, J., Spooner, W., Stabenau, A., Stalker, J., Storey, R., Ureta-Vidal, A., Woodwark, K. C., Cameron, G., Durbin, R., Cox, A., Hubbard, T., and Clamp, M. (2004) An overview of Ensembl. *Genome Res* 14, 925–928.

13

BGI-RIS V2
Rice Information System at the Beijing Genomics Institute

Ximiao He and Jun Wang

Summary

Rice serves as both a staple for over half of the world's population and a model organism for plants of the grass family. Beijing Genomics Institute (BGI) has long been engaged in rice genomic research: sequencing, assembly, information analysis and integration. Such intensive research results in public releases and biological applications. In order to facilitate obtaining and operating on the rice genomic data, as well as to provide a genomic groundwork for comparative, functional or evolutionary research on important cereal crops, BGI has established and updated the Rice Information System (BGI-RIS V2), an integrated information resource and comparative analysis workbench for rice genomes. BGI-RIS V2 offers not only genomic sequences, which combine the genomic data of *Oryza sativa L. ssp. indica* (by BGI) with *Oryza sativa L. ssp. japonica*, but also most detailed annotation data, including genetic markers, Bacterial Artificial Chromosome (BAC) end sequences, gene contents, cDNAs, oligos, tiling arrays, repetitive elements, and genomic polymorphisms. As a basic platform, BGI-RIS V2 also offers graphical interfaces and a series of tools and services for gene finding, genomic alignment and genomic assembly. This database is available through the web server (http://rise.genomics.org.cn or http://rice.genomics.org.cn) and the File Transfer Protocol (FTP) server (ftp://ftp.genomics.org.cn/pub/database/rice).

Key Words: Rice; cereal crops; genome; comparative genomics; database; information system.

1. Introduction

Rice is one of the most important cereal crops and the principal food for more than half of the world's population. This species has the smallest genome size among major cereal crops, estimated at 430 Mb *(1)*. Evolutionary trees have

From: *Methods in Molecular Biology, vol. 406: Plant Bioinformatics: Methods and Protocols*
Edited by: D. Edwards © Humana Press Inc., Totowa, NJ

proved that these crops diverged from a common ancestor some 60 million years ago *(2)*. Whole genome organization exhibits a high degree of synteny *(3–7)*. Thus, rice is the most suitable model organism for cereal genome analysis. The genome sequences of rice *(8,9)* provide firm foundations for integrating other biological information. These foundations include genetics, gene expression, physiology, development and evolution, which extend to cereal crops, monocots and even general plants. Accordingly, it is feasible and highly desirable to construct a robust, versatile workbench or specific tools to facilitate biological research in rice and other cereal crops *(10,11)*.

As the major genome research institute in China, the Beijing Genomics Institute (BGI) has been carrying out the SuperHybrid Rice Genome Project with best endeavors to comprehend the genome biology of rice *(8)*. We have released a 4.2× draft genome sequence, obtained with the whole-genome shotgun (WGS) scheme *(12)* for *93-11*, which is a cultivar of the *Oryza sativa L. ssp. indica* subspecies grown widely in China and Southeast Asia. An improved version was later reported in which we brought the coverage of the *93-11* data set up to 6.28 *(8,13)*. In order to make thorough use of our updated knowledge about rice genomics, BGI conceived of the Rice Information System (BGI-RIS V2) as a highly integrated information resource for the storage, retrieval, visualization and analysis of rice data *(14)*. The current version of BGI-RIS V2 focuses on rice genomic assembly, which anchors contigs/scaffolds onto chromosomes, based on mapped genetic markers, BAC-based physical maps and annotations. Oligos and tiling arrays are also used to confirm previously predicted genes. We use the rice genome as a framework to organize data for other cereal crops such as wheat and barley, and expand it to *Arabidopsis thaliana* and other plant species. A special emphasis is directed toward comparative analysis among different subspecies of rice and, in the future, among rice, other cereal crops and *A. thaliana*. BGI-RIS V2, together with its most updated database, search engine, species-specific map viewer, comparative genomics viewer, and analysis tools, provides both a comprehensive information resource and a comparative analysis workbench for genome research of rice, other cereal crops and plants.

2. Materials

2.1. Brief Data Overview

2.1.1. Genomic Sequence

WGS (*see* **Note 1**) sequences for the genomes of *indica* (*93–11*) and *japonica* (Syngenta) are available. All the genomic sequence assembles are listed in **Table 1**. Reads produced by WGS are assembled into contigs, scaffolds and

Table 1
Brief Data Content of BGI-RIS V2: Sep 24, 2005

Data type	Data statistics (item numbers)			
	93–11 vs. Syngenta		*93–11* vs. RGP	
	93–11	Syngenta	*93–11*	RGP
Genomic sequences				
Contigs				
Total size (Mb)	410.8	374.2	407.8	–
Number of pieces	50,231	35,047	47,664	–
Scaffolds				
Total size (Mb)	411.7	375.1.	408.8	–
Number of pieces	39,922	26,160	37,393	–
Super-scaffolds				
Total size (Mb)	426.3	391.1	433.9	450.8
Number of pieces	149	119	231	BACs: 3,315
Chromosomes (Mb)[a]	374.5	353.4	352.2	363.2
Annotation data				
Genetic markers	1,408	1,539	1,416	1,343
BAC ends	63,495	70,543	–	–
Full-length cDNAs mapped	25,645	25,645	26,359	25,591
FGeneSH predictions	49,088	45,824	47,905	43,635
BGF predictions	49,710	46,453	48,833	44,665
Genomic polymorphisms				
Total in Chr. level	4,723,468	4,723,468	5,019,016	5,019,016
SNPs	3,936,020	3,936,020	4,249,158	4,249,158
InDels	787,448	787,448	769,858	769,858
Total in cDNA Level	53,833	53,833	54,743	54,743
SNPs	49,471	49,471	49,946	49,946
InDels	4,362	4,362	4,797	4,797
Tiling arrays	6,539,432	6,539,432	–	–
Oligos	–	–	58,404	–
Homology regions	65,171	65,171	50,811	50,811

BAC, Bacterial Artificial Chromosome; BGF, Beijing Gene Finder; InDel, insertion/deletion; SNPs, Single Nucleotide Polymorphisms.

[a]The statistics of chromosome (Mb) not including ChrUn.

93–11: the genomic assembly of *indica* (*93–11*); Syngenta: the genomic assembly of *japonica* (Nipponbare) sequenced by Syngenta; RGP: the genomic assembly of *japonica* (Nipponbare) sequenced by the International Rice Genome Sequencing Project (IRGSP); *93–11* vs. Syngenta: the comparative genomic assemblies referred to each other; 93–11 vs. RGP: like 93–11 vs. Sygenta, the assemblies referred to each other.

super-scaffolds, and mapped to chromosomes according to information of homology and genetic markers.

1. Contig: The result of joining an overlapping collection of usable reads, in which each base is safely recognized *(15)*.
2. Scaffold: The result of connecting contigs by linking information (from paired-end reads, known messenger RNAs, etc.), in which contigs are ordered and oriented with respect to one another.
3. Super-scaffold: The scaffolds are further assembled into super-scaffold. Scaffold and super-scaffold both have gaps where their lengths are known but base contents are unknown.
4. Chromosome: All of the above WGS sequences are assembled into 12 chromosomes. Those that cannot be assembled finally are combined into chromosomeUn, a virtual chromosome for convenience of further analysis.

2.1.2. Annotation Data

Data are annotated to the genomic sequences on different levels: chromosome level, scaffold level or cDNA level, by different standards according to the type of annotation data.

1. Genetic marker: These DNA sequences associated with a particular gene or trait have been located onto the 12 chromosomes, the known location of which are precise in both genetic distance in centiMorgans (cM, unit of distance in genetic maps) and physical distance in base pairs (bp, unit of distance in genomic sequences). Users can retrieve the above information as well as the genomic sequence of the marker.
2. BACends: Both ends of BAC are mapped onto each chromosome by sequence alignment tools such as Basic Local Alignment Search Tool (BLAST) (*see* **Note 2**), and the BAC-end information also plays an important role in the process of the genome assembly.
3. Full-length cDNA: The complementary DNA copies of mRNA, which cover the open reading frame (ORF) of the gene, are mapped to the chromosome level using strict criterion in the BLAST Like Alignment Tool (BLAT) (*see* **Note 3**). Users can acquire the information of the genomic sequence, the corresponding protein, its location in the chromosome, and the structure of the cDNA, i.e., each exon, coding sequence, start and end etc.
4. FGeneSH predictions: The gene prediction results of the FGeneSH (*see* **Note 4**), generated by an *ab initio* gene finding tool for rice, are available. The annotations of FGeneSH are in the chromosome level for the 12 chromosomes and the scaffold level for chromosomeUn. Only entire predicted gene structures (i.e., including initial exon through terminal exon) are present, retained and loaded into BGI-RIS V2.
5. Beijing Gene Finder (BGF) predictions: The gene prediction results of the BGF (*see* **Note 5**), generated by an *ab initio* gene finding tool, which was powered by BGI and designed specially for the rice genome, are also available. Like the

FGeneSH, the procedure was carried out in the chromosome and scaffold levels, and only the complete genes are returned.

6. Repeats: The genomic sequence was annotated on the scaffold level by running the program RepeatMasker (http://www.repeatmasker.org/), a program that screens DNA sequences for interspersed repeats and low complexity DNA sequences. Repeat sequences, location in the scaffold, and the type, such as transposable elements (TEs), long interspersed nuclear element (LINE), and short interspersed nuclear element (SINE), are available.

7. Genomic polymorphisms: Genomic polymorphisms including single-nucleotide polymorphisms (SNPs) and insertion/deletion polymorphisms (InDels) between the *indica* (*93–11*) and *japonica* (Syngenta, RGP) genomes are detected on the chromosome or cDNA levels. Users can get further information of the effects on amino acids and ORFs from SNPs and InDels in the cDNA level.

8. Tiling array: The tiling microarrays are designed using two independent sets of 36-mer probes, with 10-nucleotide intervals (*16*), tiled throughout both strands of each chromosome. The signal oligos are aligned according to their chromosomal coordinates, and the oligo index scores (reflecting the intensity of each signal oligo) are given.

9. Oligos: Oligo microarray was designed for the gene sets, including nr-KOME cDNAs, FGeneSH predictions and BGF predictions. Users can acquire the oligo sequence, melting temperature (TM), gene ontology (GO) and InterPro information.

10. Homology: The homology sequence regions between *indica* (*93-11*) and *japonica* (Syngenta, RGP) are identified by alignment tools and programs developed by us. Users can see a comparative analysis focusing on the gene, marker, etc. within these regions.

2.1.3. Integration with Other Database

BGI-RIS V2 has integrated other widely used databases, including GO, InterPro and GenBank, for user access to trace files, or to get more detailed information from the data origin.

1. GO: The GO (*see* **Note 6**) annotation of full-length cDNAs, FGeneSH predictions and BGF predictions are shown in the report respectively and users can click the GO identifier to switch to the GO database (*see* chapter 24) to see more detailed information.

2. InterPro: The InterPro (*see* **Note 7**) annotation of full-length cDNAs, FGeneSH and BGF gene predictions are also shown in the report pages, and users can click the InterPro identifier to jump to the InterPro database to see more detailed information.

3. GenBank: The genomic sequences of contig and super-scaffold have been submitted to GenBank (*see* **Note 8**), each has the access number for GenBank and can be linked to GenBank in the reports respectively.

2.2. Database Type and Software

BGI-RIS V2 has three components: a world-wide web server, a database server, and a sequence analysis/homology search engine. Tomcat, Oracle 9i and BLAST/BLAT/BGF are running concurrently under the Sun Solaris operating system (Solaris OS).

1. Jakarta Tomcat: BGI-RIS V2 runs Jakarta Tomcat which supports servlet and JSPs. With its own HTTP server, it can be run on any operating system that has a Java Virtual Machine. The release is Tomcat 5.0.28.
2. Oracle 9i: BGI-RIS V2 uses the Oracle (*see* **Note 9**), the most significant and prevalent database management system (DBMS), to build up the relational database of RIS at the back end in the database server. The release used in BGI-RIS V2 is Oracle9i Database Release 2: 9.2.0.1.
3. BLAST/BLAT/BGF: This sequence analysis software is running on separate servers, and more detailed information is presented in **Subsection 3.2**.
4. Sun Solaris: The computer operating system running on the BGI-RIS V2 server is the Solaris OS, which is based on open-source UNIX, developed by Sun Microsystems. The release used in BGI-RIS V2 is Solaris 8.
5. Model-View-Controller (MVC) system: The most important function models of the BGI-RIS V2 search engine and visualization system are both based on the MVC system (*see* **Note 10**). This system consists of a "model" where the business logic resides, a "View" that is generated by JSP pages, and a "Controller" that is a servlet or a collection of servlets to provide centralized process handling.

2.3. Hardware

BGI-RIS V2 and the associated environment are running on a supercomputer: Sun 10k. Parameters are as follows:

1. CPU: Scalable Processor Architecture (SPARC) (*see* **Note 11**) 400 MHz × 64.
2. Hard disk: 6 TB (1 TB =1024 GB).
3. Memory: 16 GB.
4. Network: 1000M Network Interface Card.

2.4. Data Outline

1. Type: Most data are stored in databases as tables. Genomic sequence is stored in FASTA format using flat text. In the web server, users can see the files in the view systems (MapView and CompView) in the BGI-RIS V2, which are in Portable Network Graphics (PNG) format. To handle the large amount of complex rice genome data, we developed our own standard set of genome-based Bio-XML format that lays the foundation for our research work and allows BGI-RIS V2 to accommodate the fast accumulating data and to integrate new data types when encountered.

2. Source: BGI-RIS V2 integrates our own genomic data on *O. sativa* L. ssp. *indica* (*93-11*) with genome sequences of *Oryza sativa* L. ssp. *japonica* from other institutions, as well as EST sequences of rice from our own productions and public data of rice and other cereal crops, such as maize, wheat and barley (http://www.ncbi.nlm.nih.gov/dbEST/). Additional related information from rice include genomics such as BACs (ftp://ftp.genome.arizona.edu/pub/stc/rice/), genetics, such as genetic markers (http://www.gramene.org/resources/), and cDNAs such as the nr-KOME data set of non-redundant cDNAs from the knowledge-based Oryza Molecular Biological Encyclopedia (*17*) (ftp://cdna01.dna.affrc.go.jp/pub/data/). Wherever publicly available, data are carefully curated and integrated into BGI-RIS V2.

3. Organization: Due to the complexity and the large-scale nature of the genomic data, the strategy of comprehensive organization and effective management are essential for successive analyses. In BGI-RIS V2, we organize the genomic data at three vertical levels as different modules: chromosome level, contig/scaffold level and genetic element level, which are in accordance with the main tables in the database schema, and link the data of the three different levels through the genome-oriented MapView and CompView for comparative analysis.

4. Volume: The total volume of the database data, genomic file and running software is about 60 GB.

5. Updates: BGI-RIS V2 updates the rice genome sequence information and annotation data bimonthly, constantly incorporating more data, once they become available from other plant genomes, and different types of biological data, such as tRNA, mRNA, SAGE and microarray data. To assist users, we have introduced into BGI-RIS V2 a version system and a frame of reference around different versions of the rice data. In the near future, it will be possible for users to retrieve data from different versions.

3. Methods

3.1. From Search to Viewer

3.1.1. Search Engine

The BGI-RIS V2 provides users with identifier-based, keyword-based and genetic location-based subject searches for querying the major data types housed in the database, including identification numbers of scaffold, contigs, genes, cDNAs, repeats, genomic polymorphisms and markers (**Fig. 1**).

1. Scaffolds: Users can access a certain scaffold through the exact identifier of scaffold (e.g., Scaffold000024) in BGI-RIS V2, and a group of scaffolds through the part identifier of a scaffold (e.g., Scaffold0024), by fuzzy searching. BGI-RIS V2 also provides the users with genetic location searches to focus on a batch of scaffolds in a specific region on the chromosome (e.g., Chr2:2000000-400000, i.e., the region from 2 Mbp to 4 Mbp in chromosome 2).

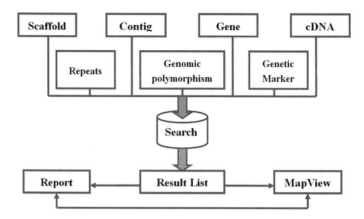

Fig. 1. Outline of the work flow for the Search Engineer.

2. Contigs: Besides the BGI-RIS V2 identifier and genetic location search, users can also retrieve contigs through the access number from the National Center for Biotechnology Information (NCBI) (e.g., NM_000024).

3. Genes/cDNAs: For genes/cDNAs, BGI-RIS V2 provides users with all of three search means: identifier-based, genetic location-based and keyword-based search. Users can also access a gene through a GO identifier (e.g., GO: 000124) and Interpro identifiers (e.g., IPR00000124).

4. Repeats: For repeats, operations are similar to other data types; users can search the BGI-RIS V2 identifiers. But for genetic location, two levels are specified: chromosome level (e.g., Chr12:1234-567890) and scaffold level (e.g., Scaffold000012:0-5000000). Users can also address a kind of repeats selecting the repeats type (e.g., TEs, LINE and SINE) listed in the drop list box.

5. Genomic polymorphisms: Users can search the genomic polymorphism identifier, the genomic location and genomic polymorphism types. For genomic location, there are two levels: chromosome level and cDNA level (e.g., OsJRFA059764:0-10000). Three genomic polymorphism types are specified, that is, "S" stands for SNPs, "I" stands for insertion, and "D" stands for deletion. Users can also combine the genomic location with the genomic polymorphism types within a search.

6. Genetic markers: Like the above data types, users can access the genetic markers through an identifier-based search, chromosome-level genomic location search and keyword-based search.

3.1.2. A Detailed Example

Suppose that a user is interested in the genes or proteins related to proteolysis in *indica*, the user could retrieve the information through the Search Engine by the following steps: (1) Select "*indica (93-11)*" in the search bar and "Gene" for the search data type; (2) input the keyword "proteolysis" in the text frame; (3) click the button "Go" to submit the request. Then, the result pages with gene list are returned (**Fig. 2A**). Here, 914 items satisfied and the first 10 ones are shown in the first page by default. Users can change the display pattern, the number of items displayed in one page, and switch to the other pages. For more information, users can click the identifier of a gene (e.g., OsIFCC000064) to see a detailed gene report page (**Fig. 2B**), and click the "mapviewer" to visualize the map information related to the gene, here switch to GeneView (**Fig. 2C**) (For more description about the visualization tool MapView such as GeneView, *see* **Subsection 3.1.3.**). If users are interested in the comparison of gene/cDNA regions between *indica* and *japonica*, they can go to the CompView, and input the specified chromosome and location (in this case, chromosome 1: 520,000-680,000) to see the genomic comparative information (**Fig. 2D**) (For more about CompView, *see* **Subsection 3.1.4.**).

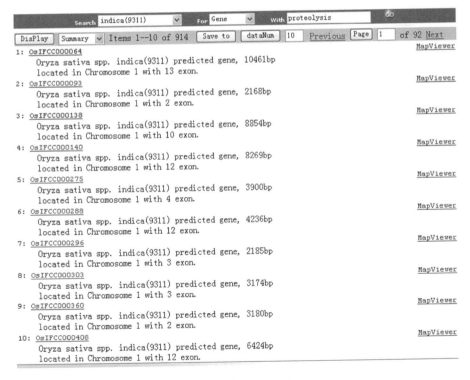

Fig. 2A. (*Continued*)

OsIFCC000064   <u>MapViewer</u>

DESCRIPTION Oryza sativa spp. indica(9311)predicted gene by FGeneSH. 10461bp.
            located in Chromosome1 from 549336 to 559796 with 13 exons.
VERSION   OsIFCC000064.1
SOURCE   Oryza sativa spp. indica(9311)
   ORGANISM Oryza sativa
            Eukaryota; Viridiplantae; Streptophyta; Embryophyta; Tracheophyta;
            Spermatophyta; Magnoliophyta; Liliopsida; Poales; Poaceae;
            Ehrhartoideae; Oryzeae; Oryza.
FEATURES

     EXON with 13 exons.

            Exon (< Exon_start...Exon_end >, < direction >)
                 phase size score
            1   (<550852...550961>,   <+>)   0 110 15.15
            2   (<551038...551257>,   <+>)   2 220 22.85
            3   (<552129...552329>,   <+>)   0 201 15.80
            4   (<552439...552509>,   <+>)   0 71 3.98
            5   (<552582...552741>,   <+>)   2 160 10.65
            6   (<554631...554957>,   <+>)   0 327 19.25
            7   (<554994...555224>,   <+>)   0 231 16.75
            8   (<555334...555714>,   <+>)   0 381 16.90
            9   (<556038...556113>,   <+>)   0 76 2.99
            10   (<556201...556264>,   <+>)   1 64 4.91
            11   (<556799...556934>,   <+>)   2 136 6.99
            12   (<557848...557892>,   <+>)   0 45 3.60
            13   (<558033...558098>,   <+>)   0 66 4.59

     PROMOTER promoter_start=549336
         CDS < 550852...558098 >

     Ontology <u>GO:0008236</u>:serine-type peptidase
              <u>GO:0006508</u>:proteolysis and peptidolysis
     Interpro <u>IPR004106</u>Prolyl oligopeptidase, N-terminal beta-propeller domain

BASECOUNT A:535 C:419 G:518 T:616
SEQUENCE
      1 ATGGGTTCTG TCGCCGGCGA CGCCGCCCGC CTTTCTTATC CACCCACTCG
     51 CCGCGACGAT TCCGTCGTCG ACATGTACCA TGGCGTCCCT GTCACCGACC
    101 CTTACCGTTG GCTGGAGGAC CCAGAGTCGG AAGATACCAA AGAATTTGTG
    151 GCAAGTCAGG TGGAGCTAGC AGAGTCTGTG CTTGCGGGCT GCTTTGACAG
    201 GGAGAACCTT CGCCATGAGG TCACTCGCCT CTTCGACCAC CCGCGCCACG
    251 GGGCGCCGTT CCGTCGCGGC GACAAGTACT TCTACTTTCA CAACTCCGGA

                        Fig. 2B. (*Continued*)

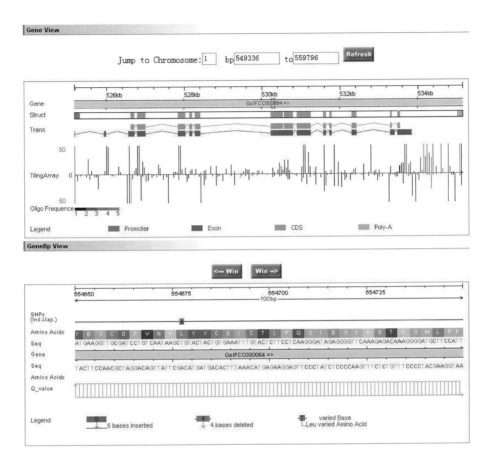

Fig. 2C. (*Continued*)

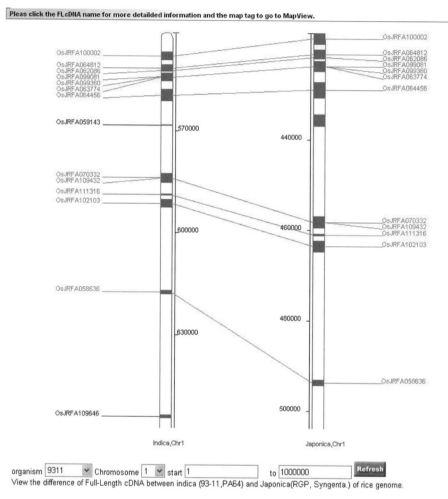

Fig. 2D. An example of a search for genes related to "proteolysis". Screenshots include (**A**) the result list of the search, (**B**) the detailed report of "OsIFCC000064", (**C**) the GeneView, and (**D**) the FL-cDNA Compview of the related regions of the gene.

3.1.3. MapView

As an important and efficient visualization tool in BGI-RIS V2, MapView is composed of three main types of subviewer, in hierarchical architecture *(18,19)*: ChroView/OverView, ContigView, and GeneView/cDNAView. They are in accordance with the organization of three vertical levels of complex genomic data.

1. ChroView: In this model, we show users the outline of a certain chromosome: position of centromere, statistics of distributing trends of SNPs, FGeneSH/BGF predictions, full-length cDNAs, genomic repeats, and GC content, in the chromosome level.
2. OverView: Focusing on about 100-kbp region (by default) or whatever region the user specified, this model shows users low-resolution physical map with sequence super-scaffolds/scaffolds aligned to, mapped genetic markers, BGF predicted genes, and the distribution of SNPs. Homologous regions between *indica* (*93-11*) and *japonica* are also marked out in this model.
3. ContigView: In this model, users can highlight an area in OverView to browse the annotated information for the chosen scaffolds/contigs (**Fig. 3**). The annotation with distinct color coding, includes anchored BAC ends, BGF/FGeneSH predicted genes, full-length cDNAs, classes of repeats, SNPs that oligo frequency of tiling array, and GC content. A factual report for each element contained in the visualization system is displayed automatically by clicking. For predicted genes and full-length

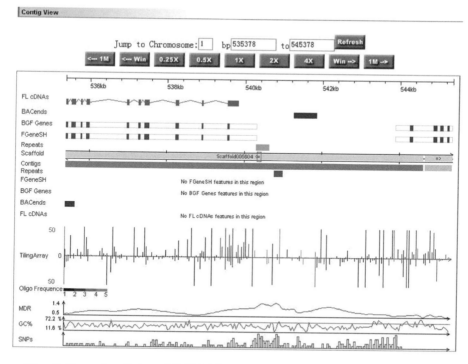

Fig. 3. Example of the of BGI-RIS V2 viewer for *Oryza sativa* L. ssp. *indica* (*93-11*). Screenshots of the ContigView.

cDNAs, users can also go to GeneView/cDNAView to see a more detailed view, such as exon-intron structures, CDS and protein sequence etc.

4. BaseView: In this model, users can focus on a region of 100-bp or investigate the Q-value for each nucleotide. Genomic polymorphisms (SNPs and InDels) are also shown, and protein sequence and the ORF of genes in the region are shown in the GeneBpView/cDNABpView, a similar model to the GeneView/cDNAView.

5. GeneView: In this model, we focus on the structure of a certain gene, the promoter, exons and introns, CDS and poly A. For the gene prediction, the cDNAs are also aligned in the same model. General information, such as location and orientation on the chromosome are shown. A more detailed report for the gene is displayed automatically by clicking on the map.

6. cDNAView: Very similar to the GeneView, this model focuses on the information of structure, and includes the location and orientation on the chromosome. In the cDNABpView model, users can see the peptides and SNP information. A link to the report referring to the cDNA is also available on the map.

3.1.4. CompView

CompView is another interactive visualization tool that is being developed for identifying and visualizing conserved syntenic blocks (homologous chromosome segments and gene homologs) across multiple-related genomes simultaneously. It is designed to allow users to switch between CompView and MapView and to start with a gene or region of interest to search for related information, and will provide users with timely genomic information across species beyond genera and families.

1. Marker CompView: Rice genetic markers are identified on a certain chromosome of *indica* (*93-11*) and *japonica* (Syngenta or RGP) in this model, and users can view comparative analysis between the two subspecies by focusing on the same genetic marker in different genetic locations on the chromosome. Users can also switch to a MapView of the genetic marker by clicking on the chromosomal coordinate tag in the chromosome map, and get a more detailed report by clicking the genetic marker name. In order to facilitate the user to see the difference in a compact figure, we use a different coordinate system for the two chromosomes and mark respectively.

2. FL-cDNA CompView: Similar to the Marker CompView, users can view the comparative information on a certain chromosome of *indica* (*93-11*) and *japonica* (Syngenta or RGP). The links to the MapView and reports are also available, and again, the coordinate systems for both chromosomes are different.

3.2. Analysis Tools

BGI-RIS V2 offers users a series of tools and services to analyze the genetic sequence, aimed at gene finding (i.e., BGF), genomic homolog search (i.e., BLAST/BLAT), and assembly of sequenced reads/contigs [i.e., Repeat-masked

Phrap with Scaffolding (RePS), *see* **Note 12**]. The services of BLAST, BLAT and BGF are also packaged into a grid service in order to optimize the resource of computer machines.

3.2.1. BGF: To Find New Genes

Users can submit genomic sequences in FASTA format to get the predicted gene results in the BGF page. Two methods are offered for the sequence input either typing sequences in the Sequence Text Frame, or uploading sequence files. For longer sequences, more than 500 Kb, BGF recommends users to provide an e-mail address and will ultimately return the prediction results by e-mail in order to save time waiting for the result pages.

Here is an example: Users input or paste a sequence of 9867 bps in the Sequence Text Frame, which starts with ">" and is named "MySeq01234", select "Oryza sativa" for the species, leaving the e-mail address empty as the sequence is not more than 500 Kb (**Fig. 4A**), then click the "Submit" button to get the result. The predicted genes are listed in the new page, with the gene structure, predicted proteins and other related information. Here, three new genes are predicted (**Fig. 4B**).

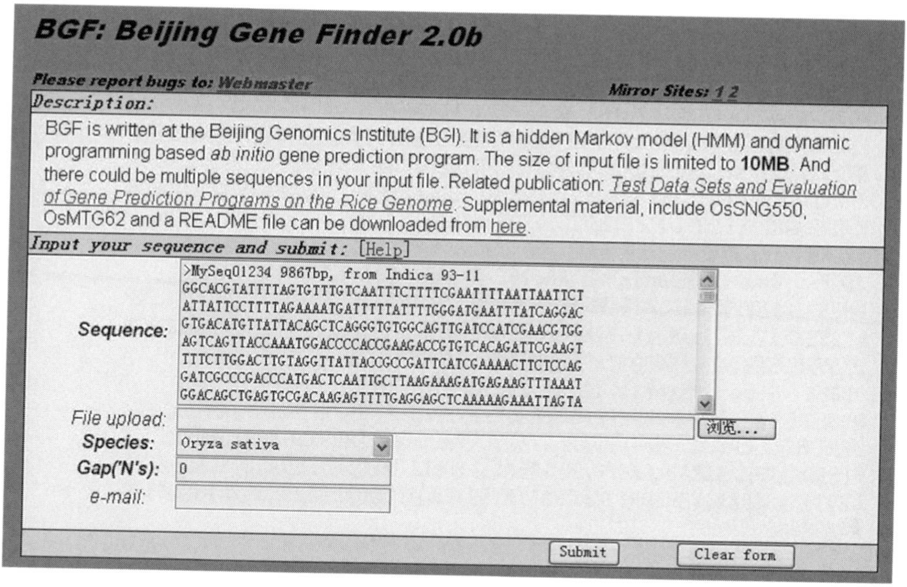

Fig. 4A. (*Continued*)

```
Program    : BGF
Version    : 2.0.20
Time       : Fri Jan 13 18:20:33 2006
Parameter  : Rice
Sequence   : MySeq01234 9867bp, from Indica 93-11
Length     : 9867
GC%        : 41.63
Total Genes:   3 (  1 in + strand &   2 in - strand)
Total Exons:   6 (  1 in + strand &   5 in - strand)
```

Gene#	S	Exon#	Type	Start		End	ORF_S		ORF_E	Score	Len
=======	===	=======	======	=======	===	=======	=======	===	=======	=======	======
1	+	1	Sngl	264	-	1388	264	-	1388	5.68	1125
1	+		PolA	1549	-					-0.92	
2	-		PolA	1637	-					-3.69	
2	-	1	Term	1683	-	1910	1683	-	1910	0.85	228
2	-	2	Init	2299	-	2559	2299	-	2559	7.76	261
2	-		Prom	3836	-					-2.54	
3	-		PolA	5582	-					-1.44	
3	-	1	Term	6010	-	6422	6010	-	6420	5.40	413
3	-	2	Intr	6723	-	6834	6724	-	6834	10.26	112
3	-	3	Init	7937	-	8158	7937	-	8158	4.50	222

```
Predicted protein(s):
>BGF:  Gene:1 Exon(s):1 AA:374 Chain+ H+T+
MTQLLKKDEKFKWTAECDKSFEELKKKLVSAPVLILPDQMKDFQVYCDASRHGLGCVLMQ
EGRVVAYASRQLRPHEGNYPTHDLELAAVVHALKIRRHYLIGNRCEVYMDHKSLKYIFTQ
PDLNLRQRRWLELIKDYDMSIHYHPEKANVVADALSRKSYCTALGIEGMCEELRQELERL
NVGIVEHGFVAALEARPTLADQVRAAQVNDPEIAELKKNMRVRKARDFHEDEHGTIWLGE
RLCVPDDKELKDLILTEAHQTQYSIHPGSTKMYQDLKEKFWWVSMRREIAEFVALCDVCQ
RVKAEHQRPVGLLQPLQIPEWKWEEIGMDFITGLPRTSSGHDSISVVNVTPTVLKRVCIP
IRLCRGVILSSGRL
>BGF:  Gene:2 Exon(s):2 AA:162 Chain- H+T+
MMFEDTDVEVPMPNVDLRTSTNGAKGSTKKSSNYTCKEDIQLCISWQSISSDPIIGNEQP
GKAYLKRITEHCHANRDYESVGNMGPGAEWGRVCPVPSACSRACLPVHESTKVSMELRGN
DRDVPRVSRRTALGSTWGQTTALLPDEGMDDSSTQNMWAPIQ
>BGF:  Gene:3 Exon(s):3 AA:248 Chain- H+T+
MDSDDHPPEFIIHVVDDLAPPPPPPPPRAQPALVPPRILPLPPPAFHGAAVQRRRPRPPPR
DEATNNRTLLFVFQVCMFTMMVINLSIAIYLTVHPSPEESIADERSGVRMTFERAVEKCR
FERRDEAVPDAAARAELGRGPGARNGWERAIDEDELLDALDGEDLERAHEQALVLVHEAG
GEVTAWASSRAAQVSESDMPTGRCSASTRTSEVERQTLRQRVESAAVRAASLTGRQETIR
RRARRGGG
```

Fig. 4B. An example of using the Beijing Gene Finder (BGF) to find new genes. Screenshots of (**A**) submission and (**B**) the result page to the submitted sequence.

input

input section

* **database** database sequence
SynVs9311: 9311 Scaffold Nucleotide

* **query** input of query sequence
>MySeq007 2167bp, from Indica 93-11
GGCACGTATTTTAGTGTTTGTCAATTTCTTTTCGAATTTTAATTAA
TTCT
ATTATTCCTTTTAGAAAATGATTTTTATTTTGGGATGAATTTATCA
GGAC

浏览...

R PSI-TBLASTN checkpoint file, Optional

浏览...

* **p** Program Name, input should be one of "blastp", "blastn",
"blastx", "tblastn", or "tblastx".
blastn

output

output files automatically

output BLAST report Output File
myBlast.txt

O SeqAlign file [File Out] Optional

Job Status

SoftName : SynVs9311Blast

● **Your job ID:**
20060113164543_1648

Task Assignment

● **Your Command Line:**
-o "myBlast.txt" -d "SynVs9311: 9311 Scaffold Nucleotide" -i "20060113164543_1648_query.file" -p "blastn"

● **Your job is running on:**
BGI(Beijing) 61.50.158.108:8081

● **Time on queue:**
0 second

● **Time on excecuting:**
14 seconds

● **Job status:**
Job ended (terminated execution).

● **Your input file:**
20060113164543_1648_query.file

● **Job standard output:**
6668.sun-do.OU

● **Job standard error:**
6668.sun-do.ER

● **Your output file:**
myBlast.txt

Fig. 5A and B. (*Continued*)

Job Results

/public/20060113164543_1648/output/myBlast.txt

```
BLASTN 2.2.5 [Nov-16-2002]


Reference: Altschul, Stephen F., Thomas L. Madden, Alejandro A. Schaffer,
Jinghui Zhang, Zheng Zhang, Webb Miller, and David J. Lipman (1997),
"Gapped BLAST and PSI-BLAST: a new generation of protein database search
programs", Nucleic Acids Res. 25:3389-3402.

Query= MySeq007 2253bp, from Indica 93-11
        (2167 letters)

Database: /netgrid/database/public/rice/SynVs9311/9311/Sequence/Scaffo
ld/Indica_scaffold.seq
           39,922 sequences; 411,710,190 total letters

Searching.....................................................done

                                                            Score    E
Sequences producing significant alignments:               (bits) Value

Scaffold011234 2003-04-20 BGI                               3370    0.0
>Scaffold011234 2003-04-20 BGI
          Length = 9867

 Score = 3370 bits (1700), Expect = 0.0
 Identities = 1700/1700 (100%)
 Strand = Plus / Plus


Query: 1     ggcacgtattttagtgtttgtcaatttcttttcgaattttaattaattctattattcctt 60
             ||||||||||||||||||||||||||||||||||||||||||||||||||||||||||||
Sbjct: 1     ggcacgtattttagtgtttgtcaatttcttttcgaattttaattaattctattattcctt 60


Query: 61    ttagaaaatgattttatttgggatgaatttatcaggacgtgacatgttattacagctc 120
             ||||||||||||||||||||||||||||||||||||||||||||||||||||||||||||
Sbjct: 61    ttagaaaatgattttatttgggatgaatttatcaggacgtgacatgttattacagctc 120


Query: 121   agggtgtggcagttgatccatcgaacgtggagtcagttaccaaatggaccccaccgaaga 180
             ||||||||||||||||||||||||||||||||||||||||||||||||||||||||||||
Sbjct: 121   agggtgtggcagttgatccatcgaacgtggagtcagttaccaaatggaccccaccgaaga 180
```

Fig. 5C. An example of using the Basic Local Alignment Search Tool (BLAST) to
search for a genomic homolog. Screenshots of (**A**) Submission, (**B**) the Job Status, and
(**C**) the BLAST result page for the submitted sequence.

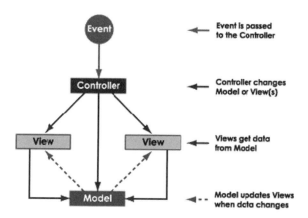

Fig. 6. Outline of the control flow in a Model-View-Controller application.

3.2.2. BLAST/BLAT: To Search for a Genomic Homolog

BGI-RIS V2 provides BLAST services, a popular alignment tool, to search for a genomic homolog in some of the rice genomic sequence data. The sequence databases formatted for BLAST in BGI-RIS V2 include the genomic assembly at the chromosomal level, scaffold level, predicted genes and relevant proteins data set, full-length cDNAs and relevant proteins data set.

Like BGF, users can input the sequences either by typing sequences in the Query Text Frame or by uploading sequence files. Suppose that we have a 2167-bp DNA sequences of rice *indica* 9311 and want to know the genomic location of it. We can select "SynVs9311:9311 ChrAll Nucleotide" for the database type, input or paste the sequence in the Query Text Frame, run "BLASTn" for the program, set the output file with the name "myBlast.txt", and then click the button "run" to submit it (**Fig. 5A**). Then the job ID and job status are shown in a new page (**Fig. 5B**), and users can click outfile (here it is myBlast.txt) to see the alignment result (**Fig. 5C**).

Because of its shorter time consumption, BLAT is another popular alignment tool offered in BGI-RIS V2. The sequence databases formatted for BLAT are very similar to BLAST. The version of BLAT offered in BGI-RIS V2 is BLAT 27.

4. Notes

1. WGS: Whole Genome Shotgun sequencing, first introduced by J. Craig Venter in 1994, using the method of Dideoxy Sequencing invented by Fred Sanger in 1982 *(20)*, is one of most important sequencing methods. In the year 2000, Celera scientists, in collaboration with the publicly funded *Drosophila* Genome Project,

published the WGS assembly of the *Drosophila* genome *(21)*, with descriptions of the paired end sequencing strategy, and new algorithms *(12)* which made WGS the prevailing genome sequencing approach. The procedure of WGS sequencing involves (1) physically break up the DNA into millions of fragments, (2) inserting these fragments into cloning vectors in order to amplify the DNA to the required levels for a sequencing reaction, (3) sequence the fragments to get the sequences of both ends of the fragments, called pairs of reads, and (4) assembly of the fragments by algorithms that involve information of the overlaps in the fragments and pairs of reads. The common step is reads–contigs–scaffolds–chromosomes.

2. BLAST: BLAST, developed in 1990 by Altschul, Gish, Miller, Myers and Lipman *(22)*, is the most popular sequence alignment tool hosted by NCBI (http://www.ncbi.nlm.nih.gov/BLAST/). It integrates a set of sequence comparison algorithms optimized to search sequence databases for optimal local alignments of a query sequence. The basic algorithm is applied in a variety of contexts including straightforward DNA and protein sequence database searches, motif searches and gene identification searches, and in the similarity analysis of multiple regions in long DNA sequences. The current version provided in BGI-RIS V2 is BLAST 2.2.12, 28 Aug. 2005.

3. BLAT: BLAT, developed by W. James Kent *(23)* in 2002, is a powerful tool for the mRNA/DNA and cross-species protein alignment of vertebrate genomes. Based on index of non-overlapping K-mers in the genome, it proves more accurate and faster than BLAST (i.e., 500 times faster than popular mRNA/DNA alignments tools, and 50 times faster in protein alignments). This widely used alignment tool is hosted by UCSC (http://www.genome.ucsc.edu/cgi-bin/hgBlat). The current version is BLAT 32, 18 Feb. 2005.

4. FGeneSH: FGeneSH, developed by Asaf, Victor and their gene finding group, is a program predicting multiple genes in genomic DNA sequences. Based on the Hidden Markov Model *(24,25)*, it is one of the fastest and most accurate gene finders. It is a commercial software owned by Softberry Inc. and online testing is available at http://www.softberry.com/.

5. BGF: BGF, developed by the BGF team in BGI, is a gene prediction tool *(26)* based on a Hidden semi-Markov model and dynamic programming. BGF is written in C++ using the Standard Template Library. It has been trained for the rice (*indica* and *japonica*) genome and silkworm *Bombyx mori* genome. Users can access the BGF either in BGI-RIS V2 (http://bgf.genomics.org.cn) or a mirror site in Fudan University (http://tlife.fudan.edu.cn/bgf/). The current version is BGF 2.0b, Aug. 2005.

6. GO: The GO project launched by the Gene Ontology Consortium established in 1998 aims to provide a set of structured vocabularies for specific biological domains that can be used to describe gene products in any organism *(27–30)*. Three ontologies have developed to describe attributes of gene products or gene product groups: molecular function, biological process and cellular component. Users can

address the GO Database (GOD) through the GO Browsers at the GO Web site (http://www.geneontology.org). The GO terms, definitions, and ontologies are updated monthly by FTP and updated every 30 min on the GO web site download page (http://www.geneontology.org/GO.downloads.shtml).

7. InterPro: The InterPro database, established in 1999 when the InterPro Consortium was formed, is an integrated documentation resource for protein families, domains, and functional sites, in which identifiable features found in known proteins can be applied to unknown proteins *(31–34)*. It was created to integrate the major protein signature databases, including PROSITE, Pfam, PRINTS, ProDom, SMART, TIGRFAMs, PIRSF and SUPERFAMILY. The database is aiming to update every 2 to 3 months, and the current release is InterPro release 11.0.

8. GenBank: GenBank, hosted and maintained by NCBI and established in 1988 as a national resource for molecular biology information, is a comprehensive genetic sequence database, an annotated collection of all publicly available DNA sequences *(35–37)*. Its data are obtained primarily through submissions from individual laboratories and batch submissions from large-scale sequencing projects. As a part of the International Nucleotide Sequence Database Collaboration, GenBank exchange data on a daily basis with the other two members: the DNA DataBank of Japan (DDBJ) and the European Molecular Biology Laboratory (EMBL). GenBank is aiming at making a new release every 2 months, and the current release is NCBI-GenBank Flat File Release 149.0, Aug. 15 2005.

9. Oracle: Oracle database, the product of the Oracle Corporation, which was founded in 1977 by Larry Ellison and changed to this name in 1983, is commonly referred to as the Oracle Relational Database Management System (RDBMS), which is a DBMS based on the relational model. Oracle RDMBS stores data logically in the form of tables, and physically in the form of files. The current release is Oracle Database 10g Release 2: 10.2.0.1. The g stands for "grid", emphasizing a marketing thrust of presenting 10g as "grid-computing ready" *(38)*.

10. MVC: MVC, first described in 1979 by Trygve Reenskaug, is a software architecture that separates an application's data model, user interface and control logic into three distinct components, so that modifications to one component can be made with minimal impact to the others. Generally, constructing an application using MVC architecture involves defining three classes of modules (**Fig. 6**). (1) Model: the domain-specific representation of the information on which the application operates. (2) View: renders the model into a form suitable for interaction, typically a user interface element. (3) Controller: responds to events, typical user actions, and invokes changes on the model or view as appropriate.

11. SPARC: SPARC, originally designed in 1985 by Sun Microsystems, is a pure big-endian, Reduced Instruction Set Computing (RISC) microprocessor architecture. RISC is a microprocessor CPU design philosophy that favors a smaller and simpler set of instructions that all take about the same amount of time to execute.

12. RePS: RePS, developed by BGI in 2002, is a program for assembling shotgun DNA sequence data *(39)*. In the process of assembly, RePS explicitly identifies exact 20-mer repeats from the shotgun DNA sequence data and masked the repeats. Phrap (http://www.phrap.org/phredphrapconsed.html), another established assembling software, is used to compute meaningful error probabilities for each base. It combines the clone-end-pairing information to construct the scaffolds, in which the contigs are ordered and oriented. It has been successfully used in the assembly of the rice *indica* genome in BGI. Users can download the software from our FTP site, and install and run RePS on their computer system. The current version is RePS v2.01, Aug. 2004.

Acknowledgments

This work was sponsored by Chinese Academy of Sciences, Commission for Economy Planning, Ministry of Science & Technology, National Natural Science Foundation of China, Zhejiang University and China National Grid. The permission of using Oxford University Press publication for the source of the material was appreciated. We thank Hui Song for some constructive suggestions.

References

1. Zhao, W., Wang, J., He, X., Huang, X., Jiao, Y., Dai, M., Wei, S., Fu, J., Chen, Y., Ren, X. (2004) BGI-RIS: an integrated information resource and comparative analysis workbench for rice genomics. *Nucleic Acids Res.* 32, D377–D382.
2. Chen, M., SanMiguel, P., de Oliveira, A.C., Woo, S.S., Zhang, H., Wing, R.A., Bennetzen, J.L. (1997) Microcolinearity in sh2-homologous regions of the maize, rice, and sorghum genomes. *Proc. Natl. Acad. Sci. USA* 94, 3431–3435.
3. Bevan, M. and Murphy, G. (1999) The small, the large and the wild: the value of comparison in plant genomics. *Trends Genet.* 15, 211–214.
4. Feuillet, C. and Keller, B. (2002) Comparative genomics in the grass family: molecular characterization of grass genome structure and evolution. *Ann. Bot. (Lond)*, 89, 3–10.
5. Wicker, T., Stein, N., Albar, L., Feuillet, C., Schlagenhauf, E., Keller, B. (2001) Analysis of a contiguous 211 kb sequence in diploid wheat (*Triticum monococcum* L.) reveals multiple mechanisms of genome evolution. *Plant J.* 26, 307–316.
6. Shimamoto, K. and Kyozuka, J. (2002) Rice as a model for comparative genomics of plants. *Annu. Rev. Plant Biol.* 53, 399–419.
7. McCouch, S.R. (2001) Genomics and synteny. *Plant Physiol.* 125, 152–155.
8. Yu, J., Hu, S., Wang, J., Wong, G.K., Li, S., Liu, B., Deng, Y., Dai, L., Zhou, Y., Zhang, X. (2002) A draft sequence of the rice genome (*Oryza sativa* L. ssp. *indica*). *Science* 296, 79–92.

9. Goff, S.A., Ricke, D., Lan, T.H., Presting, G., Wang, R., Dunn, M., Glazebrook, J., Sessions, A., Oeller, P., Varma, H. (2002) A draft sequence of the rice genome (*Oryza sativa* L. ssp. *japonica*). *Science* 296, 92–100.

10. McCouch, S. (1998) Toward a plant genomics initiative: thoughts on the value of cross-species and cross-genera comparisons in the grasses. *Proc. Natl. Acad. Sci. U. S. A.* 95, 1983–1985.

11. Bennetzen, J. (2002) The rice genome. Opening the door to comparative plant biology. *Science* 296, 60–63.

12. Myers, E.W., Sutton, G.G., Delcher, A.L., Dew, I.M., Fasulo, D.P., Flanigan, M.J., Kravitz, S.A., Mobarry, C.M., Reinert, K.H., Remington, K.A. (2000) A whole-genome assembly of Drosophila. *Science* 287, 2196–2204.

13. Yu, J., Wang, J., Lin, W., Li, S., Li, H., Zhou, J., Ni, P., Dong, W., Hu, S., Zeng, C. (2005) The genomes of *Oryza sativa*: a history of duplications. *PLoS. Biol.* 3, e38.

14. Zhu, J.H., Stephenson, P., Laurie, D.A., Li, W., Tang, D., Gale, M.D. (1999) Towards rice genome scanning by map-based AFLP fingerprinting. *Mol. Gen. Genet.* 261, 184–195.

15. Lander, E.S., Linton, L.M., Birren, B., Nusbaum, C., Zody, M.C., Baldwin, J., Devon, K., Dewar, K., Doyle, M., FitzHugh, W. (2001) Initial sequencing and analysis of the human genome. *Nature* 409, 860–921.

16. Li, L., Wang, X., Xia, M., Stolc, V., Su, N., Peng, Z., Li, S., Wang, J., Wang, X., Deng, X.W. (2005) Tiling microarray analysis of rice chromosome 10 to identify the transcriptome and relate its expression to chromosomal architecture. *Genome Biol.* 6, R52.

17. Kikuchi, S., Satoh, K., Nagata, T., Kawagashira, N., Doi, K., Kishimoto, N., Yazaki, J., Ishikawa, M., Yamada, H., Ooka, H. (2003) Collection, mapping, and annotation of over 28,000 cDNA clones from japonica rice. *Science* 301, 376–379.

18. Ashurst, J.L., Chen, C.K., Gilbert, J.G., Jekosch, K., Keenan, S., Meidl, P., Searle, S.M., Stalker, J., Storey, R., Trevanion, S. (2005) The Vertebrate Genome Annotation (Vega) database. *Nucleic Acids Res.* 33, D459–D465.

19. Hubbard, T., Barker, D., Birney, E., Cameron, G., Chen, Y., Clark, L., Cox, T., Cuff, J., Curwen, V., Down, T. (2002) The Ensembl genome database project. *Nucleic Acids Res.* 30, 38–41.

20. Sanger, F., Coulson, A.R., Hong, G.F., Hill, D.F., Petersen, G.B. (1982) Nucleotide sequence of bacteriophage lambda DNA. *J. Mol. Biol.* 162, 729–773.

21. Adams, M.D., Celniker, S.E., Holt, R.A., Evans, C.A., Gocayne, J.D., Amanatides, P.G., Scherer, S.E., Li, P.W., Hoskins, R.A., Galle, R.F. (2000) The genome sequence of *Drosophila melanogaster*. *Science* 287, 2185–2195.

22. Altschul, S.F., Gish, W., Miller, W., Myers, E.W., Lipman, D.J. (1990) Basic local alignment search tool. *J. Mol. Biol.* 215, 403–410.

23. Kent, W.J. (2002) BLAT – the BLAST-like alignment tool. *Genome Res.* 12, 656–664.

24. Krogh, A., Larsson, B., von, H.G., Sonnhammer, E.L. (2001) Predicting trans-membrane protein topology with a hidden Markov model: application to complete genomes. *J. Mol. Biol.* 305, 567–580.

25. Krogh, A., Mian, I.S., Haussler, D. (1994) A hidden Markov model that finds genes in *E. coli* DNA. *Nucleic Acids Res.* 22, 4768–4778.

26. Li, H., Liu, J.S., Xu, Z., Hao, B.L. (2005) Test data sets and evaluation of gene prediction programs on rice genome. *J. Comput. Sci. & Technol.* 20, 446–453.

27. Ashburner, M., Ball, C.A., Blake, J.A., Botstein, D., Butler, H., Cherry, J.M., Davis, A.P., Dolinski, K., Dwight, S.S., Eppig, J.T. (2000) Gene ontology: tool for the unification of biology. The Gene Ontology Consortium. *Nat. Genet.* 25, 25–29.

28. Harris, M.A., Clark, J., Ireland, A., Lomax, J., Ashburner, M., Foulger, R., Eilbeck, K., Lewis, S., Marshall, B., Mungall, C. (2004) The Gene Ontology (GO) database and informatics resource. *Nucleic Acids Res.* 32, D258–D261.

29. Camon, E., Magrane, M., Barrell, D., Binns, D., Fleischmann, W., Kersey, P., Mulder, N., Oinn, T., Maslen, J., Cox, A. (2003) The Gene Ontology Annotation (GOA) project: implementation of GO in SWISS-PROT, TrEMBL, and InterPro. *Genome Res.* 13, 662–672.

30. Ashburner, M., Ball, C.A., Blake, J.A., Butler, H., Cherry, J.M., Corradi, J., Dolinski, K., Eppig, J.T., Harris, M., Hill, D.P., Lewis, S., Marshall, B., Mungall, C., Reiser, L., Rhee, S., Richardson, J.E., Richter, J., Ringwald, M., Rubin, G.M., Sherlock, G., Yoon, J. (2001) Creating the gene ontology resource: design and implementation. *Genome Res.* 11, 1425–1433.

31. Mulder, N.J., Apweiler, R., Attwood, T.K., Bairoch, A., Bateman, A., Binns, D., Bradley, P., Bork, P., Bucher, P., Cerutti, L. et al. (2005) InterPro, progress and status in 2005. *Nucleic Acids Res.* 33, D201–D205.

32. Mulder, N.J., Apweiler, R., Attwood, T.K., Bairoch, A., Barrell, D., Bateman, A., Binns, D., Biswas, M., Bradley, P., Bork, P. et al. (2003) The InterPro Database, 2003 brings increased coverage and new features. *Nucleic Acids Res.* 31, 315–318.

33. Apweiler, R., Attwood, T.K., Bairoch, A., Bateman, A., Birney, E., Biswas, M., Bucher, P., Cerutti, L., Corpet, F., Croning, M.D. et al. (2000) InterPro–an integrated documentation resource for protein families, domains and functional sites. *Bioinformatics* 16, 1145–1150.

34. Apweiler, R., Attwood, T.K., Bairoch, A., Bateman, A., Birney, E., Biswas, M., Bucher, P., Cerutti, L., Corpet, F., Croning, M.D. et al. (2001) The InterPro database, an integrated documentation resource for protein families, domains and functional sites. *Nucleic Acids Res.* 29, 37–40.

35. Burks, C., Cinkosky, M.J., Gilna, P., Hayden, J.E., Abe, Y., Atencio, E.J., Barnhouse, S., Benton, D., Buenafe, C.A., Cumella, K.E. et al. (1990) GenBank: current status and future directions. *Methods Enzymol.* 183, 3–22.

36. Benson, D.A., Karsch-Mizrachi, I., Lipman, D.J., Ostell, J., Wheeler, D.L. (2005) GenBank. *Nucleic Acids Res.* 33, D34–D38.

37. Bilofsky, H.S., Burks, C., Fickett, J.W., Goad, W.B., Lewitter, F.I., Rindone, W.P., Swindell, C.D., Tung, C.S. (1986) The GenBank genetic sequence databank. *Nucleic Acids Res.* 14, 1–4.

38. Stephens, S.M., Chen, J.Y., Davidson, M.G., Thomas, S., Trute, B.M. (2005) Oracle Database 10g: a platform for BLAST search and Regular Expression pattern matching in life sciences. *Nucleic Acids Res.* 33, D675–D679.

39. Wang, J., Wong, G.K., Ni, P., Han, Y., Huang, X., Zhang, J., Ye, C., Zhang, Y., Hu, J., Zhang, K. et al. (2002) RePS: a sequence assembler that masks exact repeats identified from the shotgun data. *Genome Res.* 12, 824–831.

14

GrainGenes

A Genomic Database for Triticeae and Avena

Helen O'Sullivan

Summary

The GrainGenes website hosts a wealth of information for researchers working on Triticeae species, oat and their wild relatives. The website hosts a database encompassing information such as genetic maps, genes, alleles, genetic markers, phenotypic data, quantitative trait loci studies, experimental protocols and publications. The database can be queried by text searches, browsing, Boolean queries, MySQL commands, or by using pre-made queries created by the curators. GrainGenes is not solely a database, but serves as an informative site for researchers and a means to communicate project aims, outcomes and a forum for discussion. This chapter describes the type of information available on the site and database, and the options available to access this data.

Key Words: Database; genetic maps; Triticeae; *Avena*; comparative mapping; physical mapping; quantitative trait loci; germplasm; genetic stocks.

1. Introduction

The GrainGenes 2.0 database is an extensive web-based resource available online for researchers working on barley, wheat, rye, oat and their wild relatives. The aim of the resource is to aid the flow of data between researchers, pathologists, molecular biologists, geneticists and small-grain breeders to promote the development of new and improved crop varieties *(1)*.

The original database was created in 1992 by the United States Department of Agriculture (USDA). It is a large collaborative project with data provided by

From: *Methods in Molecular Biology, vol. 406: Plant Bioinformatics: Methods and Protocols*
Edited by: D. Edwards © Humana Press Inc., Totowa, NJ

a broad range of scientists working in small grain research. A list of the current contributors can be found at http://greengenes.cit.cornell.edu/contrib.html. Financial support is provided by the International Triticeae Mapping Initiative, through a grant from the USDA/DOE/NSF Joint Program on Collaborative Research in Plant Biology *(2)*.

Researchers are encouraged to contribute to the site by submitting data. Templates are available for multiple data types including genetic probe, map and locus data, quantitative trait loci (QTL) information, germplasm records, and images. There are also guidelines for assigning unique identifiers to GrainGenes references, for example, each reference entered into GrainGenes is required to have a unique alpha-numeric identifier.

The database is hosted at the GrainGenes website http://wheat.pw.usda.gov/GG2/index.shtml and is updated daily. A major element of the GrainGenes database is the integration of a vast quantity of Triticeae and *Avena* genetic data gathered over decades. Information on the genes includes map location, alleles, key references, occurrence in cultivars and disease symptoms *(1)*.

In addition to the large quantity of data stored in the GrainGenes database, sections of the GrainGenes website are dedicated to collaborative work and forums for discussion. For example, coordinating worldwide participation in projects such as expressed sequence tag–simple sequence repeat (EST–SSR) and expressed sequence tag–single nucleotide polymorphism (EST–SNP) molecular genetic marker development. Users of GrainGenes are invited to join the website's mail group, which is a forum for open discussion, requests and announcements.

The website also hosts a number of tools to search, extract and analyse the stored information. It is beyond the scope of this chapter to illustrate all the tools available in GrainGenes and therefore **Subsection 3** describes some of the most useful tools. These include methods for data retrieval, including simple text-based queries that return all objects containing the search term, pre-made quick queries (QQ) and advanced querying methods, and a specific example of querying deletion mapping data.

2. Materials

The GrainGenes website hosts a wealth of information including databases, tools, meeting announcements, employment information, forums and links to useful web resources. Specific data types include genetic and physical maps; genetic markers, including probes, primers and genes; gene expression sequences from the Barley 1 Genechip; genetic stocks and germplasm collections; phenotypes; sequences; QTL; genotypes; and pedigree information.

A number of tools are available on the site to enable a researcher to search, view and compare results. These include cMAP, which allows users to view and compare genetic maps. The vast array of data contained in GrainGenes is organized into classes. All classes of the database can be browsed through the 'browse GrainGenes' button on the menu on the left-hand side of the webpage (*see* **Note 1**). Currently, there are 27 classes of data (*see* **Fig. 1**). Records in

GrainGenes Class Browser

Query (optional) [_____] in Class [All ▾] [GO]

Results: **1755662** Records in **27** Classes

Class	Records
Allele	1164
Author	21181
Colleague	2307
Gene	2861
Gene Class	532
Gene Product	3368
Germplasm	32991
Image	2367
Journal	1166
Keyword	22124
Library	38
Locus	44100
Map	1387
Map Data	157
Marker	57154
Pathology	450
Polymorphism	2799
Probe	26036
Protein	18588
QTL	1056
Rearrangement	838
Reference	13016
Sequence	1497152
Species	1800
Trait	240
Trait Study	192
2 Point Data	598

Fig. 1. The GrainGenes class browser showing 1755562 records contained in 27 classes of the database (June 2006).

these classes frequently link to external plant databases such as the National Center for Biotechnology Information (NCBI) database (*see* Chapters 2 and 3), Gramene (see **Note 2** and Chapter 15) and TIGR (*see* Chapter 5).

The main access point to the GrainGenes database is through a web browser (http://wheat.pw.usda.gov/GG2/index.shtml). As of June 2006, GrainGenes 2.0 is the current version of the database. Prior to GrainGenes 2.0, the database was known as 'GrainGenes Classic' and was maintained and accessed using ACEDB (*see* **Note 3**) software. GrainGenes classic is still available and is located at http://wheat.pw.usda.gov/ggpages (*see* **Note 4**). For users accustomed to ACEDB it is possible to download the GrainGenes ACEDB database and run a local installation on your own Windows95/NT, Unix or Macintosh computer (*see* **Note 5**) *(3)*. The GrainGenes classic database is updated daily or as required. Standalone versions can be updated by replacing any previous release with the newer version (*see* **Note 6**). However, using the online version is preferential as it ensures that the most recent version of the GrainGenes database is automatically deployed.

GrainGenes version 2.0 is a relational version of the GrainGenes classic database, implemented using the MySQL relational database management system *(4)*. The method of the database conversion from ACEDB to MySQL is described at http://wheat.pw.usda.gov/ggmigration/. The process of reorganizing the data in GrainGenes 2.0 resulted in a more robust database that is superior to the original GrainGenes classic database *(5)*.

3. Methods

3.1. Organization and Navigation of the GrainGenes Homepage

The GrainGenes database can be accessed through the following homepage http://wheat.pw.usda.gov/GG2/index.shtml. The database is only one component of the information available from the homepage. A toolbar is present on the left-hand side of the website (*see* **Fig. 2**). This displays the tools available, the data types contained in the database that can be queried, other web resources contained on the website, and user services. The centre of the homepage displays a featured tool that changes upon refreshing. In **Fig. 2.**, the tool displayed is the Quick Queries tool, which allows users to carry out a complex query through a click of a button. This is described further in **Subsection 3.3**.

A list of 'hot topics' and 'up and coming meeting announcements' can also be found on the homepage (as shown on the right hand side of **Fig. 2**). The bottom right of the homepage contains a 'featured link' of relevant web resources for users. For example, the site 'Wheat Pedigree and Identified Alleles of Genes

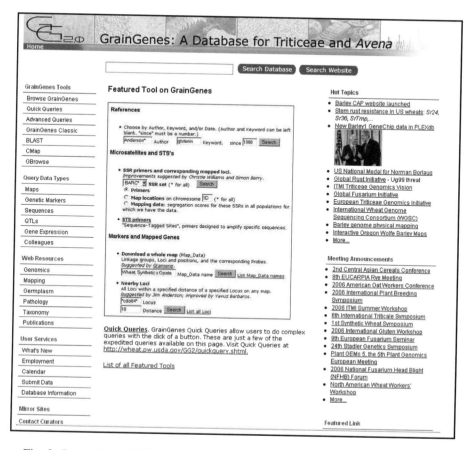

Fig. 2. Screenshot of GG homepage demonstrating the toolbar on the left hand side and 'Hot topics' and meeting announcements on the right.

online' *(6)* contains genealogies and identified alleles of genes for 72,848 wheat accessions, defined by 3034 information sources (as of June 2006).

3.2. Querying the GrainGenes Database

The GrainGenes database, like many databases, can be accessed through text searches and Boolean queries (*see* **Note 7**), as well as by browsing *(1)*. To browse the classes of the database, press the 'Browse GrainGenes' button on the toolbar. This allows you to click on the required class. However, this is not the most direct approach for retrieving information. Another way is to query the data types (also on the toolbar). By browsing, the user can retrieve information on maps, genetic markers, sequences, QTL, gene expression and contact information for GrainGenes members.

Simple text-based queries can be carried out by typing text into the search box at the top of the homepage. Wild cards are accepted, use '*' as the wild card character for searches. This allows for an expanded search using fewer characters. The user can choose to search either the database indices or the website (*see* **Note 8**) by pressing the relevant button. This returns links to all objects containing the search term (*see* **Note 9**).

The database curators allow direct access to the GrainGenes database through SQL queries using a web interface (http://wheat.pw.usda.gov/cgi-bin/graingenes/sql.cgi). It is rare to provide unrestricted SQL access to databases such as this, and it may involve some risk. However, the curators encourage users to perform such queries as this is a very powerful way of accessing data and the perceived level of risk is low (*see* **Note 10**). Useful information on how to use the GrainGenes SQL interface is available at (http://wheat.pw.usda.gov/GG2/SQLhelp.shtml).

3.3. Quick Queries (QQ)

GrainGenes QQ allow users to carry out complex queries with a click of a button and provide rapid access to some of the most frequently asked questions from GrainGenes users (*see* **Note 11**). Several pre-made queries are available at http://wheat.pw.usda.gov/GG2/quickquery.shtml.

The QQ section is a particularly useful place to start when a user is seeking access to specific information. Many aspects of the database are covered by the pre-made queries. A list of the QQ and the type of information that they yield are summarized in **Table 1**.

QQ functions are intuitive and easy to use and it is possible to restrict the search results based on many options. Once the user has selected a query, it will execute immediately and display the results. An example of one of the QQ is shown in **Fig. 3**. In this scenario, the user wants to identify the map location for all available BARC microsatellite markers on chromosome 1D (*see* **Note 16**). To do this, the user must select the BARC* option for the SSR set, click the 'Map locations' button and press search. Many other options for searching microsatellites markers are available to the user. Entering the '*' character will search for all microsatellites markers or all map locations. User can also obtain primer information and mapping data. **Figure 4** shows the result page for this QQ. Records can be browsed by clicking on the links. The top of the page displays the SQL query and from here it is possible to edit and submit a revised query (*see* **Note 17**).

Table 1
Summary of Pre-Made 'Quick Queries' Available to Access the GrainGenes Database

Quick query search type	Information that can be obtained with quick query
References	Research articles. Search via author, keyword and/or date (*see* **Note 12**)
Microsatellites and sequence-tagged sites	Primers, map locations and mapping data
Markers and mapped genes	Maps, mapping scores, linkage groups, loci and positions, and the corresponding probes, nearby genes (*see* **Note 13**), loci between two markers, trait markers (*see* **Note 14**), genes mapped to a specified chromosome group and probes mapped to a specified chromosome group
Sequences	Mapped sequences, sequenced genes, cloned genes mapped in a specific species (*see* **Note 15**) and mapped sequence tag sites
QTL	Can search for QTL for a specified trait or lists all QTL traits
Genes	Barley genes
Polymorphisms	Allele differences between two germplasm lines
Address book	A list of plant breeders and researchers who have submitted their details or contributed to the website

QTL, quantitative trait loci.

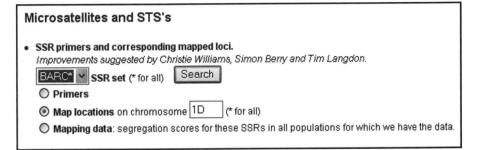

Fig. 3. Example of the GrainGenes quick query tool.

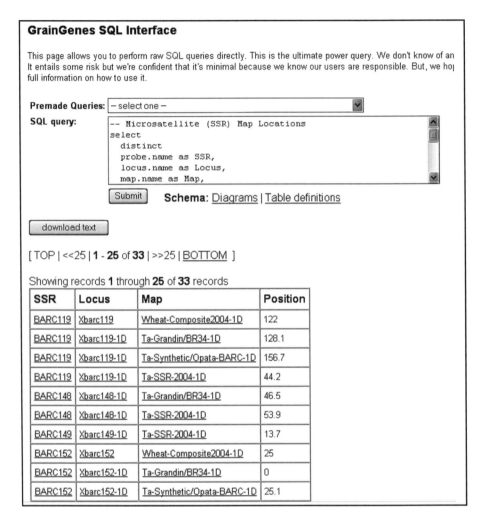

Fig. 4. Results from the quick query tool, searching for BARC microsatellite markers on chromosome 1D.

3.4. Advanced Queries

There are several other advanced modes for searching the GrainGenes database. For a field-based search, users can search the database by a combination of fields. This results in a list of matching records.

If the user is familiar with the SQL querying language, this is the most powerful way to access the database. This allows users to perform custom

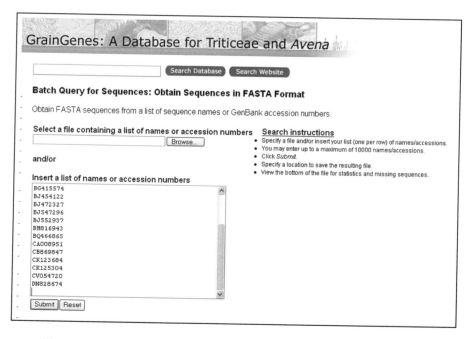

Fig. 5. Example of a Batch query to obtain sequences in FASTA format.

SQL queries directly on the database. The GrainGenes Batch SQL Interface accepts lists of names and can restrict query results to those items. The pre-made queries at the bottom of the menu are helpful in designing queries (*see* **Note 18**).

A very useful aspect of the advanced query facility is that a batch query may be used to retrieve lists of sequences in FASTA format. Enter a list of sequence names or accessions, either by browsing for a file or inserting a list (*see* **Note 19**). Click submit and choose a location to save the results (*see* **Fig. 5**). The resulting file is a concatenated list of FASTA files. The data can be downloaded by specifying a location to save the file. The bottom line of the file contains the information related to the success of the query, for example, 19 sequences requested, 19 obtained.

3.5. BLAST Facility in GrainGenes

Over recent years, large-scale sequencing projects have produced an enormous amount of EST sequence information for Triticeae species. The sequence data class of the GrainGenes database holds sequences from EST projects, contig assemblies and genome sequencing efforts. Sequence data are obtained from NCBI with quarterly updates. As of March 2006, GrainGenes contained 1,190,000 sequences from *Triticum, Aegilops, Hordeum, Secale, Avena* and relatives.

The BLAST tool *(7)* is available on the GrainGenes website and it is possible to either use all GrainGenes sequences or a subset of the sequence data class for BLAST searches. The BLAST tool can be accessed through the toolbar or directly at http://wheat.pw.usda.gov/GG2/blast.shtml. Users can choose the specific BLAST program (*see* **Note 20**) and the database to search. The sequence databases available are listed on http://wheat.pw.usda.gov/blast/blast_databases.html. This page can be accessed by clicking the 'Database' button on the BLAST page. With the exception of 'All GrainGenes sequences' the databases are available for download files in FASTA format.

Examples of databases available for BLAST search include the following:

(1) Poaceae EST–SSRs: Species-specific contigs from microsatellite-containing ESTs from wheat, barley, rice, maize and sorghum. This tool enables researchers to identify microsatellite markers in their genes of interest.
(2) Another very useful database to use is the 'Mapped wheat ESTs'. This is a collection of ESTs that have been mapped in the Chinese Spring deletion lines by the NSF wheat EST project. ESTs were mapped to chromosome locations defined by deletion line breakpoints and other cytogenetic stocks *(8)*. This tool enables a researcher to determine the chromosomal location in wheat for genes of interest.

In order to carry out the BLAST search, the user should choose the BLAST program and database and enter the sequence of interest in FASTA format (*see* **Note 21**). The parameters given are the BLAST default parameters and should be suitable for most applications. Simply press search and the individual BLAST hits resulting from the search will be shown.

The following is an example of using the BLAST facility in GrainGenes to identify the chromosomal location of a wheat EST.

1. Go to http://wheat.pw.usda.gov/GG2/blast.shtml or click on 'BLAST' on the tool bar.
2. For program, choose 'BLASTn'. For the database, choose 'Mapped Wheat ESTs'.
3. Enter your query sequence in FASTA format (*see* **Note 22**).
4. Press search.
5. This returns a list of sequences producing significant alignments. Contained in the list is the chromosomal location of the EST (*see* **Fig. 6**).
6. If no significant matches are found this indicates that the corresponding EST was not mapped in the NSF project.

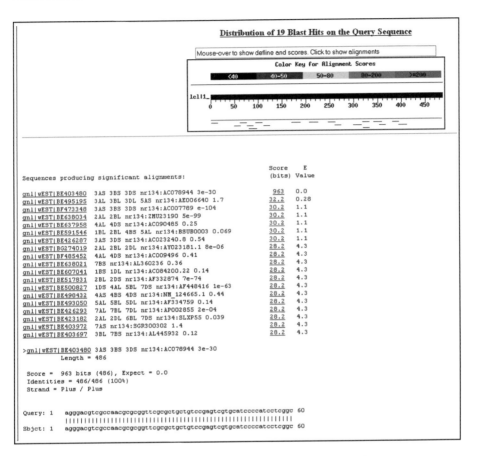

Fig. 6. A list of sequences producing significant alignments with the query sequence.

7. The top of the list is the sequence with the most significant alignment score (*see* **Fig. 6**). In this example, the query sequences matches exactly with a sequence in the database (E-value = 0.0, 100% identity).

8. Clicking on the link opens a webpage containing further information on the mapped EST (*see* **Fig. 7** for the results table). Information includes the chromosome the EST maps to, the bin reference and the images from the mapping experiments (*see* **Note 23**). It is possible to view more information in the results table, such as BLAST descriptions. This can be done by selecting the relevant columns to view and pressing submit (*see* **Note 24**).

9. Results can be downloaded as text, or the user can obtain a sequence file in FASTA format for the relevant probe ESTs.

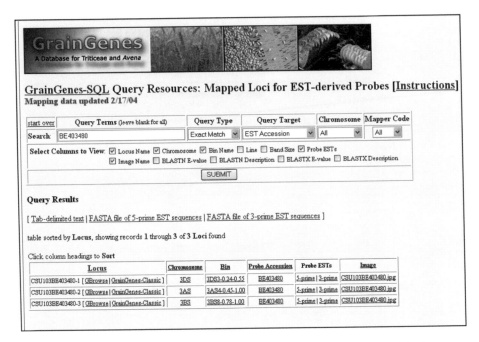

Fig. 7. Information on the mapped expressed sequence tag (EST) loci.

4. Notes

1. For example, following the link to 'Library' returns a list of all records of all the cDNA libraries in the database. Contained within each link are the details of the library, including the type, species, germplasm, developmental stage, tissue, treatment, vector and source.
2. Gramene is closely related to the GrainGenes database, with the focus of Gramene being inter-species comparisons among the Poaceae.
3. ACEDB is an acronym for A *Caenorhabditis elegans* Database. It refers to a database system originally developed for data on *C. elegans*.
4. This homepage is no longer updated and will not contain current information.
5. This can be obtained by ftp from www.graingenes.org.
6. It is possible to maintain a copy of GrainGenes classic by incremental updating. The software package 'Mirror' can be used for this purpose and downloaded from the following site ftp://www.graingenes.org/pub/GrainGenes/more/.
7. Boolean operators; 'AND', 'OR', 'NOT' can be used in the search field.
8. Search database: use this to search the GrainGenes MySQL database. This search returns data associated with the general classes in the database. Search website:

use this to confine the search to the webpages of GrainGenes, the database will not be searched.

). From the results page it is possible to limit the search by either searching just the title or the body and title.

). If the user experiences problems creating queries, the GrainGenes curators may be able to assist and help develop the query for you. They can be contacted through curator@pw.usda.gov.

1. The curators are keen to receive suggestions of queries you may want to explore, and can be contacted at matthews@greengenes.cit.cornell.edu.

2. GrainGenes contains some references that are not found in other databases such as PubMed or Web of Science. Therefore, it is worthwhile conducting literature searches on the site as well as using other sources.

3. It is also possible to search for all loci within a specified distance of a specified locus on any map. This section also allows the user to search for nearby genes, which will return genes near the specified locus.

4. Linkages between useful genes and molecular markers.

5. The default setting is barley.

6. There is not an option to choose a species in which to search for specific microsatellite markers. However, if the user wants to specify the species, searching based on a specific chromosome, for example, 1H Barley or 1A, 1B or 1D in hexaploid wheat, can do this.

7. This is only recommended for users who have some experience of using MySQL commands.

8. The QQ are also implemented using the SQL interface.

9. It is possible to enter up to a maximum of 10,000 names/accessions.

0. The following is a list of programs available for the BLAST search:

    a. BLASTn: Compares a nucleotide query sequence against a nucleotide sequence database.

    b. TBLASTn: Compares a protein query sequence against a nucleotide sequence database dynamically translated in all reading frames.

    c. TBLASTX: Compares the six-frame translations of a nucleotide query sequence against the six-frame translations of a nucleotide sequence database. Please note that tblastx program cannot be used with the nr database on the BLAST Webpage.

    d. BLASTp: Compares a protein query sequence against a protein sequence database.

    e. BLASTX: Compares a nucleotide query sequence translated in all six reading frames against a protein sequence database.

1. A sequence in FASTA format begins with a single-line description, followed by lines of sequence data. The description line is distinguished from the sequence data by a greater than ('>') symbol in the first column. It is recommended that all lines of text be shorter than 80 characters in length.

22. In order to follow this example, the user will need to first get the sequence. Go to 'Browse GrainGenes' (on the toolbar). In the query box type the following accession number 'BE403480'. In the class box, select 'sequence'. Press go. This returns a GrainGenes sequence report 'BE403480'. The sequence for this EST is at the bottom of the page. Highlight and copy it. Return to the BLAST page, choose 'BLASTn' for the database, choose 'Mapped wheat ESTs', then paste the sequence into the space provided.

23. The image file name is linked to a view of the Southern blot autoradiograph image, along with a table of the mapped loci represented on that image.

24. From this location the user can restrict the search to Chromosome/Arm, Chromosome group or Genome. The user can also search for mapped loci based on the Locus Info (Locus Name, Bin Name, Deletion Line(s) and Mapper remarks), or Probe EST Info (EST Name, EST GenBank Accession and EST Best BLASTN/X result description).

5. References

1. Matthews, D.E., Carollo, V.L., Lazo, G.R., and Anderson, O.D. (2003) Grain-Genes, the genome database for small-grain crops. *Nucleic Acids Res.* 31, 183–186.
2. Matthews, D.E. (1999) GrainGenes: A Database for Triticeae and *Avena*. http://wheat.pw.usda.gov/ggpages/aboutgg.html.
3. Matthews, D. (1997) ACEDB Genome Database Software FAQ. http://www.faqs.org/faqs/acedb-faq/.
4. Carollo, V., Matthews, D.E., Lazo, G.R., Blake, T.K., Hummel, D.D., Lui, N., Hane, D.L., and Anderson, O.D. (2005) GrainGenes 2.0. An improved resource for the small-grains community. *Plant Physiol.* 139, 643–651.
5. Hummel, D. (2004) Progress on Migration of GrainGenes from ACEDB to RDBMS. http://wheat.pw.usda.gov/ggmigration/.
6. Martynov, S.P. (2005) Wheat Pedigree and Identified Alleles of Genes online. http://genbank.vurv.cz/wheat/pedigree/.
7. Altschul, S.F., Gish, W., Miller, W., Myers, E.W., and Lipman, D.J. (1990) Basic local alignment search tool. *J. Mol. Biol.* 215, 403–410.
8. The Structure and Function of the Expressed Portion of the Wheat Genomes. http://wheat.pw.usda.gov/NSF/.

15

Gramene
A Resource for Comparative Grass Genomics

Doreen Ware

Summary

Grasses are one of the largest agricultural crops, providing food, industrial materials and renewable energy sources. Due to their large genome size and the number of the species in the taxa, many of the genomes are not targeted for complete sequencing. Gramene seeks to provide basic researchers, industry and educators with a resource that can be used as a tool for knowledge discovery across grass species. This chapter briefly outlines system requirements for end users and database hosting, outlines data types and basic navigation within Gramene and provides an example of how a maize researcher would use Gramene to leverage rice genome organization and phenotypic information to support targeted experimental research in maize.

Key Words: Rice; maize; wheat; barley; genome sequence; comparative genomics.

1. Introduction

Gramene seeks to provide the plant community with a resource for the comparative genomics of grass species where reference genomes and their respective annotations, including sequence and phenotypes, serve as a scaffold species where the complete genome sequence is unavailable.

Because of the high degree of conservation of gene order in grass species and conserved gene function between closely related species, the information available for a reference grass genome is useful to make informed decisions about genome organization, gene function and potential phenotypes in related species. In comparison, the genome information for *Arabidopsis* may provide

From: *Methods in Molecular Biology, vol. 406: Plant Bioinformatics: Methods and Protocols*
Edited by: D. Edwards © Humana Press Inc., Totowa, NJ

researchers with detailed information on potential gene function, though not necessarily genome organization.

The Gramene Genome Browser *(1,2)* hosts an implementation of the EnsEMBL Genome Browser *(3)* containing collated information on rice genes. This information includes predicted function as assigned by GO annotations *(4,5)* and Pfam domain similarity *(6)*, single-nucleotide polymorphisms (SNPs) and rice phenotypes [as quantitative trait loci (QTL)]. Comparative genome information is provided by the alignment of plant expressed sequence tags (ESTs) with the best rice ortholog, and maize synteny information is based on the maize physical map. The front Web page acts as portal to internal pages in Gramene as well as external links to other databases (*see* **Table 1**). Internal links within Gramene include Mapviewer (an implementation of the CMap comparative genetic viewer developed as part of the GMOD consortium), markers, genes, QTL and an ontology browser. Contextual links to external sources include reference resources such as GenBank *(7)* (*see* Chapters 2 and 3), model organism databases such as MaizeGDB *(8)* (*see* Chapter 16) or GrainGenes *(9,10)* (*see* Chapter 14) and project specific databases such as Barleybase *(11)* (*see* Chapter 17) and Panzea *(12,13)* (*see* **Table 1**).

2. Materials

2.1. Hardware and Software Requirements for Users

A computer with internet access and a standard web browser such as Mozilla/Firefox, Netscape 6 and above, or Safari.

2.2. Gramene System Components

Gramene is a web-based application that allows users to search and view biological data, making use where appropriate of graphics viewers such as the EnsEMBL genome browser or the CMap genetic and comparative map package. Data are maintained in distinct relational databases (MySQL), and users connect to the site using a standard web browser. User queries for static (HTML) and dynamic content are negotiated by the Apache web server and a middleware layer written in Perl. Bulk downloads of data are provided through an FTP site.

2.3. Local Installation of Gramene

Local installation of Gramene requires a computer running a Unix operating system (such as Linux, FreeBSD and Solaris), Perl (5.6.1 or greater), Apache 1.x (1.3.26 or greater, *see* **Note 1**), MySQL (4.1 or greater), Perl modules (Comprehensive Perl Archive Network, CPAN), CMap (0.15) and the EnsEMBL

Table 1
Links to External Websites, Databases

Resource	URL	Data type
EBI (InterPro)	http://www.ebi.ac.uk/interpro	Protein families
EMBL (*see* Chapter 1) (SMART)	http://smart.embl-heidelberg.de	Protein domains
ExPASy (PROSITE)	http://www.expasy.ch/prosite	Protein domains
GenBank (*see* Chapters 2 and 3)	http://www.ncbi.nlm.nih.gov/	Protein, gene
Gene Ontology Consortium (*see* Chapter 24)	http://www.geneontology.org	Protein, gene, Ensembl genes
Georgetown University (PIR)	http://pir.georgetown.edu/pirwww/index.shtml	Protein domains
Graingenes (*see* Chapter 14)	http://www.graingenes.org	ESTs
GRIN	http://www.ars-grin.gov/	Protein
INRA (ProDom)	http://protein.toulouse.inra.fr/prodom/current/html/home.php	Protein domains
IRRI (*see* Chapter 22)	http://www.iris.irri.org/	Protein
KEGG (*see* Chapter 21)	http://www.genome.ad.jp/	Protein
KOME	http://cdna01.dna.affrc.go.jp/cDNA	cDNA sequences
Laboratory for Genomics & Bioinformatics	http://www.fungen.org/Sorghum.htm	ESTs, clustered ESTs

(Continued)

Table 1 (Continued)

Resource	URL	Data type
MaizeGDB (*see* Chapter 16)	http://www.maizegdb.org	ESTs, clustered ESTs, gene
Maize Oligonucleotide Array Project	http://www.maizearray.org	Microarray target sequences
Manchester University (SPRINT)	http://umber.sbs.man.ac.uk/dbbrowser/sprint	Protein domains
MGOS	http://www.mgosdb.org	SAGE tag sequences
MRC-LMB (Superfamily)	http://supfam.mrc-lmb.cam.ac.uk/SUPERFAMILY	Protein domains
NASC (*Arabidopsis*) (*see* Chapter 9)	http://atensembl.arabidopsis.info	Gene models
NCBI (GenBank, dbEST)	http://www.ncbi.nih.gov/dbEST/index.html	ESTs
NCBI (GenBank, dbGSS) (*see* Chapters 1 and 2)	http://www.ncbi.nih.gov/dbGSS/index.html	Genome Survey Sequences, Bacterial Artifical Chromosome (BACs), Methyl-filtered/Hi-cot selected reads, FSTs
NCBI (GenBank, dbPLN)	http://www.ncbi.nih.gov/Genbank/index.html	CDS sequences, cDNA sequences, Genomic clone sequences
NSF Rice Oligo Array Project	http://www.ricearray.org	Microarray oligonucleotide sequences

Database	URL	Content
Oryzabase	http://www.shigen.nig.ac.jp/rice/oryzabase/top/top.jsp	Gene
PlantGDB (*see* Chapter 25)	http://www.plantgdb.org	Clustered ESTs, FST sequences
PLEXdb (*see* Chapter 17)	http://www.plexdb.org	Microarray oligonucleotide sequences, Microarray target sequences
Sanger (Pfam)	http://www.sanger.ac.uk/Software/Pfam	Protein domains
TIGR (Gene Indices) (*see* Chapter 5)	http://www.tigr.org/tdb/tgi/plant.shtml	Clustered ESTs
TIGR (Rice Genome Annotation)	http://www.tigr.org/tdb/e2k1/osa1	Gene models
TIGR (TIGRFAM)	http://www.tigr.org/TIGRFAMs	Protein domains
TIGR (TIGR Maize Database)	http://maize.tigr.org	Clustered Genome Survey Sequences
Uniprot (*see* Chapter 4)	http://us.expasy.org	Protein
University of Delaware Massively Parallel Signature Sequencing (MPSS) (*see* Chapter 19)	http://mpss.udel.edu/	MPSS tag sequences

EST, expressed sequence tag; FST, flanking sequence tag.

genome browser. The current Gramene server (http://www.gramene.org/) is running on a Fedora Core 1 Linux machine with dual CPUs, 6 GB memory and 600 GB of disk space available.

3. Methods
3.1. Basic Navigation of the Gramene Website

Gramene contains many different data types (*see* **Table 2**) that are maintained as multiple modules within the database. The main entry point for the system is through the front Web page (http://www.gramene.org/) (*see* **Note 2**). Every Gramene page contains the main navigation bar as well as module specific navigation bars, a general search, a link to the site map and a feedback page. The main navigation bar is found at the top of each Gramene page and is the

Table 2
Major Data Types in Gramene

Data type	Description
Sequence	Genome assembly such as rice pseudomolecules, maize BACs, Arabidopsis assembly, Genomic survey sequences such as BAC ends, mRNAs such as full-length cDNA, ESTs
Map	Sequence, genetic, physical, Maize Bin, Deletion, QTL
Marker	Gramene has a broad definition for Marker data type, which include all features located on chromosomes, eg, SSR, RFLP, AFLP, RAPD markers, BAC end sequences, BAC Clones, cDNAs, ESTs, Gene Primers, STS, Tos17 FSTs
Gene	Genetically defined phenotype genes, sequenced genes with or without experimental evidence, tRNAs, rRNAs, pseudogenes
Protein	Swiss-Prot/TREMBL, from predicted gene models, families and domains
Diversity	Variations with species on the same allele, e.g., SNPs, allelic data of SSR, genes
Pathway	Biochemical or genetic pathways based EcoCyc
Germplasm	Related to the species and other data in Gramene
Controlled Vocabularies	GO, TO, EO, PO, GRO, cross-reference to other data in Gramene
Reference	Publications associated with the data types and methods associated with data integration

BAC, bacterial artificial chromosome; EST, expressed sequence tag; QTL, quantitative trait loci.

main entry point to the search modules (Search), genome browsers (Genomes), software and data sets (Downloads), general information (Resources), help documents (Help), and information about the project and participants (About). In addition to the navigation bar, there is a general search which can be refined to interrogate the different data modules, as well as the Feedback link. The Feedback link is set up to provide the user with a comment page where the URL from the page the user was viewing at the time of the response is automatically included in the message. From the main navigation bar, a dropdown menu is available from each of the main headings containing refined subjects. For example, from the Help menu, entry points are available to the site map, help documents, release notes, tutorials, frequently asked questions (FAQs), workshop exercises, mailing list, rice gene nomenclature and ask us feedback links. By selecting the site map, a user is provided with detailed set of options available within each one of the search modules, as well as information available from reference pages, data analysis, help documents and downloads within Gramene. The interfaces within Gramene are interactive, providing the user with links to external reference databases (*see* **Table 1**) as well as links to internal modules with in Gramene.

3.2. Example Use of Gramene

Within the constraints of this chapter it is will not be possible to go through all of the Gramene interfaces. Instead, this example examines a sample question and walks through how to use Gramene to obtain information to facilitate genomic research. In this example, a maize line expressing a mutant phenotype has identified candidate genes for the mutation based on microarray gene expression data. We would like to identify candidate rice genes orthologous to the differentially expressed maize genes, as well as related genes from other cereals, ascertain the potential function of these rice genes based on annotation of the region in rice, and use synteny to identify candidate genomic position in maize.

In this example, we will use the BLAST search module as an entry point to the Rice Genome Browser.

From the Gramene home Page select Search from the main navigation bar and select "Sequences-BLAST". This will take the user to the BLAST home page where a sequence can be entered either pasting it directly into the sequence box or uploading a local file (*see* **Fig. 1**). Once the sequence has been entered we must identify the sequence type by selecting "DNA queries" or "peptide queries". In this example, we are using a DNA sequence and check "DNA queries". Next we must determine which sequence database to query. The

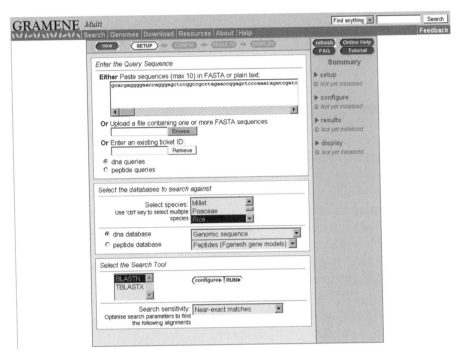

Fig. 1. The Gramene BLAST home page.

Gramene sequence databases are composed of reference genomes available in the Genome Browser module (*Arabidopsis*, rice and maize) as well as the sequence of features aligned to these reference genomes. Sequences are organized first by species and then by sequence type. In this example we select the species "Rice" and then "Genomic sequence". The search tool used will be "BLASTN" and the search sensitivity "Near-exact matches".

The Gramene BLAST result page contains three views, the alignment locations vs. karyotype, the alignment locations vs. query sequence and the alignment summary. All three views are interactive and allow the user to view details of the alignment and link to other pages within Gramene.

In the alignment location vs. karyotype view (*see* **Fig. 2**), the top scoring hit is highlighted by mousing over the glyph, we are presented with a drop down box containing links to the BLAST alignment with the genome (align), the alignment in context to the submitted sequence (query sequence) or genomic sequence (genomic sequence), and a link to the rice genome browser (ContigView).

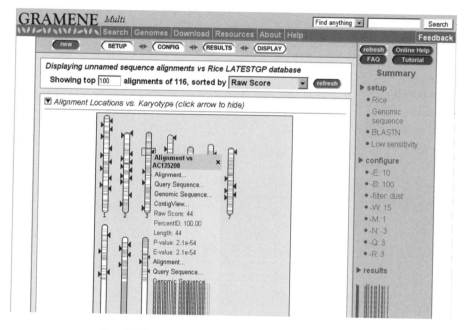

Fig. 2. The alignment location vs. karyotype view.

By selecting ContigView we leave the BLAST search module and enter the rice genome browser (*see* **Note 3**).

The ContigView of the EnsEMBL genome browser contains four views, the Chromosome or Clone view, Overview, Detailed view and Basepair view. All views contain a small red box that positions the user for each successive view. Views can be turned off or on by right clicking on the "+" or "−" symbol. In nearly complete genomes such as rice and *Arabidopsis*, the highest level view is of a chromosome, whereas in emerging genome sequences that use a bacterial artificial chromosome (BAC) based sequencing approach (such as maize), the highest level sequence view is of a clone or BAC.

The Chromosome or Clone view provides users with a representation of the complete pseudomolecule (a chromosome in the case of rice) and the position (in a red box) represented by the Detailed view.

The Overview provides the user with information on long range-derived features such as QTL, and synteny in the context of smaller sequence features such as genes and markers (*see* **Fig. 3**).

The Detailed view (*see* **Fig. 4**) provides semantic zooming, customizable tracks and links to internal pages, as well as contextual links to data sources

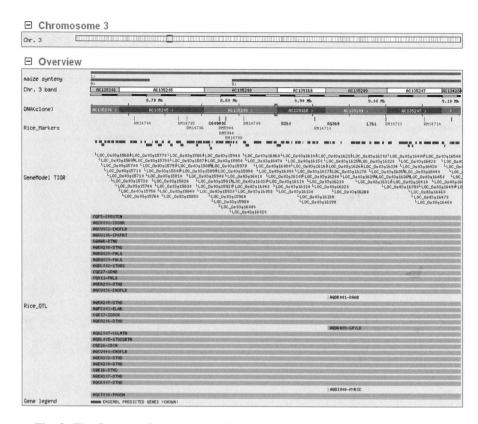

Fig. 3. The Gramene Overview provides information on long range features such as quantitative trait loci (QTL) and synteny.

outside of Gramene (*see* **Table 1**). Feature types include genes, expressed sequences, genomic survey sequences, QTL, gene expression microarray probes and small RNAs (*see* **Table 2**) and provide internal links within Gramene and contextual links to reference databases such as Genbank, Model Organism databases such as MaizeGDB, or project specific databases such as BarleyBase *(6)* or Panzea *(12,13)*.

The Basepair view displays the rice genome sequence for both the forward and reverse strands as well as predicted amino acid sequences representing the three potential reading frames for each strand (*see* **Fig. 4**).

From the Chromosome view and Overview we can see that the maize sequence with the highest BLAST hit aligns with rice chromosome 3, about one quarter of the way down the chromosome at ~8.87 MB. The BLAST hits are

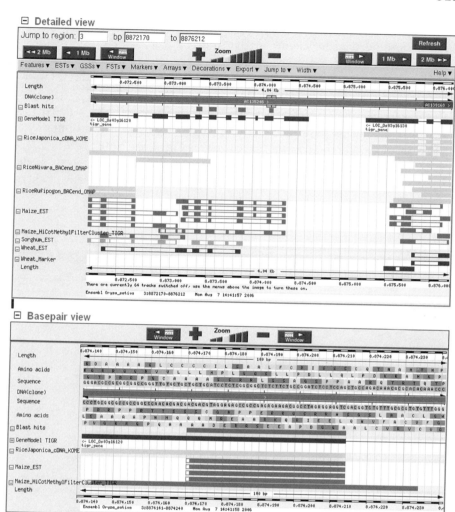

Fig. 4. The Detailed view provides semantic zooming, customizable tracks, links to internal pages as well as links to external data. The Basepair view displays the rice genome sequence for both the forward and reverse strands as well as predicted amino acid sequences representing the three potential reading frames for each strand.

displayed as the first feature track in the Detailed view (*see* **Fig. 4**, *see* **Note 4**). This region is annotated with rice QTL and is syntenic with maize chromosomes 1 and 9. The region also contains 21 different QTL reflecting 11 traits (grain number, soluble protein content, 1000 seed weight, seed number, spikelet fertility, seed dormancy, chalky endosperm, spikelet number, root number, and days to heading panicle length) in five trait categories (yield, biochemical, vigor, quality, development, and anatomy) (*see* **Fig. 3**). For more detailed information we can link directly to the QTL database for details of the QTL and information on the studies and trait descriptions.

In this example, the BLAST hit overlaps the rice gene model LOC_os03g16120. The cereal alignments show that the rice gene model has sequence similarity to ESTs from wheat, sorghum as well as maize (*see* **Fig. 4**, *see* **Note 5**).

The cereal features are a useful validation for the rice gene model structure and for the development of species-specific molecular genetic markers.

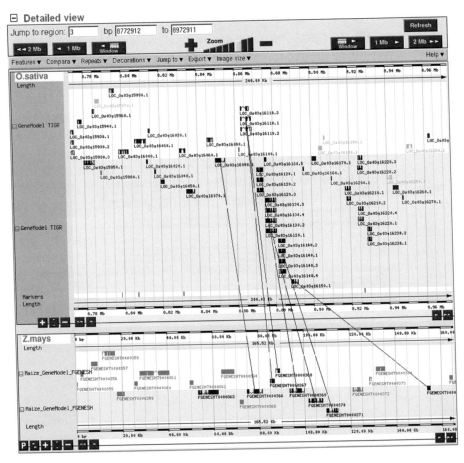

Fig. 5. MulticontigView between rice and maize.

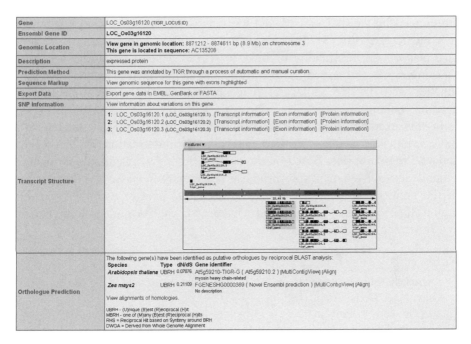

Transcript/Translation Summary

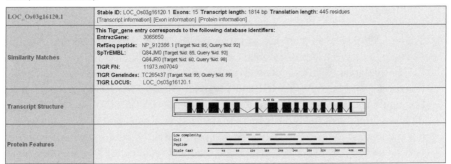

Fig. 6. Gene model page displaying gene annotation.

Sequence features also provide links to the alignment, reference databases such as GenBank, or Model Organism databases such as MaizeGDB, or project specific databases such as BarleyBase or Panzea. A graphical representation of this analysis is available from a Multicontig view (*see* **Fig. 5**). A link to the Multicontig view is available from the gene model page (*see* **Fig. 6**).

In this example the user is able to view several rice and maize genes that have their respective best reciprocal hits within the same genomic region, as represented by the maize BAC and the rice assembly, suggesting this region is syntenic (*see* **Fig. 5**). In addition to the ortholog information, the user may also obtain information or links to related genomic location, transcript structure, SNPs and protein. This information can be used to suggest potential function and develop molecular reagents for marker assisted selection or genome walking.

In this example, we entered Gramene through the BLAST module and traversed to the genome browser view to view information on the related genomic region in rice. Based on the position in rice we were able to identify syntenic locations in maize on chromosomes 1 and 9. We assessed the function of the predicted orthologous rice gene and were able to ascertain that this region of the genome in rice is associated with yield, vigor, protein content, and flowering time phenotypes.

Notes

1. Apache 2.x is not supported yet due to significant differences in the persistent Perl interpreter module (mod_perl).
2. As Gramene matures in the next 18 months it is anticipated that much of the steps described here will become part of GrameneMart and will reduce the independent steps that a user would need to answer this question within Gramene.
3. Alternatively, we could enter the ContigView by selecting "[c]" in the alignment summary.
4. The tracks displayed can be adjusted by the user. For the purpose of this example, we have selected features: GeneModel_TIGR, Rice_CDS and Rice_QTL, EST; RiceJaponica_cDNA_Kome, maize, wheat, rice and sorghum and GSS; MaizeBacEnd Maize_WGS_JGI and Sorghum_Methylfiltered_Orion.
5. For genomes that are not completely sequenced, ESTs may provides the best representation of orthologous genes. In the case of a genome that is sequenced or has sequencing in progress, ESTs provide potential orthologous gene sets based on best reciprocal hit, many best reciprocal hits, and reciprocal hit based on synteny around a best reciprocal hit.

References

1. Jaiswal, P., Ni, J., Yap, I., Ware, D., Spooner, W., Youens-Clark, K., Ren, L., Liang, C., Zhao, W., Ratnapu, K., Faga, B., Canaran, P., Fogleman, M., Hebbard, C., Avraham, S., Schmidt, S., Casstevens, T.M., Buckler, E.S., Stein, L., and McCouch S. (2006) Gramene: a bird's eye view of cereal genomes. *Nucleic Acids Res.* 34, D717–D723.
2. Ware, D., Jaiswal, P., Ni, J., Pan, X., Chang, K., Clark, K., Teytelman, L., Schmidt, S., Zhao, W., Cartinhour, S., et al. (2002) Gramene: a resource for comparative grass genomics. *Nucleic Acids Res.* 30 (1), 103–105.

3. Birney, E., Andrews, T. D., Bevan, P., Caccamo, M., Chen, Y., Clarke, L., Coates, G., Cuff, J., Curwen, V., Cutts, T., Down, T., Eyras, E., Fernandez-Suarez, X. M., Gane, P., Gibbins, B., Gilbert, J., Hammond, M., Hotz, H.R., Iyer, V., Jekosch, K., Kahari, A., Kasprzyk, A., Keefe, D., Keenan, S., Lehvaslaiho, H., McVicker, G., Melsopp, C., Meidl, P., Mongin, E., Pettett, R., Potter, S., Proctor, G., Rae, M., Searle, S., Slater, G., Smedley, D., Smith, J., Spooner, W., Stabenau, A., Stalker, J., Storey, R., Ureta-Vidal, A., Woodwark, K. C., Cameron, G., Durbin, R., Cox, A., Hubbard, T., and Clamp, M. (2004) An overview of Ensembl. *Genome Res.* 14, 925–928.

4. Lomax, J. (2005) Get ready to GO! A biologist's guide to the Gene Ontology. *Brief. Bioinform.* 6, 298–304.

5. Clark, J.I., Brooksbank, C. and Lomax, J. (2005) It's all GO for plant scientists. *Plant Physiol.* 138, 1268–1278.

6. Bateman, A., Coin, L., Durbin, R., Finn, R.D., Hollich, V., Griffiths-Jones. S., Khanna, A., Marshall, M., Moxon, S., Sonnhammer, E.L.L., Studholme, D.J., Yeats, C., Eddy, S.R. (2004) The Pfam protein families database. *Nucleic Acids Res.* 32, D138–D141.

7. Benson, D.A., Karsch-Mizrachi, I., Lipman, D.J., Ostell, J., and Wheeler, D.L. (2006) Genbank. *Nucleic Acids Res.* 34, 16–20.

8. Lawrence, C.J., Dong, Q., Polacco, M.L., Seigfried, T.E., and Brendel, V. (2004) MaizeGDB, the community database for maize genetics and genomics. *Nucleic Acids Res.* 32, D393–D397.

9. Matthews, D.E., Carollo, V.L., Lazo, G.R., and Anderson, O.D. (2003) GrainGenes, the genome database for small-grain crops. *Nucleic Acids Res.* 31, 183–186.

10. Carollo, V., Matthews, D.E., Lazo, G.R., Blake, T.K., Hummel, D.D., Lui, N., Hane, D.L., and Anderson, O.D. (2005) GrainGenes 2.0. An improved resource for the small-grains community. *Plant Physiol.* 139, 643–651.

11. Shen, L., Gong, J., Caldo, R.A., Nettleton, D., Cook, D., Wise, R.P., and Dickerson, J.A. (2005) BarleyBase–an expression profiling database for plant genomics. *Nucleic Acids Res.* 33 (1), D614–D618.

12. Du, C.G., Buckler, E., and Muse, S. (2003) Development of a maize molecular evolutionary genomic database. *Comp Funct Genomics* 4 (2), 246–249.

13. Zhao, W., Canaran, P., Jurkuta, R., Fulton, T., Glaubitz, J., Buckler, E., Doebley, J., Gaut, B., Goodman, M., Holland, J., Kresovich, S., McMullen, M., Stein, L., and Ware, D. (2006) Panzea: a database and resource for molecular and functional diversity in the maize genome. *Nucleic Acids Res.* 34 (1), D752.

16

MaizeGDB
The Maize Genetics and Genomics Database

Carolyn J. Lawrence

Summary
MaizeGDB is the community database for biological information about the crop plant *Zea mays*. Genetic, genomic, sequence, gene product, functional characterization, literature reference, and person/organization contact information are among the data types stored at MaizeGDB. At the project's website (http://www.maizegdb.org) are standardized custom interfaces enabling researchers to browse data and to seek out specific information matching explicit search criteria. In addition, pre-compiled reports are made available for particular types of data, and bulletin boards are provided to facilitate communication and coordination among members of the community of maize geneticists.

Key Words: Maize; database; genetics; genomics; genome; model organism database.

1. Introduction
MaizeGDB is the repository for and interface to maize genetics and genomics data. Its content is comprised of records previously stored at the MaizeDB *(1)* and ZmDB *(2)* repositories (which are no longer in operation), as well as sequence data provided by workers at PlantGDB *(3)*, information gleaned from primary literature and entered into the database through manual curation, and data provided directly by the maize researchers who generated it. Because not all maize data are housed within MaizeGDB, contextual links are embedded throughout the data interface to enable navigation to other sites of interest (*see* **Table 1** for a list of linked sites).

From: *Methods in Molecular Biology, vol. 406: Plant Bioinformatics: Methods and Protocols*
Edited by: D. Edwards © Humana Press Inc., Totowa, NJ

Table 1
Contextually Linked Site List

Site name	Web address	Connection context
NCBI Map Viewer (see Chapter 3)	http://www.ncbi.nlm.nih.gov/mapview/	Maps
Gramene (*see* Chapter 15)	http://www.gramene.org/	Loci, maps
PlantGDB (*see* Chapter 25)	http://www.plantgdb.org/	Loci, sequences, Sequence contigs
DDBJ	http://www.ddbj.nig.ac.jp/	Sequences
EMBL (*see* Chapter 1)	http://www.ebi.ac.uk/	
TIGR (*see* Chapter 5)	http://maize.tigr.org/	
CerealsDB (SNPs)	http://www.cerealsdb.uk.net/cgi-bin/maize_snip.pl	
GenBank (*see* Chapter 2)	http://www.ncbi.nih.gov/Genbank/	Sequences, probes
Maize Mapping Project (WebFPC)	http://www.genome.arizona.edu/fpc/WebAGCoL/maize/	Probes
BioCyc	http://www.biocyc.org	Gene products
KEGG (*see* Chapter 21)	http://www.genome.ad.jp/kegg/	
SwissProt/TrEMBL (*see* Chapter 4)	http://www.expasy.org/	
AmiGO (*see* Chapter 24)	http://www.godatabase.org/cgi-bin/amigo/go.cgi	
PubMed	http://www.pubmed.gov/	Person/Organization, References
GRIN	http://www.ars-grin.gov/npgs/	Stocks

SNP, single-nucleotide polymorphism.

In addition to storing and making available maize data, workers at MaizeGDB also provide services to the community of maize geneticists. Bulletin boards for news items, information of interest to cooperators, lists of websites for projects that focus on the scientific study of maize, an editorial board's recommended reading list, and educational outreach items, are among the webpages made available through the MaizeGDB site (*see* **Table 2**). In addition, workers at MaizeGDB provide technical support for the Maize Genetics Executive Committee and the Annual Maize Genetics Conference.

Information about the history of MaizeGDB and the technical aspects of project's operation are described elsewhere (see references 4 and 5, respectively). Reported here are the types of data that are made available at

Table 2
Bulletin Boards and Static Pages

Page title and web address	Content description
News Column (http://www.maizegdb.org/)	News bulletins are displayed in the right margin. Older items are accessible through a link near the bottom
Tutorial (http://www.maizegdb.org/tutorial/)	An online step-by-step tutorial explains how to use the MaizeGDB website
Data Contribution 'How To' Guide (http://www.maizegdb.org/ data_contribution.php)	Displays sources of currently stored data and how researchers can contribute their own data
Editorial Board (http://www.maizegdb.org/ editorial_board.php)	A list of noteworthy references selected monthly by the MaizeGDB Editorial Board
Cooperators' Page (http://www.maizegdb.org/cooperators.php)	Page of links to resources supporting the cooperative spirit shared among maize researchers
Maize Genetics Cooperation – Newsletter (http://www.maizegdb.org/mnl.php)	Makes accessible online, copies of the MNL, and provides information on how to receive hard copies
Maize Genetics Executive Committee (http://www.maizegdb.org/mgec.php)	Explains the membership, goals, function, and history of the MGEC
Maize Genetics Conference (http://www.maizegdb.org/maize_meeting/)	Online access to information about when and where the Maize Genetics Conference will take place and access to online forms for submitting abstracts, and so on
Maize Research Projects List (http://www.maizegdb.org/maizeprojects.php)	A list of maize projects and links to their respective project sites
Educational Resources (http://www.maizegdb.org/education.php)	Makes accessible, materials for maize educational outreach

MaizeGDB, some generalized search strategies that can be applied across various data types, and a number of specialized example usage cases. Mechanisms for adding data to the database are also described in detail.

2. Materials

2.1. Genetic Data

1. Loci including (but not limited to) genes, chromosomal segments, centromeres, introns, probed sites, and quantitative trait loci (QTL).
2. Variations including the set of alleles at a given locus, chromosomal structural variations, cytoplasmic variations, DNA polymorphisms, rearrangements, transpositions, etc.
3. QTL experiment environmental conditions, parental stocks, traits of interest, locus summaries, and raw data files.
4. Maps, over 1200, including high-resolution genetic maps and cytogenetic (cytological) maps, along with associated data including mapping panel descriptors, population size, source information (usually a researcher's name), and related maps.
5. Seed stock descriptors consisting of a unique identifier (the stock name) and known synonyms, the stock source (e.g., an individual researcher's name or an organization name, such as the "Maize Genetics Cooperation Stock Center"), and associated focus linkage group assignments, genotypic variations, karyotypic variations, phenotypes, and parental stock identifiers.

2.2. Genomic and Sequence Data

1. Sequences and sequence contig membership data.
2. Molecular probes including (but not limited to) bacterial artificial chromosomes (BACs), Yeast Artificial Chromosomes (YACs), cDNAs, and expressed sequence tags (ESTs).
3. Molecular probe preparation methods for the amplification of, for example, overgos, Simple Sequence Repeats (SSRs), Random Amplified Polymorphic DNAs (RAPDs), Restriction Fragment Length Polymorphisms (RFLP), Amplified Fragment Length Polymorphisms (AFLPs), and other genomic DNAs.

2.3. Gene Product and Functional Characterization Descriptions

1. Gene products with associated Enzyme Commission numbers, expression induction conditions, subcellular localization data, metabolic pathway, known metabolic cofactors, mass (kDa), and links to loci that encode them.
2. Phenotypic descriptions that include trait descriptions and affected tissue types/organs (body parts) alongside mutant images.

2.4. Literature References and Person/Organization Records

1. References from primary literature, the Maize Genetics Cooperation Newsletter, and abstracts from the Annual Maize Genetics Conference; associated with virtually all other data types.
2. Contact information records for cooperators, authors, and organizations.

2.5. Terms, Controlled Vocabularies, and Ontologies

1. Terms and term definitions that describe stored data of various types.
2. Controlled vocabularies, the set of terms that describe a given process or data type. For example, terms of type "Developmental Stage" make up one controlled vocabulary.
3. Ontologies, hierarchically related controlled vocabularies that serve to enable communication across different databases and data sets. Within MaizeGDB are the Gene Ontologies *(6)* and the Plant Structure Ontology *(7)*. Nascent ontologies, such as the Trait and Environment Ontologies, which are being developed by Gramene in collaboration with the Plant Ontology Consortium (http://www.plantontology.org/ and *see* Chapter 24) *(8)*, are also incorporated and utilized as they emerge.

3. Methods

Navigating data to find specific, useful pieces of information is not always a simple task. Learning to use the tools that will enable facile data navigation is, therefore, a good use of time. By learning the general methods for browsing and searching MaizeGDB, the time required to locate information will be decreased, allowing for more time to be spent testing hypotheses at the bench.

In each of the following sections, general techniques for efficiently and effectively navigating the MaizeGDB interface are described. Following each general description is an example usage case that explains how to access relevant and useful data to meet a specific research need.

3.1. Embedded Search and Feedback Mechanisms

The fastest and easiest way to navigate to data of interest at MaizeGDB is by using the search bars. Search bars are available at the top and bottom of each page within a horizontal green band (*see* **Fig. 1**). Also present at the top of the page are the MaizeGDB icon (*see* **Note 3**) and links to bulletin boards including a job board, a list of upcoming events, and others. At the bottom of the page are the search bar and a link to send feedback to MaizeGDB personnel (*see* **Note 4**). The top and bottom search bars function identically. Virtually, all data types can be queried simultaneously by selecting a type of record to search using the drop-down menu and entering a query word or phrase in the text box, then pressing the button marked "Go!" (*see* **Note 5**). Described below is a protocol describing how to use the search bars to locate records of interest. Here, instructions are given to find information about available probes that mark the *bronze1* (*bz1*) locus as well as information about how to obtain a cloned sequence of *bz1* for experimental use.

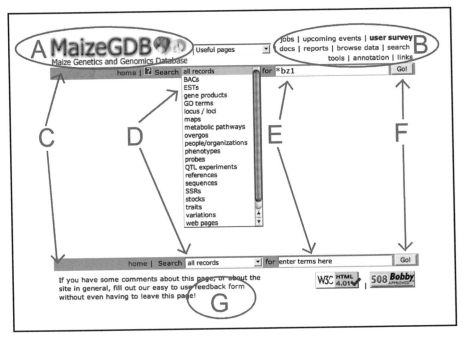

Fig. 1. Search bars available at the top and bottom of all MaizeGDB pages. (A) The MaizeGDB icon is located at the top left of each page. Clicking the icon causes a return to the home page (http://www.maizegdb.org). (B) Links to bulletin boards and static pages are listed at the top right. (C) Within bands located at the top and bottom of the page host the search bars (colors as seen on the Web page). To carry out a search, choose a data type (D) from the drop-down menu, type a search term (E) into the typing field, and press the button (E) marked "Go!".

1. Go to http://www.maizegdb.org.
2. Locate the top search bar and click on the button marked with a question mark within the green band.
3. Read the note that appears in the popup window (*see* **Note 1**).
4. Specify criteria to locate records about *bronze1* by selecting "locus / loci" from the drop-down menu and by typing *bz1 into the field to the right. Click the button marked "Go!" (*see* **Note 2**).
5. The locus page for the *bz1* gene is first in the results list (*see* **Note 6**). Click the link to the *bz1* gene.
6. On this page, scroll down to see the list of BACs, overgos, and other probes known to mark the *bz1* locus.
7. Click on the link to BAC a0020G06.

8. This BAC is listed as having been made available by Andrew Paterson. Click this link to access information about how to contact him to make arrangements to get the cloned a0020G06 BAC for experimental use.

3.2. Interrogation Tools

Questions asked by biologists are complex, so tools that query the database must enable complex queries to be made. In the left margin of the main page (http://www.maizegdb.org) are links to various Data Centers (including Gene Products, Locus/Loci, Maps, Metabolic Pathways, etc.). Each Data Center name is linked to a page that makes available a Simple Search (similar in function to the search available through the search bar described in **Subsection 3.1.**), an Advanced Search (which will be discussed more fully in this section), and a Discussion of the Data Type (written at a level comprehensible by the general public).

The use and functionality of the various Data Centers' Advanced Search tools are best explained by the following example. Two SSRs are known to flank a QTL of interest on maize chromosome 10 in bin 10.04. The two flanking SSR repeat patterns are (AG)28 and (TC)12. First, we determine whether the repeat patterns are present and mapped (at the level of the chromosome bins). Second, BACs that contain the SSRs are identified and BACs that contain both SSRs are selected. Finally, information on how to obtain the BACs for experimental use is accessed.

1. Go to http://www.maizegdb.org and look in the left bar for the Data Center called Molecular Markers and click on it (*see* **Note 7**).
2. Toward the top of this page is a link to specific information for BACs, ESTs, overgos, and SSRs. Click the link for SSRs.
3. In the Advanced Search Tool (the green box labeled "SSR Browser") use the checkbox and pull-down menus to limit the search to SSRs found in bin 10.04. Check the box to limit to a given repeat sequence and specify the repeat sequence as AG. Click the submit button toward the bottom of the green box.
4. In the results page, at least six results containing AG repeats are identified, but only one (p-umc2163) is anywhere near 28 units in length (the others are significantly shorter). Click the link labeled "p-umc2163".
5. Beneath the heading "Related Probes", five BACs are listed: b0045D05, b0161N11, b0256M13, b0187L03, and b0187C05. Make a note of these BAC identifiers (on paper or by other means).
6. Go back to the SSR Data Center (*see* **Note 8**).
7. Search for the SSR that is on the other side of your QTL by once again limiting your search within the green box to bin 10.04 and by specifying the repeat pattern (this time use TC). Click the submit button.

8. Of the results, the SSR p-MZETC34 has TC repeated 12 times. Click on "p-MZETC34" and scroll down to see the list of BACs that contain the repeat.
9. This probe detects four BACs: b0187L03, b0161N11, b0045D05, and b0256M13. Compare these BAC names to your notes (from **step 4**) to find that all four are also present in the first result sets, so all could be used for sequencing to find out more about the QTL.
10. To find out how to order the BACs, click on the first BAC name, b0187L03. It is made available by CUGI (the Clemson University Genomics Institute).
11. Toward the top of this page, note the heading that reads, Want this clone? You can order it from CUGI using their BAC ordering system. Be sure to request clone "b0187L03"! Click the link within this heading to access the "BAC ordering system". A form for ordering the BAC directly from CUGI appears.

It should be noted that in instances where the SSRs of interest are known to be mapped, it is also possible to locate information using the "Mapped SSRs" browser toward the top of the SSR Data Center page. More specifically, if those SSRs are known to be mapped onto IBM2, it is possible to check for BACs anchoring those SSRs using the "Mapped & Anchored SSRs" browser, which is also accessible toward the top of the same page.

3.3. Analysis Tools

In addition to methods enabling data access, tools for data analysis are also made available through MaizeGDB. Because the maize genome is currently being sequenced, sequence data are of particular interest at present. The sequence data analysis tool BLAST *(9)* is accessible via a link on the left of the main page (http://www.maizegdb.org) and through links available on all sequence pages. BLAST enables the identification of sequences stored in the database that are similar to a sequence of interest.

Because the MaizeGDB sequence data set is updated on a monthly basis, the BLAST service at MaizeGDB should not be considered to be the one-and-only place for searching for similar maize sequences! A sampling of other useful sites that host a BLAST service and that could be searched include NCBI (the site that hosts the most up-to-date sequence set, *see* Chapters 2 and 3), PlantGDB (which enables maize-specific BLAST searches and hosts a searchable assembly of the maize genome, *see* Chapter 25), Panzea (where sequence similarity searches can be carried out against sequence from a diverse set of maize lines), the *Mu* Transposon Information Resource (where a sequence set tagged by the Robertson's *Mutator* elements is made available), and others. All these BLAST services are accessible through their own sites, and links to them exist at the MaizeGDB BLAST page (http://www.maizegdb.org/blast.php).

Two distinct MaizeGDB BLAST services are available: one against the standard maize sequence set made available through NCBI (available at http://www.maizegdb.org/blast.php) and another against the "Cornsensus Sequence Set" *(10)* which is made accessible through a link toward the top of the standard MaizeGDB BLAST page. These two services are kept distinct in an effort to distinguish those sequences that exist in nature (the former) from the "Cornsensus Sequences" which represent alignments of similar sequences that are not guaranteed to have been derived from a single locus or even the same background of maize.

The MaizeGDB BLAST services are especially unique and useful in the following respect: result sets display the map locations of identified similar sequences and are linked to other records stored at MaizeGDB and offsite. The following example usage case shows these useful aspects of the MaizeGDB BLAST search utility using a sequence from B73 (the maize line that is currently being sequenced; see http://www.nsf.gov/awardsearch/showAward.do?AwardNumber=0527192) as the query. Here, a sequence record of interest is located using its GenBank identifier. Contextual links are followed to arrive at the MaizeGDB BLAST page, and a BLAST search against maize nucleotides is carried out. Within the result set, links to known map locations of matches are available. By following those links, maps are accessed and displayed, with the locus of interest highlighted.

1. Using the search bar at the top of the page (as described in **Subsection 3.1.**), use the drop-down menu to select "sequences" and type CG247295 into the field to the right. Click the submit button.
2. On the sequence page that appears, find the heading "Bioinformatics Tools" toward the bottom of the green bar on the right. Beneath the heading is a link to "BLAST against MaizeGDB". Click it.
3. On this page, scroll about half way down the page and fill in a sequence name, move right to choose the BLAST program (for this example, the default "BLASTn" is correct), continue right to choose "Maize Nucleotide" as the database to search, and to the right of that select an E-value (the default 0.0001 is appropriate for this example).
4. Note that the sequence query is already present (filled in from the sequence page from **step 2**) and click the button marked "Run BLAST" (*see* **Note 9**).
5. Scroll down the result set page (shown in **Fig. 2**) to view the results summary. Beneath the "Detailed Results Summary" for sequence AY772455.1 is a table consisting of three columns, the names of maps charting the position of this sequence's locus, the coordinate of the sequence on a given map, and the name of the marker or locus that was mapping (for AY772455.1, the locus listed is *umc95*). Click the link for the map named UMC 93 9, which is the second map from the top.

Results Summary

BLASTN 2.2.4 [Aug-26-2002]

Reference: Altschul, Stephen F., Thomas L. Madden, Alejandro A. Schaffer, Jinghui Zhang, Zheng Zhang, Webb Miller, and David J. Lipman (1997), "Gapped BLAST and PSI-BLAST: a new generation of protein database search programs", Nucleic Acids Res. 25:3389-3402.

Accession # / Link to Detailed Result	Map Locations	Links to Full Sequence Record	Sequence Title	Score	E-value
AY772455.1	75	MaizeGDB | DDBJ | EMBL | GenBank	gi|54111453|gb|AY772455.1|AY772455 PLN Zea mays RFLP probe umc95...	1308	0.0
G10872.1	76	MaizeGDB | DDBJ | EMBL | GenBank	gi|984505|gb|G10872.1|G10872 STS umc95 R maize DGrant6 Zea mays ...	389	e-107
G10873.1	76	MaizeGDB | DDBJ | EMBL | GenBank	gi|984936|gb|G10873.1|G10873 STS umc95 F maize DGrant6 Zea mays ...	351	4e-96

Detailed Result Summaries

AY772455.1

Retrieve this record at MaizeGDB \| DDBJ \| EMBL \| GenBank

Maps: AY772455.1 has known map locations on these **75** maps:

Map	Coordinate	Marker
BNL 93 9	81	umc95
UMC 93 9	103.3	umc95
UMC 89 9	88.2	umc95
chapalote/Z.mexicana 9	43	umc95

Fig. 2. BLAST results show map locations and are linked to other data. Within the "Results Summary" is a table showing the list of similar sequences matched by BLAST alongside those sequences' map locations, links to records at MaizeGDB and offsite, and the similarity scores and E-values. Below the table are "Detailed Results Summaries" which list the maps containing the sequences identified. Clicking on the map name results in access to a map display, where the locus name for the sequence match identified is highlighted.

6. Note that on the UMC 93 9 map page that appears, the locus *umc95* is highlighted and that the page scrolls to show this locus. Note also that the epithet "CBM 9.05" is appended to the locus name. This connotes that the locus is the core marker for bin 9.05 (i.e., core bin marker 9.05).

3.4. Ways to Add Data to the Database

MaizeGDB's sequence data set comes from PlantGDB. That data set is updated automatically on a monthly basis. Other data sets are classified as one of three types: large data sets, small data sets, and notes. Large data sets are generally added to the database in bulk by members of the MaizeGDB Team and are contributed by researchers directly. To contribute a large data set to the project (or to find out whether the data set you have generated constitutes a "large" or "small" data set), use the feedback button at the bottom of any MaizeGDB page to make an inquiry.

Researchers can add "notes" to records (like the one contributed by Hugo Dooner which is shown in **Fig. 3**). To add such a note, log in to the site using the "annotation" link displayed at the top right of any MaizeGDB page. Once

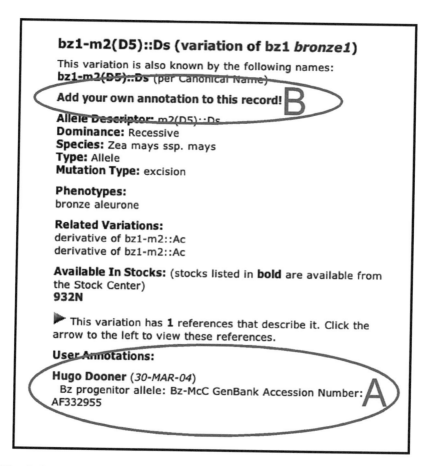

Fig. 3. Researchers can add notes to records at MaizeGDB. (A) Hugo Dooner has annotated the *bz1-m2(D5)::Ds* allele as the *bz1* progenitor allele from the maize line McC. (B) Almost all data types stored can be annotated by following links to "**Add your own annotation to this record**" (*see* **Note 10**).

logged in, click the link to "Add your own annotation to this record" shown at the top of virtually all data displays (*see* **Note 10**). Small data sets also can be added to the database by researchers directly by way of the MaizeGDB Community Curation Tools. The method for adding a small data set is explained below, using a newly published reference as the example usage case. The citation for our pretend reference is as follows:

Lawrence, CJ. (2005) How to use the reference curation module at MaizeGDB. *Plant Physiology* 9:3–4.

1. Click on the "annotation" link at the top of any MaizeGDB page. Click the link to "Create an Annotation Account" and fill out all information required. Be sure to check the box to become a MaizeGDB curator before clicking the submit button.
2. A confirmation email along with a Community Curation manual will be sent once the new account has been activated.
3. To begin adding data to the database, click on the link marked "tools" toward the top right of any MaizeGDB page.
4. Toward the bottom of this page click the link marked "Playground Community Curation Tools" (*see* **Note 11**).
5. Log in using the newly created username and password.
6. Click the link toward the center of the page to download the curation tools' user manual for future reference.
7. In the left bar, click the link marked "Reference".
8. Fill in the title and select "article" as the reference type. (Because you are working at the "Playground Community Curation Tools" feel free to make up pretend information for the purposes of this exercise). When in doubt of what information to put into a given field, click on the buttons labeled with a question mark (*see* **Note 12**).
9. Fill in the year, volume, and pages information. For the "In Journal" field note the label "Lookup Field – Enter a Search String". Fill in the journal title.
10. Click the link beneath the "Author" heading to "Add Authors".
11. Half way down the page is a typing field where the name of an author can be typed to locate a person record to associate with the new reference. For this example, type Lawrence.
12. Lawrence is the first (and only) author on this imaginary publication, so leave the drop-down menu with "Author" selected and type the number **1** into the box labeled "Order". Press the "Submit & Continue" button.
13. Note that "Lawrence, CJ" is available in the drop-down menu. Select this item from the drop-down menu that has replaced the typing field for Author, scroll to the bottom of the page, and click the button labeled "Add to List of Authors".
14. Note that "Lawrence, CJ" now appears in the list of authors at the top of the page. Click the button marked "Author List Complete".

15. Scroll to the bottom of the page and press the button marked "Submit & Continue". Returned in place of the "In Journal" search string are available instances of matching journal names. Select the records "Plant Physiol" (*see* **Note 13**).

16. Click the button at the bottom of the page marked "Insert into Database".

The newly created record enters a queue for approval by a worker at MaizeGDB. Once the record has been approved, it will become available through the MaizeGDB interface after the next database update. Other curation tool modules function similarly, and a detailed manual is available through the curation tools.

4. Notes

1. A wildcard is appended to the right ends of all queries automatically. This means that if bz1 is typed into the search window, a search is carried out for all instances matching the pattern bz1*. Asterisks or percent signs can be appended manually to the left ends of search strings to enable matching for instances where the string bz1 is preceded by other characters.

2. In general, searches carried out through the search bars are simple searches that are limited to record names and synonyms. Names of probe records are those assigned by researchers and are not subject to any standardized naming convention. For this reason, searching through probe records for *bz1 is not guaranteed to yield a list of all probes that mark the *bz1* locus. Searching for the *bz1* locus record and subsequently browsing through associated probes is a better method to follow if the desire is to find all probe records associated with the *bz1* locus.

3. Clicking the icon within the top bar on any page results in a return to the main page, http://www.maizegdb.org.

4. When feedback is sent, the web address for the page from which the feedback was sent is included in the message automatically to enable workers to better understand how or why questions or concerns were submitted.

5. Instead of hitting "Go!" or "submit" buttons, it is also possible to simply press the "enter" or "return" key on the computer's keyboard to submit queries.

6. Had the asterisk not been appended to the left end of the query, the "Probed Sites" and the one "YAC" would not have been found because the search phrase is preceded by other characters in those records' names. The gene *zp15* is found by the search because its synonym *bz15* contains the search string.

7. To return to the SSR Data Center, click on the upper left MaizeGDB icon then select the SSR Data Center from the left bar on the main page.

8. SSRs are a probe type. All probes can be searched by name through the "Probe" Data Center, but to search SSRs by sequence pattern, the SSR Data Center subset should be searched directly.

9. The search could be limited to find only those sequences with known map locations by selecting the check box just above the "Run BLAST" button.

10. This link is not displayed unless you are logged in to the site.
11. It is advisable to work first with the "Playground Community Curation Tools" before entering data into the real database. The "Playground Community Curation Tools" function just like the "Real Community Curation Tools", but access a different copy of the database intended for testing and training purposes. Using the "Playground Community Curation Tools" allows the entry of fictitious data while learning to use the tools, without the possibility of compromising MaizeGDB's content.
12. All fields required to create a complete reference record are marked with a red asterisk.
13. If the content of the drop-down menu returned in place of the field where a search string was entered does not contain the item desired, scroll to the bottom of the page and use the link to "Re-edit Search Fields". If by making edits the item is still not found, use the link at the bottom of the left bar to "Email an Expert Curator".

Acknowledgments

The author thank Trent Seigfried and Darwin Campbell for their work maintaining the MaizeGDB interface and database, respectively; Mary Schaeffer, Ed Coe, and Marty Sachs for their curatorial work; PlantGDB partners Volker Brendel, Qunfeng Dong, and Matthew Wilkerson for building and maintaining the sequence update pipeline; Michael Brekke for technical support; and Sanford Baran and Jason Carter for developing the MaizeGDB curation toolsets. This work was supported by the USDA-ARS.

References

1. Polacco, M., and Coe, E. (1999) *Bioinformatics: Databases and Systems* (Letovsky, S. I., Ed.), pp. 151–162, Kluwer Academic Publishers, Norwell, MA.
2. Dong, Q., Roy, L., Freeling, M., Walbot, V., and Brendel, V. (2003) ZmDB, an integrated database for maize genome research. *Nucleic Acids Res.* 31, 244–247.
3. Dong, Q., Lawrence, C. J., Schlueter, S. D., Wilkerson, M. D., Kurtz, S., Lushbough, C., and Brendel, V. (2005) Comparative plant genomics resources at PlantGDB. *Plant Physiol.* 139, 610–618.
4. Lawrence, C. J., Dong, Q., Polacco, M. L., Seigfried, T. E., and Brendel, V. (2004) MaizeGDB: The maize genome database, the community database for maize genetics and genomics. *Nucleic Acids Res.* 32, D393–D397.
5. Lawrence, C. J., Seigfried, T. E., and Brendel, V. (2005) The maize genetics and genomics database. The community resource for access to diverse maize data. *Plant Physiol.* 138, 55–58.
6. Gene Ontology Consortium (2001) Creating the gene ontology resource: design and implementation. *Genome Res.* 11, 1425–1433.

7. Bruskiewich, R., Coe, E., Jaiswal, P., McCouch, S., Polacco, M., Stein, L., Vincent, L., and Ware, D. (2002) The plant ontology[tm] consortium and plant ontologies. *Comp Funct Genomics* 3, 137–142.

8. Ware, D. H., Jaiswal, P., Ni, J., Yap, I. V., Pan, X., Clark, K. Y., Teytelman, L., Schmidt, S. C., Zhao, W., Chang, K., Cartinhour, S., Stein, L. D., and McCouch, S. R. (2002) Gramene, a tool for grass genomics. *Plant Physiol.* 130, 1606–1613.

9. Altschul, S. F., Madden, T. L., Schaffer, A. A., Zhang, J., Zhang, Z., Miller, W., and Lipman, D. J. (1997) Gapped blast and psi-blast: a new generation of protein database search programs. *Nucleic Acids Res.* 25, 3389–3402.

10. Gardiner, J., Schroeder, S., Polacco, M., Sanchez-Villeda, H., Fang, Z., Morgante, M., Landewe, T., Fengler, K., Useche, F., Hanafey, M., Tingey, S., Chou, H., Wing, R., Soderlund, C., and Coe, E. (2004) Anchoring 9,371 maize expressed sequence tagged unigenes to the bacterial artificial chromosome contig map by two-dimensional overgo hybridization. *Plant Physiol.* 134, 1317–1326.

17

BarleyBase/PLEXdb

A Unified Expression Profiling Database for Plants and Plant Pathogens

Roger P. Wise, Rico A. Caldo, Lu Hong, Lishuang Shen, Ethalinda Cannon, and Julie A. Dickerson

Summary

BarleyBase (http://barleybase.org/) and its successor, PLEXdb (http://plexdb.org/), are public resources for large-scale gene expression analysis for plants and plant pathogens. BarleyBase/PLEXdb provides a unified web interface to support the functional interpretation of highly parallel microarray experiments integrated with traditional structural genomics and phenotypic data. Users can perform hypothesis building queries from multiple interlinked resources, e.g., a particular gene, a protein class, EST entries, and physical or genetic map position—all coupled to highly parallel gene expression, for a variety of crop and model plant species, from a large array of experimental or field conditions. Array data are interlinked to analytical and biological functions (e.g., Gene and Plant Ontologies, BLAST, spliced alignment, multiple alignment, regulatory motif identification, and expression analysis), allowing members of the community to access and analyze comparative expression experiments in conjunction with their own data.

Key Words: Expression profiling; microarray; MIAME; plant ontology; cluster analysis.

1. Introduction

1.1. Mission Statement

PLEXdb (Plant Expression Database) aims to provide a unified web interface to connect a wide range of microarray data sets from both model and crop species. The primary objective is to apply parallel gene expression resources

From: *Methods in Molecular Biology, vol. 406: Plant Bioinformatics: Methods and Protocols*
Edited by: D. Edwards © Humana Press Inc., Totowa, NJ

to accelerate comparative analysis of genes important in plant biology and translate these findings to agriculture. The integrated tool sets of PLEXdb make it possible for individual investigators to capitalize on the commonalities of plant biology as a comparative approach to functional genomics through large-scale expression profiling data sets.

1.2. Background

Microarray analysis commonly consists of a data-driven exploratory approach that relies on searching for differentially expressed or co-regulated genes. Subsequently, the investigator is left with the task of searching for connections between the genes that showed interesting activity, by searching through annotation from BLAST hits, specialized genome databases [such as Gramene *(1)* (*see* Chapter 15), Graingenes *(2)* (*see* Chapter 14), or TAIR *(3)* (*see* Chapter 8)], and protein information and pathway links. This process is very labor intensive and time consuming as well as risking missing subtle linkages due to the sheer amount of information available. In addition, each database has a different format, making it difficult for the user to perform uniform data retrieval. PLEXdb automates this procedure for the biologist by extracting key information such as physical location, functionality, potential pathways, homologs in related species, and grouping the diverse information in meaningful ways.

For example, an investigator has a list of genes associated with a particular treatment. The investigator then wishes to see where all genes that are co-regulated or that belong to a particular gene family map on the genome. This would be particularly beneficial to investigators interested in high-throughput quantitative trait loci (QTL) analysis. For sequenced genomes, such as rice or Arabidopsis, and soon Medicago, tomato, and maize, this will be possible by a simple look-up of the pre-calculated coordinates in a genome browser. However, the problem is complex for species without fully sequenced genomes, such as wheat or barley. In this case, it would be desirable to identify syntenic positions on the most closely related model genome (in this example, rice). These map locations could then be used to search for associations with trait loci to integrate gene expression data with phenotype data or further determine how genes associated with a particular phenomenon are co-regulated. For example, are there conserved promoter motifs that may be recognized by regulatory factors that are responsible for the coordinated regulation?

Investigators may also be interested in that are associated with pathways their list of co-regulated genes. Then one could look at conserved genes or pathways in another organism (e.g., Arabidopsis) to build hypotheses regarding function or compare expression profiles of similar experiments in the other organism.

While tasks in the hypothetical case above can be achieved with available software, the integration of data sources, visualizations, and analytical tools at PLEXdb will greatly facilitate this process. Thus, PLEXdb focuses on providing data integration and tool development for both specialized and comparative interpretation of plant microarray data.

2. Materials

2.1. Array Types

PLEXdb fully supports Affymetrix GeneChip data from Plants and Plant Pathogens (22K Barley1, 22K Arabidopsis, 16K grape, 61K wheat, 51K rice, 18K maize, 8K sugarcane, 10K tomato, 61K soybean/*Phytophthora sojae*/ soybean cyst nematode, 61K *Medicago truncatula/M. sativa/Sinorhizobium meliloti*, 61K poplar, and 8K *Fusarium graminearum*) (*see* **Note 1**). Annotation information and experiment submission is also available for long oligo arrays, e.g., NSF 45K rice and NSF 58K maize. Data sets are contributed by users in the community and also retrieved from archives such as ArrayExpress (*4,5*), NCBI-GEO (*6*) (*see* Chapter 3), TAIR (*3*) (*see* Chapter 8), or NASC Arrays (*7*) (*see* Chapter 9) and uploaded by the PLEXdb curators.

2.2. Available Data Formats

For Affymetrix GeneChips, PLEXdb requires raw CEL and EXP data files; EXP and DAT files are recommended for submission. All submissions are normalized in a standard way to enable cross-experiment comparison. Long oligo array data submission requires the scanner data and raw data files. After the submitter uploads the experiment data, the curator checks the data integrity and computes the normalized expression measures, summary statistics, and graphs. The raw experimental data are available for download as a compressed archive of CEL files or in a comma separated value (CSV) form which links each probe pair to the raw data. PLEXdb generates SOFTmatrix format files. The SOFT files can be submitted to NCBI's GEO repository (http://www.ncbi.nlm.nih.gov/projects/geo/info/soft2.html).

2.3. Server Type

PLEXdb uses open-source tools or tools free for academic institutions. The server uses RedHat Linux. The website is powered by an Apache server, PHP and Javascript for dynamic Web pages, and a MySQL 4.0 relational database as the back end. The data pre-processing uses R *(8,9)*, Bioconductor, and Perl. R is an open platform for statistical computation, and Bioconductor is a project written in R for microarray data analysis. Robust Multichip Average (RMA) *(10)* in the Affy package of Bioconductor and Affymetrix MAS 5.0 are used to compute normalized expression measures from the raw expression values. As newer methods of normalization are developed, PLEXdb will reassess which methods of normalization are optimal for searches and data analysis. Current methods being considered for inclusion are GCRMA and PLIER (from Affymetrix) *(11)* (*see* **Note 2**).

2.4. Gene Expression and Visualization

Typically in parallel expression analysis, an individual will need to manually download expression data from a repository site; check for conformity to standards that allows cross-experiment comparisons; map the respective array tags to genes and those genes to genomic locations and orthologs in other species; install local software for expression analysis; rely on disparate resources to view associated data, such as EST coverage or SAGE and MPSS data for genes of interest; and develop their own tools to post-process results (e.g., sequence data for promoter motif-finding software). To preclude these difficulties, all microarray data sets in PLEXdb are interlinked with extensive filtering and analysis tools. These tools include multiple BLAST options for each gene on a chip, probe sequences and alignments, microarray platform translator, for comparing gene sequences between species, and extensive annotation options on the respective species exemplar annotation pages (e.g., http://www.plexdb.org/modules/PD probeset/annotation.php?genechip=Barley1). Tools to facilitate analysis of hybridization data include expression profile plots, individual probe plots, and gene list creation based on a variety of statistical filtering and analysis. Data queries are integrated with analysis and visualization tools to allow users to explore their experimental data, ranging from simple assessment of data quality to derivation of expression profiles and annotation of a particular expression cluster.

2.5. Integration With Other Databases

PLEXdb features MIAME compliant experiment annotations *(12)* as well as required Plant Ontology terms through PLEX Express, its user-friendly, web-based submission tool *(13)*. In collaboration with the Plant Ontology Consortium *(14)*, PLEXdb utilizes the plant and trait ontology terms during data submission so that experiments can accurately describe developmental stages and plant tissue types. Using these terms allows cross-species comparisons based upon the common identifiers, facilitating interoperability between existing plant databases to determine genes that may have exhibited similar expression profiles *(15)*. Currently, interconnecting links with PlantGDB *(16)* (*see* **Note 3** and Chapter 25), Gramene *(1)* (*see* Chapter 15), Graingenes *(2)* (*see* Chapter 14), TIGR *(17)* (*see* Chapter 5), or TAIR *(3)* (*see* Chapter 8) (e.g., http://www.plexdb.org/modules/PD probeset/annotation.php?genechip=Barley1) allow PLEXdb users to perform gene predictions using the probe sets located on any of the GeneChip or oligo arrays in the database or cross-species comparisons linked to respective genome databases [e.g., Barley1 probe sets *(18)* on rice and other cereals at Gramene; http://www.gramene.org/Oryza sativa/contigview?bottom=%7C Barley Arrayconsensus Affy22 K%3Aon&w=1938&C=1%3A41692749.5%3A1 &h=%28130-1064%29%7Cbarley1 01637]

2.6. Tutorials/Help Files

To ensure ease of use of PLEXdb, a variety of FAQ, tutorials, help files, and downloadable presentations from National and International meetings are posted (see "Documentation" on the front page at http://plexdb.org/ or more specifically http://www.plexdb.org/modules/documentation/documentation.php).

3. Methods
3.1. Basic Visualization Tools

In microarray experiments, the number of interesting genes identified by statistical analysis and data mining is large, usually hundreds to thousands. Each individual GeneChip hybridization on the new generation of chips generates approximately 170 Mb of raw data, and for a replicated multifactor experiment (genotype × treatment × time) it is not uncommon to use 50–100 chips, generating 8–17 GB of data *(19)*. It is imperative that the user appreciates the reliability of an experiment before proceeding further. Therefore, one of the features of PLEXdb is to facilitate the visualization of the quality of all public data sets. If an investigator is new to the field of parallel expression analysis, they may want to utilize existing data sets to build a set of hypotheses before actually physically performing their own experiments.

1. To view, click on the "+" to the left of "Data Access" and then "Experiments" on the main menu. Public experiments can be viewed without Login. Login is required to view private and group managed experiments.

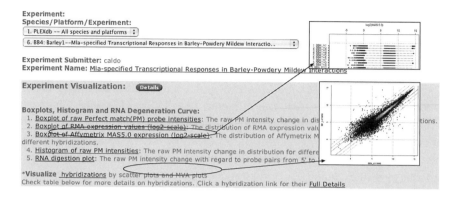

Fig. 1. Visualization tools. To check data quality, users browse experiments such as BB4. From there, choices include batch download of the complete data sets in raw, processed CSV, and MAGE-ML formats; view expression box plots; visualize hybridizations or treatment means with scatter plots and MvA plots; or continue on to perform gene-centric data analysis.

2. Select "hybridization/visualization" to check hybridizations, treatments or replications in box plots, density plots, or scatter plots. Choose "sample information" to view all information associated with the experimental sample, e.g., tissue, stage of development or growth, or plant ontology.

3. For example, by selecting accession "BB4" from the pull-down menu, the user would arrive at the screen shown in **Fig. 1** (http://www.plexdb.org/modules/PDbrowse/expsummary.php). This page allows the user to visualize, in multiple formats, the quality of the data and number of replications. From there, the user can perform any number of probe set/gene queries or download the complete data set for further analysis (*see* **Note 4**).

3.2. Experiment Submission (http://www.plexdb.org/modules.php?name=Your_Account)

PLEXdb is committed to making data from plant and plant pathogen-based microarrays publicly available to researchers worldwide. Researchers are encouraged to submit their data to the repository through PLEX Express, a MIAME compliant and Plant Ontology enhanced online submission tool *(13)*. An on-line tutorial is available to assist users with data submission (http://www.plexdb.org/modulesdocumentation/submissionhelp.php). Submitted data can be made public immediately or be kept private, and an accession number can be obtained up to 6 months prior to submission for publication. Submitters can also grant access to their data to designated individuals or groups, including manuscript reviewers.

Submissions will be checked for structure and content by the PLEXdb curators. Submissions are prioritized in the following order. (1) Submissions that are to be made public immediately regardless of publication status. (2) Submissions that require database accession information in page proof. (3) Private submissions that are related to a paper that has not yet been accepted. (4) Private submissions that are related to a paper that has not been submitted.

1. Select "Data Submission". Log in to submit an experiment. User registration is needed to obtain a username and password.
2. Select submission type. Protocols used for RNA isolation and array processing should be entered before you submit your experiment. Select "Extraction protocol" to browse submissions. Add a new extraction protocol if your protocol is not on the list. Fill in the text boxes with information about the protocol and type or paste the protocol in the text box for description. Submit the protocol. Perform this same submission procedure for the labeling, hybridization, scanning, and washing protocols.
3. Select "Experiment Submission". Fill in the information (Title, Experiment types, Experiment factors, Experiment Descriptions, Quality control steps/description, and Publication) about the design of the experiment and array used. Also fill in experimenter information. Save the information (*see* **Note 5**).
4. Provide the number of experimental factor levels. An example of a factor could be time points, treatment, or tissue sampled. Save the information (*see* Note 5).
5. Describe the treatment design. Fill in the names of factor levels and number of replication(s). Save the information. The number of hybridizations should coincide with the number of replications × treatment factors × treatment factor levels.
6. Upload the data in the Java UploadApplet. Browse the location of data files. Select data to upload and click "add files to upload". Sometimes it is necessary to select "Overwrite all" for the options for existing files to start uploading.
7. The required fields are the sample name (provide unique name), organism (genus and species), developmental stage (based on the plant ontology definition), organism part (anatomy ontology of plants or fungi), and protocol binding. All protocols previously submitted should be bound to each hybridization. Go to step 1 and save the information.
8. Continue the submission by associating data files with treatment. Select the files associated with particular experimental factors and replications. Submit information on data association.
9. Select "finalize submission page". It is necessary to check if the submission is correct and complete before finalizing the experiment. Click the "edit/update" button to edit submission, otherwise click the "finalize" button. An e-mail will be sent to the submitter acknowledging the submission with the appropriate database

number. The curator will check the submission and data quality, so it will take 1–2 weeks to access the experiment along with fully curated tools for analysis.

3.3. Genelist Analysis

Gene lists may be derived from statistical analysis of expression, or alternatively, based on function, e.g., biochemical pathways or groupings based on enzyme activities. To access the analysis tools in PLEXdb, gene lists should be imported or created in the database. These tools are currently only available for data from Affymetrix GeneChips although analysis support for long oligo arrays is planned in the near future.

3.3.1. Create Gene List

1. Select "Data Access" and then "Gene Lists".
2. In Wizard mode there are eight major steps to create and analyze a gene list. First, choose an appropriate action either to You may create a new gene list or create a new gene list from saved gene lists. Click the "next" button to go to the next page.
3. Select an approach to create a gene list. Choose from either single or multiple experiments where you would want to obtain your gene list. Then, select an approach to create a gene list: expression profile, keywords, importing your own list, identifying co-expressed probe sets, sequence BLASTing, or gene ontology (GO). Not all of these methods are available for all organisms. To analyze the expression profile of a certain experiment, create a gene list by expression profile and a single experiment.
4. Choose an expression estimation method, either MAS 5.0 or RMA, and an experiment. Then query the experiment by type, experimental factors, experimenter, or institution. Click the search button. Select experiment if it is on the list, or if not, go back to change the query. Click the button "add experiment to data folder", and then go to the next page for filtering genes. If the user has the experiments in the data folder, choose experiment from that folder.
5. Select filters for the expression profile using single or multiple parameters. Choose filtering parameters from absolute value, MAS 5.0 detection call, variation, fold-change, and statistical analysis. Absolute value filter is based on the level of expression, MAS 5.0 call filter is based on the presence or absence of detection calls, variation filter is based on absolute or relative variation, fold-change filter is based on expression increase or decrease, and statistics filter is based on statistical analysis.
6. Select the parameters for the selected filter and specify the number of arrays where the filters be applied.
7. Check the results of filtering You may choose to save the genelist at PLExds or download it to your computer.

3.3.2. Analyze Gene List

Analysis can be done after creating a gene list or from a saved gene list. In some algorithms, data transformation is needed to improve analysis performance. Log transformation, mean centering, median centering, and standardization are some of the common data transformations.

3.3.2.1. CLUSTER ANALYSIS

Cluster analysis can be applied in expression profiling experiments to group genes with similar expression patterns. It encompasses different algorithms and methods to sort genes into clusters, so that the degree of similarity in one group is high, and low between groups. Gene associations are revealed based on expression patterns, which could be due to their involvement in a similar biochemical pathway or function.

1. Select a clustering method from hierarchical clustering, K-medoids, and K-means, self-organizing map (SOM), principal component analysis (PCA), or Sammon's non-linear mapping *(20)*.
2. For hierarchical clustering, perform $\log_2$ and mean centering data transformations by clicking the boxes. Select the linkage method for combining clusters (single linkage, average linkage, complete linkage, Ward's, and centroid) and distance measure (correlation-centered Pearson, non-centered Pearson, Euclidean distance, maximum, Manhattan, and Canberra) to use. Selection should be based on the user's objective in clustering gene expression profiles. The results are displayed as a heat map color spectrum and a hierarchical tree. Different clustering method–distance combinations give different results.
3. K-medoids and K-means partitioning. Select the number of clusters to search for in the data. Transform data by clicking the boxes for $\log_2$ and mean centering. Select a distance metric from either correlation-centered Pearson, non-centered Pearson, Euclidean distance, maximum, Manhattan, or Canberra. For K-means, use the Euclidean distance metric. Identify the maximum number of iterations to be used.
4. Self organizing maps (SOMs) are another type of supervised cluster analysis. Transform the data by selecting the $\log_2$ transformation. The SOM algorithm is implemented using the SOM package in R and default parameters are used. User identifing the number of grids to separate the results into.
5. In Principal Components Analysis (PCA), select which method to use. The methods are (1) no transformation of data matrix and singular value decomposition (SVD) is carried out on the sum of squares and cross-product matrix; (2) observations are centered to zero mean and SVD uses the variance-covariance matrix; (3) observations are standardized by centering to mean 0 and variance 1, and SVD is carried out on correlation matrix; (4) observations are normalized by being range divided,

and then the variance-covariance matrix is used in SVD; (5) SVD is carried out on Kendall (rank order) correlation matrix; (6) SVD is carried out on Spearman (rank order) correlation matrix; (7) SVD is carried out on the sample covariance matrix; (8) SVD is carried out on sample correlation matrix. This is implemented using the statistics package in R.

6. Sammon's non-linear mapping. Use the default parameters. Select $\log_2$ and mean centering transformations.

3.3.2.2. GENE ONTOLOGY ANNOTATIONS

GOs are used as common identifiers (*see* Chapter 24), facilitating interoperability between existing databases to determine genes that may have exhibited similar expression profiles. This analysis is useful for sequenced genomes that have existing GO information. Currently this capability is only available for Arabidopsis.

3.3.2.3. MAPPING LISTS OF EXPRESSED GENES TO SYNTENOUS POSITIONS ON SEQUENCED GENOMES

The Model Genome Mapper tool helps the user obtain the syntenous position of genes on sequenced model genomes. For example, the Rice Genome Mapper currently allows batch mapping Barley1, wheat, maize, or rice probe sets onto the Rice Genome Browser at Gramene (*see* **Fig. 2**). Subsequently, the user can retrieve FASTA outputs of rice genome sequence, 5' and 3' regulatory regions or specific exons and introns. These FASTA outputs can be used for input into motif-finding software (*see* **Note 6**). All the relations between the data are captured in tables in the relational database, and updates implemented by changes to only the requisite tables (e.g., ortholog designations will be stored in separate tables linked by the unique gene identifiers as common keys). Future updates will include capability for mapping on other sequenced genomes (e.g., soybean or Medicago).

1. Model Genome Mapper queries a list of GeneChip probe set locations on the sequenced genomes of rice and Arabidopsis genome. It can be accessed by selecting "Tools" from the main menu and then choosing "Rice Genome Mapper".
2. To use Model Genome Mapper, select a microarray platform. There are currently five choices for monocots and five for dicots.
3. Paste the query probe set names for the chosen platform into the text box (*see* **Fig. 2A**). Example input probe names are provided to illustrate the correct format for different GeneChips.
4. Set parameters as follows:

   a. Output format: Output results can be sorted by GeneChip order, or unsorted. The default is to sort by gene name.

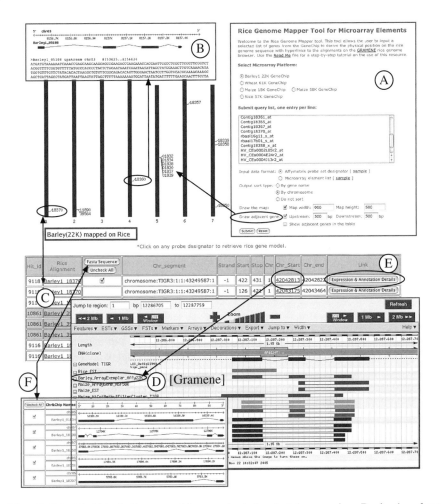

Fig. 2. Model (Rice) Genome Mapper tool for batch mapping Barley1, wheat, maize, or rice probe sets onto the Rice Genome Browser at Gramene. This tool allows a user to enter a list of genes derived from any cereal GeneChip (or maize oligo array) experiment (**A**) and immediately visualize their positions on the rice genome (**B**). Using the table that is generated below the map, the user can then get details on position and alignment of orthologous ESTs by a direct link to Gramene (**C**, **D**), or annotation in terms of predicted function (**E**). Subsequently, the user can retrieve FASTA outputs of rice genome sequence, or 5' and 3' regulatory regions, or specific exons and introns (**B**, **F**). These FASTA outputs can be used for input into motif-finding software.

b. Draw the map (default): A map will be displayed illustrating the queried gene locations in the rice genome. You can determine the picture size by setting "Map width" and "Map height". The default values are 900 and 500.

5. Submit data

a. Click "Submit Query" to display the results in your browser.
b. The queried genes are shown in red on the map and as hyperlinks in the table. The adjacent genes are shown in purple in both the picture and the table. For each row, the first column is the queried GeneChip or oligo designator and is hyperlinked to Gramene (*1*) where detailed information on this alignment is displayed. Column 2 allows the retrieval of multiple rice gene models and accompanying FASTA files. Columns 3–14 contain information on the alignment (*see* **Fig. 2C**): alignment orientation (PLUS/PLUS or PLUS/NEGATIVE), alignment start and end, the chromosome where the gene is located, and alignment start and end in the rice genome (*see* **Fig. 2D**). The 15th column contains a link to the PLEXdb annotation page, providing detailed information for a probe set, such as predicted function, sequence, and expression (*see* **Fig. 3**). The last column (not shown in **Fig. 2**) is an abbreviated gene annotation.

6. Check detailed gene structure and retrieve sequence.

a. Click the probe set designator on the output picture and a detailed alignment between this gene and the rice gene model will be displayed (*see* **Fig. 2B**).
b. Chromosomal number and scale represent the position of this gene on the rice genome. Blue rectangles represent exons, and the leftmost arrow designates the orientation.
c. Click exons (blue rectangle) or introns and the corresponding sequence will be displayed (*see* **Fig. 2B**).
d. Click the arrow at the ends of the scale to retrieve 3 kb upstream and downstream sequences of this gene.
e. For FASTA outputs of multiple gene sequences, check selected boxes in the table below the map. Click "FASTA sequence" above the third column. As shown in **Fig. 2F**, representative rice gene models and FASTA files will be displayed for input into motif-finding software (*see* **Note 6**).

3.3.2.4. ANNOTATION OF GENES

Gene and exemplar annotation is essential for understanding the results of microarray experiments. As shown in **Fig. 3**, for every exemplar on every chip in PLEXdb, the following information is provided on the species annotation page (e.g., http://www.plexdb.org/PD probeset annotation.php?genechip=Barley1).

1. Probe sequence and alignment: This function displays the position of each probe on the consensus gene sequence as well as its individual nucleotide sequence.

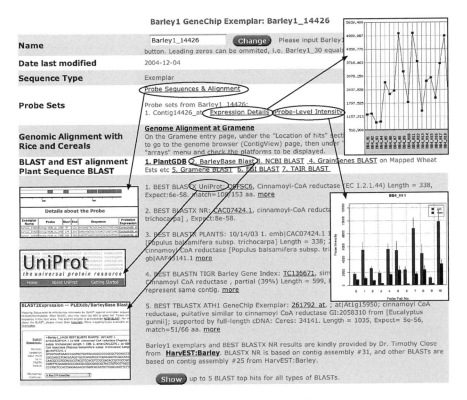

Fig. 3. Gene alignments, expression, and annotation.

2. Expression data

   a. Clicking the "expression details" link provides the user with expression plots for a particular probe set in any public experiment in PLEXdb that used a platform with that probe set.

   b. The expression page has further links to all hybridization data for that probe set, including individual probes within a probe set.

3. Genome alignment with model sequenced genome: This function has links to syntenous positions on sequenced model genomes, e.g., barley genes aligned to the rice genome in Gramene.

4. EST BLAST alignments are provided from a variety of plant-oriented databases. Examples are shown below.

   a. PlantGDB (http://www.plantgdb.org/search/display/data.php?Seq_ID=Barley1_01637) (*see* **Note 3**).

b. Graingenes
   (http://www.graingenes.org/cgi-bin/ace/custom/goBlast/graingenes?
   name=Barley1_01637&class=Sequence).
c. Gramene (http://143.48.220.116/Multi/blastview).
d. EBI (http://www.ebi.ac.uk/blastall/) uses NCBI-BLAST 2 software.
e. TAIR (http://www.arabidopsis.org/Blast/).
g. BarleyBase
   (http://www.plexdb.org/modules.php?name=PD_dataSelection
   &page=blast2expression.php).

5. BLAST searches are updated frequently to provide predicted protein functions.

   a. BLASTX results using UniProt *(21)* provide users with better annotation
      including GO terms, domains, and other key information to understand
      a protein and its function (http://www.pir.uniprot.org/cgi-bin/upEntry?id=E13
      B_HORVU).
   b. TIGR gene index (http://www.tigr.org/tigr-scripts/tgi/tc_report.pl? tc=TC139945
      &species=barley).
   c. Other GeneChips: Microarray Platform Translator allows you to enter a gene list
      from one GeneChip and extract similar genes from another species GeneChip.

6. Nucleotide/amino acid sequence is provided for each gene such that users can
   develop probes or primers for further analysis.

3.4. Conclusions

Many biological phenomena are conserved across taxonomic boundaries; in
order to take advantage of conserved genes and pathways, we need efficient
ways of data presentation and dissemination. With the advent of increasing
numbers of DNA chips for many crop and crop pathogen species, the rationale
for the development of PLEXdb as a hypothesis building information resource
is to provide efficient access to highly parallel expression data with seamless
portals to related genetic, physical, and trait data.

4. Notes

1. New arrays will be added to PLEXdb as they are developed, e.g., cotton and
   Brassica arrays will be released in the coming year.
2. Additional information on normalization can be found at
   http://www.biomedcentral.com/1471-2105/6/214 *(22)*.
3. Probe set annotation links to PlantGDB-assembled Unique Transcripts (PUT)
   assemblies. The probe locations for the Affymetrix GeneChip and long oligo array
   designs in PLEXdb are linked to the PlantGDB PUT assemblies and to the original
   assemblies. These links allow investigators to examine the historical data used to

design the GeneChips as well as the most up-to-date EST assemblies that support the probe sets.

4. "Notes" and "Help" files are provided at each stage of the inspection process, such that the user has assistance in interpreting what they see. We will point out, however, that whatever the quality of the data or database, there is no substitute for a thorough interaction with your local statistics faculty when you are planning a large microarray experiment.

5. Factorial submission and association each sample description with each hybridization makes possible the full function of downstream analysis tools, e.g., gene list creation, expression graphs, barplots, heatmaps, and so on.

6. An initial analysis step would be through promoter-specific motif finders, such as the Signal Search Analysis server at ISREC, http://www.isrec.isb-sib.ch/ssa/ssa.html. Directed searches for known motifs can be performed using the FINDPATTERNS program (GCG). Broad searches for plant promoter motifs can be performed by using the PLACE database of published motifs found in plant cis-acting regulatory DNA elements (http://www.dna.affrc.go.jp/PLACE/signalup.html) or NSITE-PL (http://www.softberry.com/berry.phtml?topic=nsitep&group=programs&subgroup=promoter) for recognition of plant regulatory motifs with statistics.

More in-depth analysis may involve an *ab initio* motif-finding program that looks for over-represented sequence strings as compared to a random group such as MotifSampler http://www.esat.kuleuven.ac.be/~thijs/Work/MotifSampler.html or GLAM http://zlab.bu.edu/glam/. GLAM, in particular, will process iteratively data sets and eliminate incorrect results inherent in stochastic methods. Additionally, GLAM will eliminate problem sequences, whereas some other *ab initio* programs will force a sequence to fit, resulting in an incorrect identification.

Statistical tests should be used to determine if motifs of interest are significantly over-represented in the co-regulated gene sets and their orthologs. For example, using the frequencies of potentially conserved motifs from all rice and Arabidopsis genes as a reference, individuals can calculate *p* values using the Poisson distribution to describe the likelihood of the observed sequence being present by chance.

Acknowledgments

The authors thank Nick Lauter for critical review of the manuscript. BarleyBase/PLEXdb was initiated with funds from USDA-CSREES North American Barley Genome Project and is currently supported by USDA-National Research Initiative grant no. 02-35300-12619 and National Science Foundation Plant Genome grant no. 0500461 to RPW and JAD.

References

1. Ware, D., Jaiswal, P., Ni, J., Pan, X., Chang, K., Clark, K., Teytelman, L., Schmidt, S., Zhao, W., Cartinhour, S., et al. (2002) Gramene: a resource for comparative grass genomics. *Nucleic Acids Res.* 30 (1), 103–105.

2. Matthews, D.E., Carollo, V.L., Lazo, G.R. and Anderson, O.D. (2003) GrainGenes, the genome database for small-grain crops. *Nucleic Acids Res.* 31 (1), 183–186.

3. Rhee, S.Y., Beavis, W., Berardini, T.Z., Chen, G., Dixon, D., Doyle, A., Garcia-Hernandez, M., Huala, E., Lander, G., Montoya, M., et al. (2003) The Arabidopsis Information Resource (TAIR): a model organism database providing a centralized, curated gateway to Arabidopsis biology, research materials and community. *Nucleic Acids Res.* 31 (1), 224–228.

4. Brazma, A., Parkinson, H., Sarkans, U., Shojatalab, M., Vilo, J., Abeygunawardena, N., Holloway, E., Kapushesky, M., Kemmeren, P., Lara, G.G., et al. (2003) ArrayExpress–a public repository for microarray gene expression data at the EBI. *Nucleic Acids Res.* 31 (1), 68–71.

5. Parkinson, H., Sarkans, U., Shojatalab, M., Abeygunawardena, N., Contrino, S., Coulson, R., Farne, A., Garcia Lara, G., Holloway, E., Kapushesky, M., et al. (2005) ArrayExpress–a public repository for microarray gene expression data at the EBI. *Nucleic Acids Res.* 33 (1), D553–555.

6. Barrett, T., Suzek, T.O., Troup, D.B., Wilhite, S.E., Ngau, W.-C., Ledoux, P., Rudnev, D., Lash, A.E., Fujibuchi, W. and Edgar, R. (2005) NCBI GEO: mining millions of expression profiles–database and tools. *Nucleic Acids Res.* 33 (1), D562–566.

7. Craigon, D.J., James, N., Okyere, J., Higgins, J., Jotham J. and May, S. (2004) NASCArrays: a repository for microarray data generated by NASC's transcriptomics service. *Nucleic Acids Res.* 32 (1), D575–577.

8. Ihaka, R. and Gentleman, R. (1996) R: A language for data analysis and graphics. *J. Comput. Graph. Stat.* 5, 299–314.

9. R Development Core Team (2003) *R: A Language and Environment for Statistical Computing.* Vienna, Austria, R Foundation for Statistical Computing.

10. Gautier, L., Cope, L., Bolstad, B. and Irizarry, R. (2004) Affy–analysis of Affymetrix GeneChip data at the probe level. *Bioinformatics* 20 (3), 307–315.

11. Irizarry, R.A., Wu, Z. and Jaffee, H. A. (2005) Comparison of Affymetrix GeneChip expression measures. *Bioinformatics* 1 (1), 1–7.

12. Brazma, A., Hingamp, P., Quackenbush, J., Sherlock, G., Spellman, P., Stoeckert, C., Aach, J., Ansorge, W., Ball, C.A., Causton, H.C., et al. (2001) Minimum information about a microarray experiment (MIAME)—toward standards for microarray data. *Nat. Genet.* 29 (4), 365–371.

13. Tang, X., Shen, L. and Dickerson, J.A. (2005) BarleyExpress: a web-based submission tool for enriched microarray database annotations. *Bioinformatics* 21 (3), 399–401.

14. The Plant OntologyTM Consortium (2002) The Plant Ontology™ Consortium and Plant Ontologies. *Comp. Funct. Genomics* 3 (2), 137–142.

15. Shen, L., Gong, J., Caldo, R.A., Nettleton, D., Cook, D., Wise, R.P. and Dickerson, J.A. (2005) BarleyBase–an expression profiling database for plant genomics. *Nucleic Acids Res.* 33 (1), D614–618.

16. Dong, Q., Schlueter, S.D. and Brendel, V. (2004) PlantGDB, plant genome database and analysis tools. *Nucleic Acids Res.* 32 (1), D354–359.

17. Quackenbush, J., Liang, F., Holt, I., Pertea, G. and Upton, J. (2000) The TIGR Gene Indices: reconstruction and representation of expressed gene sequences. *Nucleic Acids Res.* 28, 141–145.

18. Close, T.J., Wanamaker, S.I., Caldo, R.A., Turner, S.M., Ashlock, D.A., Dickerson, J.A., Wing, R.A., Muehlbauer, G.J., Kleinhofs, A. and Wise, R.P. (2004) A new resource for cereal genomics: 22K Barley GeneChip comes of age. *Plant Physiol.* 134 (3), 960–968.

19. Caldo, R.A., Nettleton D. and Wise, R.P. (2004) Interaction-dependent gene expression in *Mla*-specified response to barley powdery mildew. *Plant Cell* 16, 2514–2528.

20. Duda, R.O. and Hart, P.E. (1973) *Pattern Classification and Scene Analysis*. New York, John Wiley & Sons.

21. Bairoch, A., Apweiler, R., Wu, C.H., Barker, W.C., Boeckmann, B., Ferro, S., Gasteiger, E., Huang, H., Lopez, R., Magrane, M., et al. (2005) The Universal Protein Resource (UniProt). *Nucleic Acids Res.* 33 (1), D154–159.

22. Zakharkin, S.O., Kim, K., Mehta, T., Chen, L., Barnes, S., Scheirer, K.E., Parrish, R.S., Allison, D.B. and Page, G.P. (2005) Sources of variation in Affymetrix microarray experiments. *BMC Bioinformatics* 6, 214–225.

18

Reaping the Benefits of SAGE

Stephen J. Robinson, Justin D. Guenther, Christopher T. Lewis,
Matthew G. Links and Isobel A.P. Parkin

Summary

Serial analysis of gene expression (SAGE) is a powerful technique which yields a digital measure of gene expression through the sequencing of libraries of specific mRNA-derived fragments, namely SAGE tags. This chapter introduces the methods and software tools that are available for researchers to analyze gene expression through SAGE analysis. A detailed examination of SAGE analysis in *Arabidopsis thaliana* using the publicly available analysis tool, SaskSAGE, is provided. The use of this software allows the user to maximize the information gained from SAGE experiments in a model system with a fully sequenced genome.

Key Words: Serial analysis of gene expression; expression profile; novel transcripts; *Arabidopsis thaliana*; SaskSAGE.

1. Introduction

Global gene expression analysis has become widely adopted as a tool for both candidate gene identification and elucidation of the regulatory pathways controlling complex traits. Microarray analysis is the most common technology used to simultaneously sample the expression of large numbers of genes. However, microarray analyses are limited by the available gene repertoire for a particular species. Sequenced-based methodologies, such as serial analysis of gene expression (SAGE) and massively parallel signature sequencing (*see* Chapter 19), are not dependent upon prior sequence information and allow the identification of millions of transcripts from a single mRNA population derived from any species of interest.

From: *Methods in Molecular Biology, vol. 406: Plant Bioinformatics: Methods and Protocols*
Edited by: D. Edwards © Humana Press Inc., Totowa, NJ

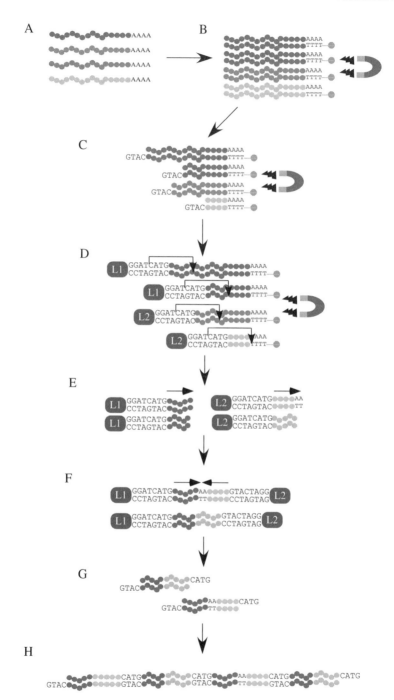

The SAGE protocol isolates a short, between 14 and 26 base pairs (bp), tag from the 'anchoring' enzyme (AE) restriction site closest to the terminal end of each transcript (*see* **Fig. 1**). The length of the tag and the targeted location of the tag together determine the specificity of the tag for each transcript within the cell. The expectation is that the frequency with which each SAGE tag is isolated is directly proportional to the abundance of their parent RNA molecules. Due to the absolute nature of the SAGE tag counts, direct comparisons can be made between libraries derived from different tissues and treatments. Since its inception, SAGE analysis *(1)* has been used in a plethora of transcript profiling experiments in a number of organisms including *Saccharomyces cerevisiae (2)*, *Homo sapiens (3)*, *Caenorhobditis elegans (4)* and *Drosophila melongaster (5)*. It is only in the last 5 years that SAGE has been widely applied to the analysis of plant transcriptomes, this delay was due to the relative size of plant genomes, the availability of sequence data, and the cost of large-scale sequencing. SAGE data have now been generated to study changes in gene expression responsible for different developmental processes and in response to a wide range of biotic, abiotic and environmental stresses for numerous plant species, including *Oryza sativa (6, 7)*, *Arabidopsis thaliana (8–10)* and *Zea mays (11)*. SAGE has also proved valuable in uncovering changes in gene expression in plant species with as yet limited genomics resources such as *Musa acuminata (12)* and *Pinus taeda (13)*.

In this chapter, we will describe the generation of SAGE libraries and the methods and available resources for analyzing SAGE tag data. Although a number of modifications have been made to the SAGE protocol, we will focus on the use of the original SAGE protocol described by Velculescu et al. *(1)*, which is still the most commonly applied method, and indicate where appropriate differences in application for the revised methods of SAGE analysis. Additionally, we will describe in detail the use of SaskSAGE software for the comprehensive analysis of SAGE tags derived from the model plant *A. thaliana*.

◄─────────────────────────────────────

Fig. 1. Diagram showing the different steps in serial analysis of gene expression (SAGE) library construction (modified from Jiang Long www.seq.ubc.ca/images/sage3b.ai). (**A**) Production of poly(A) RNA. (**B**) Synthesis of double-stranded cDNA using a biotinylated oligo $dT_{(18)}$ primer; this enables the isolation of 3′ ends. (**C**) Digestion with anchoring enzyme (AE), *Nla*III. (**D**) The cDNA population is divided and each half is ligated separately to different linkers (L1 or L2 sequence), both adapters generate the tagging enzyme (TE) (*Bsm*FI) recognition site, but each contains a different polymerase chain reaction (PCR) primer annealing sequence. (**E**) Digestion with TE, which releases the SAGE tag. (**F**) Formation of ditags. (**G**) PCR amplification of the ditags followed by digestion with the AE, which releases the ditags. (**H**) Concatamerization of the ditags, which are then cloned and sequenced.

2. Materials

2.1. Web Resources for SAGE analysis

1. AAFC SaskSAGE: http://www.brassica.ca/SaskSAGE. Access to software and comparative SAGE tag databases for analyzing *A. thaliana* SAGE data.
2. SAGENet: http://www.sagenet.org/. Information on SAGE library construction, a list of publications which have employed SAGE analysis, and access to SAGE2000 software.
3. SAGEmap: http://www.ncbi.nlm.nih.gov/SAGE/. Provides resources for SAGE tag matching, comparing between SAGE libraries and access to publicly available SAGE data.
4. IDEG6: http://telethon.bio.unipd.it/bioinfo/IDEG6_form/. Access to six different statistical tools for comparing between sequenced expression libraries, including SAGE data.
5. DiscoverySpace: http://www.bcgsc.ca/discoveryspace/. Access to graphical tools for analysis and storage of SAGE data.
6. Melbourne Brain Genome Project: http://www.mbgproject.org/MBGP_Tools.html. Maintains a list of most publicly available SAGE resources.

2.2. SAGE Analysis Software

There are a number of available resources which perform different steps in SAGE analysis. Although the majority of these software packages were developed for the analysis of mammalian transcriptomes, elements from these packages can be utilized for the analysis of plant-derived SAGE data. **Table 1** lists the different packages and compares the features available in each.

2.3. Plant Resources for SAGE Tag-To-Gene Matching

1. http://www.ncbi.nlm.nih.gov/SAGE/: Virtual SAGE tags derived from Unigene contigs for *A. thaliana, Lycopersicon esculentum (Solanum lycopersicum), Medicago trunculata, Musa acuminate, O. sativa, P. taeda, Saccharum officinarum, Triticum aestivum, Vitis vinifera*, and *Z. mays*.
2. http://www.brassica.ca/navigation/sage/sage_e.shtml Virtual SAGE tags derived from the *A. thaliana* genomic sequence (TIGR release v5: http://www.tigr.org/tdb/e2k1/ath1/).

3. Methods

3.1. SAGE Library Generation

To analyze SAGE output it is important to understand how the tags are generated, as each stage of the process may introduce experimental errors

Table 1
List of available SAGE software

Name	Website	Platform	Input file	Sequence Quality Filter	SAGE method	SAGE tag extraction	Tag-to-gene mapping	Output	Statistical Analysis		Targeted Organism
									pairwise	multi	
SaskSAGE	2.1.2	Unix Windows (cygwin) WebBrowser	FASTA PHD	y	Short Long	y	y	By tag By gene	y	y	A. thaliana
SAGE2000	2.1.2	Windows	FASTA	n	Short Long	y	y	By tag	y	n	Human
SAGEmap	2.1.3	WebBrowser	Tab delimted text	n	Short Long	n	y	By tag	y	n	Multiple
IDEG6	2.1.4	WebBrowser	Tab delimted text	n	Short Long	n	n	By tag	y	y	Multiple
Discovery Space	2.1.5	Windows	FASTA	n	Short Long	y	y	By tag	y	n	Human

which could bias the results of a SAGE experiment. The SAGE tags are obtained through a series of restriction and ligation reactions performed on cDNA molecules derived from polyadenlyated RNA, with the location of the SAGE tag within a transcript being determined by the use of a four-base restriction AE (*see* **Fig. 1**). The two most commonly used AEs are *Nla*III (CATG) and *Sau*3A (GATC), the remainder of this chapter will adopt *Nla*III as the AE. The cDNA molecules are generated using biotin-labeled oligodT$_{(18)}$ primers, this allows the 3′ end of the cDNA to be captured on streptavidin beads, and after digestion with the AE, the 5′ fragments from the cDNA molecules are discarded. The ligation of linker molecules generates both polymerase chain reaction (PCR) priming sites and a 'tagging' enzyme (TE) site which releases the SAGE tags. The TE is a Type IIS restriction enzyme, which recognizes asymmetric base sequences and cleaves DNA at a specified position up to 20 bp outside the recognition site. The released SAGE tags are blunt end cloned to generate ditags, which are then amplified by PCR before being concatenated, cloned and sequenced.

The most commonly used TE is *Bsm*FI, which produces a tag of length 14 bp. Modification of the SAGE protocols through the use of different TEs, *Mme*I (LongSAGE) *(14)* and *EcoP15*I (SuperSAGE) *(15)* have increased the length of the released SAGE, tag to 21 bp and 26 bp, respectively. These later modifications have increased the complexity of the SAGE tag, and thus the power for unambiguous tag assignment, but this comes at the price of lowering the cost efficiency per tag, as additional sequencing is required to attain the same depth of tag count.

It is necessary to determine the length of tag that is sufficient for unambiguous tag assignment in the organism under investigation, where a 14-bp tag maybe sufficient for *A. thaliana*, a more complex tag will be required for studies in organisms with larger genomes or with higher ploidy levels. Studies investigating host–pathogen interactions may also require longer tags, which would provide additional information to allow discrimination of transcripts from the two species.

3.2. Extraction of Experimental SAGE Tags

The data from a SAGE experiment can easily be extracted from the sequencing reads, as the ditags are delimited by the restriction recognition site of the AE used during the protocol (*see* **Fig. 2**). **Figure 2** shows a sequence from a SAGE clone and the typical output after SAGE tag extraction. The SAGE extraction software recognizes each AE site and extracts each ditag flanked by the AE sites. Although the ditags can vary in length, each tag

A

```
TGGGCCCTCTAGATGCATGCCAGTACCTTGTCGAAACAAGCTCATGAGAG
AGTTCAGTTATCTAAAACGCATGTTGAAGTCTCAGGCAGTAACCTTCATG
TAACATTTTTTTCAAACAGACTACATGGGGAACCGATGGATTGATCTGGC
ATGATTATGATTCGGTTACACGTTGCATGGATGCTTGTTCGGCCACACCT
GCATGGTGACGCCATCCTAATTACCATCATGCAGGATTTGATGCCGATGG
GTACATGGTGAGCAGAAGATTTGTACAAACATGTGTTTTTATGTCGCTCT
GTTATGCATGGTTCTGGCAAGATTCTACCAGCCATGTGTCCATATGGTTA
CATCTTCTTCCATGCACCTGAACGAGAGCATCGCAAACATGGGCGTTGTA
AACAAAAGAGAAAACATGGTTGAGATCCATTTTTTTTTATCATGACAAAA
TGCATCTCTTCGTCAAGCATGTTTAGAGCCAGTGGCCACACCTTCATGTC
TGTCAAGAATAGGGCTCCGAGCATGAGCTCGAAAAGAAAGGAAGGAAACA
TGAAGTGCAACACACTCCTTCAAACCATGACGTTCACGATTATTTGTACA
AACATGGGTCAGTGCTTAAGGAAGGAAACATGGAGGTGGTGAATAAAATG
AAATCATGTTTTTATTAAAGTTACAGGTGACATGGGCCTTCGCCACACTT
TTGATTCATGGGCCTTCGCTAAATCTGTCTTTCATGAAAGAAGGAGAA
```

B

Tag	Tag (id)	Tag Count	Duplicate dTags	Duplicate dDitags	Frequency
AACCTTAGTA	(0)	1	0	0	0.000012
AACCTTATCC	(0)	3	0	0	0.000037
AACCTTGAAC	(0)	4	0	0	0.000050
AACCTTGCTT	(0)	10	0	0	0.000125
AACCTTGGTA	(0)	1	0	0	0.000012
AACCTTTTTT	(0)	4	0	1	0.000050
AACGAAAAAT	(0)	3	0	0	0.000037
AACGAAACAA	(0)	2	0	0	0.000025
AACGAAATCG	(0)	5	0	0	0.000062
AACGAAATTG	(0)	1	0	0	0.000012
AACGAACAAT	(0)	4	0	0	0.000050
AACGAAGCGG	(0)	1	0	0	0.000012
AACGAATGGA	(0)	1	0	0	0.000012
AACGAATTCA	(0)	2	0	0	0.000025
AACGAATTTG	(0)	3	0	1	0.000037
AACGACAAAA	(0)	1	0	0	0.000012
AACGACAAAG	(0)	36	0	3	0.000449
AACGACAACT	(0)	1	0	0	0.000012
AACGAGCTCA	(0)	4	0	0	0.000050
AACGAGCTCT	(0)	7	0	1	0.000087
AACGAGTTGA	(0)	15	0	0	0.000187
AACGATATGC	(0)	2	0	0	0.000025
AACGATCTAG	(0)	1	0	0	0.000012
AACGATCTTC	(0)	1	0	0	0.000012
AACGATCTTG	(0)	33	0	3	0.000412
AACGATGATG	(0)	1	0	0	0.000012
AACGATTTTG	(0)	1	0	0	0.000012
AACGCAAAAA	(0)	1	0	0	0.000012
AACGCAAAGA	(0)	1	0	0	0.000012
AACGCAACAT	(0)	1	0	0	0.000012
AACGCACACC	(0)	27	0	2	0.000337

C

```
              Ditags:  31146
      Invalid Ditags:  1371
    Duplicate Ditags:  1611
   Low Quality Ditags:  3029
     Ditags Processed:  25135

Sequence Tags Processed:  50270

Sequence Tags Excluded:
       In Exclude File:  461
         Too many N's:  15
        Duplicate Tags:  0

 Valid Sequence Tags:  49794
              Unique:  13512
```

extracted is of equal length, as defined by the limits of the TE. Once the first tag is extracted from the proximal end of each ditag, the second tag is extracted as the reverse complement from the terminal end of the ditag. The total number of SAGE tags and the individual tag counts are recorded and used to calculate the tag frequency within a library. These digital expression data are directly proportional to the abundance of the transcripts from which the tags originate. There are a number of software packages that have been developed to perform this function, these software and their various features are listed in **Table 1**. Typically, tags that are observed only once, known as singletons, make up between 50–70% of the total tags detected *(9,10)*. Singletons are often excluded from further analysis, as they have little value when attempting to identify differentially regulated genes, and it has been suggested that they arise from sequencing errors. However, singletons that are unambiguously assigned to their parental transcript can further the annotation of genomic sequence and contribute to the characterization of the transcriptome.

The most pervasive source of error in SAGE analysis, which is commonly overlooked, is due to poor quality sequence data. Sequencing errors have been estimated to range between 0.7 and 1% per base for each ditag *(2)*. The effects of these errors can be diminished by applying the Phred sequencing quality characteristic algorithm to all chromatograms (*16*; http://www.phrap.org) and excluding all low quality ditags (generally with an average Phred score <20).

A more contentious issue in SAGE analyses is the exclusion of duplicate ditags from further analysis. The SAGE method generates ditags prior to PCR amplification (*see* **Fig. 1**). Due to the complexity and size of any SAGE tag population, it is expected that the formation of identical ditags would occur rarely by chance, thus it has been proposed that duplicate ditags are artifacts of the PCR amplification step. The majority of SAGE analyses ignore all but one

---

Fig. 2. (**A**) An example of a sequence from a serial analysis of gene expression (SAGE) clone, which is punctuated by the anchoring enzyme site (CATG), highlighted in bold, that flanks each ditag. The tagging enzyme *Bsm*FI generates a 10 bp/14 bp 3′ overhang which defines the minimum possible length of each tag at 10 bp. (**B**) Sample output after SAGE tag extraction; the 10-bp tags are extracted from the sequence, and the abundance of each unique tag is calculated. The number of duplicate ditags associated with each SAGE tag is also noted. (**C**) Information describing the quality of a SAGE library, with regard to total number of SAGE tags, number of excluded tags and number of duplicate ditags.

occurrence of a particular ditag. However, this may introduce bias, especially in less complex tissues, as highly abundant transcripts would engender proportionally abundant SAGE tags, which could pair to form legitimate duplicate ditags. It has been proposed that in specific instances a calculable proportion of the duplicate ditags should not be removed from the analysis as they may well represent a true reflection of gene expression *(17)*.

Digestion of each cDNA 3′ end with the TE results in a 3′ or 5′ overhang which is made blunt prior to ligation. Variations in enzyme efficiencies in each of these steps can lead to SAGE tags of variable lengths. To prevent the inadvertent extraction of chimeric SAGE tags, the SAGE software excludes ditags that are either too short or too long, the minimum and maximum lengths being specific to the TE used. Ligation of the linker molecules can result in the formation of linker dimers, that upon restriction using the TE, yield a linker-derived SAGE tag. The frequency of linker-derived contamination has been estimated to occur at a rate of 1–5% *(18,19)*. Poly (A) tags can also be excluded from further analysis, as the low complexity of the sequence prevents specific matching of such tags.

3.3. SAGE Tag to Transcript Matching

Once the tags are extracted and counted, their utility for analyzing gene expression depends on the ability to unambiguously match each tag back to its transcript of origin. Tag-to-transcript matching is achieved by comparing each tag with a data set of virtual SAGE tag libraries generated *in silico* from available nucleotide sequences for the species of interest. This analysis results in the assignment of SAGE tags to previously identified genes, to genes of unknown function, or to a transcriptional unit not present within the available data sets, allowing SAGE to identify novel expressed genes.

SAGE is designed to specifically capture the tag derived from the 3′ most AE site of each transcript, and this is referred to as the canonical tag site. The defined position of the canonical tag within the transcript aids in unambiguously assigning a SAGE tag to the transcript of origin by reducing the complexity of the virtual tag sites. Currently, the majority of the available SAGE software will only assign matches to canonical tag sites within annotated transcripts.

The dichotomy of SAGE analysis lies in the fact that SAGE tags can be isolated from any species. However, the ability to match the tags back to a functional gene unit is dependent upon the available sequence for the targeted species. The level of matching which can be achieved and also the types of transcripts that can be identified vary according to the available sequence data.

3.3.1. SAGE Tag to Transcript Matching in an Unsequenced Genome

For most species with as yet unsequenced genomes, there are collections of either expressed sequence tag (EST) or cDNA sequences available that can be exploited for SAGE tag matching. The most valuable sequences for SAGE tag matching are full-length cDNAs, which include both 5′ and 3′ untranslated regions (UTRs). Currently, there are no complete sets of full-length transcript sequences available for any organism, although a number of projects have been initiated to achieve this goal *(20–24)*. The limitation in using EST sequences for tag matching is largely due to the fact that they are single-pass sequences with an estimated error rate of 1% and an overall error rate of 10% for a 10-bp tag *(25)*. ESTs also introduce errors in tag-to-gene mapping assignments due to the presence of chimeras and truncated gene sequences. In most instances contigs of ESTs, namely Unigenes, are assembled prior to extracting the *in silico* SAGE tags. However, this process may also introduce errors, because the presence of highly similar genes can corrupt the contig alignment. NCBI provides Unigene sets for 10 different plant species (*see* Chapters 2 and 3) which is an excellent resource for SAGE tag matching.

3.3.2. SAGE Tag to Transcript Matching Within a Sequenced Genome

In addition to the complete genome sequence of *A. thaliana* *(26)* and *O. sativa* *(27,28)*, projects are underway to sequence the genomes of *M. trunculata* (http://www.genome.ou.edu/medicago.html), *L. esculentum* (*Solanum lycopersicum*) (http://www.sgn.cornell.edu/help/about/tomato_sequencing.pl) *Z. mays* (http:// www.maizegenome.org.) and *B. rapa* (http://www. brassica.info/b_rapa_sequencing_project/bac_sequencing.htm). These sequence data will improve the ability to assign SAGE tags to their gene of origin. Genomic sequence data are generally based on multiple sequence reads that reduce the presence of sequencing errors. Additionally, SAGE tags generated from low abundance molecules, alternatively processed transcripts, anti-sense transcripts and microRNAs can be identified using genomic sequence.

The SAGE protocol is designed to recover a sequence tag that is derived from the 3′ end of each transcript. However, due to the random nature of restriction site position, the tags may often be derived from the 3′ UTR sequence, and in the absence of an AE site in the coding region, the tag may be derived from the 5′ UTR sequence. Ideally, the data sets that are used to match experimental tags should include these sequences. The availability of genome sequence allows the development of default UTR sequence for genes without annotated UTR sequence. SAGE tag analysis using data sets with computationally determined

UTR sequence has been carried out for *A. thaliana* and was shown to increase the efficacy of SAGE tag matching (*10, see* **Note 1**).

3.4. Identification of Novel Transcripts

Alternative transcript processing has been estimated to inflate the number of possible transcripts detected within *A. thaliana* from between 4 and 17% (*29,10*). The unique assignment of SAGE tags to individual transcripts at both canonical and non-canonical positions enables the detection of possible alternative transcript processing. However, SAGE tag artifacts that mimic alternative transcript processing may be generated through incomplete digestion with the AE, which may complicate the identification of such novel transcripts. Tags generated through incomplete AE digestion would be detected at a frequency that is inversely proportional to their relative tag site position from the 3' end of the molecule. Genes with tag frequencies that follow this pattern can be excluded when attempting to identify alternatively processed transcripts.

The SAGE protocol defines the orientation of each tag which allows anti-sense transcripts to be identified. Anti-sense transcription has been proposed as a novel mode of gene regulation, and 24% of the *A. thaliana* annotated genes were shown to possess potential anti-sense transcripts (*30*).

3.5. Statistical Analysis of SAGE Data

SAGE experiments are designed to identify transcripts whose observed tag count is significantly different between SAGE libraries. This can be achieved by calculating the probability that the difference in individual tag counts observed between libraries occurred by chance. A thorough statistical approach would measure a number of independently sampled biological replicates that would allow a comparison of variance both within and between the treatments. However, typically, SAGE libraries only estimate gene expression levels from a single sample or from pooled biological replicates resulting in a single measurement. As the within library variance is unknown, a number of different statistical methods have been used to provide an estimate for the variance of the SAGE tag distribution. A review of these methods is provided by Man et al. (*31*). A number of the reviewed statistical methods are available at http://telethon.bio.unipd.it/bioinfo/IDEG6_form. Due to its inclusion within SAGE2000 and SAGEmap, the simulation approach of Zhang et al. (*3*) is the most widely used method for statistical comparisons between two SAGE libraries. However, for statistical analysis of SAGE libraries comprising large numbers of tags, the chi-squared test has a high level of statistical power and is effective over a wide range of tag frequencies (*31*).

More recently, SAGE experiments have employed the use of multiple libraries, and with the increasing amounts of publicly available SAGE data, there is now a need for multi-comparison statistical tests. The chi-squared test for homogeneity using a contingency table $2 \times t$ (where $t > 2$) can be used for such analyses *(31)*. For any statistical test between libraries, the probability calculated provides a threshold for type I errors (or false positives). As the test is applied repeatedly to each unique tag detected, the use of the Bonferroni correction can reduce the occurrence of these errors.

3.6. Analyzing SAGE Data in Practice

The ability to carry out comprehensive SAGE tag matching has been applied to the analysis of gene expression data in the *A. thaliana* genome *(10)*. The AAFC SaskSAGE analysis suite uses a set of Perl modules to extract experimental SAGE tags and matches these tags to data sets derived from genomic sequence. Presently targeted to *A. thaliana*, these scripts could be adapted to accommodate any sequenced annotated genome. SaskSAGE uses an iterative matching process that successively matches SAGE tags to defined data sets, which can increase the efficiency of matching experimental SAGE tags in complex genomes *(10; see* **Fig.** 3*)*. The experimental SAGE tags are matched in the following order:

1. SAGE tags are matched to the canonical tag site extracted from the sense strand of all annotated transcripts, either including experimentally defined UTR or virtual UTR sequence.
2. Unmatched tags are compared to all non-canonical sites from the same set of transcripts as in step 1.
3. Unmatched tags are compared to all possible tag sites from immature mRNA sequences, including intronic sequence derived from the same set of transcripts as in step 1.
4. Unmatched tags are compared to tags derived from the anti-sense strand in each of the different data sets.

At each stage, ambiguous matches, where the experimental tag matches more than one transcript, are indicated.

SaskSAGE uses chi-squared analysis both for a two library comparison where the data are organized into a $2 \times t$ contingency table, and for multiple libraries where the data are arranged into a $2 \times t$ contingency table.

3.7. Analyzing SAGE Data Using SaskSAGE

1. Sequence sufficient SAGE tags (*see* **Note 2**) from experimental libraries of interest (*see* **Note 3**).

A

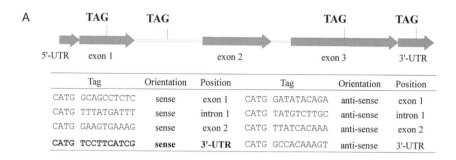

Tag		Orientation	Position	Tag		Orientation	Position
CATG	GCAGCCTCTC	sense	exon 1	CATG	GATATACAGA	anti-sense	exon 1
CATG	TTTATGATTT	sense	intron 1	CATG	TATGTCTTGC	anti-sense	intron 1
CATG	GAAGTGAAAG	sense	exon 2	CATG	TTATCACAAA	anti-sense	exon 2
CATG	**TCCTTCATCG**	**sense**	**3'-UTR**	CATG	GCCACAAAGT	anti-sense	3'-UTR

B

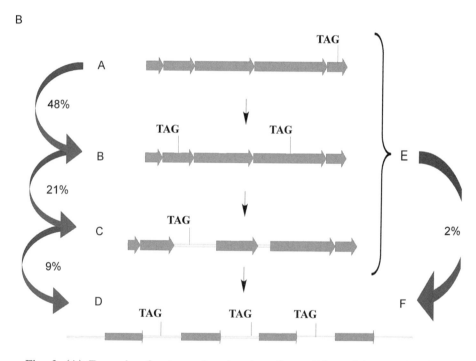

Fig. 3. (**A**) Example of a transcript showing all possible serial analysis of gene expression (SAGE) tag sites within the transcript, the canonical tag is indicated in bold. (**B**) The iterative matching process in a sequenced genome. The percentages indicate the percent unmatched tags remaining after each stage, using SAGE tag matching in *Arabidopsis thaliana* as an example: SAGE tags are matched to A, canonical tags sites from spliced transcripts; B, all non-canonical tag sites in spliced transcripts; C, all tag sites from immature transcripts; D, all possible tag sites in intergenic DNA; E, anti-sense tag sites from mature and immature transcripts; F, all remaining possible 'anti-sense' sites in the genome.

2. To enable sequence quality filtering, analyze the chromatograms using Phred (http://www.phrap.org), which assigns sequence quality values for each base and returns the data as a .PHD file.**4**

3. Install saskSAGE software http://sasksage.sourceforge.net sasksage_core_v1.1.tar.gz

   • unpack

   tar -xzvf sasksage_core_v1.1.tar.gz

   • set the environment variable SASKSAGE to point to the unpacked directory export SASKSAGE=/path/to/sasksage_core

   • saskSAGE has also been integrated into the web portal software APED (http://sourceforge.net/projects/aped) *see* **Fig. 4**

4. Organize directories such that there is an experiment directory under which there are independent SAGE libraries. In each library there should be a directory named PHD_FILES which contains the phd files for the sequencing reads for that library.

5. For each library perform the multi-level tag matching sasksage_core/bin/multi-csage.pl -p Experiment-X/control/PHD_FILES -w Experiment-X/control

Tag	SAGE control 2005	SAGE 30 mins	SAGE 120 mins	SAGE 2 day	SAGE 1 week	Gene Codes	Chi2	p-value
AAAAAAATCA	1	9	1	3	14	At1g21310.1 At2g11780.1 At2g32950.1 At5g53775.1	29.70	5.63e-06
AAAAATGGAG	15	6	1	10	20	At4g13940.1 At4g13940.1 At4g13940.1 At4g13940.1	36.77	2.01e-07
AAAACACAAA	8	14	0	14	29	At1g24140.1 At1g63400.1 At2g27920.1 At2g27920.2	52.88	9.03e-11
AAACTTAAAT	16	51	11	50	35	At1g61210.1 At4g13940.1 At1g61210.1 At4g13940.1	26.29	2.77e-05
AAACTTTATT	2	4	6	37	25	At2g05530.1 At5g17460.1 At2g05530.1 At5g17460.1	87.80	3.86e-18
AAAGATAAAA	0	1	1	6	8	At4g32220.1 At5g04480.1 At5g14010.1 At3g10620.1	24.04	7.84e-05
AAAGATATTC	0	2	2	13	16	At5g17460.1 At5g17460.1 At5g17460.1 At5g17460.1	48.70	6.74e-10
AAAGGAGGTG	0	13	10	0	1	At3g22380.1 At5g02210.1 At5g02210.1 At3g22380.1	32.20	1.74e-06
AACCACCCTG	1	0	0	1	9	At1g07920.1 At1g07930.1 At1g07940.1 At5g60390.1	40.20	3.94e-08
AACCCAGCCG	37	196	48	54	2	At3g26520.1 At3g26520.1 At3g26520.1 At3g26520.1	97.64	3.13e-20
AACGCAGTTC	22	14	2	19	4	At2g10940.1 At2g10940.2 At2g10940.1 At2g10940.2	25.67	3.69e-05
AACTAGTAGT	1	0	0	0	8	Chr3 Chr4 Chr3 Chr4	40.10	4.13e-08
AACTATCTGT	58	55	15	19	18	At2g42120.1 At2g42120.2 GI:7525093 At2g42120.1	34.41	6.14e-07
AAGAAAAAAG	16	5	0	4	6	At1g63710.1 At2g04290.1 At4g29300.1 GI:7525093	27.67	1.45e-05
AAGAACCATA	13	7	0	17	13	At1g01320.1 At1g53550.1 At1g01320.1 At1g53550.1	25.36	4.26e-05
AAGAGAGAGG	0	0	0	6	6	At4g69430.1 At2g42540.1 At2g42540.2 At2g42540.1	24.51	6.31e-05
AAGCACATAT	1	4	2	3	12	Chr1 Chr1 Chr3 Chr1	28.63	9.29e-06
AAGTTTCCGT	1	29	8	2	2	At5g40450.1 At5g40450.1 At5g40450.1 At5g40450.1	26.40	2.63e-05
AATATAGAGG	87	107	22	29	70	At4g21100.1 At4g21100.1 At4g21100.1 At4g21100.1	49.58	4.42e-10
AATCAAAAGT	56	42	18	43	12	At2g30560.1 At2g30560.1 At2g30570.1 At2g30570.2	38.83	7.55e-08
AATCAAAATT	2	0	0	13	2	At4g03824.1 At4g14690.1 At4g03824.1 At4g14690.1	39.11	6.61e-08
AATCATAATA	8	5	2	4	20	At3g42922.1 At4g31985.1 At3g21430.1 At2g21430.1	44.86	4.25e-09
AATGAGAATT	12	10	2	31	23	At1g26610.1 At3g04120.1 At5g12970.1 At1g26610.1	48.31	8.13e-10
AATGGTTAAC	2	0	0	9	11	At2g03320.1 At2g03320.1 At2g03320.1	36.21	2.62e-07
ACAAAAGAA	1	1	0	9	10	At1g16850.1 At1g65290.1 At1g69220.1 At1g69220.2	32.50	1.51e-06
ACAAGAAAAA	0	6	2	6	15	At2g02290.1 At5g23470.1 At2g02290.1 At5g23470.1	35.09	4.45e-02

Fig. 4. Screenshot of the SaskSAGE Browser showing typical data from a serial analysis of gene expression experiment, in this instance five libraries are being compared. The tag sequence is shown, along with the tag count for each of the libraries. The P value indicates whether there is a significant difference in tag count between any of the libraries, and the *Arabidopsis thaliana* gene code to which each of the tags has been matched is shown.

A

Tag	Control	30min	2hours	2days	1week	p-value	AGI-codes
CCTAAGATCT	289	86	218	141	47	2.78e-47	At5g66570.1:d:+1181:secondary
TTCAATAGTT	3	0	10	33	60	2.24e-44	At2g42530.1:d:+521:primary
AAGTACCGTA	9	17	11	15	60	1.00e-32	At2g02100.1:d:+367:primary
TTACAAGAGG	41	13	20	19	61	2.01e-22	At1g04270.1:d:+690:primary
ATTCCCTCAA	31	19	20	32	57	2.40e-16	At1g55490.2:d:+2026:primary
							At1g55490.1:d:+2053:primary
AATATAGAGG	104	85	30	30	69	8.57e-15	At4g21100.1:d:+1291:secondary
TTTGGTTATC	37	45	36	32	59	6.18e-08	At1g16000.1:d:+434:primary
							At1g43170.1:d:+1283:primary
							At1g43170.2:d:+1249:primary
TCAAAGGCCT	86	49	76	64	59	9.96e-06	At3g04120.1:d:+1130:secondary
							At5g36226.1:p:+377:secondary
							At1g13440.1:d:+1080:secondary
GCCATAAAAA	0	0	0	2	0	4.74e-06	Not matched
GAGGAGTGGT	0	0	0	2	0	4.74e-06	Chr2:+2216966:quaternary

B

```
LOCUS: AT2G42540
DESCRIPTION: cold-responsive protein / cold-regulated protein (cor15a), identical to cold-r
egulated protein cor15a (Arabidopsis thaliana) GI:507149; contains Pfam profile PF02987: La
te embryogenesis abundant protein
DATA:                     Control 30min   2hours  2days   1week   p-value       pos
SENSE COUNTS:             2       0       0       16      235     2.41e-184
GENES (3 total):
  AT2G42540.2
    SENSE COUNTS:         2       0       0       16      235     2.41e-184
    TAGS: (2 total)
    d+1  TTTAATAGTA       2       0       0       16      234     5.59e-183     577
    d+2m GCGATGTCTT       0       0       0       0       1       1.70e-01      88
  AT2G42540.1
    SENSE COUNTS:         2       0       0       16      234     5.59e-183
    TAGS: (1 total)
    d+1  TTTAATAGTA       2       0       0       16      234     5.59e-183     512


LOCUS: AT2G42530
DESCRIPTION: cold-responsive protein / cold-regulated protein (cor15b), nearly identical to
  cold-regulated gene cor15b (Arabidopsis thaliana) GI:456016; contains Pfam profile PF02987
: Late embryogenesis abundant protein
DATA:                     Control 30min   2hours  2days   1week   p-value       pos
SENSE COUNTS:             3       0       9       39      92      2.50e-45
GENES (1 total):
  AT2G42530.1
    SENSE COUNTS:         3       0       9       39      92      2.50e-45
    TAGS: (2 total)
    d+1  TTCAATAGTT       3       0       9       39      91      2.24e-44      521
    d+2m GCGATGTCTT       0       0       0       0       1       1.70e-01      76


LOCUS: AT4G25480
DESCRIPTION: encodes a member of the DREB subfamily A-1 of ERF/AP2 transcription factor fam
ily (CBF3). The protein contains one AP2 domain. There are six members in this subfamily, i
ncluding CBF1, CBF2, and CBF3. This gene is involved in response to low temperature an
DATA:                     Control 30min   2hours  2days   1week   p-value       pos
SENSE COUNTS:             0       0       50      4       3       4.23e-33
GENES (2 total):
  AT4G25480.1
    SENSE COUNTS:         0       0       50      4       3       4.23e-33
    TAGS: (2 total)
    d+1  AAGTCGACGG       0       0       50      4       3       4.23e-33      726


LOCUS: AT2G46110
DESCRIPTION: ketopantoate hydroxymethyltransferase family protein, similar to SP|Q9Y7B6 3-m
ethyl-2-oxobutanoate hydroxymethyltransferase (EC 2.1.2.11) (Ketopantoate hydroxymethyltran
sferase) {Emericella nidulans}; contains Pfam profile PF02548: Ketopantoate hydroxymet
DATA:                     Control 30min   2hours  2days   1week   p-value       pos
SENSE COUNTS:             0       0       0       0       21      1.68e-15
GENES (1 total):
  AT2G46110.1
    SENSE COUNTS:         0       0       0       0       21      1.68e-15
    TAGS: (2 total)
    d+1  GTTTCTCTGT       0       0       0       0       21      1.45e-18      1303
```

6. Combine the library csage reports into a multi-library report sasksage_core/bin/csage_combined.pl -d sage_libraries > combined_report.txt

7. Perform a 2 by t test on the independence of tag vs. treatment (a < 0.005) sasksage_core/bin/2byt_csage -f combined.csage -a 0.005 > combined_report_0.005.txt

8. SaskSAGE reports can be further manipulated using the parsetags.pl utility available from http://sasksage.sourceforge.net.

(a) Reports can be ordered by SAGE tag or by individual gene *see* **Fig. 5.** Traditionally, SAGE analyses describe the results of any experiment in relation to each detected SAGE tag. However, it may be perspicacious to combine SAGE tag counts where different observed tags are uniquely matched to the same transcript. Arranging data by gene identifier with each tag and corresponding count listed highlights the possible occurrences of alternative transcripts and/or anti-sense transcripts for the identified gene.

(b) Alternative transcription *see* **Fig. 6:** There are three arbitrary stringency levels for identifying alternative transcripts: (1) the default where any possible alternative transcripts are indicated, that is when more than one SAGE tag is

---

Fig. 5. An example of serial analysis of gene expression (SAGE) output from an experiment to investigate the effect of low temperature on the leaf transcriptome of *Arabidopsis thaliana*, organized by (**A**) tag and (**B**) gene. (**A**) The output contains information for each unique SAGE tag identified during the experiment. The data can be ordered by tag or by statistical significant difference (shown here). The output provides a normalized tag count for each library analyzed, along with the p-value. The tag to gene matching information provided includes (1) the AGI gene code the tag matches to, differentiating between known alternative transcripts (denoted by numerical suffix); (2) the data set each gene is derived from, 'd' genes have experimentally defined UTR sequence, 'v' genes have virtual UTR sequence, 'p' are annotated pseudogenes; (3) the bp position of the anchoring enzyme (AE) site within the transcript and '+' or '−' denotes the sense orientation of the match; (4) the stage the match occurred in the iterative process, 'primary' match to a canonical tag site, 'secondary' match to a non-canonical tag site, 'tertiary' match to a intronic sequence, 'quaternary' match to inter-genic sequence (chromosome number and the bp position on the pseudochromosome of the AE position). (**B**) The data for each gene are displayed separately, with the individual tag and total tag abundance indicated for each SAGE library. This output also shows the ability of SAGE data to differentiate between closely related homologues cor15a and cor15b. The 'pos' column indicates the bp position of the tag within the transcript. The data sets to which the tags are matched are indicated to the left of each tag as above. However, the step in the iterative matching process is shortened to 1, 2, 3, and 4 for primary, secondary and so on.

A
```
LOCUS: AT1G32060
DESCRIPTION: phosphoribulokinase (PRK) / phosphopentokinase, nearly identical to SP|P25697
Phosphoribulokinase, chloroplast precursor (EC 2.7.1.19) (Phosphopentokinase) (PRKASE) (PRK
) {Arabidopsis thaliana}
DATA:               Control 30min   2hours  2days   1week   p-value     pos
SENSE COUNTS:       337     154     427     499     150     3.14e-64
GENES (1 total):
  AT1G32060.1
    SENSE COUNTS:   337     154     427     499     150     3.14e-64
    TAGS: (2 total)
    d+1  GCGAAAAGGA 337     154     424     499     150     5.20e-64    1414
    d+2  CTGATTTCCC 0       0       3       0       0       3.30e-02    1182
         ----------                                                     1056
         ----------                                                     737
         ----------                                                     570
         ----------                                                     509
```

B
```
LOCUS: AT5G66570
DESCRIPTION: oxygen-evolving enhancer protein 1-1, chloroplast / 33 kDa subunit of oxygen e
volving system of photosystem II (PSBO1) (PSBO), identical to SP:P23321 Oxygen-evolving enh
ancer protein 1-1, chloroplast precursor (OEE1) (33 kDa subunit of oxygen evolving sys
DATA:               Control 30min   2hours  2days   1week   p-value     pos
SENSE COUNTS:       412     103     290     207     113     5.85e-63
GENES (2 total):
  AT5G66570.1
    SENSE COUNTS:   412     103     290     207     113     5.85e-63
    TAGS: (5 total)
    d+1  TCCTTCATCG 122     38      81      37      42      2.48e-17    1212
    d+2  CCTAAGATCT 290     65      207     169     71      2.78e-47    1181
         ----------                                                     1091
    d+2  ACCCGTCTTA 0       0       2       1       0       9.84e-02    579
         ----------                                                     396
         ----------                                                     90
```

C
```
LOCUS: AT1G15820
DESCRIPTION: chlorophyll A-B binding protein, chloroplast (LHCB6), nearly identical to Lhcb
6 protein (Arabidopsis thaliana) GI:4741960; contains Pfam profile PF00504: Chlorophyll A-B
binding protein
DATA:               Control 30min   2hours  2days   1week   p-value     pos
SENSE COUNTS:       177     38      230     138     21      8.83e-55
GENES (2 total):
  AT1G15820.1
    SENSE COUNTS:   177     38      230     138     21      8.83e-55
    TAGS: (4 total)
                                                                        1117
    d+2  AGGCTTGTTC 176     38      216     137     21      4.47e-50    1012
                                                                        895
                                                                        823
    d+2  GTCGAAGACC 1       0       11      1       0       6.93e-07    639
                                                                        565
                                                                        486
    d+2  GCCGATGGGC 0       0       3       0       0       6.04e-03    422
                                                                        109
```

Fig. 6. Output indicating possible alternative transcription. The Pos (position) column lists the bp position of all possible tag sites within the transcripts of interest, where the canonical tag is indicated first. The sequence of each tag is only displayed when there is an associated tag count among the library data. Absence of a potential tag in a library is represented by a dashed line. Three examples of possible alternative transcript processing are shown. (**A**) Two tags are uniquely assigned to gene *At1g32060*. Tags captured from the canonical site are detected at high levels, and the 3′ most non-canonical tag is detected three times in one library, the latter may be the result of incomplete digestion during the serial analysis of gene expression protocol. (**B**) Three of six possible tags are detected for gene *At5g66570*. The canonical tag is detected at high levels in all libraries. However, the 3′ most non-canonical tag is consistently more abundant than the canonical tag. This pattern is unlikely to be the result of incomplete digestion. (**C**) The canonical tag has not been detected for gene *At1g15820*. This may be due to mis-annotation, although the random pattern with which the non-canonical tags are observed suggests alternative transcription.

unambiguously matched to a single transcript; (2) if any additional match to a transcript is made to an intronic sequence it is given higher priority; (3) tags are excluded if the pattern of matching indicated they may be the result of incomplete digestion by the AE.

(c) Match orientation: the default is to display the tags matched to both strands. However, the user can specify to only display matches in a single orientation.

Notes

1. There are 17,754 (58%) annotated *A. thaliana* genes with experimentally verified 5′ and 3′ UTRs. These sequences can be used to accurately match experimental SAGE tags; however, for the remaining 42% of the annotated genes, it is necessary to construct default UTR sequence. The mean length and distribution of the 17,754 experimentally verified UTRs was calculated after removal of all intronic regions. The mean UTR length sufficient to encompass 95% of each distribution was determined to specify a default length of 250 bp and 500 bp for the 5′ UTR and 3′ UTR, respectively *(10)*. The default UTRs are extracted from the *A. thaliana* genomic sequence and appended to the annotated genes for SAGE tag matching.

2. The sample size of each SAGE library is an important parameter for the design of an experiment and is dependent on the expression profile of target genes, as there is a positive correlation between library size and the ability to detect significant changes in gene expression for rare transcripts. The percentage of novel tags discovered within a library of increasing size is presented in **Fig. 7**, with the rate of new tag discovery beginning to plateau between 60 and 80,000 tags. If this rate of novel tag discovery were maintained the library would have to be sequenced to approximately 380,000 to get below the 5% level. Therefore, it is evident that small SAGE libraries (<25, 000 tags) will reveal predominantly qualitative differences, and to yield reproducible quantitative data, large libraries (>80, 000 tags) are required. SAGE data from *A. thaliana* leaf tissue was used to estimate the effect that library size has toward accurately determining the transcript abundance. This effect may be more pronounced in species with an increased ploidy level, a larger genome or where alternative transcript processing adds to the complexity of the transcriptome. Conversely, the use of a less complex tissue type may reduce the minimum library size required to capture accurate quantitative data.

3. It is recommended that care should be taken when generating material for SAGE library construction. Where possible, plants should be grown in controlled environments or ideally under axenic conditions to ensure that tags derived from foreign mRNAs are excluded. Tags derived from alien mRNA are likely to contribute to the unmatched tags and in some instances may increase the counts for highly homologous transcripts. In either instance, the alien tags will reduce the effective library size that has been sampled, reducing the accuracy of the quantitative description of the transcriptome.

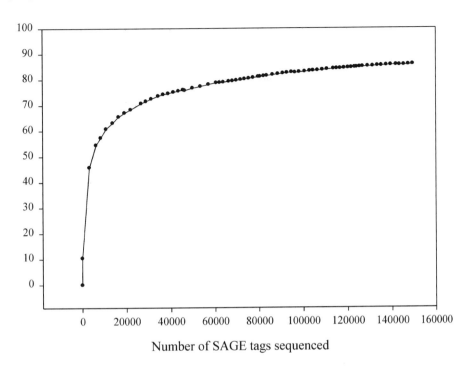

Number of SAGE tags sequenced

Fig. 7. Graph showing the percentage of unique tags which are identified during the incremental sequencing of serial analysis of gene expression (SAGE) tags for *Arabidopsis thaliana* SAGE libraries.

4. For any exploratory SAGE expression analysis, to identify candidate genes, the user must select a threshold probability value based on the acceptable level of type I statistical errors. The occurrence of type I errors are reduced by applying the Bonferroni correction. However, many of the genes that have a major or controlling effect on a particular phenotype, such as transcription factors, may be expressed at relatively low levels, and any significant differences in gene expression may be overlooked when using the Bonferroni correction. Therefore, to maximize the ability to identify candidate genes, it might be advantageous to only apply the Bonferroni correction to the highly abundant transcripts, recognizing that there will be a higher rate of false positives found among the rare transcripts.

Acknowledgments

The development of the SaskSAGE website was funded in part by the Genome Prairie project 'Functional Genomics of Abiotic Stress.' The authors would like to thank Dr Andrew Sharpe (AAFC, Saskatoon Research Centre) for critical reading of the chapter.

References

1. Velculescu, V.E., Zhang, L., Vogelstein, B., and Kinzler, K.W. (1995) Serial analysis of gene expression. *Science* 270, 484–487.
2. Velculescu, V.E., Zhang, L., Zhou, W., Vogelstein, J., Basrai, M.A., Bassett, D.E. Jr., Hieter, P., Vogelstein, B., and Kinzler, K.W. (1997) Characterization of the yeast transcriptome. *Cell* 88, 243–251.
3. Zhang, L., Zhou, W., Velculescu, V.E., Kern, S.E., Hruban, R.H., Hamilton, S.R., Vogelstein, B., and Kinzler, K.W. (1997) Gene expression profiles in normal and cancer cells. *Science* 276, 1268–1272.
4. Jones, S.J., Riddle, D.L., Pouzyrev, A.T., Velculescu, V.E., Hillier, L., Eddy, S.R., Stricklin, S.L., Baillie, D.L., Waterston, R., and Marra, M.A. (2001) Changes in gene expression associated with developmental arrest and longevity in *Caenorhabditis elegans*. *Genome Res.* 8, 1346–1352.
5. Gorski, S.M., Chittaranjan, S., Pleasance, E.D., Freeman, J.D., Anderson, C.L., Varhol, R.J., Coughlin, S.M., Zuyderduyn, S.D., Jones, S.J., and Marra, M.A. (2003) A SAGE approach to discovery of genes involved in autophagic cell death. *Curr. Biol.* 13, 358–363.
6. Matsumura, H., Nirasawa, S., and Terauchi, R. (1999) Technical advance, transcript profiling in rice (*Oryza sativa* L.) seedlings using serial analysis of gene expression (SAGE). *Plant J.* 20, 719–726.
7. Gibbings, J.G., Cook, B.P., Dufault, M.R., Madden, S.L., Khuri, S., Turnbull, C.J., and Dunwell, J.M. (2003) Global transcript analysis of rice leaf and seed using SAGE technology. *Plant Biotechnol. J.* 1, 271–285.
8. Lee, J.Y., Lee, D.H. (2003) Use of serial analysis of gene expression technology to reveal changes in gene expression in Arabidopsis pollen undergoing cold stress. *Plant Physiol.* 132, 517–529.
9. Fizames, C., Munos, S., Cazettes, C., Nacry, P., Boucherez, J., Gaymard, F., Piquemal, D., Delorme, V., Commes, T., Doumas, P., Cooke, R., Marti, J., Sentenac, H., and Gojon, A. (2004) The *Arabidopsis* root transcriptome by serial analysis of gene expression. Gene identification using the genome sequence. *Plant Physiol.* 134, 67–80.
10. Robinson, S.J., Cram, D.J., Lewis, C.T., and Parkin, I.A.P. (2004) Maximizing the efficacy of SAGE analysis identifies novel transcripts in *Arabidopsis*. *Plant Physiol.* 136, 3223–3233.
11. Poroyko, V., Hejlek, L.G., Spollen, W.G., Springer, G.K., Nguyen, H.T., Sharp, R.E., and Bohnert, H.J. (2005) The maize root transcriptome by serial analysis of gene expression. *Plant Physiol.* 138, 1700–1710.
12. Coemans, B., Matsumura, H., Terauchi, R., Remy, S., Swennen, R., and Sági, L. (2005) SuperSAGE combined with PCR walking allows global gene expression profiling of banana (*Musa acuminata*), a non-model organism. *Theor. Appl. Genet.* 111, 1118–1126.

13. Lorenz, W.W., Dean, J.F. (2002) SAGE profiling and demonstration of differential gene expression along the axial developmental gradient of lignifying xylem in loblolly pine (*Pinus taeda*). *Tree Physiol.* 22, 301–310.

14. Saha, S., Sparks, A.B., Rago, C., Akmaev, V., Wang, C.J., Vogelstein, B., Kinzler, K.W., and Velculescu, V.E. (2002) Using the transcriptome to annotate the genome. *Nat. Biotechnol.* 20, 508–512.

15. Matsumura, H., Reich, S., Ito, A., Saitoh, H., Kamoun, S., Winter, P., Kahl, G., Reuter, M., Kruger, D.H., and Terauchi, R. (2003) Gene expression analysis of plant host-pathogen interactions by SuperSAGE. *Proc. Natl. Acad. Sci. U. S. A.* 100, 15718–15723.

16. Ewing, B., Green, P. (1998) Base-calling of automated sequencer traces using phred. II. Error probabilities. *Genome Res.* 8, 186–194.

17. Vilain, C., Libert, F., Venet, D., Costagliola, S., and Vassart, G. (2003) Small amplified RNA-SAGE, an alternative approach to study transcriptome from limiting amount of mRNA. *Nucleic Acids Res.* 31, e24.

18. Velculescu, V.E., Madden, S.L., Zhang, L., Lash, A.E., Yu, J., Rago, C., Lal, A., Wang, C.J., Beaudry, G.A., Ciriello, K.M., Cook, B.P., Dufault, M.R., Ferguson, A.T., Gao, Y., He, T.C., Hermeking, H., Hiraldo, S.K., Hwang, P.M., Lopez, M.A., Luderer, H.F., Mathews, B., Petroziello, J.M., Polyak, K., Zawel, L., Kinzler, K.W., et al. (1999) Analysis of human transcriptomes. *Nat. Genet.* 23, 387–388.

19. Cheng, G., Porter, J.D. (2002) Transcriptional profile of rat extraocular muscle by serial analysis of gene expression. *Invest. Ophthalmol. Vis. Sci.* 43, 1048–1058.

20. Seki, M., Narusaka, M., Kamiya, A., Ishida, J., Satou, M., Sakurai, T., Nakajima, M., Enju, A., Akiyama, K., Oono, Y., Muramatsu, M., Hayashizaki, Y., Kawai, J., Carninci, P., Itoh, M., Ishii, Y., Arakawa, T., Shibata, K., Shinagawa, A., and Shinozaki, K. (2002) Functional annotation of a full-length Arabidopsis cDNA collection. *Science* 296,141–145.

21. Zhou, Y., Tang, J., Walker, M.G., Zhang, X., Wang, J., Hu, S., Xu, H., Deng, Y., Dong, J., Ye, L., Lin, L., Li, J., Wang, X., Xu, H., Pan, Y., Lin, W., Tian, W., Liu, J., Wei, L., Liu, S., Yang, H., Yu, J., Wang, J.V. (2003) Gene identification and expression analysis of 86,136 Expressed Sequence Tags (EST) from the rice genome. *Genomics Proteomics Bioinformatics* 1, 26–42.

22. Kikuchi, S., Satoh, K., Nagata, T., Kawagashira, N., Doi, K., Kishimoto, N., Yazaki, J., Ishikawa, M., Yamada, H., Ooka, H., Hotta, I., et al. (2003) Collection, mapping, and annotation of over 28,000 cDNA clones from japonica rice. *Science* 301, 376–379.

23. Osato, N., Yamada, H., Satoh, K., Ooka, H., Yamamoto, M., Suzuki, K., Kawai, J., Carninci, P., Ohtomo, Y., Murakami, K., Matsubara, K., Kikuchi, S., and Hayashizaki, Y. (2003) Antisense transcripts with rice full-length cDNAs. *Genome Biol.* 5, R5.

24. Castelli, V., Aury, J.M., Jaillon, O., Wincker, P., Clepet, C., Menard, M., Cruaud, C., Quetier, F., Scarpelli, C., Schachter, V., Temple, G., Caboche, M., Weissenbach, J., and Salanoubat, M. (2004) Whole genome sequence comparisons and "full-length" cDNA sequences: a combined approach to evaluate and improve *Arabidopsis* genome annotation. *Genome Res.* 14, 406–413.

25. Lash, A.E., Tolstoshev, C.M., Wagner, L., Schuler, G.D., Strausberg, R.L., Riggins, G.J., and Altschul, S.F. (2000) SAGEmap, a public gene expression resource. *Genome Res.* 10, 1051–1060.

26. Arabidopsis Genome Initiative. (2000) Analysis of the genome sequence of the flowering plant *Arabidopsis thaliana*. *Nature* 408, 796–815.

27. Yu, J., Hu, S., Wang, J., Wong, G.K., Li, S., Liu, B., Deng, Y., Dai, L., Zhou, Y., Zhang, X., et al. (2002) A draft sequence of the rice genome (*Oryza sativa* L. ssp. *indica*). *Science* 296, 79–92.

28. Goff, S.A., Ricke, D., Lan, T.H., Presting, G., Wang, R., Dunn, M., Glazebrook, J., Sessions, A., Oeller, P., Varma, H., et al., .(2002) A draft sequence of the rice genome (*Oryza sativa* L. ssp. *japonica*). *Science.* 296, 92–100.

29. Ruijter, J.M., Van Kampen, A.H., and Baas, F. (2002) Statistical evaluation of SAGE libraries, consequences for experimental design. *Physiol. Genomics* 11, 37–44.

30. Yamada, K., Lim, J., Dale, J.M., Chen, H., Shinn, P., Palm, C.J., Southwick, A.M., Wu, H.C., Kim, C., Nguyen, M., et al. (2003) Empirical analysis of transcriptional activity in the Arabidopsis genome. *Science* 302, 842–846.

31. Man, M.Z., Wang, X., and Wang, Y. (2000) POWER_SAGE, comparing statistical tests for SAGE experiments. *Bioinformatics* 16, 953–959.

19

Methods for Analysis of Gene Expression in Plants Using MPSS

Kan Nobuta, Kalyan Vemaraju, and Blake C. Meyers

Summary

Massively parallel signature sequencing (MPSS) is a technology capable of sequencing simultaneously almost all the DNA molecules in a sample. This technology is well suited for deep profiling of mRNA and small RNA by the sequencing of cDNA tags. A series of mRNA MPSS databases has been created from various libraries derived from four different species (Arabidopsis, rice, grape, and *Magnaporthe grisea*, the rice blast fungus). Our mRNA MPSS databases measure the absolute expression level of most genes in the sample and provide information about potentially novel transcripts (antisense transcripts, alternative splice isoforms, and regulatory intergenic transcripts). In addition to these data, we have recently built an extensive database from MPSS-derived Arabidopsis small RNA samples. All the databases are accessible thorough our Web interface (http://mpss.udel.edu), and the individual pages are equipped with various graphical and analytical tools. Here, we focus on a subset of these tools (e.g., Gene Analysis [GA], Chromosome Viewer [CV], and Library Analysis [LIBAN]) and describe how the users can analyze and interpret our MPSS expression data.

Key Words: MPSS; gene expression; small RNA; antisense transcript; transcription unit; alternative splice variant; unannotated transcript; transposon; repeats.

1. Introduction

1.1. mRNA Massively Parallel Signature Sequencing Technology

Massively parallel signature sequencing (MPSS) is a sequencing-based technology that uses a unique cloning and sequencing method to quantify gene

From: *Methods in Molecular Biology, vol. 406: Plant Bioinformatics: Methods and Protocols*
Edited by: D. Edwards © Humana Press Inc., Totowa, NJ

expression levels *(1,2)*. "Regular" or mRNA MPSS data measures nearly all genes expressed in a sample by generating a 17–20 nucleotide "signature" or "tag" sequence derived from the 3′-most *Dpn*II restriction site (GATC) of each polyadenylated mRNA transcript. The expression level of the transcripts is determined by counting the number of occurrences of each specific signature sequence in the sample. Although there are MPSS-specific problems such as duplicated or under-represented signatures *(3)*, one of the advantages of MPSS is that, because it does not use a hybridization intensity to determine expression levels, it eliminates the problems of background interference and cross-hybridization and provides an absolute, rather than relative, transcript level in a given tissue. The other advantage is that, unlike most microarrays, the transcripts that are analyzed are not pre-selected. This random sequencing method allows the user to identify novel transcripts such as antisense transcripts, alternative splice isoforms, or intergenic transcripts that may play an important role under certain experimental or biological conditions *(4)* (*see* **Note 1**). While MPSS is now obsolete, other short-read, high-throughput technologies can be applied in exactly the manner that we describe here, including 454's pyrosequencing. Solexa's (now Illumina's) "sequencing-by-synthesis" (or SBS), Applied Biosystem's "supported oligo ligation detection" (or SOLID) system, or others that are not yet commercialized.

1.2. Classification of mRNA Signatures

One of the first steps in mRNA MPSS data analysis is the classification of the signature sequences. These signatures can be roughly divided into three broad categories (sense-strand signatures, antisense signatures, and unannotated signatures) based on their location relative to annotated genomic features. In our approach, sense or antisense signatures are identified by comparison with annotated transcription units (TUs); sense-strand matches are further subdivided into four classes (classes 1, 2, 5, and 7). Class 1 and 5 signatures match within annotated exons and introns, respectively (*see* **Fig. 1A**). Class 2 signatures are located within an annotated 3′-untranslated region (3′-UTR) or within 500 bp after the annotated stop codon (*see* **Fig. 1A**). Class 7 signatures include an exon/intron boundary and therefore can only be defined from the spliced transcript sequences (*see* **Fig. 1A**). There are two classes (classes 3 and 6) for the antisense signatures that match to the opposite strand of annotated TUs. Classes 3 and 6 signatures are positioned antisense to exons and introns, respectively (*see* **Fig. 1A**). The last category of signatures (class 4; unannotated signatures) is located within intergenic regions (IGRs; *see* **Fig. 1A**). In our Web

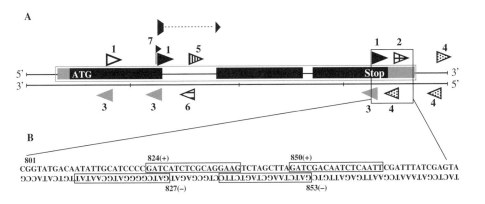

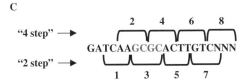

Fig. 1. Classification, extraction, and sequencing frames of massively parallel signature sequencing (MPSS) signatures. (**A**) Arrowheads and flag represent the location and direction of mRNA MPSS signatures. The numbers associated with the arrowheads correspond to the classes that are described in **Subsection 1.2**. The class 7 signature is described by the flag and the objects connected by a dotted line. See our Web interface for the color scheme and detail of the signatures. (**B**) An example of the signatures extracted from genomic DNA sequence. Boxes indicate signatures identified in the genomic sequence; a non-complementary signature on each strand is identified from each *Dpn*II site because of the palindromic nature of the site. Numbers above and below the sequence indicate the nucleotide position and the strand information stored for each signature. (**C**) MPSS uses two sequencing reactions that are performed in reading frames shifting by two bases. The 2-step frames are indicated below the example signature, and the 4-step frames are indicated above. Sequencing proceeds in sets of four bases ("words"), which are illustrated by the numbering 1–8. The "word" in the third frame is one of the "bad words" that under-perform in MPSS. The three Ns at the end of the 17-base signature indicate the bases that are not sequenced but can affect the performance (e.g., "bad words") of the seventh or eighth word.

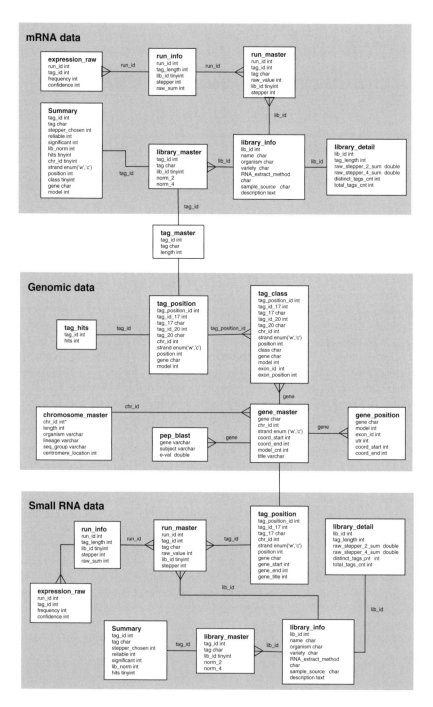

site, different classes are indicated by seven different colors. See our Web site for the true colors and details.

1.3. Genomic Database

Because the classification is based on the position of the signature sequence relative to the annotated genes, the genomic annotation data are a crucial component of the MPSS data analysis. For Arabidopsis, we have used genome annotation that was released from The Institute for Genomic Research (TIGR) (v5.0), although we regularly update our Web site to use the current TIGR version *(5)*. This annotation includes 26,207 protein-coding genes and 3,786 pseudogenes. It also includes many alternative splice variants and UTRs that were identified based on expressed sequence tag (EST) and full-length cDNA sequences. We extract all the necessary information from the XML file provided by TIGR and build a database of "genomic" signatures (*see* **Fig. 2**). In the gene_master table, the most general information (e.g., start and end coordinate, strand, etc.) of the annotated TUs is stored. The gene_position table contains the detail of each TU in the context of the coordinates of exons, introns, and UTRs. Because MPSS signatures are anchored with a *Dpn*II restriction site (GATC) and are of a specific length, we extract 17 and 20 nucleotides based on all occurrences of GATC, from both strands of the entire genomic sequence (*see* **Fig. 1B**). These sequences and additional information are stored in the tag_position table (e.g., position, the corresponding gene if one exists, etc.). After this process, the genomic signatures are then characterized as one of the seven classes described in **Subsection 1.2** and stored in the tag_class table. Although relatively few of these genomic signatures are expressed, it is necessary to store all the signatures as "potential MPSS signatures" and use the tables as lookup tables because MPSS signatures are not pre-selected for a location in the gene.

Fig. 2. Schema of massively parallel signature sequencing (MPSS) database. The database is designed with three major sets of data. The lines connecting tables indicate one-to-one (simple lines) or one-to-many (branched lines) relationships. The field name above or below the lines indicates the key that connects the tables. The detail of genomic data, mRNA MPSS data, and small RNA MPSS data can be found in **Subsections 1.3–1.5,** respectively. Note that not all the fields and the tables are listed in this figure. For example, several tables with repetitive sequence information, which are useful for small RNA MPSS data analysis, are not listed.

1.4. MPSS Expression Database

1.4.1. Sequencing Runs and Steps

Independent of the genomic database, we build an "expression" database from the raw MPSS data produced for our RNA samples (*see* **Fig. 2**). All the raw data are stored in the expression_raw table, which contains the sequence of the MPSS signatures and the number of occurrences of signature per reaction (the "abundance"). Usually, four to eight MPSS reactions (sequencing runs) are performed in two different frames ("steppers") for each cDNA library. In the Arabidopsis libraries, runs were performed using the 2 or 4 steppers (*see* **Fig. 1C**; the 2 or 3 steppers predominate in our rice data). Multiple sequencing runs in different frames function as technical replicates and help to ameliorate the impact of sequence-specific technical artifacts such as "bad words" (*see* **Fig. 1C**; characters in gray). Bad words are discussed in detail elsewhere *(3)*. Briefly, certain four-base sequences (mostly palindromes like GGCC or ATAT) perform quite poorly in the MPSS reaction.

1.4.2. Normalization

Multiple sequencing runs also help to obtain an accurate estimate of the expression level of each transcript. The following equations describe the normalization method we used for a given signature, where t_norm and f_norm are the 2-step and 4-step normalized values, respectively:

$$t\_norm = \frac{\sum_{all\ 2step\ runs} raw\_value_{signature}}{\sum_{all\_2step\_runs} raw\_value_{all\_signature}} \times 10^6$$

$$f\_norm = \frac{\sum_{all\ 4step\ runs} raw\_value_{signature}}{\sum_{all\_4step\_runs} raw\_value_{all\_signature}} \times 10^6$$

The calculated values represent the number of transcripts of a gene, which corresponds to the given signature, if a library has a million transcripts. We use TPM (transcripts per million) as the unit for these values and store the values in the library_master table along with the corresponding signatures (*see* **Fig. 2**).

1.4.3. Summary Table of Expression Database

On the basis of the calculated normalized values and the information stored in the genomic database, we build the "summary" table (*see* **Fig. 2**). This table stores the higher of the two normalized values (t_norm and f_norm, summed across all libraries), as the normalized abundance value for the given signature.

We do not average the two values of the steppers, because a "bad word" in one frame can produce a false-negative result, underestimating the abundance of a signature *(3)*. These normalized abundance values are used to identify "significant" signatures in the table. A given signature is defined as "significant" if the signature is found in any library at ≥4 TPM. We also created the "reliable" filter that determines if signatures are found in more than one of the runs in all libraries. This filter is best for removing signatures resulting from sequencing errors. The results of these two filters are included in the summary table to identify real transcripts from "noise". Additional information on the normalization and filtration techniques can be found elsewhere *(3)*.

The tag_master table connects the two databases, and all the genomic data in the summary table are derived from this table *(see* **Fig. 2**). It contains not only the signatures identified in tag_class and tag_position tables, but also the signatures identified by MPSS with no match to a given genomic sequence. Although the main source of these orphan signatures is most likely sequencing errors, a portion of these signatures could be derived from unsequenced regions of the genome like the centromeres or other gaps. We identify these unmatched signatures as "class 0" and store their abundance information in the summary table. Detail discussion about the sources of class 0 signatures can be found elsewhere *(3)*.

1.5. Small RNA MPSS Expression Database

Recently, the MPSS technology was adapted to capture and sequence small RNAs *(6)*. This new technology was applied to Arabidopsis, and small RNA MPSS libraries were generated from two tissues. Because the structure of the raw data (e.g., a signature sequence and the count) is the same between mRNA and small RNA MPSS data, we adapted the same database methods for small RNA MPSS data, although this was modified in a number of significant ways *(see* **Fig. 2**). For these libraries, the sequencing runs were performed using 2 and 3 steppers instead of the 2-step and 4-step runs that had previously populated our mRNA database. However, this does not affect our approach, as the differences between the two methods do not affect the raw abundance count of the signatures. A more substantial difference is that the total number of small RNA MPSS signatures sequenced was closer to 250,000 per stepper per run (the lower yield is typical of the longer runs required to sequence signatures that lack an initial GATC). Therefore, we modified the equations and normalized the raw data to transcripts per quarter million (TPQ) instead of TPM.

The other major difference between mRNA and small RNA MPSS signatures is that whereas all the mRNA signatures start with GATC, small RNA signatures can initiate at any nucleotide in the genome. Therefore, we could not pre-extract all the potential signatures from the genome sequence and store them in the tag_position table as we did for the mRNA signatures. No "genomic" tables can be generated in advance for the small RNAs, and these sequences can only be matched back to the genome after the expression data are produced. The smallRNA_tag_position table was created with these signatures. In addition to this table, three unique tables (e.g., chr_repeat, chr_tandem, and chr_inverted), which contain repetitive sequence information, were created (not shown), as this information is directly relevant to the matching locations of the small RNAs *(6)*. We used freely available computer programs, such as Repeat-Masker (http://repeatmasker.org) *(7)*, Etandem, and Einverted *(8)*, to extract different kind of repetitive sequence from Arabidopsis genome sequence. The information stored in these tables was used to display these repetitive regions in our Web viewer and to analyze the correlation between small RNAs and the identified regions.

1.6. Web Interface

We have developed four mRNA MPSS databases from Arabidopsis, rice, grape, and the *Magnaporthe grisea*; the Arabidopsis and rice databases currently have small RNA MPSS data. All the databases are accessible through our Web interface (http://mpss.udel.edu), which is equipped with various graphical and analytical tools that allow the user to retrieve and analyze data. In this chapter, we focus on the Arabidopsis mRNA and small RNA MPSS data, and introduce the GA, CV, and LIBAN tools. If the user wants to extract MPSS data (mRNA and/or small RNA) for a particular gene, the GA tool is most relevant. If the user is interested in a particular bacterial artificial chromosome (BAC) or a specific region of a chromosome, the CV gives a "bird's eye" view of the complexity of the given region. If the user wants to compare expression data and select certain genes (e.g., select the genes that have more than two-fold higher expression in library A compared with library B), the new LIBAN tool allows the users to define a condition and retrieve sets of data from our database. Detailed descriptions and instructions for these three tools are in **Subsection 3**.

2. Materials
2.1. Arabidopsis mRNA MPSS Data

1. This data set currently consists of 17 mRNA MPSS libraries, representing treated and untreated tissues and flower mutants. Although all the libraries have both 17- and 20-bp signature data, we mainly describe 17-bp data in this chapter.

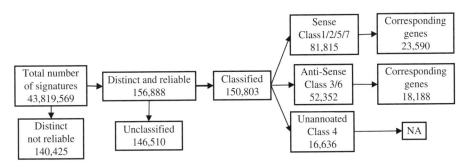

Fig. 3. Filter and classification results of 17 Arabidopsis massively parallel signature sequencing (MPSS) libraries. A total of 43,819,569 signatures sequenced from 17 Arabidopsis MPSS libraries were filtered and classified. Distinct and reliable signatures were preferentially classified by their genomic positions. Class 0 signatures and the signatures that can be classified in more than one class were put in the "unclassified" category. The signatures were grouped into three categories, sense, antisense and unannotated, and summed.

2. A total of 43,810,569 (156,888 distinct and reliable) signatures from poly(A)+ RNA were sequenced. Each distinct signature represents a different transcript (*see* **Fig. 3**), although some of these transcripts may differ only in the 3′-polyadenylation site. Among the distinct signatures, 81,815 are sense signatures (classes 1, 2, 5, and 7), which correspond to 23,590 genes annotated by TIGR (*see* **Fig. 3**). The difference between these two numbers suggests the existence of many as-yet uncharacterized alternatively isoforms in Arabidopsis.

3. There are 52,352 antisense signatures (classes 3 and 6) and 16,636 unannotated signatures (class 4) in this data set (*see* **Fig. 3**). Many class 0 signatures (145,240), which do not match to the Arabidopsis genomic sequence, have been identified.

4. The database tables are built with Oracle and transferred to MySQL for the public Web server, for which we use a Dell Poweredge 2650. The Web interface, mainly written in PHP, extracts the data requested by the user from MySQL tables and displays the query results in graphical and analytical outputs.

5. The normalized and raw data are downloadable from our Web site for the users who want to perform their own analyses.

2.2. Small RNA MPSS Data

1. This data set consists of inflorescence and seedling small RNA MPSS libraries. A total of 1,407,168 (104,800 distinct) signatures from short RNA libraries was sequenced. Known and new small RNAs were identified in this data set (*6*). Because repetitive sequences are known to be sources of short interfering RNAs (siRNAs),

we identified repetitive regions as described above. Many transposon- or retrotransposon related sequences that are often supported by small RNA signatures were identified *(6)*.

2. Perl scripts and MySQL were used to build the small RNA MPSS database. We made extensive modifications to the PHP scripts that run the mRNA MPSS Web site, and we added the graphical tools for the small RNA data as well as the repetitive sequence data. New analytical tools, which are specific for small RNA data, were also developed and added to our Web interface.

3. The normalized and raw small RNA data are downloadable from our Web site for the users who want to perform their own analysis.

4. In the Arabidopsis home page, we provide a brief description of the small RNA MPSS data and provide links to known microRNAs, inverted repeats, tandem repeats, pericentromeric regions, weakly predicted transposons, and trans-acting siRNAs (http://mpss.udel.edu/at). Users can simply click the examples to get an idea of our visualization and data access tools.

2.3. Library Analysis Tool

1. We use a Dell Optiplex with an Intel Pentium 4 processor to run this tool. This tool is written in JSP and it retrieves the data from the tables that reside in the Oracle server.

2. Users can access and utilize this tool through our public interface. This tool includes different analysis methods that are explained in the section below.

3. Methods

In this section, we focus on three tools: GA, CV, and LIBAN. We use Arabidopsis mRNA and small RNA data as examples to describe these tools. Although all the mRNA libraries have both 17- and 20-bp signature data, we mainly describe the data with 17-bp tags in this section. We encourage the readers to go to our Web site to see true color images (rather than the black and white printed in this chapter).

3.1. Gene Analysis Tool

3.1.1. Gene Image and Expression Table

1. This tool is accessible from various pages and the Arabidopsis MPSS home page (http://mpss.udel.edu/at). For example, entering At1g80830 on the home page takes the user to the GA page for this gene *(see* **Fig. 4A**).

2. In this page, the user can find the detailed structure of this gene as it was annotated by TIGR *(5)*. Light gray boxes at the beginning and at the end of this gene represent the annotated UTR regions, and black boxes represent the coding regions *(see* **Fig. 4A**). The gray box that surrounds the gene specifies the maximum region

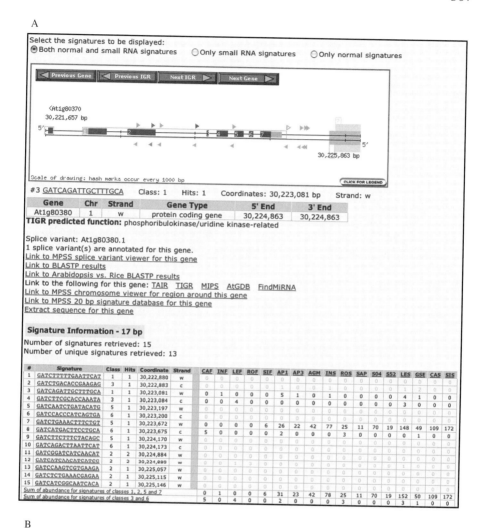

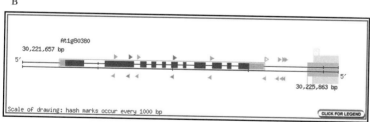

Fig. 4. Gene Analysis (GA) output page and the image from splice variant output page. (**A**) GA output page of At1g80380. Pastel-shaded boxes (on the Web interface)

occupied by this gene (*see* **Fig. 4A**). The right end of this gray line is slightly downstream of the end of 3'-UTR in **Fig. 4A** because At1g80380 has two splice variants, and we utilize the size of the largest mRNA transcript to display the size of this gene. The details of the splice variant are explained below.

3. The arrowheads above and below the gene indicate the location and direction of potential or "genomic" MPSS signature sequences (17- or 20-nucleotide sequences that start with GATC). Light gray signatures are those that are not expressed, whereas darker signatures satisfy our filter criteria for expression (significant and reliable).

4. In our Web site, the arrowheads are in different colors that represent different classes of the signature sequences as described in **Subsection 1**. The colors can be interpreted using a legend obtained by clicking on a button in the image (*see* **Fig. 4B**: "CLICK FOR LEGEND"). This legend not only describes the signature classes but also the color scheme of the repetitive regions (e.g., shaded boxes downstream of the gene in **Fig. 4A**) and small RNA signatures that are described in the next section.

5. The expression level or abundance of each signature in the 17 libraries (*see* **Note 2**) can be found in a table that is displayed at the bottom of the same GA page (*see* **Fig. 4A**). The name of the libraries is abbreviated with a three-letter code, and the details of each library can be found by clicking on these abbreviations.

6. To determine the correspondence between the arrowheads in the display window and the expression data, the user has to mouse-over the signature of interest in the display window. This action produces a brief description of the signature at the bottom of the gene image. For example, the second arrowhead from the left (class 1 signature) in the "w" strand is signature 3, and this signature is 5 and 4 TPM in the AP1 (*apetala 1-10*) and INS (inflorescence) libraries, respectively (*see* **Fig. 4A**).

7. The expression levels of highly expressed signatures are displayed in black, and those of non-expressed or very weakly expressed signatures are displayed in gray in the signature information table (*see* **Fig. 4A**). Clicking either the triangle or the signature sequence in the table takes the user to Signature Analysis page (not shown). This page allows the user to examine the detail (e.g., raw data, "bad word", etc.) of the signature chosen.

◄───

Fig. 4. with a small question mark indicate repetitive regions that are identified by one of the three programs (e.g., RepeatMasker, einverted, and etandem). A short description of each box can be obtained mousing-over the question mark. The information on the repetitive regions as well as that on mRNA and small RNA signatures can be turned on and off by clicking a set of the selection buttons above the gene image. (**B**) The other isoform of At1g80380. This isoform has more exons than the one in (**A**), and this affects the class of the signatures within the exons at the center of the gene. More details of (**A**) and (**B**) are found in **Subsection 3.1**.

8. There are five signatures (1, 3, 5, 7, and 9) associated with the sense strand of this gene (*see* **Fig. 4A**). However, according to the annotation of this gene, most of these sequences (5, 7, and 9) should be spliced out during the maturation stages of the mRNA. As mentioned in the **Subsection 1**, MPSS technology sequences 17–20 bases immediately adjacent to the 3′-most *Dpn*II restriction site in the cDNA sequence *(3,4)*. Therefore, signature 3 (second from the left) satisfies this criterion (*see* **Fig. 4A**), and the numbers listed in signature information table for signature 3 represent the expression level of this transcript (At1g80830.1).

9. An additional expressed signature is shown as a dark arrowhead and is located in the second intron of this transcript (signature 7, class 5) (*see* **Fig. 4A**). This signature is highly expressed in almost all of the libraries. This result is inconsistent with splice variant At1g80830.1 because this signature should not exist in the mature mRNA transcript. However, as stated in the middle of **Fig. 4A**, this gene has another isoform, which is accessible by clicking "Link to MPSS splice variant viewer for this gene" (*see* **Fig. 4A and B**). In the second isoform, signature 7 is no longer in intron, but it resides in the sixth exon (*see* **Fig. 4B**). This signature is the 3′-most signature of this isoform, and the numbers corresponding to this signature in the table represent the expression level of this isoform (At1g80380.2). Therefore, the MPSS data indicate that the second isoform is the predominant form and is expressed at levels substantially above those of the first isoform.

10. There are many mechanisms that regulate the expression of a gene, and the production of antisense transcripts is one such mechanism *(9–14)*. On the strand opposite to the protein coding region of this gene, there are two signatures (second and fourth from the left in the "c" strand) expressed at levels considered both "significant" and "reliable" (signatures 4 and 8 in **Fig. 4A**). These data, along with the expression patterns shown in the table, suggest the expression of antisense transcripts of this gene in the CAF (Callus, "Full" MPSS method) and LEF (Leaf, "Full MPSS method") libraries. It is possible that antisense transcription regulates the abundance of the protein-coding mRNA for this gene.

3.1.2. Navigation and Links

1. The navigation bars at the top of the image are linked to the GA page of the adjacent genes or IGRs (*see* **Fig. 4A**). The user can easily pass through to examine adjacent genes or examine the existence of expressed class 4 signatures (unannotated signatures) in the IGR. Class 4 signatures indicate the existence of unannotated transcriptional units in the specified IGR.

2. The GA page has links to our own and outside databases. All the links are listed between the gene image and the expression table (*see* **Fig. 4A**). These links provide additional information about the gene or region selected by the user.

3. As described in the above section, the "Link to MPSS splice variant viewer for this gene" displays all the annotated splice variants and the expression tables. This

viewer is a great graphical tool to compare different isoforms side by side (*see* **Fig. 4A** and **B**). In addition, the user can use this viewer and the expression table to examine the tissue specificity of expression for different splice variants.

4. We performed BLASTP with the Arabidopsis amino acid sequences against themselves and the rice amino acid sequences. We selected the subject genes that have similarity E-value $< 1 \times 10^{-10}$ against the query gene and stored these results in a genomic database (*see* **Fig. 2**). The two links ("Link to BLASTP results" and "Link to Rice vs. Arabidopsis BLASTP results") connect to the BLASTP tables and extract the genes that are similar to the query. The result of these links displays the gene images and the expression pattern of the similar genes. This function may be useful to compare the expression pattern of genes in the same gene family or orthologous genes in rice.

5. The detailed information for the annotated genes can be obtained from outside sources. We have created the links to The Arabidopsis Information Resource (TAIR) *see* Chapter 8, TIGR *see* Chapter 5, Munich Information Centre for Protein Sequence (MIPS) *see* Chapter 6, AtGDB, and FindMiRNA for Arabidopsis (*see* **Fig. 3A**). The links take the user directly to the selected gene information in these sites.

6. The "Link to MPSS chromosome viewer for region around this gene" takes the user to our CV, which displays all the genes within 50 kb upstream and downstream of the selected gene. The details of this viewer are described in the next section.

7. Although we use 17-bp signatures to describe the tools in this chapter, same or similar information are available with 20-bp signatures from our Web interface. The "Link to MPSS 20 bp signature databases for this gene" takes user to the GA page with 20-bp signature information.

   Signature 12 (the last arrowhead in the "w" strand) in **Fig. 4A** is gray and hollow because this signature has two copies in the Arabidopsis genome and the expression level is not significant. However, in the 20-bp signature database, it shows up as solid purple signature (class 2) because the additional three nucleotides form a 20-bp signature with specific and significant expression data (not shown). Although 20-bp tags are useful to reduce the number of ambiguous signatures (more than one match in the genome), at the same time, the additional three nucleotides increases the chance of "bad words" in the signature. This issue and the relative merits and costs of longer MPSS signatures are discussed elsewhere *(3)*.

8. The "Extract sequence for this gene" link displays the sense sequence, spliced coding sequence, UTR and exon sequences, and antisense sequence of the selected gene. A similar function is available for IGRs. This function is useful for the users who plan to perform their own biological experiments such as RACE (Rapid Amplification of Complementary DNA Ends) experiments to identify novel transcripts based on MPSS data.

9. Additional questions about these pages are addressed in our FAQs page (*see* **Note 3**).

3.2. Chromosome Viewer and Small RNA Data

1. This section describes the CV tool and our small RNA MPSS data. The CV page is accessible from various pages within our Web interface. At the home page, the user can define start and end coordinates of a chromosome to access the CV tool. **Figure 5A** is a sample CV page between coordinates 22,880,000 and 23,000,000 of chromosome 1.

2. This page consists of a graphical tool to observe the complexity of the Arabidopsis chromosome and that of MPSS data (*see* **Fig. 5A**). If this view appears too complex, either the small RNA or mRNA data can be turned off using radio buttons at the top of the page (*see* **Fig. 5A**).

3. The user can navigate through the chromosome by clicking one of the navigation buttons, or jump to a certain location on a chromosome by clicking the rod-shape chromosome at the top of the image (*see* **Fig. 5A**). Significant and reliable mRNA MPSS signatures are displayed as default. However, the middle navigation button "show all signatures" allows the user to change this setting to display all the potential signatures (locations of GATC/*Dpn*II sites) in this region, regardless of the expression data.

4. In our Web site, different types (e.g., protein coding, tRNA, transposon, and miRNA) of transcriptional units are indicated by specific color schemes. For example, protein-coding genes in "w" and "c" strands are represented by blue and red, respectively. The legend of the color scheme and the details of this viewer can be found below the chromosome image (not shown).

5. In addition to mRNA MPSS signatures, small RNA signatures are displayed (*see* **Fig. 5A**). The black triangles, which are pointing downward or upward, indicate the location of small RNA signatures identified by MPSS (*see* **Fig. 5A**). The solid black triangles are unique signatures and the hollow triangles indicate the same signature sequence is found at multiple locations on the genome.

6. A short description of the shaded boxes can be found at the bottom of the CV page (not shown). These boxes represent various repetitive regions predicted by the computer programs described in **Subsection 1.5**. In this specific example, there are several shaded boxes representing retrotransposons predicted by RepeatMasker. In fact, one of these boxes corresponds to At1g62075 (*see* **Fig. 5A**), which has been annotated as member of the copia-like transposon family.

7. Near the end of the chromosome image, there is a shaded box, which represents a transposon with many associated small RNA signatures (*see* **Fig. 5A**). All the signatures (except one) in this shaded area have more than one hit on the genome. Although no TU has been annotated in this region, our results indicate the existence of a transposon, which is apparently silenced by small RNAs.

8. One of the most well-characterized microRNAs, miR171c *(15)*, is displayed almost at the middle of this image (*see* **Fig. 5A**). Clicking the box takes the user to the GA page of this miRNA (*see* **Fig. 5B**). In this page, the user can observe and retrieve small RNA MPSS signatures that are associated with miR171c.

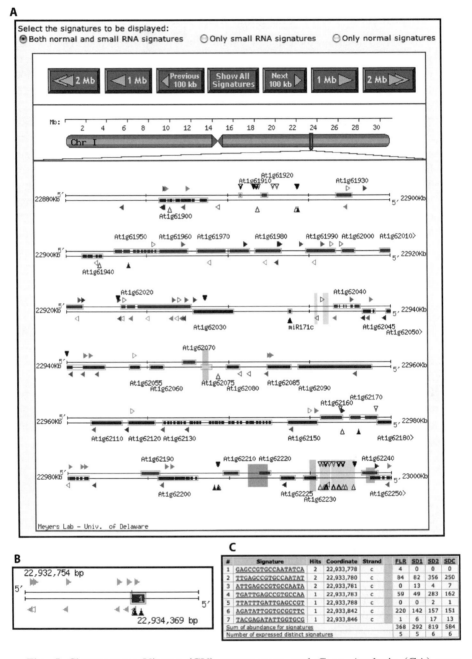

Fig. 5. Chromosome Viewer (CV) output page and Gene Analysis (GA) page with small RNA data. (**A**) The CV output page of a specific region of Arabidopsis

9. Although this GA page (*see* **Fig. 5B**) is equipped with almost all the links and tools introduced in the section above, it also has information specific to small RNAs. For example, in addition to the expression levels of mRNA signatures (not shown), those of small RNA signatures are also listed (*see* **Fig. 5C**). There are seven (four unique) small RNA MPSS signatures associated with miR171c, three of which are highly expressed in both flower and seedling libraries (*see* **Fig. 5B** and **C**).

10. Below the graphical image, there is a link to "All miRNAs associated with this gene" (not shown). This link activates a multi-GA tool and displays a graphical image and the expression data of all the genes associated with the given gene. In this particular example, the link displays miR171a, miR171b, and miR171c (not shown). Users can examine the expression of three associated microRNAs by simply sliding a scroll bar.

3.3. Library Analysis

1. The LIBAN tools are powerful methods for analyzing MPSS expression data. This page offers five different ways to analyze the data: "Basic Library Analysis", "Basic Operations on Libraries", "Inter-Comparison on Libraries", "Advanced Operations on Libraries", and "Logic Operations on Libraries". A short description of each tool can be found at the entry page of LIBAN.

2. Here, we use the Inter-Comparison tool and demonstrate how the user can identify genes that are highly expressed in the root library compared with the leaf library.

3. In the first step, the user can choose the database to search (*see* **Fig. 6A**). The "Signature-Based" and "Gene-Based" options retrieve the signatures and genes that match the user-defined conditions, respectively.

4. By default, all the signatures analyzed in LIBAN pages are "reliable" signatures. Users have a choice to change this default setting by selecting the "Include UnReliable signatures" (*see* **Fig. 6A**).

◀───────────────────────────────────────

Fig. 5. chromosome 1. Both mRNA and small RNA signatures are displayed along with repetitive regions. Different types of genes (e.g., protein coding, tRNA, etc.) have unique color schemes in our Web interface. The genes and signatures are linked to the GA and the Signature Analysis pages, respectively. **(B)** and **(C)** The gene image of miR171c and the small RNA signature information of this gene. A total of seven small RNA signatures that match to this gene were identified. The expression level of these seven small RNA signatures can be found in the table, which is displayed in **(C)**. The detail of the libraries can be found in the pop-up window, which is linked from the three letter codes of the libraries in the table.

A

Criteria	Options
Database	◉ 17 bp ○ 20 bp ○ Signature - Based ◉ Gene - Based
Filter Results	☐ Include ☑ Include UnReliable InSignificant ⑦ signatures signatures
Class	☑ Sense-strand transcripts ☐ Antisense transcripts (Clear Class Selection)	☐ Class 0 - no match to genome, thus not classified ☑ Class 1 - inside exon, same strand ☑ Class 2 - within 500 bp of 3' end of annotated gene ☐ Class 3 - inside exon but on anti-sense strand ☐ Class 4 - in intergenic region (IGR) and not in Class 2 region ☑ Class 5 - within intron, sense strand ☐ Class 6 - within intron, antisense strand ☑ Class 7 - overlaps with an exon-intron splice boundary ☐ Class 99 - multiple hits to genome, but unclassifiable
Threshold	Unit for comparison: Choose the unit for the comparison (e.g. 10% = 0.1X, or 200% = 2X) ○ Percent Values (%) ◉ Fold Values (X) Direction of the comparison: ◉ Library Group1 compared ○ (Library Group1 compared to Library Group2) AND to Library Group2 (Library Group2 compared to Library Group1) Lower [10] Upper [ ] If you need some examples to know how to use this page and understand the significance of the analyses performed please click here. (LIB2 * 10 <= LIB1)
Libraries	Group 1 Group 2 Library Description ☐ ☐ CAF - Callus - actively growing, classic MPSS ☐ ☐ INF - Infloresence - mixed stage, immature buds, classic MPSS ☐ ☐ LEF - Leaves - 21 day, untreated, classic MPSS ☐ ☐ ROF - Root - 21 day, untreated, classic MPSS ☐ ☐ SIF - Silique - 24 to 48 hr post-fertilization, classic MPSS ☐ ☐ AP1 - ap1-10 infloresence - mixed stage, immature buds ☐ ☐ AP3 - ap3-6 infloresence - mixed stage, immature buds ☐ ☐ AGM - agamous infloresence - mixed stage, immature buds ☐ ☐ INS - Infloresence - mixed stage, immature buds ☑ ☐ ROS - Root - 21 day, untreated ☐ ☐ SAP - sup/ap1 infloresence - mixed stage, immature buds ☐ ☐ S04 - Leaves, 4 hr after salicylic acid treatment ☐ ☐ S52 - Leaves, 52 hr after salicylic acid treatment ☐ ☑ LES - Leaves - 21 day, untreated

B

S. No	GENE ▲▼ hide	CAF ▲▼ hide	INF ▲▼ hide	LEF ▲▼ hide	ROF ▲▼ hide	SIF ▲▼ hide	AP1 ▲▼ hide	AP3 ▲▼ hide	AGM ▲▼ hide	INS ▲▼ hide	ROS ▲▼ hide	SAP ▲▼ hide	S04 ▲▼ hide	S52 ▲▼ hide	LES ▲▼ hide	GSE ▲▼ hide	CAS ▲▼ hide	SIS ▲▼ hide
1	At1g01190	7	-	3	138	1	1	1	-	-	83	-	-	-	1	3	15	1
2	At1g01200	21	3	-	14	5	-	1	0	2	5	-	-	-	0	0	33	-
3	At1g01290	1	2	6	7	0	2	3	-	5	11	2	-	-	1	-	15	12
4	At1g01340	86	2	23	122	3	-	-	-	-	35	-	-	-	1	-	-	6
5	At1g01640	35	1	5	169	-	-	-	-	-	75	2	-	-	1	-	0	-
6	At1g01670	-	-	-	-	-	1	1	3	-	13	-	-	-	1	-	3	-
7	At1g01710	65	43	41	65	116	31	66	38	37	35	63	27	27	2	61	0	60

Fig. 6. Analysis of massively parallel signature sequencing (MPSS) expression libraries with Library Analysis (LIBAN) tools. (**A**) and (**B**) The entry page for the "Inter-Comparison on Libraries" tool and the result page. This tool is one of the five LIBAN tools available through our Web interface. The image (**A**) shows the final view of the page, if the user wants to retrieve all the genes that are expressed more than

5. The class of the signatures can be chosen individually or as a group. For example, the "Sense-strand transcripts" option selects classes 1, 2, 5, and 7 (*see* **Fig. 6A**) whereas the "Antisense transcripts" option selects classes 3 and 6.

6. In the "Threshold" criterion, the user can specify the condition of the comparison. For example, if the user is interested in the genes that have 10-fold higher expression in library 2 than in library 1, the user should enter "10" in the "Lower" text box and select the "Fold Values (X)" and the "Library Group 1 compare to Library Group 2" options (*see* **Fig. 6A**). More complex analyses that analyze changes in both directions can be performed with the "(Library Group 1 compare to Library Group 2) AND (Library Group 2 compared to Library Group 1)" option. More information can be obtained by clicking the link at the end of the "Threshold" section (*see* **Fig. 6A**). The formula appears after the user chooses the libraries to analyze (*see* **Fig. 6A**).

7. In this example, the root library (ROS; Root, "Signature" MPSS method) is chosen as Group 1 (or LIB1 for "Library 1") and the leaf library (LES; Leaf, "Signature" MPSS method) is chosen as Group 2 (or LIB2 for "Library 2") (*see* **Fig. 6A**). Clicking the "Get Results" button retrieves the number of signatures that are expressed more than 10-fold higher in root than in leaf (not shown). The result is grouped by the signature class, and each result takes the user to the list of corresponding genes and their expression level in the libraries (*see* **Fig. 6B**).

8. The output table can be modified in many ways. For example, the user can hide specific libraries from the table or order the table by the expression levels of specific libraries (*see* **Fig. 6B**). Each gene and signature has a link to the corresponding GA and Signature Analysis pages for further analyses, respectively. The user can download the result in comma-separated MS Excel compatible format.

4. Notes

1. We are currently using laboratory-based methods to verify the expression and sequence of antisense and unannotated transcripts in the IGRs of Arabidopsis identified by MPSS.

2. Among the 17 libraries, we have two libraries each of callus, leaf, root, and silique. Two different MPSS methods ("full" and "signature") were applied to these libraries. CAF, LEF, ROF (Root, "Full" MPSS method), and SIF (Silique, "Full" method) were sequenced with the full method and CAS (Callus, "Signature"

Fig. 6. ten-fold higher in root (ROS) compared with leaf (LES). This page takes the user to an intermediate page, which displays the number of genes retrieved. Clicking one of these numbers of signatures (not shown) displays the result in table format as displayed in (**B**). *See* **Subsection 3.3** for more details.

MPSS method), LES, ROS, and SIS (Silique, "Signature" MPSS method) were sequenced with the signature method. The full method uses the "full-length" fragment from the 3'-most *Dpn*II site to poly(A) tail and sequences 17–20 bases from the *Dpn*II site. Therefore, the size of the target fragment from which the signature is sequenced varies among transcripts, and this can bias against very large fragments. By contrast, the signature method uses *Mme*I (Type 11S enzyme) to extract, prior to cloning and sequencing, a 21 or 22 nucleotide fragment including the *Dpn*II site. Therefore, all the transcripts produce the same-sized fragments to be sequenced. In **Fig. 4A**, the libraries sequenced with signature method show high expression whereas those produced with the full method show low or no expression. This discrepancy among the libraries is likely because of the methods used for sequencing. Because only the signature method was used for other organisms, this issue is specific for Arabidopsis, and we believe the signature data to be more accurate than the full data.

3. Our Web interface is equipped with a FAQs page that has answers to many general and specific questions. In addition to this page, we are in the process of creating a tutorial section for the users who are new to our MPSS site and/or do not have a biology background. This tutorial site is written in Flash, and it is interactive and easy to use.

Acknowledgments

The authors thank Mayumi Nakano for helpful comments and critical reading of the manuscript. The tools and the databases described here were developed with support from the National Science Foundation Plant Genome Research Program.

References

1. Brenner, S., Johnson, M., Bridgham, J., Golda, G., Lloyd, D. H., Johnson, D., et al. (2000) Gene expression analysis by massively parallel signature sequencing (MPSS) on microbead arrays. *Nat. Biotechnol.* 18, 630–634.
2. Brenner, S., Williams, S. R., Vermaas, E. H., Storck, T., Moon, K., McCollum, C., et al. (2000) In vitro cloning of complex mixtures of DNA on microbeads: physical separation of differentially expressed cDNAs. *Proc. Natl. Acad. Sci. USA* 97, 1665–1670.
3. Meyers, B. C., Tej, S. S., Vu, T. H., Haudenschild, C. D., Agrawal, V., Edberg, S. B., et al. (2004) The use of MPSS for whole-genome transcriptional analysis in Arabidopsis. *Genome Res.* 14, 1641–1653.
4. Meyers, B. C., Vu, T. H., Tej, S. S., Ghazal, H., Matvienko, M., Agrawal, V., et al. (2004) Analysis of the transcriptional complexity of Arabidopsis thaliana by massively parallel signature sequencing. *Nat. Biotechnol.* 22, 1006–1011.

5. Haas, B., Wortman, J., Ronning, C., Hannick, L., Smith, R., Maiti, R., et al. (2005) Complete reannotation of the Arabidopsis genome: methods, tools, protocols and the final release. *BMC Biol.* 3, 7.
6. Lu, C., Tej, S. S., Luo, S., Haudenschild, C. D., Meyers, B. C., and Green, P. J. (2005) Elucidation of the small RNA component of the transcriptome. *Science* 309, 1567–1569.
7. Smit, A. F. F., Hubley, R., and Green, P. (1996–2004) RepeatMasker open-3.0.
8. Rice, P., Longden, I., and Bleasby, A. (2000) EMBOSS: The European Molecular Biology Open Software Suite. *Trends Genet.* 16, 276–277.
9. Lehner, B., Williams, G., Campbell, R. D., and Sanderson, C. M. (2002) Antisense transcripts in the human genome. *Trends Genet.* 18, 63–65.
10. Vanhee-Brossollet, C., and Vaquero, C. (1998) Do natural antisense transcripts make sense in eukaryotes? *Gene* 211, 1–9.
11. Eddy, S. R. (2001) Non-coding RNA genes and the modern RNA world. *Nat. Rev. Genet.* 2, 919–929.
12. Kumar, M., and Carmichael, G. G. (1998) Antisense RNA: function and fate of duplex RNA in cells of higher eukaryotes. *Microbiol. Mol. Biol. Rev.* 62, 1415–1434.
13. Lavorgna, G., Dahary, D., Lehner, B., Sorek, R., Sanderson, C. M., and Casari, G. (2004) In search of antisense. *Trends Biochem. Sci.* 29, 88–94.
14. Szymanski, M., Barciszewska, M. Z., Zywicki, M., and Barciszewski, J. (2003) Noncoding RNA transcripts. *J. Appl. Genet.* 44, 1–19.
15. Llave, C., Xie, Z., Kasschau, K. D., and Carrington, J. C. (2002) Cleavage of scarecrow-like mRNA targets directed by a class of Arabidopsis miRNA. *Science* 297, 2053–2056.

20

Metabolomics Data Analysis, Visualization, and Integration

Lloyd W. Sumner, Ewa Urbanczyk-Wochniak, and Corey D. Broeckling

Summary

Metabolomics is the large-scale analysis of metabolites and as such requires bioinformatics tools for data analysis, visualization, and integration. This chapter describes the basic composition of chromatographically coupled mass spectrometry (MS) data sets used in metabolomics and describes in detail the steps necessary for extracting large-scale qualitative and quantitative information. This process involves noise filtering, peak picking and deconvolution, peak identification, peak alignment, and the creation of a final data matrix for statistical processing. Multivariate tools for comparative analysis are presented and illustrated using data for *Medicago truncatula*. Additional tools for visualizing and integrating metabolomics data within a biological context are discussed. Two tables are provided listing current metabolomics data processing and visualization software. Because metabolomics is rapidly maturing, a final section is presented concerning the need for data standardization and current efforts.

Key Words: Metabolite analysis; metabolic profiling; metabolomics; data processing; statistical analysis; data visualization; data integration; functional genomics.

1. Introduction

Metabolomics has joined the "omics" revolution and represents the large-scale, comprehensive biochemical analysis of both primary and secondary metabolites *(1–5)*. The metabolome can be viewed as the end products of gene expression, and the measurements of large numbers of cellular metabolites provide a high-resolution biochemical phenotype of an organism. This biochemical phenotype can be used to monitor and assess gene function *(4)*

From: *Methods in Molecular Biology, vol. 406: Plant Bioinformatics: Methods and Protocols*
Edited by: D. Edwards © Humana Press Inc., Totowa, NJ

or a system's response to environmental perturbations *(6–8)*. The phenotype can also be characterized by profiling mRNA/transcripts. These messages serve to transfer information from the genome to the cellular machinery for protein synthesis. However, mRNA levels do not always correlate well with protein levels *(9)* as multiple post-transcriptional events including post-translational modifications, protein sorting, protein–protein interactions, and controlled proteolysis all contribute to the regulation of active enzyme levels. Due to these factors, changes in the transcriptome or the proteome may not always lead to alterations in the metabolic phenotype. In addition, the majority of transcript and protein annotations are currently inferred based on sequence or structural similarity. For example, it is estimated that less than 10% of *Arabidopsis thaliana* annotated genes have experimental evidence supporting assigned function. Thus the accuracy of these annotations are of some uncertainty *(10,11)*. In the absence of functionally annotated database information, transcript or protein profiling often yields limited information. For example, transcriptomics or proteomics often reveal the differential accumulation of hypothetical or unannotated proteins; however, without annotation it is very difficult to infer biological context. Microarray or proteomics experiments may also yield putative or generic protein identifications such as a putative peroxidase or peroxidase-like protein. These generic annotations based on the presence of generic motifs have limited information as many of these enzymes are promiscuous and/or involved in a large number of different reactions. However, metabolomics has the ability to reveal that the accumulated peroxidase/enzyme is more specifically related to phenylpropanoid or to another specific biochemistry. Thus, profiling the metabolome may actually provide the most direct and "functional" information of the various "omics" technologies, and integrated approaches (i.e., combined transcript, protein, and metabolite profiling) offer the greatest opportunities for discovery and understanding of biological processes.

The metabolome size of various organisms varies. For example, the yeast metabolome has been estimated to be an order of magnitude smaller than its genome *(3)*, which equates to approximately 400–600 metabolites. However, the metabolomes of higher organisms are more complex, and the plant metabolome is believed to be the most complex due to the diversity of secondary metabolites in plants. For example, the current estimates for the various plant metabolomes are on the order of 15,000 metabolites within a given species and over 200,000 different metabolites within the plant kingdom *(12,13)*. Due to the chemical complexity of the plant metabolome, it is generally accepted that a single analytical technique will not provide comprehensive visualization of the metabolome, and therefore, multiple technologies are generally employed. The selection of the most suitable technology is generally a compromise between speed, chemical selectivity, and instrumental sensitivity. Tools such as nuclear magnetic resonance (NMR) spectroscopy are rapid, highly selective, and non-destructive, but have relatively lower sensitivity. Other methods such as capillary electrophoresis (CE) coupled to laser-induced fluorescence detection are highly sensitive, but have limited chemical selectivity. Chromatographically coupled MS methods such as gas chromatography-MS (GC-MS) and high performance liquid chromatography MS (HPLC-MS) offer the best combination of sensitivity and selectivity; therefore, are central to most metabolomics approaches. GC-MS has proven capability for profiling large numbers of metabolites with reports covering several hundred to slightly more than a thousand various components *(4,8,14–18)*. Liquid introduction techniques for MS such as electrospray ionization (ESI) and atmospheric chemical ionization remove the necessity for chemical derivatization *(19–22)*. Thus, aqueous biological samples can be analyzed with minimal sample processing or even directly from the tissue source *(23)*. Furthermore, these techniques allow for the analyses of more labile and larger metabolites and for the coupling of liquid separation technologies to MS. HPLC and CE are readily coupled to MS to yield powerful tools for metabolic profiling.

Although metabolomics by definition is intended to represent the comprehensive analysis of the metabolome *(2)*, we are still distant from this objective *(24)*, and many use the term metabolomics rather loosely. The basic terminology currently accepted includes the following:

1. Metabolic fingerprinting: This unbiased, cumulative approach is generally focused on classification with minimal resolution of specific individual metabolite concentration or identification.
2. Metabolic profiling: This targeted approach focuses on the qualitative and quantitative analysis of a limited number and/or related metabolites through a metabolic pathway or network.
3. Metabolomics: This unbiased approach focuses on the comprehensive, qualitative, and quantitative analysis of all metabolites.

These specific terms are reiterated here because data processing for each of these is slightly different. An overview of each is provided below; however, the primary focus of this report will be on metabolomics data processing for MS-based approaches as these are the most familiar to the authors.

2. Materials

The materials consist of various metabolic fingerprinting, metabolic profiling, or metabolomics data sets. The composition of each is described below.

Metabolic fingerprinting is performed using rapid techniques that measure the cumulative metabolic profile. This is typically performed without attempts to separate, differentiate, or identify individual metabolites, and typically produces uni-dimensional data (*see* **Note 1**). Commonly employed instrumental approaches used for metabolic fingerprinting of crude extracts include Fourier transform infrared spectroscopy (FTIR), direct infusion (i.e., no separation) analyses using high-resolution mass spectrometers such as Fourier transform ion cyclotron resonance (FTICR) or time-of-flight MS (TOFMS), matrix-assisted laser desorption ionization (MALDI-TOFMS), or high-resolution nuclear magnetic resonance (NMR). The utilization of high-resolution instruments such as FTICRMS and high-field NMR (800 MHz) provide greater differentiation power of the metabolite mixture, and these approaches can achieve very high throughputs on the orders of hundreds of samples per day due to automation and bypass of separations. The resultant metabolic fingerprinting data are exported and classification performed using multivariate techniques such as principal component analysis (PCA), hierarchical cluster analysis (HCA), discriminant function analysis (DFA), and/or self-organizational maps (SOMs). The cluster analysis generally includes samples of known response or composition such as wildtype relative to diseased. The classification of the experimental sample is determined by the relative distance or co-clustering with one of the target classes. A related approach known as metabolic footprinting is similar in concept; however, it is focused on the analysis of extracellular/secreted metabolites residual in the culture media or rhizosphere (*see* **Note 2**).

Metabolic profiling involves the qualitative and quantitative analysis of a limited number and often related metabolites. These metabolites are often chemically similar and/or related through a metabolic pathway or network. For example, methods have been reported for the GC-MS analysis of phytohormones *(25,26)*, HPLC-MS analysis of saponins *(19,27)*, HPLC-MS of phenylpropanoids *(28,29)*, and CE-MS of isoflavones *(30)*. Because highly selective quantitative analyses of target compounds are a critical necessity in a large diversity of applications, most instrument vendors offer efficient data processing software that generate quantitative output or reports for specific target metabolites. The data from these experiments are generally processed using descriptive statistics to assess differences in the sample set (*see* **Note 3**).

Metabolomics is a non-targeted approach with the goal of comprehensive, qualitative, and quantitative analysis of all metabolites. The breadth of this approach is what differentiates it from the others. In the purest accordance with the above definition, multiple technologies are usually used in aggregate for metabolomics-based approaches. This is because each technique has an inherent bias and limitation. As a result, typical metabolomics data sets are multi-dimensional because they provide additional resolution of metabolites and orthogonal data that are useful in chemical identification and/or discrimination *(24)*. The most commonly utilized techniques include chromatography coupled to MS or NMR (i.e., GC-MS, LC-MS, CE-MS, and LC-NMR). A single GC-MS metabolite profile can yield 300–500 distinct components, whereas, a typical LC-MS metabolite profile might yield 100–200 components (*see* **Note 4**).

Metabolomics presents several challenges in data processing and visualization that are discussed below in Section 3, however, it is beneficial to first describe the data composition. The above-mentioned multi-dimensional techniques consist of two orthogonal domains that include a separation or chromatography domain and a mass domain. The chromatography domain is essentially a time domain where individual metabolites are separated from each other over time, and the efficiency of the separation is dictated by the parameters of the chromatography. A recent review has been presented covering these parameters and will not be discussed further here *(20)*. As the metabolites separate over time, mass spectra are collected at predefined intervals generally, 1 to 100 scans per second (i.e., 1–100 Hz). Mass selective detection provides highly specific chemical information including molecular mass and/or characteristic fragment ion(s) that are directly related to chemical structure. This information can be utilized for compound identification through spectral matching with data compiled in libraries for authentic compounds or used for *de novo* structural elucidation. Furthermore, chemically selective MS information can be obtained from extremely small quantities of metabolites with limits of detection in the pmole and fmole level for many primary and secondary plant metabolites. Illustrations of the chromatography and mass domains are provided in **Fig. 1**.

2.1. GC-MS, HPLC-MS and UPLC-MS Metabolic Profiling

Primary metabolites are extracted from approximately 6 mg of dissected and lyophilized plant tissue (*see* **Note 5**) using a bi-phasic solvent system composed of water and chloroform as previously reported *(8)*. Polar metabolites are partitioned into the aqueous phase and non-polar metabolite partition in the non-polar phase (chloroform) (*see* **Note 6**). These metabolites are then derivatized and analyzed using GC-MS as previously described *(31)* and summarized below. Secondary metabolites such as flavonoids, isoflavonoids, and triterpenoids are difficult to analyze by GC-MS owing to the labile nature of their glycosylated forms and often lower abundance. Thus, these compounds are analyzed using ultra pressure

A. Time Domain: chromatographic separation   **B.** Mass Domain: mass spectrum

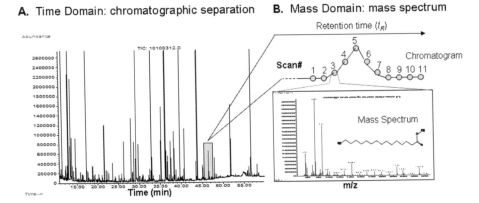

Fig. 1. Illustration of the (**A**) chromatography and (B) mass domains from a gas chromatography-mass spectrometry analysis of *Medicago truncatula* cell cultures. Chromatographic separation purifies individual components for mass measurements, which reduces matrix effects such as competitive ionization and ion suppression. Mass spectra are used for metabolite identifications through spectral matching.

liquid chromatography (UPLC) quadrupole time-of-flight mass spectrometry (QtofMS) *(32)* that offers substantially higher resolution separations than traditional HPLC and without derivatization. Secondary metabolites are extracted from approximately 20 mg of lyophilized tissue in 1.8 ml of 80% methanol containing 2 μg umbelliferone as an internal standard, for 10 h. Extracts (1.4 ml) are centrifuged at 3000 g for 60 min, and the resulting supernatant is evaporated under nitrogen to dryness. Residue is resuspended in 300 μl of 45% methanol (isoflavonoids) or 100 μl water (triterpene saponins), and samples are analyzed in negative-ion mode by HPLC ion-trap mass spectrometry (HPLC-ITMS) or UPLC-QtofMS. HPLC-ITMS (**19, 21, 27–29**) separations were achieved using an Agilent 1100 HPLC and a 250 × 4.6 mm i.d., 5 μm, reverse-phase, C18 column (J.T. Baker, Phillipsburg, NJ, USA). Samples were eluted with a linear 0.1% aqueous acetic acid : acetonitrile gradient, 95:5 to 5:95 in 90 min, at a flow rate of 0.8 ml/min. The water was adjusted with acetic acid to a final concentration of 0.1%. ITMS mass spectra were acquired using a Bruker Esquire LC equipped with an ESI source. Negative-ion ESI was performed using a source voltage of 3000 V and capillary offset voltage of −70.7 V. Nebulization was achieved using nitrogen gas at a pressure of 70 psi. Desolvation was aided by the use of a nitrogen counter current gas at a pressure of 12 psi. The capillary temperature was set at 360°C. Mass spectra were recorded over the range of 50–2200 m/z values. The Bruker ion-trap was operated under an ion current control of 20,000 and maximum acquire time of 100 ms and a trap drive setting of 60. UPLC separations were achieved using a Waters Acquity UPLC 2.1 × 100 mm, BEH C18 column with 1.7 μm particles, a flow of 600 μl/min, and a 30 min linear gradient of 5%:95% to 95%:5% for eluents A:B. Eluent A was 0.1% acetic acid in water and eluent B is acetonitrile. Mass spectra are collected on a Waters QTOFMS Premier or LECO Unique TOFMS and recorded for the range of 100–2000 m/z value.

3. Methods

3.1. Data Extraction and Alignment

A major challenge in metabolomics involves the high-throughput generation of accurate and well-aligned data matrices from the various chromatographically coupled mass spectrometry (xC-MS) data for statistical analysis. The desired data output matrix is composed of a comprehensive list of unique metabolite identifiers (noting that metabolites may or may not be chemically identifiable) and an intensity or concentration for each metabolite in each sample in the overall comparative analysis. Traditional methods for extracting quantitative information from xC-MS data sets and assembly into data matrices amenable to statistical interrogation are quite cumbersome and ineffective for large complex sample sets. Once an accurate data matrix has been created, the statistical processing of these data sets can be achieved using similar tools to those employed for microarray analysis such as descriptive statistics (averages, means, standard deviation, coefficients of variance, and t-tests), PCA, HCA, self-organizing maps, analysis of variance (ANOVA), and DFA. The primary limitation to high-throughput metabolomics is the lack of universal software to generate accurate data matrices from multiple instruments and their related file formats. Currently, several instrument manufacturers offer proprietary metabolomics data processing software including LECO's ChromaTOF, Waters MarkerLynx, and ABI Markerview. These packages represent exciting steps forward but are dedicated to the specific vendor's file formats.

Recently, several independent software packages have been developed to extract and generate data matrices for comparative analyses. These are listed in **Table 1**. Many of these are available at no cost and offer good solutions for data extraction and alignment. These solutions typically perform several of the key steps necessary for successful data processing including filtering/noise reduction, peak deconvolution to resolve co-eluting components, peak picking, peak quantification, peak identification, and alignment to compensate for experimental variation. Ideally, these processes would be efficient, automated, and high throughput to yield a truly "omics" technology.

Table 1
Available Metabolomics Software Packages

Software	Reference	URL
BinBase	(Fiehn et al., 2005) *(71)*	http://fiehnlab.ucdavis.edu/projects/binbase_setupx/
MET-IDEA Align	(Tikunov et al., 2005) *(72)*	http://www.pri.wur.nl/UK/products/MetAlign/
MET-IDEA	(Broeckling et al., 2006) *(73)*	http://www.noble.org/plantbio/MS/index.html
MZmine	(Katajamaa and Oresic, 2005)	http://mzmine.sourceforge.net/
XCMS	*(74)*	http://metlin.scripps.edu/download/

3.1.1. Filtering

Often raw xC-MS data are filtered or smoothed to remove noise and enhance signal-to-noise (S/N) ratios. A basic mechanism for doing this is a moving window method, where a "window" of fixed width, commonly three to seven scans, is used to calculate an average value. The window then moves forward one scan and repeats. As true noise is random, the average noise values have smaller amplitudes and averaging increases the S/N ratios. More elaborate variations of this method, such as the Savitsky–Golay filter, exist that use polynomials to calculate the moving window value. Alternative methods calculate descriptive statistics for the noise, and use a function of the standard deviation to subtract noise. An example of noise filtering is provided in **Fig. 2**. Because

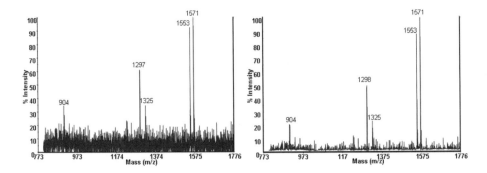

Fig. 2. Noise filtering of mass spectral data illustrating the benefit of filtering on signal-to-noise ratios.

Gaussian peak shapes are expected in chromatography, Gaussian distributions can be mandated for chromatographic peaks and also used to eliminate noise spikes characterized by significantly narrower peak widths *(33)*.

3.1.2. Peak Deconvolution, Peak Picking, and Identification

The assessment of individual components or metabolites within an xC-MS analysis is generally termed peak picking. Some instrument programs pick peaks or integrate data only within the chromatographic domain *(34)*; however, most of the more recent tools listed in **Table 1** utilize both the chromatographic and mass domain data for picking peaks. The mass domain adds additional discrimination power, as molecular mass and fragment ions are characteristic of chemical structure. Most metabolites, except isomeric compounds co-eluting in the time domain, can usually be differentiated based on the mass domain. The use of selected ion chromatograms (SICs) to resolve and differentiate co-eluting metabolites is typically referred to as deconvolution, and SICs as opposed to total ion chromatograms are commonly used for quantitative analyses.

There are two approaches for generating lists of SICs in metabolomics approaches. One simply generates a list of SICs for all masses based on fixed width m/z windows or "bins" *(35)*. An example would be 900 bins of 1 m/z width for a scan range of m/z 100 to 1000. This is the easiest method. However, this process produces highly redundant data sets as individual metabolites will often yield a large number of related ions, especially in GC-MS with electron ionization sources which produce a significant number of fragment ions (*see* **Fig. 3**, panels C–E). This is also true for LC-MS using ESI, where less fragmentation is observed. However, multiple adducts (H^+, Na^+, and Ca^+) and/or multiply charged ions are commonly observed for individual metabolites. A more selective approach is to determine unique m/z values representative of the molecular species or highly abundant ions for selected ion extraction. Furthermore, unique ions that are observed within specific elution time windows can then be used for the confident selected extraction and quantification data of both known and unknown metabolites. These ion and retention time pairs (IR_ts) are unique and characteristic of a specific compound, and these are commonly referred to as mass spectral tags *(36)*. A list of IR_ts can be generated manually, imported from a database, or through the use of specialized software such as Automated Mass Spectral Deconvolution and Identification Software (AMDIS; http://www.amdis.net/). AMDIS was developed and is maintained by the United States National Institute of Standards and Technology (NIST), is freely available, and designed for analysis of GC-MS data *(37)*. It is capable of differentiating chromatographically

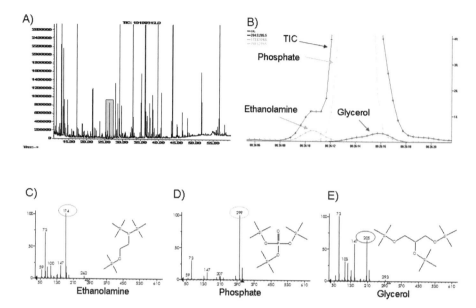

Fig. 3. Peak and spectral deconvolution are used to resolve components based on m/z data. Examples are provided for trimethylsilyl derivatives of ethanolamine, phosphate, and glycerol. Commonly, total ion chromatograms (TICs) are used to visualize data (**A**); however, co-eluting compounds are not easily differentiable in the TIC (**B**). Thus, selected ion chromatograms (**B**) based on unique ions (**C–E**; m/z 174, 299, and 205) can be used to differentiate and quantify co-eluting compounds.

unresolved and baseline-masked peaks and subsequently identifying those components by comparison with spectral libraries generated from authentic compounds. AMDIS can also be used for analysis of LC-MS and CE-MS data, albeit somewhat less effectively due to the limitations in parameter settings and lack of commercially available spectral libraries. By analyzing representative samples using AMDIS, the vast majority of components in a data set can be rapidly and efficiently converted to an IR_t list (*see* **Note 7**). Unfortunately, AMDIS does not posses quantification capabilities; however, the IR_t list can be used for high-throughput data extraction following alignment.

A moderate number of the primary metabolites can be identified using mass spectral matching to those of authentic compounds in commercial libraries such as the NIST. Unfortunately, the commercial libraries contain a low population of primary and secondary plant metabolites; therefore, many groups have generated custom mass spectral libraries *(8,16,36)*. Currently, our libraries contain over 800 primary and secondary metabolites recorded for authentic compounds on multiple instruments, and provide an additional and valuable resource for the identification of unknown compounds. However, we still have a great need for additional metabolite identifications, and an incessant effort is underway to continue the population of these spectral libraries. At present, we are able to identify approximately 100 different components from the polar, and approximately the same for the non-polar GC-MS profiles of *Medicago truncatula*, *Medicago sativa*, *Glycine max*, and *Arabidopsis* extracts. Identifiable polar metabolites include monosaccharides, disaccharides, limited sugar

Fig. 4. Linear regression alignment of both retention time and ion m/z for a series of gas chromatography-mass spectrometry metabolomics analyses using MET-IDEA (Broeckling et al., 2006). The program reports the final equations used for alignment of each file. Alignments are commonly performed based on multiple identified metabolites that are usually present in all samples such as serine, valine, phosphate, succinate, ribitol (IS), fructose, and sucrose. The metabolites illustrated have elution times distributed across the profile that help ensure representative alignment of the various regions of the chromatography domain.

phosphates, organic acids, nucleotides, amino acids, alcohols, and aromatic amines. Identifiable lipophilic metabolites include fatty acid esters, sterols, long chain alcohols, terpenoids, and others. Those metabolites not readily identifiable can still be differentiated based on their unique mass spectra and used in comparative analyses. If comparative analyses reveal these as statistically and differentially accumulated, *de novo* identification can be pursued. Mass spectral libraries for LC-MS are less efficient due to reduced fragmentation observed in LC-MS and lower instrument to instrument reproducibility. However, they are still of good utility.

3.1.3. Chromatographic Alignment

Following peak picking for a series of metabolomics analyses, the peaks must be aligned accurately across all data sets to allow proper comparative analysis. However, during a series of GC-MS, LC-MS, CE-MS, and/or LC-NMR analyses, the metabolites or features (peaks) can drift in both the chromatography and mass domains. The magnitude of the drift is dependent upon a large number of instrumental (and operator) variables, and this drift significantly complicates the alignment and generation of a data matrix. The accurate assessment and compensation for this drift are necessary to generate meaningful data. Chromatographic shifts in xC/MS can be accurately aligned using a series of exogenous retention index markers. The retention time of these internal standards are used to linearly adjust and normalize the time domain *(14,15)*. A common practice uses a series of alkanes to calculate retention indices *(16,17)*. Alternatively, commonly observed metabolites such as specific amino acids, sugars, or organic acids with predetermined retention times and/or retention indices can be used for chromatographic alignment. Linear corrections may be insufficient for analyses performed over long periods of time or with slight changes in parameters, such as use of a new separation column. In these cases, non-linear corrections such as correlation optimized warping (COW) *(38)* is best. A recently released program XCMS uses a non-linear retention correction *(35)*. Shifts also occur in the mass domain, and some software such as MET-IDEA will also provide for large-scale alignment of m/z values using a fixed value based on the average deviation, or use of a linear regression for calibration based on a single or multiple identified metabolites. Alignment in both dimensions ensures more accurate data matrix generation (**Fig. 4**).

3.1.4. Peak List and Data Matrix Creation

Once a data set has been aligned, quantitative peak areas can be extracted for all peaks in each sample based on the list of IR_ts. A peak start and stop point are usually determined based upon the changing slope of the extracted ion chromatogram, and the intensity of each scan encompassed by the peak is summed to yield a relative quantitative area measurement. These area measurements are added for each peak in each analysis to complete the data matrix. Additional filters can also be used at this point, such as a minimum S/N ratio for all peaks; however, rigorous criteria will often result in missing values in the data matrix. Alternatively, peaks may be absent due to real biological effects. In the cases where peaks are missing, many programs will still report a value based on the average or maximum intensity value observed within the retention time/index window, as opposed to a zero value. The overall output of this process is an aligned data matrix such as that provided in **Fig. 5**, which can be used for further statistical analysis. The data matrix is often exported in a universal tab delimited text file format which can then be imported into a vast variety of statistical software packages.

Result

Data folder: C:\Data\Mixed polar NETCDF\Raw NetCDF Data

FileName	176.9(11.6	174(12.325	165.9(12.3	204(12.426	174(12.611	188(12.782	219.8(13.4	220(13.466	191(13.598	281.9(14.0	188(1
22010407	587	2202	415	1325	57201	22455	230	936	515	1890	15486
22010408	1400	1992	358	2762	54651	22557	610	1531	1007	1857	15217
22010409	2298	2188	556	2097	53246	23911	517	1328	330	1500	17980
22010410	896	2014	780	1875	54172	23303	632	860	547	1936	17231
22010411	348	1896	336	955	54634	24085	206	1001	1014	1623	15386
22010412	187	2017	667	1621	51627	25895	556	995	417	1287	15302
22010413	644	1769	232	1470	54221	31473	542	887	204	1908	18602
22010414	1184	2187	399	1555	53499	29457	685	1175	0	2347	18845
22010415	553	1530	451	1462	53981	37656	0	399	860	2579	19451
22010416	334	2540	629	1613	55046	38469	606	1274	1178	1501	21990
22010417	0	2064	662	1331	54925	34094	493	957	1334	1246	17784
22010418	0	2107	1316	1111	54780	40319	500	1126	1810	1953	21648
22010419	942	2109	206	1952	53504	38782	192	953	468	2595	17634
22010420	314	2196	1109	1445	54607	38842	493	846	662	2244	18027
22010421	0	2163	1097	1060	53504	41205	506	1439	730	2126	19015
22010422	0	2144	1090	2535	53345	43535	432	890	1137	2527	20024
22010423	0	2083	727	2018	53099	43709	526	1172	1153	2984	19858
22010424	0	1871	961	1410	54061	47641	0	844	541	3673	20518
22010425	0	2226	681	1000	53129	46658	197	197	430	937	17924
22010426	798	2397	1013	2397	59713	63585	717	717	613	3193	20466
22010427	426	2333	451	2082	57425	48041	630	790	625	1723	19385
22010428	572	1717	553	1834	53838	44251	382	1168	344	1812	20515
22010429	980	1942	347	2811	54785	47586	796	796	380	3206	23617
22010430	326	1800	522	3314	55374	47668	1835	2213	513	2753	17943
22010431	974	2084	0	1728	54985	49547	997	1005	217	1598	23410

Edit Column   Correlation   Transpose   OK   Cancel

Ready....

Fig. 5. Representative metabolomics output table from MET-IDEA, listing peak areas for a series of unique ion and retention time pairs (columns) for an experimental sample series (rows). Many of the ion and retention time pairs (IR_ts) also have chemical identifiers buried in the column headers.

3.2. Data Processing

Once a data matrix has been generated then statistical analyses begin. Typically, most of the statistical processing approaches used in metabolomics are convergent (i.e., the same) with those used in other "omics" data processing. Thus, most data processing methods are not unique to metabolomics once the challenge of extracting and aligning that data has been circumvented.

Typically, the first step in data processing includes normalization to reduce the influence of experimental factors on the data set. Normalization is often performed based on the original mass of the sample and/or to an internal standard. This is performed by dividing all the peak areas for a given sample by the mass of the original sample or by the area of the internal standard calculated for a specific sample. Following normalization to the internal standard, descriptive statistical analyses are performed. These typically include the calculation of means and standard deviations. Comparative analyses of the means can then be performed by calculating a fold change, F-ratios, t-tests, or ANOVA to assess the significance of the changes *(39)*.

Due to the large number of variables and complexity of metabolomics data sets, multivariate methods are also used for comparative analyses. These commonly include unsupervised methods such as HCA, PCA, Konen's SOMs, and neural networks (NN). These are referred to as unsupervised methods as no additional information besides the raw data is required for analysis. Other supervised methods such as partial least squares and DFA require additional training information for "supervising" the data processing. The purpose of these methods is to reduce the dimensionality of the data set to better enable classification of individual samples and/or visualization of similarities and differences between samples. Multivariate analyses are available in a large number of commercial software packages, including XLSTAT (an add-on to Excel, Microsoft Corp), JMP (SAS Institute Inc.), Pirouette® (InfoMetrix, Woodinville, WA, USA), MATLAB (The MathWorks, Inc.), and various R programs (http://www.r-project.org/).

Several preprocessing steps are commonly used to allow better visualization during multivariate analyses. These include log transformation of data, which converts that data scale from a base ten to a log base. This conversion process normalizes the influence of changes in more abundant metabolites and allows visualization of changes in less-abundant metabolites. Another preprocessing step involves normalization to the mean value by dividing each peak area value by the mean peak area for that compound/peak. This helps center the data in further multivariate processing and is commonly referred to as mean centering.

PCA is one of the oldest and most widely used multivariate techniques *(40)*.
PCA is used to calculate the variance of each metabolite in the multivariate
data set and then rank the variance based on the magnitudes of the orthogonal
variables (PCs). PCA is a linear additive model, in the sense that each PC
accounts for a portion of the total variance of the data set. Usually, a small
set of PCs (two or three) can account for over 90% of the total variance, and
in such circumstances, the data from those few PCs can be used to reduce
the dimensionality of the data set. Plotting the data in the space defined by
the two or three largest PCs provides a rapid means of visualizing similarities
or differences in the data set, perhaps allowing for improved discrimination
of samples. An example of PCA analyses of a series of GC-MS metabolite
profiles collected for *M. truncatula* roots, stems, and leaves are provided in
Fig. 6.

HCA is another method of grouping samples in a data set by their similarity.
HCA involves a progressive pair-wise grouping of samples by distance. Several
distance measures can be used in HCA, such as Euclidean distance, Mahalanobis
distance, or correlation. Results vary according to which distance metric is used.
The result of hierarchical clustering is usually visualized as a dendogram or a

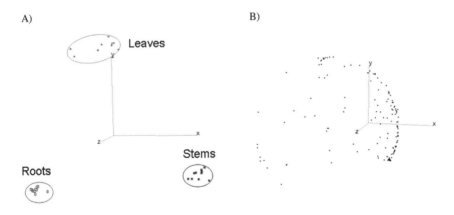

Fig. 6. Principal component analysis (PCA) showing clear segregation of *Medicago
truncatula* roots, stem, and leaves based upon their metabolic gas chromatography-mass
spectrometry profiles. The three-dimensional scores plot provided in panel (**A**) show
that the first three PCs can be used to differentiate the metabolite profiles of different
tissues. The loading plot in panel (**B**) shows the influence of individual metabolites,
represented by the various data points, on the overall samples. Those metabolites in
panel (**B**) with the maximum amplitude represent those with the largest variance and
impact on the PCA score plot (**A**).

tree. Branch lengths can be made proportional to the distances between groups. This can provide an easy visualization of the similarities of samples within data sets. Several other clustering algorithms exist such as K-means clustering, and selection of the best clustering method is generally an empirical process as the various methods often produce different results. Clustering is most useful for classifying samples in groups. A two-dimensional HCA clustering of metabolites and tissues of *M. truncatula* is provided in **Fig. 7**.

Another set of multivariate methods, called supervised, utilizes a calibration or "training" data set, i.e., a set of observations that have been classified by independent means. An example of a supervised method is the use of standards to calibrate a protein concentration assay. Supervised methods can, thus, only be carried out if one is able to provide known examples. Despite this drawback, supervised methods are usually more powerful than unsupervised methods. Raamsdonk and coworkers *(41)* have used DFA *(42)* for this purpose. Other supervised methods that could be used are feed-forward NN *(43)*, support vector machines *(44)*, genetic algorithms *(45)*, and genetic programming *(46)*. Kell, Goodacre, and coworkers have pioneered the application of supervised methods to metabolomic data *(3,47–50)*.

3.3. Data Visualization and Integration

The ultimate step in data processing is data visualization within a biological context. This process is necessary to visualize and comprehend the qualitative and quantitative changes in metabolite profiles at the pathway and system levels. This is a challenging task; however, several exciting tools are now emerging that allow the reporting of metabolomics within a pathway context. Many of these tools are discussed in focused chapters of this book and allow for the integration of metabolomics data with correlated transcriptomics and/or proteomics information. An abbreviated list of plant databases and tools is provided in **Table 2**. Others databases/tools exist such as PathwayAssist (Ariadne Genomics) and Pathway Analysis (Ingenuity Systems Inc.), but are currently focused only on mammalian and/or microbial systems. These are often useful for understanding trends in their related field, but offer limited utility to plant studies.

One of the publicly available software tools is MapMan, which is a user-driven tool that displays large data sets onto metabolic pathways or other processes *(51,52)*. The application can be used as a web-based application or downloaded from http://gabi.rzpd.de/projects/MapMan/. The MapMan software functionally categorizes genomic and metabolomic profiles in order to display quantitative color-coded data onto biochemical pathways or customized maps.

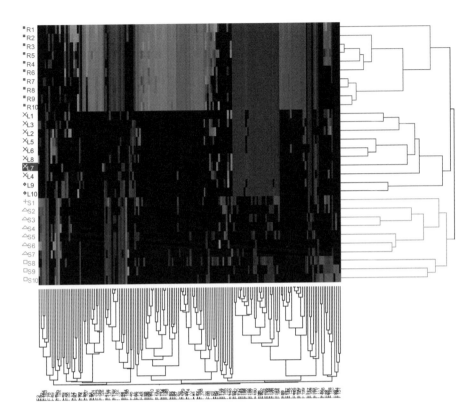

Fig. 7. Two-dimensional hierarchical cluster analysis of thirty gas chromatography-mass spectrometry metabolic profiles recorded for *Medicago truncatula* root (R), stems (S), and leaves (L). Ten replicate injections were made for each tissue and are visible on the left-hand side. A heat map and dendograms are generated based on the relative abundance for approximately 160 components observed in all the tissues. The dendogram on the right shows clustering of the similar tissue samples whereas the dendogram on the bottom reveals similarities/clustering of individual metabolites. The data illustrate the unique biochemistry of the spatially resolved tissue, and the heat map can be used to assess specific metabolite differences in the different tissues.

This approach allows the user to visualize data related to the overall plant system or to specific biological processes. MapMan was created as a generic tool that has been used to investigate responses of the model plant *Arabidopsis thaliana*. Current developments include the functional annotation of other plant genomes, e.g., tomato *(53)*, and the implementation of statistics modules that allow data pre-processing with subsequent visualization of results *(52)*.

Table 2
Current Metabolomics and Integrated Functional Genomics Databases and Visualization Tools

Database	Reference	URL
AraCyc	(Mueller et al., 2003, *(75)*)	http://www.arabidopsis.org/tools/aracyc/
Armet	*(63)*	http://www.armet.org/
DOME	*(54)*	http://mendes.vbi.vt.edu/tiki-index.php?page=DOME
Biochemical pathway maps (BioPathAtMAPS)	*(56)*	http://www.ibc.wsu.edu/research/lange/index.html.
KaPPA-view	(Tokimatsu et al., 2005) *(76)*	http://kpv.kazusa.or.jp/kappa/indexnew.jsp?g
MapMan	*(51)*	http://gabi.rzpd.de/projects/MapMan/)
DRAGON	(Bajic et al., 2005) *(77)*	http://research.i2r.a-star.edu.sg/DRAGON/ME2/
MetNet	(Yang et al., 2005) *(78)*	http://www.vrac.iastate.edu/research/sites/metnet/MetNetVR/MetNetVR.htm

Many of the above databases/tools are species specific. For example, AraCyc is dedicated to *Arabidopsis*. However, the authors of this report are interested in *Medicago* and other legumes. Currently, DOME *(54)* is the only database that houses *Medicago* specific data; however, it is envisioned that future efforts by the Legume Information System *(55)* (*see* Chapter 11) will also incorporate legume metabolomics data. In the meantime, we are currently developing a legume-specific tool aimed at the integrated analysis as well as visualization of transcripts and metabolites within the context of metabolic pathway maps using a customize version of GeneSpring (Silicon Genetics/Agilent Technologies). This effort is very similar to that described for *Arabidopsis* *(56)*. All metabolites currently detected using multiple xC-MS techniques (identified and unidentified) have been tabulated and putative biological functions assigned. In parallel, all transcript elements contained on the recently released Medicago Affymetrix chip (http://www.affymetrix.com/products/ arrays/specific/medicago.affx) were annotated and grouped into functional categories using MIPS terminology. Furthermore, the genomic annotations were utilized to predict legume-specific metabolic pathways using MetaCyc (http://MetaCyc.org), which comprises a

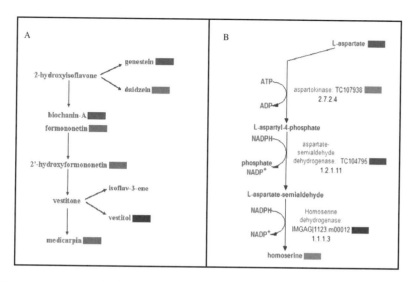

Fig. 8. Graphical displays of A) metabolites only and B) integrated metabolite and transcript data recorded for Medicago truncatula cell cultures stress responses illustrating the A) independent or B) simultaneous visualization of metabolomics data. The isoflavonoid pathway is shown in panel (**A**) and illustrates the accumulation of the phytoalexin medicarpin following elicitation. Panel (**B**) illustrates the response of aspartic acid biosynthesis following elicitation, with both transcripts and metabolite levels. Pathways can be zoomed in and out to show various levels of detail. Compounds, reactions, enzymes, and genes on a pathway page are linked to additional information.

collection of metabolic pathways and enzymes from a wide variety of organisms including microorganisms and plants. The automated alignment was manually curated, and *Medicago*-specific pathways such as isoflavonoid biosynthesis were added. The resulting pathways are being imported into GeneSpring for statistical analyses and automated visualization of integrated transcriptomics and metabolomics data. The presentation of data in this manner provides improved visualization and interpretation of the integrated data. Example data acquired by microarray and xC-MS metabolite profiling while studying the response of *M. truncatula* cell cultures to stress is provided in **Fig. 8**.

3.4. Metabolomics Data Standards

Significant effort has been directed toward the establishment of unified reporting guidelines for functional genomics. The Minimum Information About a Microarray Experiment (MIAME) *(57)* standard has been developed by the DNA micorarray community as a specification for documenting a transcriptome experiment. Similar standards have also been described for proteomics experiments. These include the Proteomics Experiment Data Repository (PEDRo) *(58)* and the Protein Standards Initiative *(59–61)*. All of the standard initiatives have focused on three main metadata areas: (1) the origin and ontology of biological samples, (2) the analytical techniques used for the analysis, and (3) the methods of data processing used to extract information from the data.

There is now a need for the standardization of metabolomics reporting *(62)*. A basic framework for the description of plant metabolomics experiments has been proposed and entitled Architecture for Metabolomics ArMet *(63,64)*. The ArMet data model could be used as a design template in data collection software tools that support the experimental process. Furthermore, the Standard Metabolic Reporting Structures (SMRS) working group has proposed minimum requirements for designing and recording the results of a metabolic study *(65)*. These included three familiar major areas: (1) experimental design, (2) description of analytical data set characteristics, and (3) data processing. The SMRS group also suggested that the minimum information about a metabolomics experiment also depends on the objective of the experiment (public database, journal supplementary data, and regulatory submissions). More recently, the National Institute of Health sponsored a workshop on metabolomics data standards (http://www.niddk.nih.gov/fund/other/metabolomics2005/). This workshop spawned an overall effort to further refine and propose a set of metabolomic standards, and a recent call for participation was put forth *(66)*.

The utility of microarray data sets for reprocessing and additional comparative analyses led to the establishment of expression profile databases, such as the Stanford Microarray Database—SMD *(67)* or NCBI-GEO *(68)* (*see* Chapter 2). Similarly, the availability of proteome data led to establishing various databases, e.g., The Universal Protein Resource UniProt *(69)* (*see* Chapter 4). An open exchange repository of metabolomics data sets is likewise desirable for plant metabolomics. One such example is the open access Golm Metabolome Database *(36,70)* that provides public access to custom mass spectral libraries, metabolite profiling experiments as well as additional information and tools, e.g., with regard to methods, spectral information, or compounds. It is envisioned that additional metabolomics databases will continue to emerge and enhance the utility of metabolomics.

4. Notes

1. Uni-dimensional data consist of data in a single dimension. Common metabolic fingerprinting uni-dimensional data include that generated by FTIR, proton NMR, and flow-injection MS of crude extracts. Commonly, flow-injection analyses utilize high-resolution mass spectrometers such as a hybrid quadrupole TOF or FTICR (which is often abbreviated as only FTMS) to improve the discrimination power

of the metabolic fingerprinting. These uni-dimensional data sets are contrasted by comparison with multi-dimensional data sets which utilize two or more orthogonal domains of chemical characterization such as GC-MS, HPLC coupled to QTOFMS (HPLC-QTOFMS), CE coupled to MS, or highly dimensional HPLC-MS/NMR. These tools provide correlated data in the chromatography, mass, and/or NMR domains providing greater information for the differentiation and identification of individual metabolites.

2. Secreted metabolomics play a key role in plant interactions. These compounds are associated with defense, alleopathy, and signaling between plants and microbes.

3. Descriptive statistics describe the data in simplified numerical terms. Examples include mean, average, median, standard deviation, and coefficient of variance. These calculations are typically available as macros in many spreadsheet programs.

4. The separation efficiencies of GC and HPLC vary based on column chemistries and architecture. Traditionally, capillary GC columns offer peak capacities of approximately 600–800 whereas traditional analytical HPLC is limited to 200–300. However, recent commercialization of ultra high-pressure ($>15,000$ psi) HPLC, also known as UPLC, is yielding peak capacities of approximately 600 by LC. The physical basis and more details on the resolving power of different techniques have been reviewed *(20)*

5. Lyophilization is used to eliminate the aqueous environment and to precipitate proteins/enzymes. This significantly reduces enzymatic activity and helps prevent further metabolic reactions and degradation. This helps ensure that the resultant metabolic profiles are indicative of the true physiological state.

6. The use of high quality glass vials and liquid transfer devices throughout is highly advised and especially so when using organic solvents. Plastic tubes often contain traces of "releasing agents" used during the molding process. The releasing agents are usually polyglycols that leach into both the aqueous and organics extracts and ultimately contaminate the samples. Thus, glass vials are highly recommended. Furthermore, we "bake" our glass (i.e., heat at $>50°C$ at least overnight) and then allow to cool to room temperature prior to use, to minimize contamination.

7. The resultant deconvoluted peak lists ($IR_t s$) generated by AMDIS are very subjective and significantly influenced by the parameters used for analysis (sensitivity, threshold, resolution, etc.). Thus, manual review and editing of these lists are usually necessary prior to further data analysis, but this can be performed in 1–2 h for a reasonably complex mixture.

Acknowledgments

The Sumner lab is supported by The National Science Foundation Plant Genome Research Program Award no. DBI-0109732, NSF 2010 MCB-0520283, NSF 2010 MCB-0520140, State of Oklahoma, and The Samuel Roberts Noble Foundation.

References

1. Sumner, L., Mendes, P., and Dixon, R. (2003) Plant metabolomics: large-scale phytochemistry in the functional genomics era. *Phytochemistry* 62, 817–836.
2. Fiehn, O. (2002) Metabolomics–the link between genotypes and phenotypes. *Plant Mol. Biol.* 48, 155–171.
3. Oliver, S., Winson, M., Kell, D., and Baganz, F. (1998) Systematic functional analysis of the yeast genome. *Trends Biotechnol.* 16, 373–378.
4. Fiehn, O., Kopka, J., Dormann, P., Altmann, T., Trethewey, R., and Willmitzer, L. (2000) Metabolite profiling for plant functional genomics. *Nat. Biotechnol.* 18, 1157–1161.
5. Trethewey, R. N., Krotzky, A. J., and Willmitzer, L. (1999) Metabolic profiling: a rosetta stone for genomics? *Curr. Opin. Biotechnol.* 2, 83–85.
6. Weckwerth, W. (2003) Metabolomics in systems biology. *Annu. Rev. Plant Biol.* 54, 669–689.
7. Fernie, A., Trethewey, R., Krotzky, A., and Willmitzer, L. (2004) Metabolite profiling: from diagnostics to systems biology. *Nat. Rev. Mol. Cell Biol.* 5, 763–769.
8. Broeckling, C. D., Huhman, D. V., Farag, M. A., Smith, J. T., May, G. D., Mendes, P., Dixon, R. A., and Sumner, L. W. (2005) Metabolic profiling of *Medicago truncatula* cell cultures reveals the effects of biotic and abiotic elicitors on metabolism. *J. Exp. Bot.* 56, 323–336.
9. Gygi, S. P., Rochon, Y., Franza, B. R., and Aebersold, R. (1999) Correlation between protein and mRNA abundance in yeast. *Mol. Cell. Biol.* 19, 1720–1730.
10. Somerville, C., and Dangl, J. (2000) Plant biology in 2010. *Science* 290, 2077–2078.
11. Somerville, C., and Somerville, S. (1999) Plant functional genomics. *Science* 285, 380–383.
12. Dixon, R. A. (2001) Phytochemistry in the genomics and post-genomics eras. *Phytochemistry* 57, 145–148.
13. Hartman, T., Kutchan, T. M., and Strack, D. (2005) Evolution of metabolic diversity. *Phytochemistry* 66, 1198–1199.
14. Roessner, U., Luedemann, A., Brust, D., Fiehn, O., Linke, T., Willmitzer, L., and Fernie, A. R. (2001) Metabolic profiling allows comprehensive phenotyping of genetically or environmentally modified plant systems. *Plant Cell* 13, 11–29.
15. Roessner, U., Wagner, C., Kopka, J., Trethewey, R. N., and Willmitzer, L. (2000) Simultaneous analysis of metabolites in potato tuber by gas chromatography-mass spectrometry. *Plant J.* 23, 131–142.
16. Schauer, N., Steinhauser, D., Strelkov, S., Schomburg, D., Allison, G., Moritz, T., Lundgren, K., Roessner-Tunali, U., Forbes, M., Willmitzer, L., Fernie, A., and Kopka, J. (2005) GC-MS libraries for the rapid identification of metabolites in complex biological samples. *FEBS Lett.* 579, 1332–1337.

17. Wagner, C., Sefkow, M., and Kopka, J. (2003) Construction and application of a mass spectral and retention time index database generated from plant GC/EI-TOF-MS metabolite profiles. *Phytochemistry* 62, 887–900.

18. Welthagen, W., Shellie, R. A., Spranger, J., Ristow, M., Zimmermannn, R., and Fiehn, O. (2005) Comprehensive two-dimensional gas chromatography-time-of-flight mass spectrometry (GC x GC-TOF) for high resolution metabolomics: biomarker discovery on spleen tissue extracts of obese NZO compared to lean C57BL/6 mice. *Metabolomics* 1, 65–73.

19. Huhman, D., and Sumner, L. (2002) Metabolic profiling of saponins in *Medicago sativa* and *Medicago truncatula* using HPLC coupled to an electrospray ion-trap mass spectrometer. *Phytochemistry* 59, 347–360.

20. Sumner, L. W. (2006) Current status and forward looking thoughts on LC-MS metabolomics, in *Biotechnology in Agriculture and Forestry: Plant Metabolomics* (Saito, K., Dixon, R.A., Willmitzer, L., Ed.), Springer-Verlag, Berlin, Vol. 57, pp. 21–32.

21. Sumner, L. W., Huhman, D. V., Urbanczyk-Wochniak, E., and Lei, Z. (2007) Methods, Applications, and Concepts of Metabolic Profiling: Secondary metabolism, in Plant System Biology, Fernie, A., Baginsky, S. Eds, Bierkenhauser-Verlag, Berlin, Germany 195–212. (ISBN 13: 978-3-7643-7261-3).

22. Tolstikov, V. V., and Fiehn, O. (2002) Analysis of highly polar compounds of plant origin: combination of hydrophilic interaction chromatography and electrospray ion trap mass spectrometry. *Anal. Biochem.* 301, 298–307.

23. Takats, Z., Wiseman, J. M., Gologan, B., and Cooks, R. G. (2004) Mass spectrometry sampling under ambient conditions with desorption electrospray ionization. *Science* 306, 471–473.

24. Bino, R. J., Hall, R. D., Fiehn, O., Kopka, J., Saito, K., Draper, J., Nikolau, B. J., Mendes, P., Roessner-Tunali, U., Beale, M. H., Trethewey, R. N., Lange, B. M., Wurtele, E. S., and Sumner, L. W. (2004) Potential of metabolomics as a functional genomics tool. *Trends Plant Sci.* 9, 418–425.

25. Birkemeyer, C., Kolasa, A., and Kopka, J. (2003) Comprehensive chemical derivatization for gas chromatography-mass spectrometry-based multi-targeted profiling of the major phytohormones. *J. Chromatogr.* A 993, 89–102.

26. Muller, A., Duchting, P., and Weiler, E. (2002) A multiplex GC-MS/MS technique for the sensitive and quantitative single-run analysis of acidic phytohormones and related compounds, and its application to Arabidopsis thaliana. *Planta* 216, 44–56.

27. Huhman, D., Berhow, M., and Sumner, L. (2005) Quantification of saponins in aerial and subterranean tissues of Medicago truncatula. *J. Agric. Food Chem.* 53, 1914–1920.

28. Frydman, A., Weisshaus, O., Bar-Peled, M., Huhman, D. V., Sumner, L. W., Marin, F. R., Lewinsohn, E., Fluhr, R., Gressel, J., and Eyal, Y. (2004) Citrus fruit bitter flavors: isolation and functional characterization of the gene encoding a 1,2 rhamnosyltransferase, a key enzyme in the biosynthesis of the bitter flavonoids of citrus. *Plant J.* 40, 88–100.

29. Liu, C., Huhman, D., Sumner, L., and Dixon, R. (2003) Regiospecific hydroxylation of isoflavones by cytochrome p450 81E enzymes from *Medicago truncatula*. *Plant J.* 36, 471–484.

30. Baggett, B. R., Cooper, J. D., Hogan, E. T., Carper, J., Paiva, N. L., and Smith, J. T. (2002) Profiling isoflavonoids found in legume root extracts using capillary electrophoresis. *Electrophoresis* 23, 1642–1651.

31. Zhang, J., Broeckling, C., Blancaflor, E., Sledge, M., Sumner, L., and Wang, Z. (2005) Overexpression of WXP1, a putative *Medicago truncatula* AP2 domain-containing transcription factor gene, increases cuticular wax accumulation and enhances drought tolerance in transgenic alfalfa (*Medicago sativa*). *Plant J.* 42, 689–707.

32. Wilson, I., Nicholson, J., Castro-Perez, J., Granger, J., Johnson, K., Smith, B., and Plumb, R. (2005) High resolution "ultra performance" liquid chromatography coupled to oa-TOF mass spectrometry as a tool for differential metabolic pathway profiling in functional genomic studies. *J. Proteome Res.* 4, 591–598.

33. Danielsson, R., Bylund, D., and Markides, K. (2002) Matched filtering with background suppression for improved quality of base peak chromatograms and mass spectra in liquid chromatography-mass spectrometry. *Anal. Chim. Acta* 454, 167–184.

34. Duran, A. L., Yang, J., Wang, L., and Sumner, L. W. (2003) Metabolomics spectral formatting, alignment and conversion tools (MSFACTs). *Bioinformatics* 19, 2283–2293.

35. Smith, C., Want, E., O'Maille, G., Abagyan, R., and Siuzdak, G. (2006) XCMS: processing mass spectrometry data for metabolite profiling using nonlinear peak alignment, matching, and identification. *Anal. Chem.* 78, 779–787.

36. Kopka, J., Schauer, N., Krueger, S., Birkemeyer, C., Usadel, B., Bergmuller, E., Dormann, P., Weckwerth, W., Gibon, Y., Stitt, M., Willmitzer, L., Fernie, A. R., and Steinhauser, D. (2005) GMD@CSB.DB: the Golm Metabolome Database. *Bioinformatics* 21, 1635–1638.

37. Halket, J., Przyborowska, A., Stein, S., Mallard, W., Down, S., and Chalmers, R. (1999) Deconvolution gas chromatography/mass spectrometry of urinary organic acids–potential for pattern recognition and automated identification of metabolic disorders. *Rapid Commun. Mass Spectrom.* 13, 279–284.

38. Nielsen, N.-P. V., Carstensen, J. M., and Smedsgaard, J. (1998) Aligning of single and multiple wavelength chromatographic profiles form chemometric data analysis using correlation optimized warping. *J. Chromatogr. A* 805, 17–35.

39. Miller, J. N., and Miller, J. C. (2000) *Statistics and Chemometrics for Analytical Chemistry*, Prentice Hall, Harlow, England.

40. Hotellin, H. (1933) Analysis of a complex of statistical variables into principal components. *J. Educ. Psychol.* 24, 417–441.

41. Raamsdonk, L., Teusink, B., Broadhurst, D., Zhang, N., Hayes, A., Walsh, M., Berden, J., Brindle, K., Kell, D., Rowland, J., Westerhoff, H., van Dam, K., and

Oliver, S. (2001) A functional genomics strategy that uses metabolome data to reveal the phenotype of silent mutations. *Nat. Biotechnol.* 19, 45–50.

42. Lachenbruch, P. A. (1975) *Discriminant Analysis*, Hafner Press, New York.

43. Cowan, J. D., and Sharp, D. H. (1988) Neural nets. *Q. Rev. Biophys.* 21, 365–427.

44. Cristianini, N., and Shawe-Taylor, J. (2000) *An Introduction to Support Vector Machines and Other Kernel-Based Learning Methods*, Cambridge University Press, Cambridge.

45. Goldberg, D. E. (1989) *Genetic Algorithms in Search, Optimization and Machine Learning.* Addison-Wesley, Reading, Mass.

46. Koza, J. R. (1992) *Genetic Programming: On the Programming of Computers by Means of Natural Selection*, MIT Press, Cambridge, Mass.

47. Goodacre, R., and Kell, D. B. (1996) Pyrolysis mass spectrometry and its applications in biotechnology. *Curr. Opin. Biotechnol.* 7, 20–28.

48. McGovern, A. C., Broadhurst, D., Taylor, J., Kaderbhai, N., Winson, M. K., Small, D. A., Rowland, J. J., Kell, D. B., and Goodacre, R. (2002) Monitoring of complex industrial bioprocesses for metabolite concentrations using modern spectroscopies and machine learning: application to gibberellic acid production. *Biotechnol. Bioeng.* 78, 527–538.

49. Shaw, A. D., Winson, M. K., Woodward, A. M., McGovern, A. C., Davey, H. M., Kaderbhai, N., Broadhurst, D., Gilbert, R. J., Taylor, J., Timmins, E. M., Goodacre, R., Kell, D. B., Alsberg, B. K., and Rowland, J. J. (2000) Rapid analysis of high-dimensional bioprocesses using multivariate spectroscopies and advanced chemometrics. *Adv. Biochem. Eng. Biotechnol.* 66, 83–113.

50. Goodacre, R. (2005) Making sense of the metabolome using evolutionary computation: seeing the wood with the trees. *J. Exp. Bot.* 56, 245–254.

51. Thimm, O., Blasing, O., Gibon, Y., Nagel, A., Meyer, S., Kruger, P., Selbig, J., Muller, L., Rhee, S., and Stitt, M. (2004) MAPMAN: a user-driven tool to display genomics data sets onto diagrams of metabolic pathways and other biological processes. *Plant J.* 37, 914–939.

52. Usadel, B., Nagel, A., Thimm, O., Redestig, H., Blaesing, O. E., Palacios-Rojas, N., Selbig, J., Hannemann, J., Piques, M. C., Steinhauser, D., Scheible, W.-R., Gibon, Y., Morcuende, R., Weicht, D., Meyer, S., and Stitt, M. (2005) Extension of the visualization tool mapman to allow statistical analysis of arrays, display of coresponding genes, and comparison with known responses. *Plant Physiol.* 138, 1195–1204.

53. Urbanczyk-Wochniak, E., Usadel, B., Thimm, O., Nunes-Nesi, A., Carrari, F., Davey, M., Blasing, O., Kowalczyk, M., Weicht, D., Polinceusz, A., Meyer, S., Stitt, M., and Fernie, A. R. (2006) Conversion of MapMan to allow the analysis of transcript data from Solanaceous species: effects of genetic and environmental alterations in energy metabolism in the leaf. *Plant Mol. Biol.* 60, 773–792.

54. Mehrotra, B., and Mendes, P. (2006) Bioinformatics approaches to integrate metabolomics and other systems biology data, in *Biotechnology in Agriculture*

and Forestry: Plant Metabolomics (Saito, K., Dixon, R.A., Willmitzer, L., Ed.), Springer-Verlag, Berlin, Vol. 57, pp. 105–115.

55. Gonzales, M. D., Arccchuleta, E., Farmer, A., Gajendran, K., Gant, D., Shoemaker, R., Beavis, W. D., and Waugh, M. E. (2005) The Legume Information System (LIS): an integrated information resource for comparative legume biology. *Nucleic Acids Res.* 33, D660–D665.

56. Lange, B., and Ghassemian, M. (2005) Comprehensive post-genomic data analysis approaches integrating biochemical pathway maps. *Phytochemistry* 66, 413–451.

57. Brazma, A., Hingamp, P., Quackenbush, J., Sherlock, G., Spellman, P., Stoeckert, C., Aach, J., Ansorge, W., Ball, C. A., Causton, H. C., Gaasterland, T., Glenisson, P., Holstege, F. C. P., Kim, I. F., Markowitz, V., Matese, J. C., Parkinson, H., Robinson, A., Sarkans, U., Schulze-Kremer, S., Stewart, J., Taylor, R., Vilo, J., and Vingron, M. (2001) Minimum information about a microarray experiment (MIAME)-toward standards for microarray data. *Nat. Genet.* 29, 365–371.

58. Taylor, C. F., Paton, N. W., Garwood, K. L., Kirby, P. D., Stead, D. A., Yin, Z., Deutsch, E. W., Selway, L., Walker, J., Riba-Garcia, I., Mohammed, S., Deery, M. J., Howard, J. A., Dunkley, T., Aebersold, R., Kell, D. B., Lilley, K. S., Roepstorff, P., Yates, J. R., Brass, A., Brown, A. J. P., Cash, P., Gaskell, S. J., Hubbard, S. J., and Oliver, S. G. (2003) A systematic approach to modeling, capturing, and disseminating proteomics experimental data. *Nat. Biotechnol.* 21, 247–254.

59. Orchard, S., Hermjakob, H., and Apweiler, R. (2003) The proteomics standards initiative. *Proteomics* 3, 1374–1376.

60. Orchard, S., Hermjakob, H., Taylor, C., Potthast, F., Jones, P., Zhu, W., Julian, R., and Apweiler, R. (2005) Further steps in standardisation. Report of the second annual Proteomics Standards Initiative Spring Workshop (Siena, Italy 17–20 April 2005). *Proteomics* 5, 3552–3555.

61. Orchard, S., Hermjakob, H., Binz, P., Hoogland, C., Taylor, C., Zhu, W., Julian, R., and Apweiler, R. (2005) Further steps towards data standardisation: the Proteomic Standards Initiative HUPO 3(rd) annual congress, Beijing 25–27(th) October, 2004. *Proteomics* 5, 337–339.

62. Bino, R. J., Hall, R. D., Fiehn, O., Kopka, J., Saito, K., Draper, J., Nikolau, B. J., Mendes, P., Roessner-Tunali, U., Beale, M. H., Trethewey, R. N., Lange, B. M., Wurtele, E. S., and Sumner, L. W. (2004) Potential of metabolomics as a functional genomics tool. *Trends Plant Sci.* 9, 418–425.

63. Jenkins, H., Hardy, N., Beckmann, M., Draper, J., Smith, A., Taylor, J., Fiehn, O., Goodacre, R., Bino, R., Hall, R., Kopka, J., Lane, G., Lange, B., Liu, J., Mendes, P., Nikolau, B., Oliver, S., Paton, N., Rhee, S., Roessner-Tunali, U., Saito, K., Smedsgaard, J., Sumner, L., Wang, T., Walsh, S., Wurtele, E., and Kell, D. (2004) A proposed framework for the description of plant metabolomics experiments and their results. *Nat. Biotechnol.* 22, 1601–1606.

64. Jenkins, H., Johnson, H., Kular, B., Wang, T., and Hardy, N. (2005) Toward supportive data collection tools for plant metabolomics. *Plant Physiol.* 138, 67–77.
65. Lindon, J., Nicholson, J., Holmes, E., Keun, H., Craig, A., Pearce, J., Bruce, S., Hardy, N., Sansone, S., Antti, H., Jonsson, P., Daykin, C., Navarange, M., Beger, R., Verheij, E., Amberg, A., Baunsgaard, D., Cantor, G., Lehman-McKeeman, L., Earll, M., Wold, S., Johansson, E., Haselden, J., Kramer, K., Thomas, C., Lindberg, J., Schuppe-Koistinen, I., Wilson, I., Reily, M., Robertson, D., Senn, H., Krotzky, A., Kochhar, S., Powell, J., van der Ouderaa, F., Plumb, R., Schaefer, H., Spraul, M., and (2005) Summary recommendations for standardization and reporting of metabolic analyses. *Nat. Biotechnol.* 23, 833–838.
66. Fiehn, O., Kristal, B., van Ommen, B., Sumner, L. W., Assuant-Sansone, S., Taylor, C., Hardy, N., and Kaddurah-Daouk, R. (2006) Establishing Reporting Standards for Metabolomic and Metabonomic Studies: A Call for Participation. *Omics* 10, 158–163.
67. Ball, C. A., Awad, I. A. B., Demeter, J., Gollub, J., Hebert, J. M., Hernandez-Boussard, T., Jin, H., Matese, J. C., Nitzberg, M., Wymore, F., Zachariah, Z. K., Brown, P. O., and Sherlock, G. (2005) The Stanford microarray database accommodates additional microarray platforms and data formats. *Nucleic Acids Res.* 33, D580–D582.
68. Barrett, T., Suzek, T. O., Troup, D. B., Wilhite, S. E., Ngau, W.-C., Ledoux, P., Rudnev, D., Lash, A. E., Fujibuchi, W., and Edgar, R. (2005) NCBI GEO: mining millions of expression profiles–database and tools. *Nucleic Acids Res.* 33, D562–D566.
69. Bairoch, A., Apweiler, R., Wu, C. H., Barker, W. C., Boeckmann, B., Ferro, S., Gasteiger, E., Huang, H., Lopez, R., Magrane, M., Martin, M. J., Natale, D. A., O'Donovan, C., Redaschi, N., and Yeh, L.-S. L. (2005) The Universal Protein Resource (UniProt). *Nucleic Acids Res.* 33, D154–D159.
70. Schauer, N., Steinhauser, D., Strelkov, S., Schomburg, D., Allison, G., Moritz, T., Lundgren, K., Roessner-Tunali, U., Forbes, M., Willmitzer, L., Fernie, A., and Kopka, J. (2005) GC-MS libraries for the rapid identification of metabolites in complex biological samples. *FEBS Lett.* 579, 1332–1337.
71. Fiehn, O., Wohlgemuth, G., Scholz, G. (2005) Setup and Annotation of Metabolomic Experiment by Intergrating Biological and Mass Spectrometric Metadata. In B. Ludascher, L. Raschid, eds, LNBI, Vol 3615. Springer-Verlag, Berlin, Germany, pp. 224–239.
72. Tikunov, Y., Lommen, A., de Vos, C. H. R, Verhoeven, H. A., Bino, R. J., Hall, R. D., Bovy, A. G. (2005) A Novel Approach for Nontargeted Data Analysis for Metabolomics. Large-Scale Profiling of Tomato Fruit Volatiles. *Plant Physiol.* 139, 1125–1137.
73. Broeckling, C., Reddy, I., Duran, A., Zhao, X., Sumner, L. (2006) MET-IDEA: Data Extraction Tool for Mass Spectrometry-Based Metabolomics. *Anal. Chem.* 78, 4334–4341.

74. Katajamaa, M., Miettinen, J., Oresic, M. (2006) MZmine: toolbox for processing and visualization of mass spectrometry based molecular profile data. *Bioinformatics* 22, 634–636.

75. Mueller, L. A., Zhang, P., Rhee, S. Y. (2003) AraCyc: A Biochemical Pathway Database for Arabidopsis. *Plant Physiology* 132: 453–460.

76. Tokimatsu, T., Sakurai, N., Suzuki, H., Ohta, H., Nishitani, K., Koyama, T., Umezawa, T., Misawa, N., Saito, K., Shibata, D. (2005) KaPPA-View. A Web-Baseed Analysis Tool for Integration of Transcript and Metabolite Data on Plant Metabolic Pathway Maps. *Plant Physiology* 138: 1289–1300.

77. Bajic, V. B., Veronika, M., Veladandi, P. S., Meka, A., Heng, M.-W., Rajaraman, K., Pan H., Swarup, S. (2005) Dragon Plant Biology Explorer. A Text-Mining Tool for Integrating Associations between Genetic and Biochemical Entities with Genome Annotation and Biochemical Terms Lists. *Plant Physiology* 138: 1914–1925.

78. Yang, Y., Engin, L., Wurtele, E. S., Cruz-Neira, C., Dickerson, J. A. (2005) Integration of metabolic networks and gene expression in virtual reality. *Bioinformatics* 21, 3645–3650.

21

KEGG Bioinformatics Resource for Plant Genomics Research

Ali Masoudi-Nejad, Susumu Goto, Takashi R. Endo, and Minoru Kanehisa

Summary

Kyoto Encyclopedia of Genes and Genomes (KEGG) is a bioinformatics resource for understanding biological function from a genomic perspective. It is a multispecies, integrated resource consisting of genomic, chemical, and network information, with cross-references to numerous outside databases and containing a complete set of building blocks (genes and molecules) and wiring diagrams (biological pathways) to represent cellular functions. KEGG consists of a suite of databases: PATHWAY, GENES/Sequence Similarity Database (SSDB), Biomolecular Relations in Information Transmission and Expression (BRITE), and LIGAND, which is a composite database of COMPOUND, DRUG, GLYCAN, REACTION, REPAIR, and ENZYME. Two new databases have been recently added to KEGG: DGENES (for draft genomes) and EGENES (for expressed-sequence tag [EST] data). EGENES is a knowledge base system for efficient analysis of organism-specific ESTs, including publicly available plant ESTs. EGENES links the genomic information with higher order functional information in a single database. The genomic information stored in EGENES is a collection of EST contigs, produced by assembling the public ESTs. In this chapter, we will introduce KEGG and discuss its importance for the plant research community by focusing on EGENES. Because all the resources in KEGG follow the same architecture and design, an appraisal of EGENES should give readers an idea of the available information stored in KEGG and how to use them efficiently.

Key Words: Plant metabolic pathway; EST assembly; plant EST clustering; functional annotation; gene index; transcriptome analysis; EGENES.

From: *Methods in Molecular Biology, vol. 406: Plant Bioinformatics: Methods and Protocols*
Edited by: D. Edwards © Humana Press Inc., Totowa, NJ

1. Introduction

The availability of different types of high-throughput experimental data in the late 1990s has expanded the role of bioinformatics and facilitated the analysis of higher order functions involving various cellular processes. For the last 10 years, automatic sequencing has had an enormous impact as it has been at the forefront of high-throughput generation of various biological data such as expressed-sequence tags (ESTs) and single-nucleotide polymorphisms (SNPs). To date, the genome sequences for over 266 different species have been published *(1)*. However, only two draft sequences among these are plant genomes. There are currently over 1000 species across various taxonomic groups for which more than 29,425,525 ESTs and 12,351,453 genome survey sequences (GSSs) have been generated (National Center for Biotechnology Information [NCBI] dbEST, release 092305, September, 2005).

An increase in the amount of large-scale sequence data does not necessarily lead to an increase in biological knowledge unless it is accompanied with new or improved tools for sequence analysis *(2)*. In general, the poorly organized nature of these data makes them difficult to interpret within a genomic context and precludes even simple comparative analyses. Common problems include significant redundancy in the data sets, contamination, poor quality, and a lack of consistent annotation between projects. An effective way to overcome these problems is to group ESTs into clusters (representing unique genes), which may be subsequently fed into downstream annotation pipelines.

To increase our understanding of cellular processes from genomic information, scientists have created pathway databases such as Kyoto Encyclopedia of Genes and Genomes (KEGG) *(3)* and Encyclopedia of Escherichia coli K-12 Genes and Metabolism (EcoCyc) *(4)* in the past decade. Whereas most databases concentrate on molecular properties (e.g., sequences, 3D structures, motifs, and gene expressions), these databases tackle complex cellular pathways, such as metabolism, signal transduction, and cell cycle, by storing the corresponding networks of interacting molecules in computerized forms, often as graphical pathway diagrams.

KEGG (http://www.genome.jp/kegg/) is a bioinformatics resource for understanding higher order functional meanings and utilities of the cell or the organism from its genome information. KEGG *(5)* integrates current knowledge on molecular interaction networks such as pathways and complexes (PATHWAY database), information about genes and proteins generated by high-throughput experiments in genomics and proteomics (GENES database/Sequence Similarity Database [SSDB]), and information about chemical compounds and reactions that are relevant to cellular processes

(LIGAND database) *(6)*. In addition, KEGG provides facilities to infer higher order functions from molecular-level information (Biomolecular Relations in Information Transmission and Expression [BRITE] database). Two new databases have been added to KEGG (release 37.0, January 2006): DGENES, for the draft genomes, and EGENES, which includes plant species and covers the plant ESTs produced by ongoing large-scale EST projects by international groups.

EGENES (http://www.genome.jp/kegg-bin/create_kegg_menu?category= plants_egenes) is a knowledge base system for efficient analysis of organism-specific ESTs, including publicly available plant ESTs. EGENES links the genomic information with higher order functional information in a single database. The genomic information stored in EGENES is a collection of EST contigs, produced from assembling the public ESTs *(7)*. EGENES also provides a gene/EST index for each genome. Functional assignments in EGENES follow the same rules for functional assignments in other KEGG resources. It is a process of linking a set of genes/transcripts in each genome with a network of interacting molecules in the cell, such as a pathway or a complex, representing a higher order biological function.

In this chapter, we will introduce KEGG and discuss its importance for the plant research community by focusing on EGENES. Because all the resources in KEGG follow the same architecture and design, an appraisal of EGENES should give readers an idea of the available information stored in KEGG and how to use them efficiently.

2. Materials

2.1. Database Design and Implementation

One of the main problems with EST data is poor quality and contamination (vectors, repeats, and low-complexity regions), which cannot be completely avoided. Many groups have respectively used different vectors and repeat databases for decontamination analysis, but none of these databases cover all contaminants. We used an improved protocol for clustering and assembling the ESTs to overcome the above-mentioned problems *(7,8)*. EST contigs for each species were used for developing EGENES and automatically annotating the KEGG pathway. EGENES provides a gene index for each plant species. The result of this stringent clustering and assembly is a set of unique, virtual transcripts, or EST contigs. These contigs can be used to functionally annotate genes, to facilitate splice alignment analysis, to link the transcripts to genetic and physical maps, and to provide a resource for comparative and functional genomic analysis.

The KEGG PATHWAY database is a collection of graphical diagrams (pathway maps) representing molecular interaction networks in various cellular processes. Each reference pathway is manually drawn and updated. In this study, we built a metabolic pathway database for 25 plants with most of the EST entries in NCBI's dbEST, using in-house software. The software allows the generation of pathway databases using automatic KEGG Orthology (KO) assignment and the Enzyme Commission (EC) numbers of the genes. The KO identifier, or the K number, is a common identifier for linking the genomic information in the GENES database and the network information in the PATHWAY database in KEGG. Organism-specific pathways were computationally generated based on the KO assignment in individual genomes. The pathway nodes represented by rectangles in the KEGG reference pathway maps are given KO identifiers, so that organism-specific pathways are automatically generated once each genome is annotated with KO's. Furthermore, the KO system includes additional classifications of orthologous groups in specific protein families, which are provided in the KEGG network hierarchy.

2.2. Content of EGENES

The current version of KEGG PATHWAY (release 37.0, January 2006) consists of 28 entries for plants. This includes automatic annotation and pathway mapping of draft genomes (DGENES) and EST contigs (EGENES). **Table 1** presents the species included in the initial version of EGENES, along with corresponding KEGG codes used, covering five classes (monocotyledon, dicotyledon, mosses, red alga, and conifers) and 12 families (mustard, rue, mallow, pea, willow, daisy, nightshade, grape, grasses, pine, mosses, and algae). Entries (genomes) with draft sequences have been allocated three character codes, and entries with ESTs have four character codes starting with "e" (e.g., "ath" represents the published *Arabidopsis thaliana* genome and "eath" the *Arabidopsis thaliana* ESTs [EGENES]). The data for each species in EGENES have been organized into two distinct parts, pathway maps and gene catalogs. Clicking on the species code provides relevant information about each species in a new window, including taxonomy data (*see* **Fig. 1A**).

2.2.1. Pathway Maps

The pathway maps for each species were derived from the automatic annotation based on KEGG's reference pathway. The latest release of KEGG contains 266 reference pathways, all of which have been manually curated (*see* **Table 2**). Each reference pathway can be viewed as a network of enzymes or a network of EC numbers. Knowledge-based prediction of metabolic pathways

Table 1
Species included in current version of the EGENES database (release 40.0, Dec. 2006)

CLASS	FAMILY	SPECIES (common name)	CODE
Dicotyledons	Mustard	Arabidopsis thaliana (thale cress)	ath
		Arabidopsis thaliana (thale cress) (EST)	eath
		Brassica napus (rape) (EST)	ebna
	Rue	Citrus sinensis (Valencia orange) (EST)	ecsi
	Mallow	Gossypium arboreum (cotton) (EST)	egar
		Gossypium raimondii (EST)	egra
	Pea	Glycine max (soybean) (EST)	egma
		Lotus corniculatus (lotus) (EST)	elco
		Medicago truncatula (barrel medic) (EST)	emtr
	Willow	Populus tremula (aspen) (EST)	eptp
		Populus balsamifera (poplar) (EST)	epba
	Daisy	Helianthus annuus (sunflower) (EST)	ehan
		Lactuca sativa (lettuce) (EST)	elsa
	Nightshade	Lycopersicon esculentum (tomato) (EST)	eles
		Solanum tuberosum (potato) (EST)	estu
	Grape	Vitis vinifera (wine grape) (EST)	evvi
Monocotyledons	Grass	Oryza sativa japonica (Japanese rice)	osa
		Oryza sativa (rice) (EST)	eosa
		Hordeum vulgare (barley) (EST)	ehvu
		Triticum aestivum (wheat) (EST)	etae
		Zea mays (maize) (EST)	ezma
		Saccharum officinarum (sugarcane) (EST)	esof
		Sorghum bicolor (sorghum) (EST)	esbi
Others	Pine	Pinus taeda (loblolly pine) (EST)	epta
		Picea glauca (white spruce) (EST)	epgl
	Mosses	Physcomitrella patens (EST)	eppa
		Physcomitrella patens subsp. patens (EST)	eppp
	Red alga	Cyanidioschyzon merolae	cme

involves the matching of genes in the genome against enzymes in the KEGG reference pathway. Once enzyme genes are identified in the genome based on sequence similarity and positional correlation of genes, and the EC numbers and KO are properly assigned, organism-specific pathways could be constructed computationally by correlating genes/sequences in the genome with gene products (enzymes) in the reference pathways, according to the matching KO identifier and EC numbers. **Figure 1B** shows a typical pathway diagram in KEGG. Each pathway consists of nodes and edges. Here, a node is an enzyme (gene product) represented by a box labeled with its EC number. An edge connects two enzymes via the compound that is both a product of one enzyme and a substrate of the other. Because a network of enzymes is equivalent to a network of genes coding for the enzyme in each organism, this graphical representation is most useful in superimposing the genomic information onto the knowledge of metabolic pathways and helps to deduce metabolism for each organism. Because the metabolic pathway, especially for intermediary metabolism, is well conserved among most organisms, it is possible to manually draw one reference pathway and consequently computationally generate many organism-specific pathways. Using KO identifiers along with EC numbers in the metabolic pathways helps to distinguish between multiple genes that match one EC number: for example, different subunits of an enzyme complex, different genes expressed under different conditions, or paralogs with similar catalyzing function but catalyzing different reactions.

A

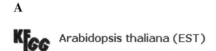

Fig. 1A Typical information stored in the Kyoto Encyclopedia of Genes and Genomes (KEGG) databases. (**A**) The entry point for each species in the EGENES database. This includes pathway data, gene indices (Gene catalogs), taxonomy information, and source sequences used for sequence assembly and analysis. (**B**) A typical pathway diagram in KEGG (glycolysis pathway for *Arabidopsis thaliana* based on expressed-sequence tag [EST] data).

B

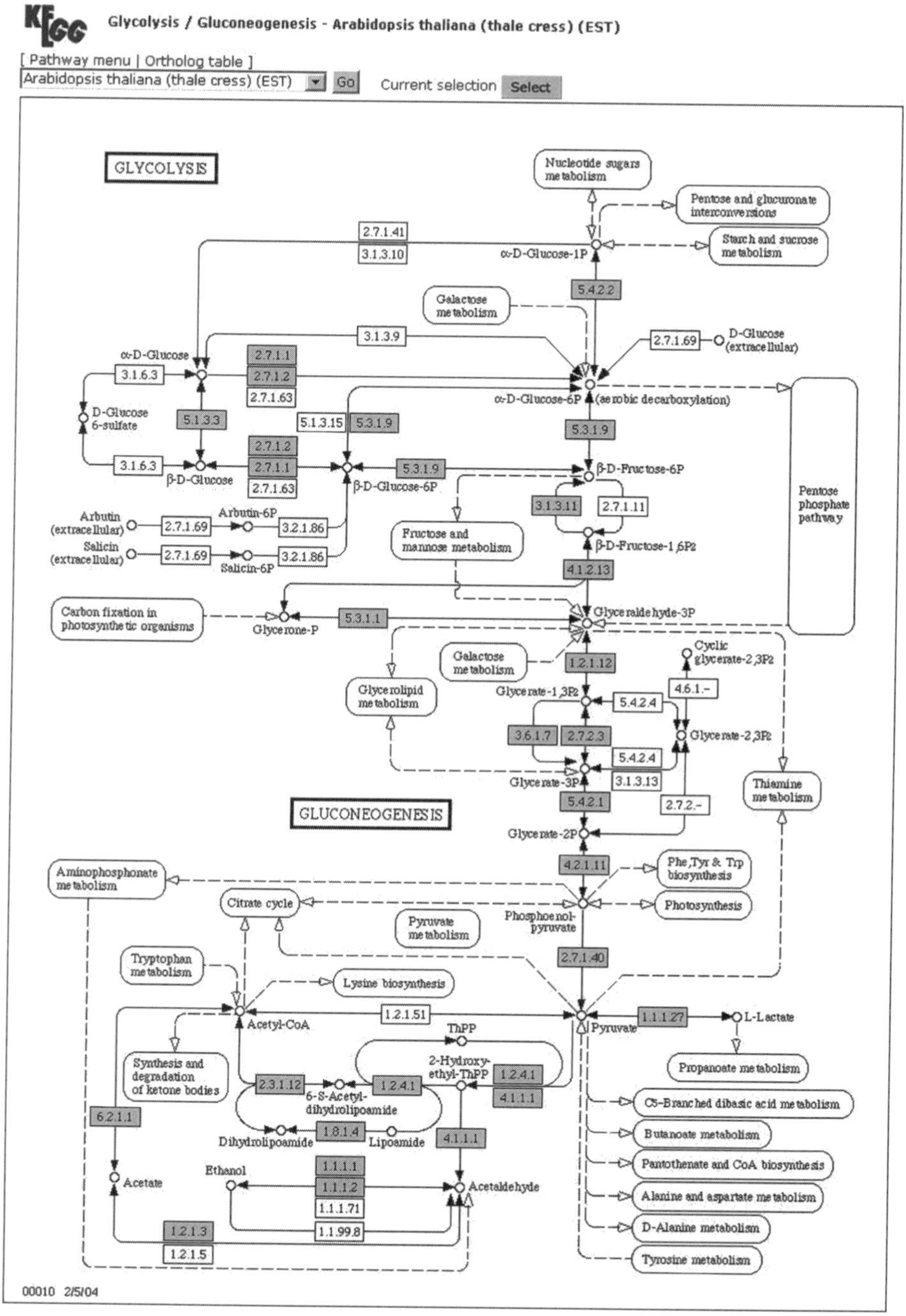

KEGG Glycolysis / Gluconeogenesis - Arabidopsis thaliana (thale cress) (EST)

[Pathway menu | Ortholog table]

Arabidopsis thaliana (thale cress) (EST) ▼ Go   Current selection Select

GLYCOLYSIS

Nucleotide sugars metabolism

Pentose and glucuronate interconversions

2.7.1.41
3.1.3.10   α-D-Glucose-1P

Starch and sucrose metabolism

5.4.2.2

Galactose metabolism

3.1.3.9

2.7.1.69   D-Glucose (extracellular)

α-D-Glucose
3.1.6.3

2.7.1.1
2.7.1.2
2.7.1.63

α-D-Glucose-6P (aerobic decarboxylation)

D-Glucose 6-sulfate

5.1.3.3   5.1.3.15   5.3.1.9

5.3.1.9

3.1.6.3   β-D-Glucose

2.7.1.2
2.7.1.1
2.7.1.63

5.3.1.9   β-D-Glucose-6P

β-D-Fructose-6P

Arbutin (extracellular)   2.7.1.69   Arbutin-6P   3.2.1.86

3.1.3.11   2.7.1.11

Salicin (extracellular)   2.7.1.69   Salicin-6P   3.2.1.86

Fructose and mannose metabolism

β-D-Fructose-1,6P2

4.1.2.13

Carbon fixation in photosynthetic organisms   Glycerone-P   5.3.1.1

Glyceraldehyde-3P

Pentose phosphate pathway

Galactose metabolism

Cyclic glycerate-2,3P2

1.2.1.12

Glycerate-1,3P2   5.4.2.4   4.6.1.-

Glycerolipid metabolism

3.6.1.7   2.7.2.3   5.4.2.4   Glycerate-2,3P2

GLUCONEOGENESIS

Glycerate-3P   3.1.3.13

5.4.2.1   Thiamine metabolism

2.7.2.-

Glycerate-2P

Aminophosphonate metabolism

4.2.1.11   Phe,Tyr & Trp biosynthesis

Citrate cycle   Pyruvate metabolism   Phosphoenol-pyruvate   Photosynthesis

Tryptophan metabolism   2.7.1.40

Lysine biosynthesis

Acetyl-CoA   1.2.1.51   ThPP   Pyruvate   1.1.1.27   L-Lactate

Synthesis and degradation of ketone bodies

2-Hydroxy-ethyl-ThPP

2.3.1.12   1.2.4.1   1.2.4.1

6.2.1.1   6-S-Acetyl-dihydrolipoamide   4.1.1.1

Propanoate metabolism

C5-Branched dibasic acid metabolism

1.8.1.4   Lipoamide   4.1.1.1   Butanoate metabolism

Dihydrolipoamide

Pantothenate and CoA biosynthesis

Ethanol   1.1.1.1
1.1.1.2
1.1.1.71
1.1.99.8   Acetaldehyde

Alanine and aspartate metabolism

D-Alanine metabolism

Acetate   1.2.1.3
1.2.1.5

Tyrosine metabolism

00010 2/5/04

Fig. 1B

Table 2
KEGG database status (release 40.0, Dec. 2006)

Number of pathways	25,733	(PATHWAY database)
Number of reference pathways	266	(PATHWAY database)
Number of ortholog tables	87	(PATHWAY database)
Number of organisms	307	(GENOME database)
Number of genes	940,473	(GENES database)
Number of ortholog clusters	38,655	(SSDB database)
Number of KO assignments	6,603	(KO database)
Number of chemical compounds	12,874	(COMPOUND database)
Number of glycans	11,061	(GLYCAN database)
Number of chemical reactions	6,469	(REACTION database)
Number of reactant pairs	6,421	(RPAIR database)

Pathway maps have been hierarchically organized into four main categories, each category of which consists of many subcategories.

1. Metabolism
2. Genetic Information Processing
3. Environmental Information Processing
4. Cellular Processes

Each subcategory is in fact a collection of pathway diagrams, consisting of many enzymes, reactions, and compounds (http://www.genome.jp/kegg-bin/show_organism?menu_type=pathway_maps&org=eath). **Table 3** is a summary of the pathways in the current version of EGENES. In total, 185 unique diagrams were created for all plants, organized into 134 in Metabolism, 19 in Genetic Information Processing, 21 in Environmental Information Processing, and 11 in Cellular Processes.

To assess whether the coverage of EST contigs is sufficient for computationally predicting pathways, we checked the number of pathway nodes present in the same pathway between the manually curated pathways for Arabidopsis (ath) and rice (osa), and the predicted pathways for these plants based on ESTs (eath and eosa). In many cases, we found that EST coverage is better than that of genomic DNA. In total, we found 251 pathway nodes unique to the EST-based rice metabolic pathway (eosa) but not found in the rice pathway based on genomic DNA. For Arabidopsis, we found 295 unique pathway nodes found only in the EST-based predicted pathway. **Figure 2A and B** shows the diagrams of genomic (ath) and EST-based (eath) Reductive Carboxylate Cycle (CO_2 fixation) pathways for Arabidopsis. The diagram for the genomic-based

Table 3
Summary of pathway in EGENES (release 40.0, Dec. 2006)

1. Metabolism	**134**
1.1 Carbohydrate Metabolism Overview of biosynthetic pathways	17
1.2 Energy Metabolism	8
1.3 Lipid Metabolism	11
1.4 Nucleotide Metabolism	2
1.5 Amino Acid Metabolism	16
1.6 Metabolism of Other Amino Acids	9
1.7 Glycan Biosynthesis and Metabolism Composite Structure Map	15
1.8 Biosynthesis of Polyketides and Nonribosomal Peptides	9
1.9 Metabolism of Cofactors and Vitamins	11
1.10 Biosynthesis of Secondary Metabolites	16
1.11 Biodegradation of Xenobiotics	20
2. Genetic Information Processing	**19**
2.1 Transcription	7
2.2 Translation	5
2.3 Folding, Sorting and Degradation	6
2.4 Replication and Repair	1
3. Environmental Information Processing	**21**
3.1 Membrane Transport	4
3.2 Signal Transduction	10
3.3 Ligand-Receptor Interaction	7
4. Cellular Processes	**11**
4.1 Cell Motility	3
4.2 Cell Growth and Death	2
4.3 Cell Communication	3
4.4 Immune System	1
4.5 Development	1
4.6 Behavior	1

predicted pathway has only seven nodes (EC: 4.1.1.31, 6.2.1.1, 4.2.1.3 [twice], 1.1.1.42, 4.2.1.2, and 1.1.1.37), whereas the same pathway for EST sequence has 11 nodes present (EC: 4.1.1.31, 6.2.1.1, 4.2.1.3 [twice], 1.1.1.42, 4.2.1.2, 1.1.1.37, 2.7.9.2, 4.1.3.6, 6.2.1.5, and 1.3.99.1). These last four enzymes are unique to "eath", and not found in "ath".

In some cases, the number of pathway hits was similar, but consisted of hits unique to one or the other, such as the Citrate Cycle (TCA) in rice, where both "eosa" and "osa" have 16 hits, but these are unique to one and not the other. In Arabidopsis, both "eath" and genomic-based "ath" have 22 hits, but both contain six unique hits not found in each other. This may be because either the sequences or the annotations of these genomes are still incomplete.

We also found some other cases where genomic-based prediction gave more hits than EST-based prediction. For example, in the pathway for the metabolism of carbon fixation, we have 30 hits, with three hits unique to "ath" and 27 common to both "ath" and "eath" (http://www.genome.jp/dbget-bin/

A

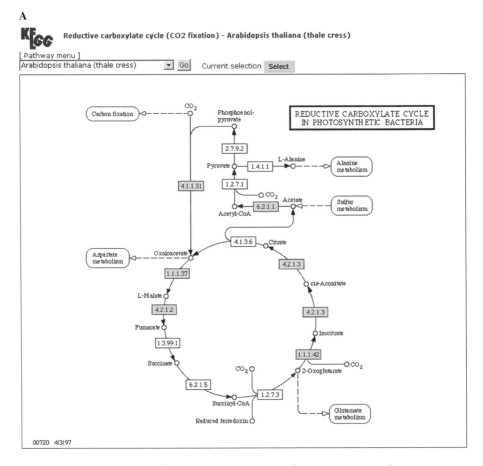

Fig. 2A. Comparison of the pathway coverage between expressed-sequence tags (ESTs) and genomic data. Coverage of the data for the Arabidopsis Reductive Carboxylate Cycle pathway between genomic and ESTs data, respectively (**A** and **B**).

B

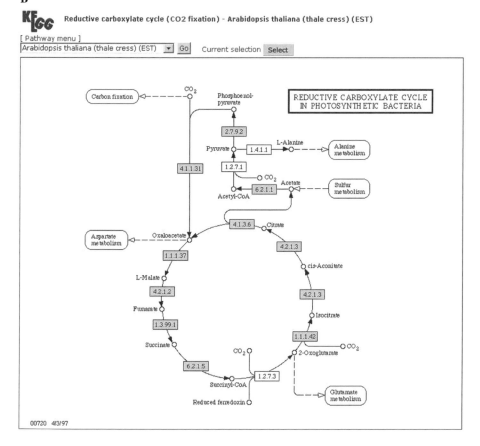

Fig. 2B

get_pathway?org_name=eath&mapno=00710 and http://www.genome.jp/ dbget-bin/get_pathway?org_name=ath&mapno=00710). This is because the numbers of sequenced ESTs, the size of the cDNA library, and the type of tissue where ESTs have been produced are insufficient to cover all pathway nodes.

2.2.2. Gene Catalogs

The gene catalog is the KEGG functional classification, which includes functional hierarchies of KEGG pathways, ortholog groups (KO), and protein families. Historically, the integration of pathway information and genomic information was first achieved in KEGG using EC numbers. Once the EC numbers were correctly assigned to enzyme encoding genes in the genome, organism-specific pathways could be generated automatically by matching against the networks of EC numbers (enzymes) in the reference metabolic pathways. The EC system uses four numbers that define the function of an enzyme in a hierarchical manner. However, there can be different enzymes that have the same EC number and function in different pathways. This means that, in cases where EC numbers refer to more than one enzyme, we have to determine which enzyme is appropriate to particular KEGG pathways. However, to incorporate non-metabolic pathways and to overcome various problems inherent in the enzyme nomenclature, we introduced a new scheme based on the KEGG ortholog replacing the EC numbers.

KO is based on computational analysis, as well as manual curation, decomposing all genes in the complete genomes into sets of orthologs. Here, two genes are considered as orthologs, or belonging to the same KO group, when they are mapped to the same KEGG pathway node. KO will be used to explore unknown pathways as well as to characterize all known pathways. Thus, EC numbers are used for mapping enzyme genes to metabolic pathways whereas KO is used for classifying all genes.

KO is classified according to the KEGG network hierarchy; thus, all KO identifiers are placed at the fourth level in this hierarchy. The KO identifier is a common identifier for linking the genomic information in the GENES database and the network information in the PATHWAY database. The pathway nodes represented by rectangles in the KEGG reference pathway maps are given KO identifiers, so that organism-specific pathways can be automatically generated once each genome is annotated with KO. There are different types of hierarchies often utilized for specific groups of proteins, such as the EC number hierarchy for enzymes. These functional/structural classifications of proteins are considered orthogonal to the fourth level of the network hierarchy and are being compiled separately in KEGG as protein families (http://www.genome.jp/kegg-bin/show_organism?menu_type=gene_catalogs&org=eath).

3. Methods

3.1. User Interface

The KEGG database has a hierarchical structure, so one can navigate KEGG either by selecting a topic of interest from the hierarchical table of contents, by clicking on the pathway maps, or through several structured database hyperlinks or searches. As one follows the path down into the resource he/she is presented with increasing levels of detail associated with the pathway, its constituent reactions, the participating complexes and macromolecules, and the relationships among the pathways in various species. At each level, there is information designed for human browsing, as well as machine-readable information. As an example, starting from **Table 1** (KEGG organism table for plants, http://www.genome.jp/kegg-bin/create_kegg_menu?category=plants_egenes), we can start with Arabidopsis EST-based data (eath). Clicking on each species' character code opens a new window, which is the starting point for exploring the information specific to that species (*see* **Fig. 1A** for details).

3.2. Exploring the Pathway Maps

Clicking on the "Pathway maps" as in **Fig. 1A** opens a hierarchical context window, which displays all the pathways for which a reference map is present for the current species (http://www.genome.jp/kegg-bin/show_organism?menu_type=pathway_maps&org=eath). Each number to the left of the pathway map name is the pathway map ID. Clicking on each pathway will show the diagram for that pathway including all nodes, edges, and compounds. For example, clicking on "Oxidative Phosphorylation" under Energy Metabolism will show the corresponding pathway diagram (*see* **Fig. 3A**). Colored boxes are the pathway nodes (enzyme) that are present in Arabidopsis genes (ESTs contigs). White boxes (EC: 3.6.3.10 and 2.7.4.1)

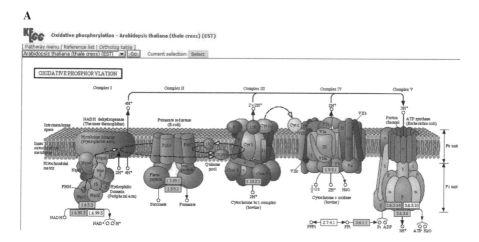

Fig. 3A. Pathway maps hierarchy in Kyoto Encyclopedia of Genes and Genomes (KEGG). **(A)** The pathway diagram for oxidative phosphorylation of Arabidopsis based on expressed-sequence tag (EST) data. **(B)** The information stored for pathway node (1.9.3.1) and the EST hit, contig 1542. **(C)** The graphical alignment of members of contig 1542.

B

KEGG Arabidopsis thaliana (thale cress) (EST): 1542

Entry	1542 Contig A.thaliana_est
KO	KO: K02265 cytochrome c oxidase subunit Vb
Pathway	PATH: eath00190 Oxidative phosphorylation
Class	(Gene catalog)
Other DBs	NCBI-GI: 49305513 8729045 747703 49252273 49307066 49302542 19740943 (Alignment) 49154042 49294919 8716518 37426662 8679559 8685632 37426664 8725190 683512 19874737
LinkDB	(PDB) (All DBs)
NT seq	849 nt (NT seq) +upstream[0] nt +downstream[0] nt

cgttaaaacaattttttgaaaatcacgtataaacacatgagcacaaggagaaagaacaaag
gtagcatctcaattatttggcaaaaaatggcgacgagagcctgaagatggaccatctatt
ttcggcggtttaaggcaatcactgaagcaattggtagagaactgagggtagaattaaaac
caaaaaaaaaaaacaagacgaggctgcttcacatttagcagaagccacactatttctta
ttacaatagattgtttttccttggatcagtgatggtcatcaccatgtccgtcaggaggtcc
tccaggaggccgaccacttcaagcttgaagtactgtgtacacacagggcattcaaaagactt
gcctttctccagccaaaaccagacaacatcatgctcatcctctccttcacctccagggca
accaactattctcatgtcgtagtaggatttgactacagcaggagcttcctttgttccaaa
aggcccttcaggaaagtctatatcgtccagcttcctcccctccaattcggcttgtagttc
ctctttctcgtgtgaccggttgcaataggcatcacatcttcaacacgcttcttcacagcagt
atccgctgcagcggagccgattaaacgagagaatatggaggaatatgcgggaatggagga
tgcacgagtcgagagagaaaggtgagtggactctacgacggcgtgcctgcacgacggagc
cgcgcaggacccgacggcggagatggatttaagatgcgatgaaacgattctcctccacat
aatggtctcgattcttcgggaaaattaggtttgtcgtttccgatctcagatgattctcag
tctccgact

Fig. 3B

C

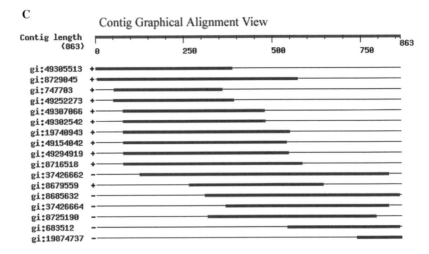

Fig. 3C

indicate that there are no Arabidopsis ESTs similar to that gene/enzyme in other species. Here, users have two options for further exploration.

1. Clicking on any colored boxes (e.g., EC: 1.9.3.1) shows all Arabidopsis EST contigs similar to that enzyme. For example, choose the contig 1542 in **Fig. 3B** for further analysis. The main window has seven sub-windows including KO annotation (gene catalogs), EST members of the contig with links to graphical and text alignment, links to pathway and other databases, and the contig's sequence.
2. Clicking on the "Alignment" button shows the text alignment of EST sequences that form this contig produced by the CAP3 *(8)* assembly program (http:// www.genome.jp/dbget-bin/show_eg_align?a.thaliana_est+1542). The graphical view of the alignment is provided to illustrate how the contig members overlap (*see* **Fig. 3C**). On the left side, the GenBank number (gi) of each single EST is shown. Clicking on this will show the corresponding GenBank data (http://www.ncbi.nih.gov/entrez/viewer.fcgi?db=Nucleotide&val=49305513).

3.3. Exploring the Gene Catalogs

The Gene catalog entry point for *Arabidopsis thaliana*, based on EST data, is shown in this link: http://www.genome.jp/kegg-bin/show_organism?menu_type=gene_catalogs&org=eath. Clicking on "KO", the pathway-based classification of ortholog groups, shows in a new window the KO's first level (*see* **Fig. 4A**), containing Metabolism, Genetic Information Processing, Environmental Information Processing, and Cellular Processes,

A

 KEGG Orthology (KO) - Arabidopsis thaliana (thale cress) (EST)

[1st Level | 2nd Level | 3rd Level | 4th Level | Text Search]

▶ 01100 Metabolism

▶ 01200 Genetic Information Processing

▶ 01300 Environmental Information Processing

▶ 01400 Cellular Processes

▶ 01500 Human Diseases

Fig. 4A Gene catalogs and Kyoto Encyclopedia of Genes and Genomes (KEGG) Orthology (KO) structure. **(A)** The hierarchical KO level for Arabidopsis expressed-sequence tag (EST) data with links to other databases. The REACTION and COMPOUND information links for each contig (**B** and **C**).

B

KEGG REACTION: R01600

Entry	R01600 Reaction
Name	ATP:D-glucose 6-phosphotransferase
Definition	ATP + beta-D-Glucose <=> ADP + beta-D-Glucose 6-phosphate
Equation	C00002 + C00221 <=> C00008 + C01172
RPair	RP: A00003 C00002_C00008 main RP: A01627 C00221_C01172 main RP: A06627 C00002_C01172 trans
Pathway	PATH: rn00010 Glycolysis / Gluconeogenesis
Enzyme	2.7.1.1 2.7.1.2
Ortholog	KO: K00844 hexokinase KO: K00845 glucokinase
LinkDB	(All DBs)

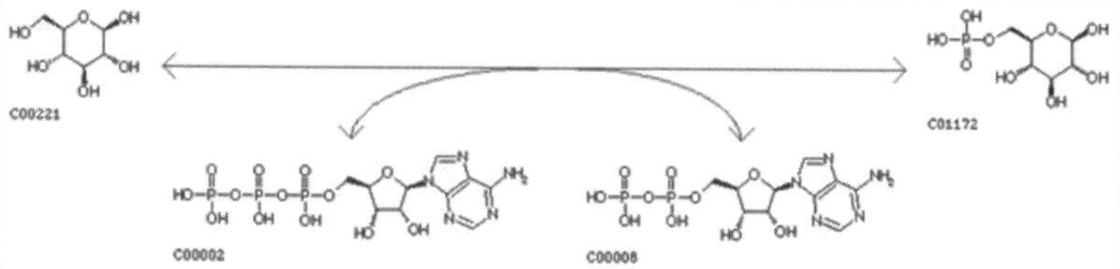

Fig. 4B

C

KEGG COMPOUND: C00002

Entry	C00002 Compound
Name	ATP; Adenosine 5'-triphosphate
Formula	C10H16N5O13P3
Mass	506.9955
Structure	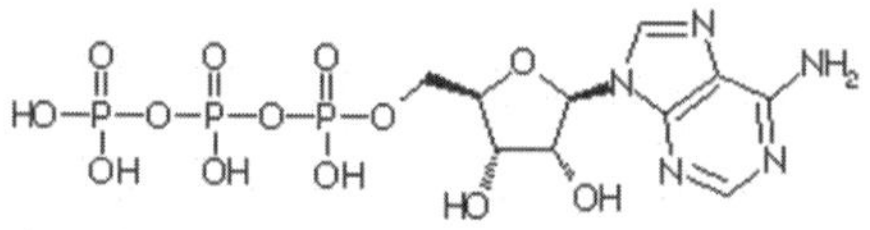 C00002 (Mol file) (KCF file) (DB search)

Fig. 4C

relevant to plants. Clicking on the link http://www.genome.jp/dbget-bin/get_htext?A.thaliana_est.kegg+-f+F+D shows the second, third, and fourth levels of the hierarchy of Arabidopsis KO. The 4th level not only shows the KO hierarchy but also the list of Arabidopsis contigs orthologous to each enzyme. For example, the first contig that appears is 16002. Clicking on this shows the information related to the sequence in a new window, and clicking on the "Gene catalog" button will display in new windows the complete information for this contig in different pathways (http://www.genome.jp/dbget-bin/www_bget?eath:16002 and http://www.genome.jp/ dbget-bin/hfind_www_sub?htext=A.thaliana_est.kegg&keywords=16002&option=-a). An ortholog of this contig is Glucokinase enzyme [EC: 2.7.1.2], present in Glycolysis [PATH: eath00010]; Galactose metabolism [PATH: eath00052] and Starch and Sucrose metabolism [PATH: eath00500] all belong to Carbohydrate Metabolism; Streptomycin biosynthesis [PATH: eath00521] belongs to the Biosynthesis of Secondary Metabolites pathway. To see information on the Glucokinase enzyme, clicking on it will show the associated information, such as enzyme Class, Reaction, Substrate, Product, Pathway, and KO number. Reaction includes the reactions in which this enzyme participates. By clicking on any reaction identifier, users will open new windows showing the associated reactions and compounds (*see* **Fig. 4B**). Substrate shows the compound on which this enzyme acts (*see* **Fig. 4C**).

Another entry point for EGENES (http://www.genome.jp/kegg-bin/show_organism?menu_type=gene_catalogs&org=eath) is Sequence Similarity Search in the KEGG database using BLAST and FASTA. For example, to find all the sequences similar to Arabidopsis cDNA clone gi: 37426855 (*see* **Fig. 5A**), click on the BLAST button, copy and paste the sequence into the form, choose BLASTN and the database to search (here KEGG EGENES), and press the "compute" button. All the similar contigs from EGENES (including species other than Arabidopsis) will be shown. From here, various analyses on the BLAST results are available, such as performing multiple alignments (clustalW and MAFFT) and graphical alignments of BLAST results, drawing a phylogenic tree (*see* **Fig. 5B**) and others.

3.4. Other KEGG Tools and Resources

3.4.1. SSDB

SSDB contains precomputed similarity scores for all protein-coding genes in KEGG GENES, together with best hit information in pairwise genome comparisons. SSDB can be used for searching orthologs and paralogs, as well as conserved gene clusters, with additional consideration of positional correlations on the chromosome.

3.4.2. EXPRESSION

The EXPRESSION database is a Web-based system for the integration and analysis of gene expression profiles, with KEGG pathways and KEGG genomes. As a database, it contains microarray data for Synechocystis PCC6803, *Bacillus subtilis*, and *Escherichia coli*. The associated Java application, KegArray, may also be used to analyze your own data.

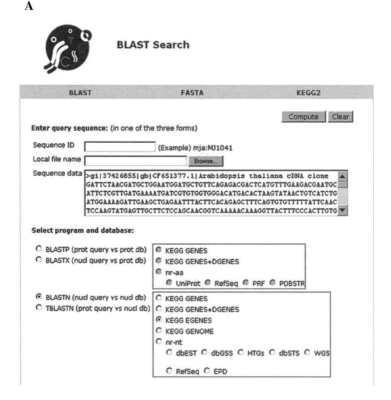

Fig. 5A. Homology search and phylogenic analysis based on expressed-sequence tag (EST) data stored in EGENES. (**A**) The BLAST and FASTA search form in KEGG. (**B**) The phylogenic tree based on homolog sequences for the query in EGENES.

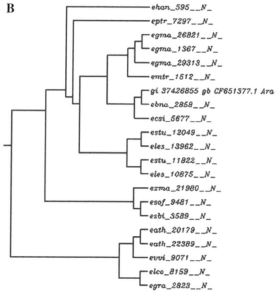

Fig. 5B

3.4.3. BRITE

BRITE is a database of functional hierarchies and binary relations between proteins or other biological molecules. BRITE aims to enhance deductive capabilities from the database on the biological relationships of genes and gene products. In view of the developments in experimental technologies for protein–protein interactions, there will be a huge amount of biological binary relation data that will be part of the BRITE database. BRITE is also designed to accumulate current knowledge on the function of biological molecules in hierarchical form, as an alternative to the PATHWAY database.

3.4.4. LIGAND

LIGAND *(6)* is a composite database of COMPOUND, DRUG, GLYCAN, REACTION, REPAIR, and ENZYME. ENZYME stores information on enzymatic reactions and enzyme molecules according to the up-to-date classification of the EC numbers. COMPOUND is a collection of about 6000 chemical compounds, most of which are metabolites in the metabolic pathways. DRUG is a derivative of COMPOUND and stores chemical structures and therapeutic categories of drugs. ENZYME and COMPOUND are linked from the KEGG reference pathways for metabolism, thus providing molecular details of network information. The future role of LIGAND is to integrate information about the environment of the network.

3.4.5. GENOME

The GENOME database is a collection of genome maps containing information about chromosomal locations of genes for completely sequenced genomes. Each entry of GENOME is an entry point for various information on the genome, including pathway maps and gene catalogs, with lineage information on the genome and related diseases, etc.

3.4.6. Simple Object Access Protocol Application Program Interface

The KEGG Application Program Interface (API) provides valuable means for accessing the KEGG system using computer programs, for searching and computing biochemical pathways in cellular processes or analyzing the universe of genes in the completely sequenced genomes. Users can access the KEGG API server using Simple Object Access Protocol (SOAP) technology over the HTTP protocol. The SOAP server also comes with the Web Services Description Language (WSDL), which makes it easy to build a client library for a specific programming language. This enables users to write their own programs for many purposes and to automate the accessing of KEGG data.

3.5. Discussion and Future Directions

In 1996–2006, bio informatics has become an integral part of research and development in the biomedical sciences. Bioinformatics now has an essential role both in deciphering genomic, transcriptomic, and proteomic data generated by high-throughput experimental technologies and in organizing information gathered from traditional biology. With the complete genome sequences for an increasing number of organisms at hand, bioinformatics is beginning to provide both conceptual bases and practical methods for detecting systemic functional behaviors of the cell and the organism. The ultimate goals of bioinformatics will be to abstract knowledge and principles from large-scale data, to present a complete representation of the cell and the organism, and to computationally predict systems of higher complexity, such as the interaction networks in cellular processes and the phenotypes of whole organisms.

With the release of the fully sequenced genomes of Arabidopsis and rice, and the initiation of sequencing projects and large-scale ESTs sequencing for many other plant species, there is a fast growing desire to place these information on the genomes in a metabolic context. Efforts to computerize our knowledge on cellular functions, at present either by the controlled vocabulary of Gene Ontology or by graphical representation in KEGG, will both facilitate the computational mapping of genomic data to complex cellular properties and detect empirical relationships between genomic and higher properties.

The EGENES project in KEGG is an attempt to capture all reactions and pathways thought to occur in organisms, including plants, based on both genomic and expressed portions of the genome. This is, however, only a starting point. We see the role of EGENES as providing a framework of possible reactions that, when combined with expression, genetics, physical location, and enzyme kinetic data, provide the infrastructure for the true integration of biological knowledge and data. Integration in this respect does not simply involve methodology, such as links and common interfaces, but rather involves biology. We are working to make EGENES data available in various standard formats, including Text and XML. This will also enable data exchange with other pathway databases, such as AraCyc and LIGAND, and other molecular interaction databases.

In summary, KEGG/EGENES provides summaries of fundamental biological processes in a form that is as useful to students working on a single gene/protein as it is to bioinformaticists striving to make sense of large-scale data sets. KEGG/EGENES provides biologist-friendly visualization of biological data and is an open-source project.

Acknowledgments

A JSPS Post Doctoral award from Laboratory of Plant Genetics and COE Post Doctoral award from Bioinformatics Center of Kyoto University to Ali Masoudi-Nejad are acknowledged. We also like to express our gratitude to Dr. Kiyoko F. Aoki-Kinoshita for critical reading of the manuscript.

References

1. Berna, A., Ear, U., and Kyrpides, N. (2001) Genomes OnLine Database (GOLD): a monitor of genome projects world-wide. *Nucleic Acids Res.* 29, 126–127.
2. Bork, P., and Koonin, E.V. (1998) Predicting functions from protein sequences where are the bottle necks? *Nat. Genet.* 18, 313–318.
3. Kanehisa, M. (1997) A database for post-genome analysis. *Trends Genet.* 13, 375–376.
4. Karp, P.D., Riley, M., Paley, S.M., Pellegrini-Toole, A., and Krummenacker, M. (1998) EcoCyc: encyclopedia of Escherichia coli genes and metabolism. *Nucleic Acids Res.* 26, 50–53.

5. Kanehisa, M., Goto, S., Kawashima, S., Okuno, Y., and Hattori, M. (2004) The KEGG resource for deciphering the genome. *Nucleic Acids Res.* 32, D277–D280.
6. Goto S., Okuno Y., Hattori M., Nishioka T., and Kanehisa M. (2002) LIGAND: database of chemical compounds and reactions in biological pathways. *Nucleic Acids Res.* 30, 402–404.
7. Masoudi-Nejad, A., Jauregui, R., Kawashima, S., Goto, S., Kanehisa, M., and Endo, T.R. (2004) The kingdom of Plantae EST Indices: a resource for plant genomics community. Genome Informatics 2004. PP-102. The 15th International Conference on Genome Informatics December 16–18, 2004, Yokohama Pacifico, Japan.
8. Huang, X., and Madan, A. (1999) CAP3: a DNA sequence assembly program. *Genome Res.* 6, 829–845.

22

International Crop Information System for Germplasm Data Management

Arllet Portugal, Ranjan Balachandra, Thomas Metz, Richard Bruskiewich, and Graham McLaren

Summary

Passport and phenotypic data on germplasm and breeding lines are available from worldwide sources in various electronic formats. These data can be collated into a single database format to enable strategic interrogation to make the best use of data for effective germplasm use and enhancement. The International Crop Information System (ICIS; http://www.icis.cgiar.org) is an open-source project under development by a global community of crop researchers and includes applications designed to achieve the storage and interrogation of pedigree and phenotypic data.

Key Words: Germplasm; phenotype; crop; database; software.

1. Introduction

The International Crop Information System (ICIS; http://www.icis.cgiar.org) is an "open-source" and "open-license" generic crop information system under development since the early 1990s by researchers from the Consultative Group on International Agricultural Research (CGIAR; www.cgiar.org) and non-CGIAR partners from the academic, government, and private sectors, worldwide *(1–4)*.

ICIS is designed to fully document germplasm genealogies with associated meta-data such as nomenclature, passport data, and characterization data, and to accurately cross-link germplasm entries with associated experimental observations from evaluations undertaken in the field, greenhouse, or laboratory.

From: *Methods in Molecular Biology, vol. 406: Plant Bioinformatics: Methods and Protocols*
Edited by: D. Edwards © Humana Press Inc., Totowa, NJ

ICIS is successfully deployed for many crops, including rice, wheat, barley, maize, common bean, chick pea, cow pea, sugarcane, potato, sweet potato, and a wide range of vegetables (*see* **Table 1** for some representative installations). ICIS is described in greater detail elsewhere *(4)*, but here we provide a general overview of its applications.

1.1. Genealogy Management System

Proper management of germplasm information is essential for the elucidation of genotype–phenotype associations. Management goals include systematic tracking of germplasm origin (passport and genealogy information), recording

Table 1
Representative Installations of ICIS

Crop	Status	*Database* (Acronym) [Web link] Institute
Rice	*Public*	*International Rice Information System (IRIS)* [www.iris.irri.org] International Rice Research Institute (IRRI), *(1,2,4)*
Wheat	*Public*	*Global Wheat Information System (GWIS)* [http://gwis.lafs.uq.edu.au], University of Queensland, Dept. of Land, Crop and Food Sciences, Brisbane, Australia
	Public	*International Wheat Information System (IWIS3)*, Centro Internacional de Mejoramiento de Maíz y Trigo (CIMMYT);
	Private	*International Wheat Information System (IWIS)* Semiarid Prairie Agricultural Research Centre (SPARC), Agriculture and Agri-Foods Canada, Saskatchewan.
Lentil and Chickpea	*Public*	*Australian Temperate Field Crops Collection (ATFCC)* DPI, Horsham Victoria, Australia
	Public	*International Centre for Research in the Semi-Arid Tropics (ICRISAT)*
	Public	*International Centre for Agricultural Research in the Dry Areas (ICARDA)*
Barley	*Public*	*International Barley Information System*, International Centre for Agricultural Research in the Dry Areas (ICARDA)
Vegetables	*Private*	Nunhems Netherlands B.V., Voort 6, 6083 AC Nunhem, the Netherlands

of alternate germplasm names, accurate linkage of experimental results to applicable genotypes, and proper material management of germplasm inventories.

An important aspect of any good germplasm information system is the separation of the management of nomenclature from identification. Users must be free to name germplasm as they like, and the system must make sure the names are bonded to the right germplasm. A key to the effective management of such variable germplasm information is the assignment of a unique germplasm identifier (GID) to each distinct germplasm sample, seed package or clone, that needs to be tracked ("bar coded"). The acid test is to ask whether or not mixing two germplasm samples together will result in an unacceptable loss of biological or management information. If the answer to this question is "yes", then each sample should be assigned a distinct GID. The GID is the essential reference point for managing all meta-data about the germplasm, for accurately attributing all experimental observations made about that sample, and for cross-linking related germplasm samples with one another, for example, the parents (sources) and progeny of the given sample, documenting methods of generation, specified as a controlled vocabulary, and membership of the sample in global "management neighborhoods", which are defined sets of related germplasm.

Once assigned, a GID is never destroyed but rather persists in the crop database long after the associated sample has become unavailable (after being fully consumed, becoming non-viable, or otherwise lost). In this manner, historical information about germplasm may be efficiently integrated with information about extant descendents of that germplasm. Although a given GID is generally a database primary key defined locally to a given database, it should be convertible into a globally unique identifier within a community of germplasm databases. This requirement is not unique to GID usage. In fact, most biological data to be shared by a distributed research community should be assigned a global identification in this manner.

The ICIS Genealogy Management System (GMS) *(4)* meets the need for global identification of germplasm and related data objects by maintaining globally unique information about the local database installation, and user who created the entry, as the authority for the information assigned to a given object identifier. This entry may eventually be published in a central ICIS repository and receive a second new "public" identifier. ICIS object identifiers are cross-linked to the original identifier. These ICIS object identifiers, such as GIDs, are not names, and although they do contain some information on domain and authority, no one will generally use them as names for germplasm.

1.2. Data Management System

ICIS stores experimental data and meta-data in its Data Management System (DMS) *(4)*. The variables of a DMS Study are classified as either a Factor or a Variate. A Factor is a known (independent, experimentally applied) variable containing discretely specified "levels". Factors can include germplasm identifiers such as GID and name, trial location identifiers, experimental design (i.e., plot layouts), applied treatments, and other information available before a trial. A Variate is a (dependent) variable measured or observed in the Study. Variates include parameters such as seed weight, yield, and disease scores. Variate data may be discrete enumerations, character strings, or continuous numeric values.

Both Factors and Variates are documented according to the general concept lying behind the variable (the "Property"), the measurement units to express the variable values (the "Scale"), and documentation about the mode of measurement (the "Method"). The Property, Scale, and Method of variables are specified in ICIS as a controlled vocabulary. Factors and Variates are combined into "observation units", which represent a single aggregate experimental observation in the Study. Such observation units may be indexed by various combinations of their (discrete) Factor levels.

2. Materials

2.1. Operating System

Most of the original (mostly data entry) ICIS tools are constrained to run under Microsoft Windows. Some newer software for ICIS, in particular for Web-based interfaces, is primarily being programmed using Java language technology and as such should run on any operating system platform supporting Java.

2.2. Database Type

ICIS is somewhat database platform independent. Within Microsoft Windows, applications are decoupled from the back-end databases using a "Dynamic Linked Library" of functions, which themselves invoke the well-established ODBC protocol to connect to any SQL 92 compliant database. Although smaller installations of ICIS can use Microsoft Access to store and exchange their data sets, larger installations of ICIS typically use either a commercial database, such as Oracle, or an open-source database platform, such as MySQL or PostgreSQL.

A partial Java language port of ICIS is also available and is now used as the foundation for (mainly read-only) Web-based applications. These Java-based applications use common open-source frameworks such as Apache Tomcat and open-source technology such as Hibernate to connect to databases. Many Web installations typically run on a MySQL-based ICIS database.

2.3. Software

ICIS software may be downloaded from http://www.icis.cgiar.org. Development versions of the software are hosted on a project server at http:// cropforge.org. ICIS technical discussions can also be found at a project Wiki site, http://cropwiki.irri.org/icis. The ICIS software suite consists of a series of software applications (*see* **Fig. 1**), most of which will be introduced below. For further detailed tutorial information and discussion about ICIS applications, refer to http://cropwiki.irri.org/icis/index.php/TDM_ Application_ Programs.

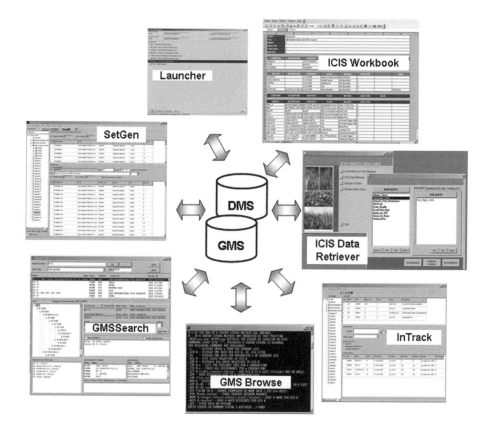

Fig. 1. The International Crop Information System (ICIS) software suite is a set of Windows-based tools for managing crop data. The purpose of each tool is summarized in the text.

2.3.1. Program Launcher

The ICIS Launcher provides a master control panel under Microsoft Windows for the launching of Windows tools. It also provides a user interface for specifying the ODBC connection parameters for both the GMS and DMS connections to point applications to both a "central" (public, shared) and a "local" (private) installation of the database.

2.3.2. SetGen

The Set Generation Module (SetGen) is an application to produce lists of germplasm entries for breeding, evaluation, distribution, management, or any other purpose. List entries need not exist already in the GMS; if they do not, they may be added through the external pedigree input tool integrated within SetGen. Germplasm entries can be selected from existing lists or directly from the GMS. Details of the new germplasm generated by the list process are stored in the GMS. Related information such as evaluation data can be retrieved from the DMS. Those data can also be used to generate new lists of entries.

2.3.3. GMS Search

The GMS Search program performs name searches in the GMS. A search string may contain wildcard characters "_" and "%". The underscore character ("_") is used to represent any single character at a specific position in the name. The percent sign (%) is used to represent any string starting where the "%" is located. All germplasm names that match the search are shown in a list. The pedigree tree, selection history, and characteristics of the highlighted germplasm in the list are displayed at the bottom. Displayed characteristics include preferred name, method of genesis, name date and location, germplasm date and location, and cross expansion. Any existing attributes and other associated names are also displayed.

2.3.4. GMS Browse

The GMS Browse program is a console application for viewing and changing GMS records. It can search for names, given a string with or without wildcards, similar to GMS Search. The other commands of Browse start with a dot (.), and these commands include computation of coefficient of parentage, display of pedigree tree, display of mendelgrams, tracing of relatives, updating a germplasm record, and deleting a germplasm record.

2.3.5. InTrack

The ICIS Inventory Tracker (InTrack) is an application that manages inventory information for any entities, including genetic resources and breeders' seed stocks. It tracks individual entities, where they are stored, what quantities are in storage, and what quantities are available for use. Lists of entries to deposit or reserve are created in SetGen, and the amounts are posted and committed in InTrack.

2.3.6. ICIS Workbook

Field or evaluation data are stored using the ICIS Workbook, an application that facilitates loading and retrieving a DMS Study. Germplasm entries created in SetGen and stored in the GMS are usually factors of a crop study; thus, the ICIS Workbook has a command to directly retrieve lists of such germplasm from the GMS.

Under the DMS, a user creates templates pertaining to studies. A Study can contain data for single experiments or a group of experiments. The user defines Factors and Variates for a given Study in the Workbook. The variables specified on one sheet are then used to generate an observation sheet to record trait values in tabulated form. Data can then be entered on this linked observation sheet. The resulting data are then readily imported into the ICIS DMS relational database schema by activating embedded Excel macros. These are organized in menus that are added in when the Workbook is opened. Templates can be reused for trials pertaining to future years, thereby achieving significant time-savings in data management. In addition, the Workbook allows validation rules to be set to ensure cleanliness of data input.

2.3.7. ICIS Data Retriever

Evaluation data of a list of germplasm entries across several studies can be retrieved using the ICIS Data Retriever. It is a Microsoft Access application that extracts data from the DMS by the specified combination of property, scale, and method. The list of entries stored in the GMS can be used as one of the criteria of the query. Queries are defined and stored in the Retriever. It also has tools for creating subsets, aggregating the values, and linking the queries with each other. The result of the query can also be stored as a list of entries in the GMS.

2.3.8. Web Query Interface

The Web-based query interface of ICIS provides for searching germplasm records in the GMS by identifiers and other germplasm attributes. Once a list of matching germplasm records is retrieved, the genealogy of each germplasm entry can be viewed in a "pedigree tree" along with other pertinent passport data. The interface also provides access to associated field studies from the DMS. Starting from the other direction, trait-driven searches of the DMS may be made to retrieve germplasm with matching characteristics. The Web interface for ICIS is evolving rapidly to incorporate other germplasm-associated data types, such as genotyping and genomic data.

2.3.9. Future Directions

The ICIS community (http://www.icis.cgiar.org) holds development and training workshops annually. The community attracts a wide range of funding sources that drive its development forward. In particular, the CGIAR-hosted Generation Challenge Programme (http://www.generationcp.org) is funding general technology development in the areas of shared domain model and ontology standard specification, platform development, and network integration of crop information platforms (http://pantheon.generationcp.org). In fact, ICIS is seen as evolving into one reference implementation of the GCP platform, which is primarily based on Java language technology already embraced by the project.

3. Methods

A detailed description of all facets of ICIS would be impossible within a single chapter. To acquaint readers with one basic method, the following example demonstrates the use of the ICIS Workbook application to load data from an experiment where traits such as yield and grain weight are measured for different lines (*see* **Note 1**).

3.1. Running the Workbook

Start the ICIS Workbook from the launcher. A dialog box may appear displaying the following warning:

The workbook you are opening contains macros. Some macros contain viruses that may be harmful to your computer. If you are sure this workbook is from a trusted source, click "Enable Macros". If you are not sure and want to prevent any macros from running, click "Disable Macros".

If the warning appears, click "Enable Macros". Otherwise, the ICIS Workbook appears automatically.

3.2. Defining a Study

From the "Study" menu select "New Study". Enter the name of the study "U03WSHB" next to the heading STUDY in the first row. Use only alphanumeric characters. A study name must be unique within the DMS. You should add a title, a study objective, and start and end dates as in **Fig. 2.** You must add a Study Type – "E" for experiment, and you may add a Project Management Key (PMKEY) to link the data to a Project Database if you have one.

If sections of the Description sheet for FACTORS and VARIATES are not visible as in **Fig. 2** you may need to unprotect the worksheet (with the "Tools, Protection, Unprotect sheet" menu option) and switch to the editor's view with the "Utilities, Workbook View" menu option. There are speed buttons on the Menu Bar for most menu options described here.

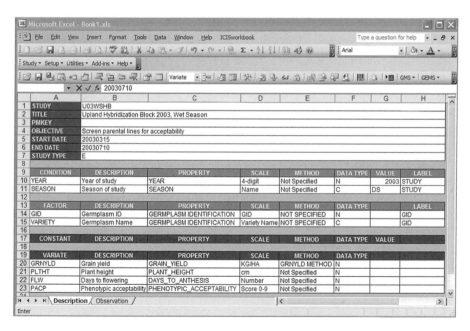

Fig. 2. The description Sheet in the DMS Workbook is the default sheet that is activated and displayed upon launching the DMS Workbook application. Users are able to define the factors and variates pertaining to an experiment or a group of experiments.

3.3. Defining Factors and Variates

Click on the word CONDITION in the first column, then click the "Setup" menu and select "Add Variable" then "Condition". Do this twice to make two Condition rows appear. In the first column of the first row under CONDITION type YEAR and in the cell below, SEASON. Conditions are factors which have a single level for all data in the current workbook.

Click on the word FACTOR in the first column and from the "Setup" menu select "Add Variable" then "Factor" to add a factor row. Do this twice and enter names GID and VARIETY in the first column of the created rows.

Click on the word VARIATE in the first column and from the "Setup" menu select "Add Variable" then "Variate" to add a variate row. Do this four times and enter names GRNYLD, PLTHT, FLW and PACP in the first column of the created rows.

CONSTANTS are variates with a single value for all observation units in the data set. We do not have any to specify in this example.

3.4. Setting Properties, Scales and Methods

We must now select Properties, Scales and Methods for our Conditions, Factors and Variates. From the "Setup" menu select "Variable Section" then "Custom Setup". A dialogue window will open with the condition names in a selection window. Highlight YEAR, add a description such as "Year of study" as in **Fig. 2.** From the pull down list next to Property select the appropriate property for Year of study. The list will depend on the particular crop implementation of ICIS being used, but there is probably a property called YEAR. If you don't find it leave it blank for the moment. Similarly pick the Scale (**Note 2**), Method and Data Type (N for numeric, C for character, **Note 3**) form the pull down lists or leave them blank if suitable ones are not found. The YEAR value applies to all data in the study, so it is a Study Label. Hence in the Label box select STUDY. When complete click on the next condition, SEASON, and perform the same operations. Season is also a Study Label.

Having completed this process for Conditions, choose FACTOR from the top left pull down list of the Custom Setup dialogue window. (If you have closed the Custom Setup dialogue window after processing Conditions, simply open it again from the Setup menu). Repeat the process for the factors. GID is chosen to be the key Factor so it is a Label of itself. In the Label box select GID. Variety, however, does not give new indexing information to the rows of data once GID is known, Hence it is a Label of GID. Select GID in the

Label box for Variety. Do this again for VARIATES. Then click "OK" on the Custom Setup dialogue window.

Type new codes in the appropriate cells of the Description Sheet for any Properties, Scales and Methods which were left blank during the previous operations. These will be added to the ontologies when the data are loaded.

The DESCRIPTION sheet will now look like **Fig. 2** depending on the particular Properties, Scales and Methods available for the crop

3.5. Assigning Factors and Variates

The factors and variates for a trial now need to be registered. From the "Setup" menu select "Observation Sheet". The variables specified in the Description sheet will now appear as column headings in the Observation sheet as in **Fig. 3**.

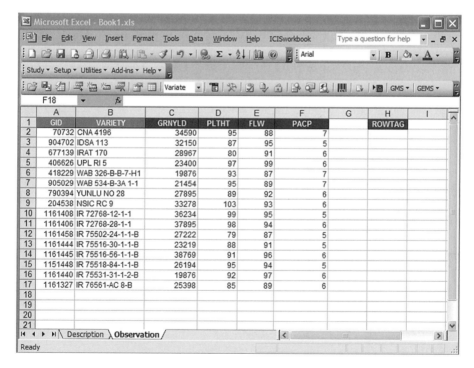

Fig. 3. The observation sheet in DMS Workbook is used to enter values in columns corresponding to the description sheet.

3.6. Entering Data

Data can be typed or pasted into the Observation sheet. Missing values should be blank, both for numeric and character data. It is not necessary to finish entering data in one session. You can save the file as an excel file at any time and continue entering your data at another time.

3.7. Loading data to DMS

Once the data have been proofread they can be loaded into the local DMS. Select "Load" from the "Study" menu and a Load Study dialogue box appears. This dialogue allows the user to run a number of data and ontology integrity checks which are described in the Workbook users manual at http://cropwiki.irri.org/icis/index.php/TDM_ICIS_Workbook_5.4. Since these checks depend on data and ontologies in the particular crop implementation being used you should disable all the checks for this exercise by un-checking all the items on the Load Study dialogue menu. Click "OK" to load the data into the local DMS.

After the data has been loaded, the excel file that you used to enter your data is no longer needed, but should be archived for back-up purposes.

3.8. Retrieving data from the DMS

Data can be retrieved from the DMS with the Workbook by selecting "Retrieve" from the "Study" menu, selecting the study of interest and then selecting the particular dataset of that study to be retrieved.

3.9. Loading Data to DMS

This process will ensure that data will be stored permanently in a database.

Data loaded into the DMS is assumed to be already clean. Do not load data into the DMS without having proofread the data first. After proofreading, click on "Study" and "Load". This loads the data into the local DMS. Exit the Excel application.

After the data have been loaded, the Excel file that you used to enter your data may be deleted as the data would already have been entered into the database, though for back-up purposes, you may wish to keep the Excel file.

3.10. Retrieving Data from the DMS

Once data have been loaded into the DMS, the Excel file where the data were entered is no longer needed. Data can be retrieved from the DMS by simply running the workbook.

1. Run the ICIS Workbook by selecting it from the Launcher.
2. Click on "Study", "Retrieve", and "Dataset". A list of existing DMS studies is displayed.
3. Select the desired study by highlighting the study name.
4. Select a data set of the study and then click the OK button. The associated data are retrieved and displayed.

Notes

1. Details provided in this example are likely to change with time. The authors recommend readers review the International Crop Information System Technical Development Manual (http://cropwiki.irri.org/icis/index.php/ICIS_Technical_Documentation) for recent changes.
2. Different scales can be entered for single traits where data are loaded from different sources measured in different units. For example, Yield can be measured in "tons per hectare" or "kilograms per hectare", and these can be entered as two different scales for the same trait.
3. A data type can only be a number or a character variable. Represent Boolean variables using either a 0/1 (numeric) or Y/N (character).

References

1. Bruskiewich, R., Cosico, A., Eusebio, W., Portugal, A., Ramos, L.R., Reyes, T., Sallan, M.A.B., Ulat, V.J.M., Wang, X., McNally, K.L., Sackville Hamilton, R. and McLaren, C.G. (2003) Linking genotype to phenotype: The International Rice Information System (IRIS). *Bioinformatics* 19(1), i63–i65.
2. Bruskiewich, R., Metz, T. and McLaren, G. (2006) Bioinformatics and crop information systems in rice research. *International Rice Research Newsletter* 31(1), 5–12.
3. Fox, P.N. and Skovmand, B. (1996) The International Crop Information System (ICIS) – connects Genebank to breeder to farmer's field in *Plant Adaptation and Crop Improvement* (eds M. Cooper and G.L. Hammer), CAB International, pp. 317–326.
4. McLaren, C.G., Bruskiewich, R.M., Portugal, A.M. and Cosico, A.B. (2005) The International Rice Information System (IRIS): a platform for meta-analysis of rice crop data. *Plant Physiology* 139, 637–642.

23

Automated Discovery of Single Nucleotide Polymorphism and Simple Sequence Repeat Molecular Genetic Markers

Jacqueline Batley, Erica Jewell, and David Edwards

Abstract

Molecular genetic markers represent one of the most powerful tools for the analysis of genomes. Molecular marker technology has developed rapidly over the last decade, and two forms of sequence-based markers, simple sequence repeats (SSRs), also known as microsatellites, and single nucleotide polymorphisms (SNPs), now predominate applications in modern genetic analysis. The availability of large sequence data sets permits mining for SSRs and SNPs, which may then be applied to genetic trait mapping and marker-assisted selection. Here, we describe Web-based automated methods for the discovery of these SSRs and SNPs from sequence data. SSRPrimer enables the real-time discovery of SSRs within submitted DNA sequences, with the concomitant design of PCR primers for SSR amplification. Alternatively, users may browse the SSR Taxonomy Tree to identify predetermined SSR amplification primers for any species represented within the GenBank database. SNPServer uses a redundancy-based approach to identify SNPs within DNA sequence data. Following submission of a sequence of interest, SNPServer uses BLAST to identify similar sequences, CAP3 to cluster and assemble these sequences, and then the SNP discovery software autoSNP to detect SNPs and insertion/deletion (indel) polymorphisms.

Key Words: Single nucleotide polymorphism; simple sequence repeat; molecular genetic marker; microsatellite; SNPServer; autoSNP; SSRPrimer; SSR Taxonomy Tree.

From: *Methods in Molecular Biology, vol. 406: Plant Bioinformatics: Methods and Protocols*
Edited by: D. Edwards © Humana Press Inc., Totowa, NJ

1. Introduction

Single nucleotide polymorphisms (SNPs) and simple sequence repeats (SSRs) are used routinely in agriculture as markers in crop-breeding programs *(1)*. They also have many uses in human genetics, such as the detection of alleles associated with genetic diseases and inferences of population history *(2,3)*. Furthermore, these markers are invaluable as a tool for genome mapping, offering the potential for generating very high density genetic maps, that can be used to develop haplotyping systems for genes or regions of interest *(4)*.

1.1. SSR Discovery

SSRs, also known as microsatellites, have been proved to be one of the most powerful genetic markers in biology. They are common, readily identified DNA features consisting of short (1–6 bp), tandemly repeated sequences, widely and ubiquitously distributed throughout eukaryotic genomes *(5)*. Furthermore, SSRs have been found in all prokaryotic and eukaryotic genomes that have so far been analyzed *(6)*. SSRs are highly polymorphic because of the mutation affecting the number of repeat units. This hypervariability among related organisms makes them informative and excellent markers for a wide range of applications including high-density genetic mapping, molecular tagging of genes, genotype identification, analysis of genetic diversity, paternity exclusion, phenotype mapping, and marker-assisted selection of crop plants *(7,8)*.

SSRs are a source of abundant, non-deleterious mutations that provide variation in the face of stabilizing selection, and their recognized role in the process of evolutionary adaptation is predicted to increase as our knowledge of them expands *(9)*. SSR stability may be correlated with overall levels of genomic stability *(10)* as mutations that affect SSR stability, such as those involved in DNA mismatch repair, can also influence genomic stability. The nature of SSRs gives them a number of advantages over other molecular markers: (1) multiple SSR alleles may be detected at a single locus using a simple PCR-based screen, (2) they are co-dominant, (3) very small quantities of DNA are required for screening, and (4) analysis may be semi-automated. Furthermore, SSRs demonstrate a high degree of transferability between species, as PCR primers designed to an SSR within one species frequently amplify a corresponding locus in related species, making them excellent markers for comparative genetic and genomic analysis.

The potential biological function and evolutionary relevance of SSRs are currently under scrutiny, which can lead to a greater understanding of genomes and genomics *(11)*. Initial suggestions that the majority of DNA was either "junk" or had no biological function are being challenged by the discovery of new functions for these sequences. Various functional roles have now been attributed to SSRs. For example, SSRs are believed to be involved in gene expression, regulation, and function *(9,12)*, and there are numerous lines of evidence suggesting that SSRs in non-coding regions may also be of functional significance *(9)*.

A common method for the discovery of SSR loci is to construct genomic DNA libraries enriched with SSR sequences, followed by DNA sequencing *(13)*. This production of enriched libraries is time consuming and the specific sequencing required is expensive. Where abundant sequence data are already available, it is more economical and efficient to use computational tools to identify SSR loci. Flanking DNA sequences may then be analyzed for the presence of suitable forward and reverse PCR primers to assay the SSR loci. Several computational tools are currently available for the identification of SSRs within sequence data as well as for the design of PCR primers suitable for the amplification of specific loci. Here, we describe the application of a package, SSRPrimer *(14)*, that integrates two such tools, enabling the simultaneous discovery of SSRs within single or bulk sequence data, and the design of specific PCR primers for the amplification of these marker loci. The Web-based version of SSRPrimer permits the remote use of this package with any sequence of interest. SSR Taxonomy Tree demonstrates the application of this package to the complete GenBank database, with the results organized as a taxonomic hierarchy for browsing or searching for SSR amplification primers for any species of interest.

1.2. SNP Discovery

SNPs and small insertions/deletions (indels) are the most frequently found DNA sequence variations *(15)*. SNPs are far more prevalent than SSRs and therefore may provide a high density of markers near a locus of interest. This abundance of SNPs offsets the disadvantage of bi-allelism, compared with the polyallelic nature of SSRs. SNPs have many uses in crop improvement programs. They can be identified within a gene of interest, or within close proximity. Although the SNP may not be responsible for the mutant phenotype, it can be used for positional cloning of the gene *(1)*. Furthermore, the information provided by SNPs is useful when several SNPs define haplotypes in the region of interest. Only a small subset is required to define the haplotype and therefore need to be assayed.

SNPs have uses in many applications, including high-resolution genetic map construction, genetic diagnostics, genetic diversity analysis, cultivar identification, phylogenetic analysis, and characterization of genetic resources *(4)*. The applications of SNPs in crop genetics have been extensively reviewed by Rafalski *(4)* and Gupta et al. *(1)*. However, these reviews highlight that for several years SNPs will co-exist with other marker systems. The use of SNPs will become more widespread with the increasing levels of genome sequencing in crops and the drop in price of SNP assays.

The development of high-throughput methods for the detection of SNPs from bulk sequence data has led to a revolution in their use as molecular markers *(16)*. However, high-throughput sequencing remains prone to inaccuracies as frequent as one base in every hundred. The false calling of these bases thereby hampers the electronic filtering of sequence data to identify potentially biologically relevant polymorphisms. The challenge of *in silico* SNP discovery is thus not the identification of polymorphic bases but the differentiation of true SNP polymorphisms from the, often more abundant, sequence errors.

Several different sources of error need to be considered when differentiating between sequence errors and true polymorphisms. The principle source of sequence error is found in the automated reading of raw chromatogram data. Here, a balance exists between the desire to read as much sequence as possible and the confidence that bases are called correctly. Phred is the most widely adopted software to call bases from chromatogram data *(17,18)*. One benefit of this software is that it provides a statistical estimate of the accuracy of calling each base and therefore provides a primary level of confidence that a sequence difference represents true genetic variation. There are several software packages that take advantage of this feature to estimate the confidence of sequence polymorphisms within alignments. Where sequence trace files are available for comparison to filter out polymorphisms in traces of dubious quality, software such as PolyBayes and Polyphred are the most efficient means to differentiate between true SNPs and sequence error (*see* **Note 1**). Unfortunately, complete sequence trace file archives are rarely available for large sequence data sets collated from various sources. Furthermore, sequence quality-based SNP discovery does not identify errors in sequences that were incorporated prior to the base calling process. The principle cause of these prior errors is the inherently high error rate of the reverse transcription process required for the generation of cDNA libraries for expressed sequence tag (EST) sequencing. Similar errors are also inherent, though to a lesser extent, in any PCR amplification process that may be part of a sequencing protocol. In cases where trace files are unavailable, the identification of sequence errors can be based on two further methods to determine SNP confidence: redundancy of the polymorphism in an alignment and co-segregation of SNPs with haplotype.

The frequency of occurrence of a polymorphism at a particular locus provides a measure of confidence in the SNP representing a true polymorphism and is referred to as the SNP redundancy score. By examining SNPs that have a redundancy score of two or greater, i.e. two or more of the aligned sequences represent the polymorphism, the vast majority of sequencing errors are removed. Although some true genetic variation is ignored because of its presence only once within an alignment, the high degree of redundancy within the data permits the rapid identification of large numbers of SNPs without the requirement for sequence trace files.

Although redundancy-based methods for SNP discovery are highly efficient, the non-random nature of sequence error may lead to certain sequence errors being repeated between runs because of conserved, complex DNA structures. Therefore, errors at these loci would have a relatively high SNP redundancy score and appear as confident SNPs. This source of error requires an additional method to differentiate them from true polymorphisms. Although sequencing errors may occur with regularity at certain positions within a sequencing read due to conserved sequence complexity, the probability of these errors being repeated between sequence reads remains random. True SNPs that represent divergence between homologous genes co-segregate to define a conserved haplotype, whereas non-random sequence errors do not co-segregate with haplotype. A co-segregation score based on whether a SNP position contributes to defining a haplotype is a further independent measure of SNP confidence. The SNP score and co-segregation score together provide a valuable means for estimating confidence in the validity of SNPs within aligned sequences independent of sequence trace files or the source of the sequence error.

Two methods currently apply a combination of redundancy and haplotype co-segregation: autoSNP *(15,16)* and SNPServer *(19)*. SNPServer is based on autoSNP and provides a real-time Internet-based SNP discovery tool combining redundancy-based SNP discovery and haplotype co-segregation scoring. Sequences may be submitted for assembly with CAP3 *(20)* or submitted preassembled in ACE format. Alternatively, a single sequence may be submitted for BLAST comparison *(21)* with a sequence database. Identified sequences are then processed for assembly with CAP3 and subsequent redundancy-based SNP discovery. SNPServer has an advantage in being the only real-time Web-based software, which allows users to rapidly identify SNPs in sequences of interest using public data.

2. Materials

The Web-based tools described in this chapter are hosted on Linux and UNIX architecture ranging from a single processor Linux server to a 64-node Linux cluster and can be accessed through a standard Web browser. The tools are reliant on public software maintained on the host server, including BLAST *(21)*, CAP3 *(20)*, SPUTNIK *(22)*, PRIMER3 *(23)*, and custom scripts coded in Perl and Bioperl *(24)*.

2.1. SSRPrimer

SSRPrimer is a Web-based tool that may also be run on the command line. Access to the Web server version requires an Internet connection and a standard Web browser. The input is in the form of single or multiple FASTA format DNA sequences. PRIMER3 options are default with the following exceptions selected to increase primer specificity. One set of primer pairs are designed at least 10 bp distant from either side of the identified SSR. Optimum size for the primers is 21 bases with a maximum of 23 bases. Optimum melting temperature is 55°C with a minimum of 50°C and maximum of 70°C. The maximum GC content is set to 70%. While these options may be modified on the SSRPrimer submission page, the authors suggest maintaining these strict criteria to ensure robust PCR amplification. All SSR PCR primers maintained within the SSR Taxonomy Tree were identified using these parameters.

SSRPrimer requires SPUTNIK and PRIMER3 as well as Perl5 and custom Perl scripts. The SSR Taxonomy Tree additionally requires MySQL. For SSR discovery, input sequences are subdivided into groups of 10. Each group is parsed to SPUTNIK, which uses a recursive algorithm to search for repeated patterns of nucleotides of length between 2 and 5. The output of SPUTNIK is then parsed to PRIMER3 for PCR primer design. The results from both SPUTNIK and PRIMER3 are combined, appended to a results file, and displayed on the Web output of SSRPrimer or parsed to a MySQL database to create the SSR Taxonomy Tree. Input of Web form data is limited to 256 KB or approximately 200,000 bp.

2.2. SNPServer

SNPServer is based on the command line SNP discovery tool autoSNP and can be run either as a command line tool or as a Web server. Access to the Web server version requires an Internet connection and a standard Web browser. The host Web server requires BLAST as well as a selection of BLAST formatted databases, CAP3, Bioperl, and the custom SNPServer perl scripts.

Parameters for BLAST and CAP3 can be selected via the Web interface, and users are encouraged to assess different parameters as optimal parameters are strongly affected by the haplotype complexity of the gene of interest (*see* **Note 2**).

3. Methods
3.1. Identification of SSRs Within DNA Sequences and the Design of PCR Primers for Their Amplification

With the increase in DNA sequence information available for marker discovery, an automated process to identify and design PCR primers for the amplification of SSR loci is a useful tool for molecular plant-breeding programs. We describe an application that integrates SPUTNIK, an SSR repeat finder, with Primer3, a PCR primer design program, into one pipeline tool, SSRPrimer. On submission of multiple FASTA-formatted sequences, the script screens each sequence for SSRs using SPUTNIK (*see* **Fig. 1**). Results are parsed to Primer3 for locus-specific primer design. The script makes use of a Web-based interface enabling remote use.

1. First identify the sequences you wish to search for SSR molecular markers. In this example, we will use *Brassica* genes annotated with the term "drought".
   Identify these genes initially from GenBank at the National Center for Biotechnology Information (NCBI) (http://www.ncbi.nlm.nih.gov/).
   Select "Nucleotide" as the database and "brassica [orgn] AND drought" as the query (*see* **Fig. 2**). In this example, 82 sequences were identified.
   To display the sequences as FASTA format, select FASTA from the display dropdown menu and show 100 as TEXT. All 82 sequences should now be seen in FASTA format. Select all (control A) and copy onto the clipboard.

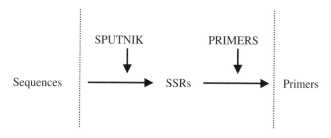

Fig. 1. SSRPrimer process pipeline.

Fig. 2. Retrieval of *Brassica* sequences from GenBank which are annotated with the term "drought".

2. Open the SSRPrimer Web site (http://hornbill.cspp.latrobe.edu.au/cgi-binpub/ ssrprimer/indexssr.pl). Within the sequence window, paste the *Brassica* sequences from the clipboard (*see* **Fig. 3**) and click on "Find SSR Primers".
3. SSRs were identified and PCR primers designed for four of the 82 sequences (*see* **Fig. 4A** and **B**). The gene name derived from the FASTA header, repeat type, SSR sequence, and position within the entry sequence is listed. Column titles "Left Sequence" and "Right Sequence" contain PCR primers designed to amplify the identified SSR. Scrolling the Web page to the right (*see* **Fig. 4B**) provides in further columns detailed information on the position of the PCR primers within the entry sequence, and primer characteristics derived from the PRIMER3 software. The data table can be selected and copied into an Excel file for further manipulation.

3.2. Identification of PCR Amplification Primers for Rosaceae SSRs from the SSR Taxonomy Tree

SSR amplification primers are often transferable between related species, enabling comparative genetic mapping and access to molecular markers for application in species with limited genomic resources. To provide access to

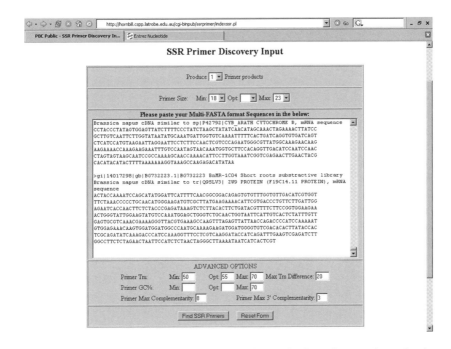

Fig. 3. The sequence entry page for SSRPrimer. Options for number of primers, primer size, and characteristics may be set here.

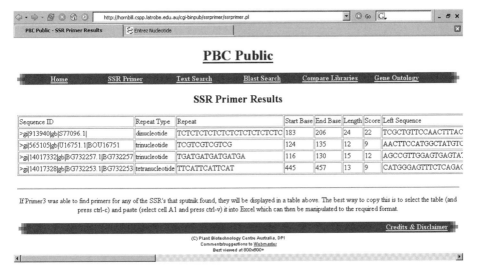

Fig. 4. (**A**) Example results from SSRPrimer.

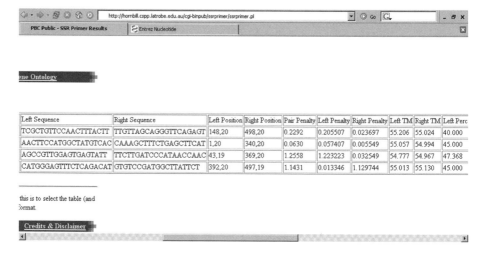

Fig. 4. (**B**) Example results from SSRPrimer.

SSR amplification primers for a broad range of species, we have applied the SSRPrimer tool for the discovery of SSRs within the complete GenBank database (currently updated to release 148) and have designed PCR amplification primers for almost 14 million SSRs. The SSR Taxonomy Tree tool provides Web-based searching and browsing of species and taxa for the visualization and download of these SSR amplification primers. The tool provides two access methods. Users may enter a search string of either a common name or taxonomic key. The tool also permits browsing between related taxa. SSRs can be viewed or downloaded for specific species or taxonomic groups.

1. Open the SSR Taxonomy Tree Web site (http://bioinformatics.pbcbasc.latrobe .edu.au/cgi-bin/ssr_taxonomy_browser.cgi).
2. Within the search box, type "rosaceae" and click on "Search" (*see* **Fig. 5A**). Two matches are identified, Rosaceae and the sub-taxa Rosaceae incertae sedis (*see* **Fig. 5B**). Click on "Rosaceae". All identified SSR primer pairs may be downloaded by clicking on "Download" (*see* **Fig. 5C** and **Note 3**).
3. To identify SSR primers for the sub-taxa *Fragaria* (strawberry), browse the taxonomic tree by clicking on "Rosoideae," then *"Fragaria"* (*see* **Fig. 6A** and **Note 4**). A total of 704 SSR primer pairs are available for several species of *Fragaria* (*see* **Fig. 6B**).
4. To view these SSR primer pairs, click on "View". Alternatively, the information may be downloaded by clicking on "Download". A list of SSRs is presented.

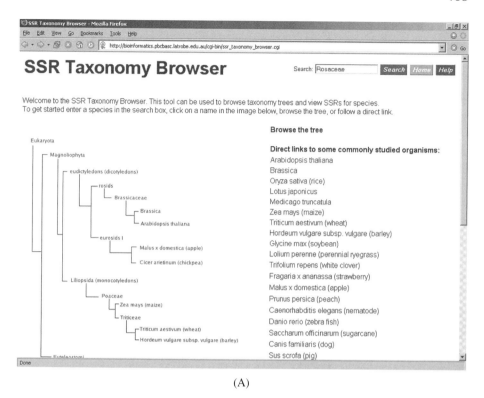

(A)

Fig. 5. (**A**) The search page for SSR Taxonomy Tree with direct links to results for commonly studied organisms.

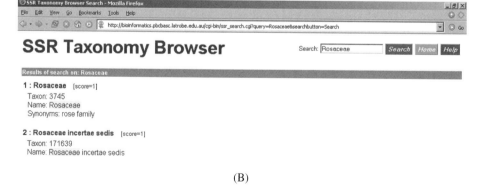

(B)

Fig. 5. (**B**) Results from an SSR Taxonomy Tree search with the query "Rosaceae".

Fig. 5. (**C**) List of Rosaceae sub-taxa results.

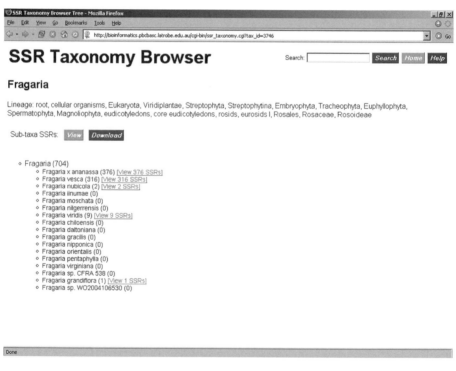

Fig. 6. (**A**) List of *Fragaria* sub-taxa results.

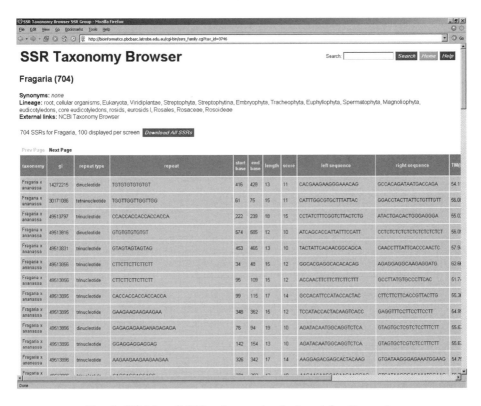

Fig. 6. (**B**) List of SSR primer pairs designed for *Fragaria*.

3.3. SNPServer: Identification of SNPs Within a Region of a Starch Synthase Gene of Maize

The real-time autoSNP Web server, SNPServer, acts as a Web interface and wrapper for the three programs, BLAST, CAP3, and autoSNP, that make up the SNP discovery pipeline (*see* **Fig. 7**). The complete pipeline accepts a single sequence as input. This entry sequence is compared with a specified nucleotide sequence database using BLAST to identify related sequences. The resulting sequences may then be selected for assembly with CAP3 and subsequent SNP discovery using autoSNP. Alternatively, users may enter a list of sequences in FASTA format for assembly, or a pre-calculated sequence assembly in ACE format. Complete options for BLAST sequence comparisons, CAP3 assembly, and SNP discovery may be specified at the user interface. SNP discovery is performed using a redundancy-based approach with a modified version of the autoSNP PERL script. Alignment data generated by CAP3 (or from a user-

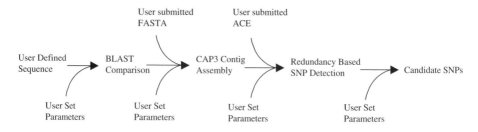

Fig. 7. SNPServer process pipeline.

>gi|32918437|gb|CF023249.1|CF023249 QBQ26e12.xg QBQ Zea mays cDNA clone QBQ26e12,
mRNAsequence

GACAACACAACACATATCTTTTAGTAACCTTTCAATAGGCGTCCCCCAAGAACTAGTAAACATGGCAAA
ACTGTTGAGCCTGTTCTTAGCCCTCTCCTTTGTAGCAGCTATGTTTGCCATAGGCTCTCATGCTGATGG
AGACTGTGAGAATGACCTCCAAAGCCTGATAAGTGACTGCAAGACATACGTGATGTTCCCAGCTAATCC
AAAGGTCCCTCCTTCGGATGCTTGCTGTGGTGTCATCAAGAAGGCAAACGTTTCTTGCTTGTGCTCCAA
GGTCACCAAAGAAACTGAGAAGGTTGTGTGCATGGAGAAGGTTGTGTATGTTGCTGAGCAGTGTAAGAG
GCCATTTGAGCATGGCTTCCAGTGCGGAAGCTACAAGGTTCCAGCAAACTGATGATATCACAAGAGGGA
AGCGCTTGACGTAGTCTTTAGATTTCAATTGTTGCAGCCGATCAACCTTCTTAGGGTTGTATAATATGA
AGTCAATGTTGAGACTAATGTGAGGTGGGAGAGGCCTATTGATGCATCAAAAGTATTATGAGCTAATAA
TAAAGCCTTC

Fig. 8. Example sequence for SNPServer assessment.

submitted ACE file) are used to load the sequences in each assembly into a 2D array. Spacing characters (-) added during sequence alignment are considered as a fifth element in addition to the four nucleotides A, C, G, and T. This permits the identification of indel polymorphisms between sequences. Each row (representing a single base locus in the assembly) is assessed for differing nucleotides. Minimum redundancy scores specified by the user and associated with alignment width (the number of sequences included in the contig) determine the number of different nucleotides at a base position required for classification as SNP. Where a SNP is recorded, a SNP score is allocated equal to the minimum number of reads that share a common polymorphism. Where several SNPs are present in an alignment, a co-segregation score is calculated for each SNP. The co-segregation score is a measure of whether the SNP contributes to defining a haplotype. This figure is then normalized to the number of sequences at that SNP position in the alignment, to produce a weighted co-segregation score. HTML format files are generated to allow the user to input data, select comparison, assembly, and SNP discovery parameters, and browse the SNP results.

1. First, obtain the sequence of CF023249, an EST sequenced by Genoplante representing a region of a maize starch synthase gene, from GenBank at NCBI

(http://www.ncbi.nlm.nih.gov/). Specify the database search as "Nucleotide" from the drop-down menu, type "CF023249" into the search bar and press "GO". This should return one match. To display the sequence as FASTA format, select FASTA from the display drop-down menu (*see* **Fig. 8**). Select and copy the resulting FASTA format sequence to the clipboard.

2. Open the SNPServer Web page (http://hornbill.cspp.latrobe.edu.au/cgi-binpub/ autosnip/index_autosnip.pl) and select BLAST.
3. Paste the FASTA format sequence into the box (*see* **Fig. 9A**). From the drop-down menus, select library "Maize" and press "Blast away". When the BLAST is complete, a list of matching sequences is displayed (*see* **Fig. 9B**).
4. Press "Run SNP detection on all sequences". The next page permits the selection of CAP3 assembly parameters and SNP discovery parameters (*see* **Fig. 9C**). For routine SNP discovery, use the default parameters and press "Start".

(A)

Fig. 9. (**A**) The sequence entry page for SNPServer. Options for sequence library, BLAST E-value, and maximum number of records returned may be set here.

(B)

Fig. 9. **(B)** Example results from a SNPServer BLAST search.

(C)

Fig. 9. **(C)** The SNPServer sequence assembly and SNP discovery parameter page. CAP3 options for sequence assembly and minimum SNP redundancy scores may be set here.

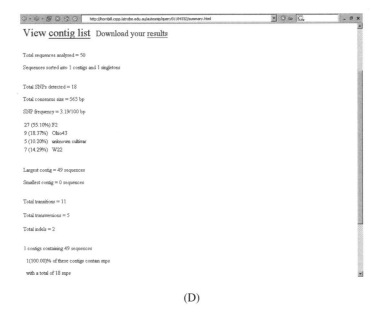

(D)

Fig. 9. (**D**) Example overview results for SNPServer sequence assembly and SNP discovery.

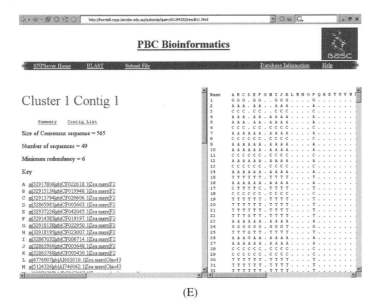

(E)

Fig. 9. (**E**)

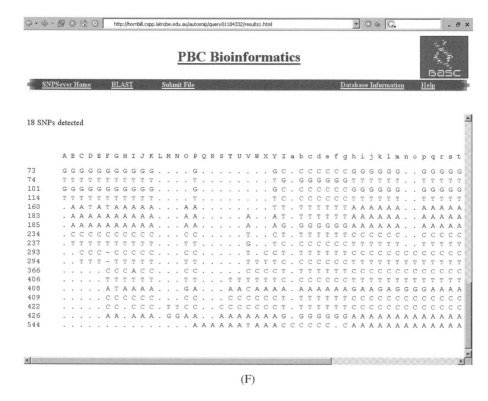

(F)

Fig. 9. **(E)** and **(F)** Example SNPServer results. The right frame shows the assembled sequences running vertically with base No. 1 at the top. The left frame displays the details of the assembled sequences along with a summary of identified SNPs **(F)**.

5. When sequence assembly and SNP discovery are complete, a summary page is displayed (*see* **Fig. 9D**). In this case, of the 50 sequences identified in the BLAST query, 49 sequences assemble together. CAP3 could not assemble the 50th sequence and this was discarded. Further information on the number of transitions, transversions, and indels is also listed. To view the sequence assembly, press "contig list" and then "Contig1." Alternatively, you can download the complete results to a local computer as a compressed file.

6. The main results page lists three types of information within two frames (*see* **Fig. 9E**). Each frame can be expanded by dragging the central bar. The right-hand frame provides the complete sequence assembly with sequences running vertically down the page. The base position of the assembly is to the left of this frame, and predicted SNPs are highlighted in color with a note of the SNP redundancy score and the SNP type.

7. The left frame provides a list of the assembled sequences coded alphabetically and by case. Plant variety or cultivar names are appended to the name to ease interpretation of the results. Each name is linked to the original sequence maintained at NCBI.

8. Below the list of sequence names within the left frame is a summary of the SNPs identified within the assembly, with the respective base-pair position in the assembly to the left and the alphabetic code for each sequence at the top (*see* **Fig. 9F**). The calculated co-segregation scores for each SNP is to the right.

9. Within this example, clear haplotypes can be observed (*see* **Note 5**). The majority of the predicted SNPs differentiate between the Ohio43 cultivar and the other cultivars. The SNP predicted at position 408 is of particular interest, as this position appears to differentiate W22 from the other cultivars. Several unknown cultivars are represented in the assembly, of these, N and Z are predicted to be Ohio43 like, h is similar to W22, and j and l are similar to F2.

4. Notes

1. PolyPhred integrates Phred base calling and peak information, within Phrap-generated sequence alignments *(25)*, with alignments viewed and marked for inspection using Consed *(26)*. More recently, this approach has been extended to include Bayesian statistical analysis. PolyBayes *(27)* is a fully probabilistic SNP detection algorithm that calculates the probability that discrepancies at a given location of a multiple alignment represent true sequence variations as opposed to sequencing errors. The calculation takes into account the alignment depth, the base calls in each of the sequences, the associated base quality values, the base composition in the region, and the expected *a priori* polymorphism rate.

2. The optimal parameters for SNP discovery are dependent on the available sequence data and the final application of the SNPs. Where submitted sequence quality is known to be near-perfect, the redundancy score may be reduced to 1 to display all polymorphic sites, even if they occur only once in an alignment. Where the object is to assess the genetic diversity of a gene between distantly related species, less stringent CAP3 sequence assembly parameters may be applied. It is recommended that users first apply the default SNPServer parameters and return to modify the BLAST, CAP3, and SNP discovery parameters and compare the results.

3. SSRs for species and sub-taxa may be viewed and downloaded using the SSR Taxonomy Browser. When sets are smaller than 3 Mb (approximately 21,000 SSRs), they may be viewed or downloaded. When sets are smaller than 5 Mb (approximately 34,000 SSRs), they may be downloaded. Some larger sets have been compressed and can be downloaded from the Large Downloads page. If you require a set larger than 5 Mb to be included in Large Downloads, contact Dave Edwards (dave.edwards@acpfg.com.au).

4. Alternatively, repeat the search using the term *Fragaria* or strawberry.

5. SNPs 73, 74, 101, 114, 183, 185, 234, 237, 406, 409, 422, 426, and 544 co-segregate and differentiate between the Ohio43 cultivar and the other cultivars. SNPs 160, 293, 294, and 366 would co-segregate with the bulk of the SNPs if it were not for minor sequence changes. SNP 160 is unique because of sequences D, F, and Y. Examination of the sequence surrounding SNP 160 shows a string of mononucleotides suggesting that these differences within sequence D and F may be sequence errors. Sequence F presents a 14-bp deletion within the region of predicted SNPs 293 and 294. This deletion prevents these two SNPs from co-segregating with the bulk of the SNPs. It is unclear from examining this data whether this 14-bp indel deletion is due to sequencing error or whether it reflects the presence of a second copy of this gene in the maize line F2. The SNP at 366 bp appears to be triallelic, though this is likely to be due to a sequence error in sequence I. The SNP at base 408 appears to differentiate between W22 and the other cultivars due to the presence of additional sequence variation within lines G and V.

References

1. Gupta, P.K., Roy, J.K. and Prasad, M. (2001) Single nucleotide polymorphisms: a new paradigm for molecular marker technology and DNA polymorphism detection with emphasis on their use in plants. *Curr. Sci.* 80, 524–535.
2. Brumfield, R.T., Beerli, P., Nickerson, D.A. and Edwards, S.V. (2003) The utility of single nucleotide polymorphisms in inferences of population history. *Trends Ecol. Evol.* 18, 249–256.
3. Collins, A., Lau, W. and De la Vega, F.M. (2004) Mapping genes for common diseases: the case for genetic (LD) maps. *Hum. Hered.* 58, 2–9.
4. Rafalski, A. (2002) Applications of single nucleotide polymorphisms in crop genetics. *Curr. Opin. Plant Biol.* 5, 94–100.
5. Tóth, G., Gáspári, Z. and Jurka, J. (2000) Microsatellites in different eukaryotic genomes: survey and analysis. *Genome Res.* 10, 967–981.
6. Katti, M.V., Ranjekar, P.K. and Gupta, V.S. (2001) Differential distribution of simple sequence repeats in eukaryotic genome sequences. *Mol. Biol. Evol.* 18, 1161–1167.
7. Tautz, D. (1989) Hypervariability of simple sequences as a general source for polymorphic DNA markers. *Nucleic Acids Res.* 17, 6463–6471.
8. Powell, W., Machray, G.C. and Provan, J. (1996) Polymorphism revealed by simple sequence repeats. *Trends Plant Sci.* 1, 215–222.
9. Kashi, Y., King, D. and Soller, M. (1997) Simple sequence repeats as a source of quantitative genetic variation. *Trends Genet.* 13, 74–78.
10. Ross, C.L., Dyer, K.A., Erez, T., Miller, S.J., Jaenike, J. and Markow, T.A. (2003) Rapid divergence of microsatellite abundance among species of *Drosophila*. *Mol. Biol. Evol.* 20, 1143–1157.

11. Subramanian, S., Mishra, R.K. and Singh, L. (2003) Genome-wide analysis of microsatellite repeats in humans: their abundance and density in specific genomic regions. *Genome Biol.* 4, R13.

12. Gupta, M., Chyi, Y-S., Romero-Severson, J. and Owen, J.L. (1994) Amplification of DNA markers from evolutionarily diverse genomes using single primers of simple-sequence repeats. *Theor. Appl. Genet.* 89, 998–1006.

13. Edwards, K.J., Barker, J.H.A., Daly, A., Jones, C. and Karp, A. (1996) Microsatellite libraries enriched for several microsatellite sequences in plants. *Biotechniques* 20, 758–760.

14. Robinson, A.J., Love, C.G., Batley, J., Barker, G. and Edwards, D. (2004) Simple sequence repeat marker loci discovery using SSRPrimer. *Bioinformatics* 20, 1475–1476.

15. Barker, G., Batley, J., O'Sullivan, H., Edwards, K.J. and Edwards, D. (2003) Redundancy based detection of sequence polymorphisms in expressed sequence tag data using autoSNP. *Bioinformatics* 19, 421–422.

16. Batley, J., Barker, G., O'Sullivan, H., Edwards, K.J. and Edwards, D. (2003) Mining for single nucleotide polymorphisms and insertions/deletions in maize expressed sequence tag data. *Plant Physiol.* 132, 84–91.

17. Ewing, B. and Green, P. (1998) Base-calling of automated sequencer traces using phred. II. Error probabilities. *Genome Res.* 8, 186–194.

18. Ewing, B., Hillier, L., Wendl, M.C. and Green, P. (1998) Base-calling of automated sequencer traces using phred. I. Accuracy assessment. *Genome Res.* 8, 175–185.

19. Savage, D., Batley, J., Erwin, T., Logan, E., Love, C.G., Lim, G.A.C., Mongin, E., Barker, G., Spangenberg, G.C. and Edwards, D. (2005) SNPServer: a real-time SNP discovery tool. *Nucleic Acids Res.* 33, W493–W495.

20. Huang, X. and Madan, A. (1999) CAP3: a DNA sequence assembly program. *Genome Res.* 9, 868–877.

21. Altschul, S.F., Gish, W., Miller, W., Myers, E.W. and Lipman, D.J. (1990) Basic local alignment search tool. *J. Mol. Biol.* 215, 403–410.

22. Abajian, C. (1994) SPUTNIK, http://abajian.net/sputnik/.

23. Rozen, S. and Skaletsky, H.J. (2000) Primer3 on the WWW for general users and for biologist programmers. In: Krawetz, S. and Misener, S. (eds) *Bioinformatics Methods and Protocols: Methods in Molecular Biology*. Humana Press, Totowa, NJ, pp. 365–386.

24. Stajich, J.E., Block, D., Boulez, K., Brenner, S.E., Chervitz, S.A., Dagdigian, C., Fuellen, G., Gilbert, J.G.R., Korf, I., Lapp, H., Lehvaslaiho, H., Matsalla, C., Mungall, C.J., Osborne, B.I., Pocock, M.R., Schattner, P., Senger, M., Stein, L.D., Stupka, E., Wilkinson, M.D. and Birney, E. (2002) The Bioperl Toolkit: Perl modules for the life sciences. *Genome Res.* 12 (10), 1611–1618.

25. Green, P. (1994) Phrap, unpublished. http://www.phrap.org/.

26. Gordon, D., Abajian, C. and Green, P. (1998) Consed: a graphical tool for sequence finishing. *Genome Res.* 8, 195–202.
27. Marth, G.T., Korf, I., Yandell, M.D., Yeh, R.T., Gu, Z.J., Zakeri, H., Stitziel, N.O., Hillier, L., Kwok, P.Y. and Gish, W.R. (1999) A general approach to single nucleotide polymorphism discovery. *Nat. Genet.* 23, 452–456.

24

Methods for Gene Ontology Annotation

Emily Dimmer, Tanya Z. Berardini, Daniel Barrell and Evelyn Camon

Summary

The Gene Ontology (GO) is an established dynamic and structured vocabulary that has been successfully used in gene and protein annotation. Designed by biologists to improve data integration, GO attempts to replace the multiple nomenclatures used by specialised and large biological knowledgebases. This chapter describes the methods used by groups to create new GO annotations and how users can apply publicly available GO annotations to enhance their datasets.

Key Words: Ontology; functional annotation; controlled vocabulary; biological database; gene product classification; automatic annotation.

1. Introduction

In recent years there has been a rapid growth in the amount of biological data generated by sequencing projects and large-scale experiments. With this increase it has become necessary for scientific databases to be able to represent information in a well-defined and standardized manner.

The Gene Ontology (GO) has become the recognised standard vocabulary used by the scientific community to describe the function and subcellular location of gene products. This resource has been widely used to associate functional information with biological database entries, and exploited by data analysis tools in biomedical research *(1,2)*.

From: *Methods in Molecular Biology, vol. 406: Plant Bioinformatics: Methods and Protocols*
Edited by: D. Edwards © Humana Press Inc., Totowa, NJ

GO consists of three controlled vocabularies of terms that represent the function of gene products in a species-independent manner. Terms are arranged in a hierarchical structure (an ontology) where each term is explicitly related to its parent and child terms.

The GO Consortium (GOC) was founded in 1998 by three model organism groups – FlyBase, SGD and MGI *(3)*. Over the last seven years the GOC has grown considerably and now consists of 15 model organism and multi-species groups tasked with both maintaining and developing the GO vocabulary and providing annotations (associations of genes or gene product identifiers with GO terms). GOC members include the Arabidopsis Information Resource (TAIR, *see* Chapter 8), The Institute for Genomic Research (TIGR, *see* Chapter 5), Gramene (*see* Chapter 15) and GO Annotation (GOA) (which supplies GO annotations to the UniProt KnowledgeBase, *see* Chapter 4) groups, which carry out GO annotation of plant species. As of April 2007, gene products in over 120,000 species have some GO annotation (including 31,808 plant species), comprising over 16 million annotations *(4)*.

GO annotations are made publicly available by the GOC and contain essentially four pieces of information: a gene or protein identifier, a GO term, a reference, and a code indicating the type of evidence found in the reference to support the association. When using GO annotations it is important to be aware of the kind of information they contain and the techniques and the evidence codes used.

The GO community is very active and the GOC is keen to assist new groups interested in contributing GO annotations *(5)*. A wide range of tools have been developed by both GOC members and external groups to facilitate the searching and analysis of GO, as well as the creation and use of its annotations. The vast majority of these tools and resources are freely available and can be located from the GOC Web site (*see* **Table 1**).

GO does not try to capture all known information about a gene or protein. Other groups, such as the Plant Ontology Consortium (POC, *6*) are developing controlled vocabularies to describe attributes such as phenotypes, structures, or developmental stages. Many of these vocabularies can be found at the OBO Foundry Ontologies Web site (*see* **Table 1**). To be included on this site, an ontology must be freely available, share a common format, and be orthogonal to the other ontologies already listed.

This chapter describes the structure of GO, how manual and electronic annotations to this vocabulary are made, some of the software tools available, and how one can apply existing GO annotations to enhance the functional information displayed by a database or analysis tool.

Table 1
Table of Useful URLs

GO Groups	URL
Gene Ontology Consortium	http://www.geneontology.org
GO SourceForge site	http://sourceforge.net/projects/geneontology
Open Biomedical Ontologies	http://obo.sourceforge.net
GOA	http://www.ebi.ac.uk/GOA
Gramene	http://www.gramene.org/plant_ontology/index.html
TAIR	http://www.arabidopsis.org
TAIR FTP site	ftp://ftp.arabidopsis.org/home/tair/Ontologies/ Gene_Ontology
TIGR	http://www.tigr.org
GO Browsers	
QuickGO	http://www.ebi.ac.uk/quickgo
AmiGO	http://www.godatabase.org
TAIR's Keyword Browser	http://arabidopsis.org/servlets/Search? action=new_search& type=keyword
TAIR's GO annotation search, functional categorization and download tool	http://arabidopsis.org/tools/bulk/go/index.jsp
Manual annotation tools	
TAIR's Preformated Excel spreadsheet	http://www.arabidopsis.org/
Manatee (TIGR)	http://manatee.sourceforge.net
PubSearch	http://www.pubsearch.org
iHOP	http://www.ihop-net.org/UniPub/iHOP
Whatizit	http://www.ebi.ac.uk/Rebholz-srv/whatizit/
CiteExplore	http://www.ebi.ac.uk/citations
Electronic annotation tools	
InterProScan	http://www.ebi.ac.uk/InterProScan
TargetP	http://www.cbs.dtu.dk/services/TargetP
Additional Annotation and Ontology files	
NCBI Gene2GO	ftp://ftp.ncbi.nih.gov/gene/DATA
TAIR GO slim	http://www.arabidopsis.org/help/helppages/ go_slim_help.jsp

2. Methods

2.1. The Gene Ontologies

GO terms are organised into three ontologies that characterise a gene product's biological activities or location within a generic cell. These are Molecular Function, Biological Process and Cellular Component. Molecular Function terms describe the discrete biochemical activities that a protein may carry out (e.g. 'kinase activity' or 'DNA binding'), Biological Process terms describe the broader pathways and activities that a gene product's functions are organised into (e.g. 'mitosis' or 'signal transduction'), and Cellular Component terms describe the subcellular components where mature gene products act (e.g. 'nucleus' or 'plasma membrane'). GO terms only describe normal cellular activities or locations, therefore those processes, functions or localisations that are unique to mutants or diseases are not included (e.g. oncogenesis).

GO terms possess a number of distinct features. Significantly each term has a name, a unique seven-digit identifier prefixed by 'GO:' (the number is arbitrary and does not have any meaning or relation to the position of the term), and a position within the ontology. Of the 23,029 terms (as of April 2007) in the three GO ontologies, 96.5% have a definition and 54.6% contain at least one synonym. Terms and relationships are added and updated regularly by members of the GO editorial team based at the European Bioinformatics Institute (EBI) and by other members of the GOC.

GO is organised in a Directed Acyclic Graph (DAG) structure. DAGs allow GO terms to be structured in a tree-like organization where broader terms at the top (parents) progressively link down to more specific terms (children) (*see* **Fig. 1**). Therefore GO can include both general (e.g. metabolism), and very specific concepts (e.g. L-methylmalonyl-CoA biosynthesis). Parent-child relationships between terms mean that every GO term inherits the properties of its parent term(s) (called the 'true path rule') and so when a GO term is assigned to a gene product all of the parent terms must also apply. Any term that breaches this rule should be brought to the notice of the GO editorial office where the entry can be corrected (*see* **Note 1**). However, unlike a tree-like hierarchy, a GO term may possess more than one relationship with any of its children and/or parent(s). Presently in GO there are two different types of relationships a parent term can have with a child term: a child term can be described as either being a type of ('is_a') or a 'part_of' a parent term. For instance, in the Cellular Component ontology, the term 'plastid' (GO:0009536) (*see* **Fig. 1**) has two parents, as it 'is_a' specific type of 'intracellular membrane-bound organelle' (GO:0043231) and also 'is_a' type of 'cytoplasmic part' (GO:0044444). Additionally, the parent term

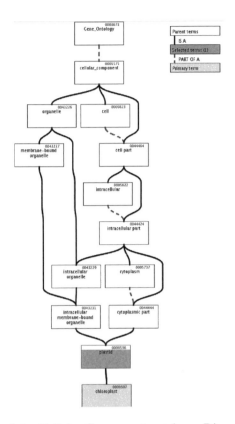

Fig. 1. A section of the Cellular Component ontology. Diagram showing relationships between a few terms in the Cellular Component ontology displaying the relationships that GO terms can have with each other. Terms describing more general concepts are located at the top of the ontology, which progressively link down to more specific terms. The 'is_a' relationships (continuous line) indicate the child term is a type of the parent term, while the 'part_of' relationship type (dashed line) indicate the child term is part of the parent term. A term may have multiple parents and/or children.

'cytoplasmic_part' is_a component (part_of) of the 'cytoplasm' (GO: 0005737). Editors of the GO have provided 'is a' parentage to the root of the ontology for all GO terms, allowing users to view the 'is a' and 'part of' trees independently. This 'is a' completeness is particularly useful when only a subset of the whole GO is needed (*see* **Subsection 2.6**).

The terms in the GO are intended to be as species-independent as possible, however when there is a possibility that a concept could be interpreted differently by different research communities, a term name is selected that is neutral but as descriptive as possible, and an extensive set of synonyms are provided for any community-specific name, ensuring that all users are able to successfully search the GO.

If a GO term is found to be wrongly defined and the correction proposed is agreed to by all groups, then the term will become obsolete. This means the term is tagged as with 'is_obsolete: true' in the ontology and will often contain an explanation in the comments section of the term's entry where the reason for the obsoletion is stated and suitable alternative terms are often suggested. Therefore, although the obsoleted term is still present within the ontology for reference purposes, its obsolete tag indicates to users that it should no longer be used in annotation.

2.1.1. Accessing the Gene Ontologies

GO can be downloaded in a number of different formats from the GOC Web site (*see* **Table 1**). OBO flat files and XML formats are updated daily and GO flat file MySQL formats are updated weekly. The OBO formats is preferred as the content is more easily parsed, is more human readable, contains minimal redundancy, and is easier to maintain as GO grows.

2.1.2. Web Sites for Browsing GO

To enable the scientific community to effectively use the GO vocabularies and annotations, a number of web-based tools have been developed by both members of the GOC and third parties to search, browse and view GO terms. All such tools are listed on the GOC Web site (*see* **Table 1**) and include the official GOC browser AmiGO, QuickGO (produced by the GOA group at the EBI), and TAIR's Keyword Browser (*see* **Table 1**). Each browser has a number of unique features and it is worth trying a couple of different tools initially to compare their functionality.

2.1.3. Worked Example on the AmiGO Browser

See **Fig. 2** for screenshots of the AmiGO browser. When the AmiGO browser is first opened, the site just shows a simple query box − allowing the user to start by searching for a particular GO term or an annotation to a particular gene or protein. Users can also browse the GO (by clicking on the 'Browse' option in the top menu. When browsing, only the roots of the three ontologies are initially displayed: Biological Process, Cellular Component and Molecular Function. Users can open up the GO hierarchy further by clicking on the '+' (which expands a node, showing child terms) and '−' (which closes the node,

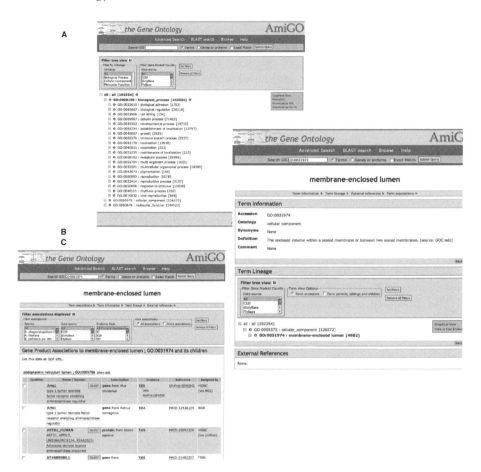

Fig. 2. Screenshots from the AmiGO browser. (**A**) The Browse page of the AmiGO browser, containing a partially opened GO hierarchy. The top-most terms under Biological Process have been expanded in this view. The unexpanded Cellular Component and Molecular Function ontologies are also shown. Numbers in parentheses after terms indicate the number of gene or gene product identifiers that have been manually annotated to the term or one of its children. Search options are located in the blue menu at the top of the page. (color as seen on the Web site) (**B**) AmiGO page showing details of one GO term, including the term's description, any synonyms, external database cross-references and position within the GO hierarchy (**C**) AmiGO page showing the annotation detail for gene/gene products which have been associated with the GO term 'membrane-enclosed lumen'.

hiding child terms) icons next to parent nodes, shown to the left of a GO term. The number in parentheses after a GO term indicates the number of gene/gene product identifiers that have been manually annotated to the term or its children.

Clicking the round red, green and blue pie chart icon, shown on the right-hand side of term names in the main browse AmiGO page, will display a pie chart showing the percentage of gene products annotated to the child terms that are being displayed.

All the information available for a GO term is shown upon clicking a term. The page that subsequently appears displays the term's definition, any synonyms, its position in the GO hierarchy and external database references. By clicking on the 'Term associations' link, users will be provided with all genes/gene products manually annotated either to the term itself or to the term and its children. In this view, clicking on a gene symbol will link you to the AmiGO gene product detail page, which shows the information held in the GO database about the particular gene/gene product, including all of its GO annotations and the protein sequence (if available).

The advanced search in AmiGO allows the user a number of additional options, including searching for multiple GO terms or gene names, searching particular fields in the GO ontology and filtering results based on species or source database.

The BLAST search option enables uers to perform a sequence search using either BLASTP or BLASTX (from the WU-BLAST package), to find the GO terms which have been associated to similar, annotated sequences (located when clicking on the 'get annotation summary' radio button. This function can also be called when looking at a particular annotation, where clicking on the adjacent 'BLAST' button will submit the sequence of the annotated identifier to the BLAST tool.

2.2. GO Annotations

Annotations are basically associations of GO terms with gene or protein identifiers. An identifier can be annotated to multiple GO terms at any position in each of the three GO categories. The majority of publicly available GO annotations are produced by species-specific databases (e.g. TAIR or FlyBase) or multi-species resources (e.g. GOA, TIGR and Gramene).

There are two ways of creating annotations – either by an annotator reading through scientific literature and manually creating each annotation, or by applying computational techniques. The following sections describe how both manual and electronic annotations are created and how these annotations can be applied to enhance the coverage of other datasets. Readers interested in only using existing GO annotations are advised to read the following sections, as an understanding of how these annotations are made is crucial for their correct interpretation and utility.

2.2.1. Evidence Codes

There are twelve GOC-agreed evidence codes (*see* **Table 2**) that are used to class the type of experiment or analysis upon which an annotation was made. These codes provide a broad classification of how an annotation to a particular GO term is supported. Codes include 'IEA', to tag information obtained from

Table 2
Go evidence codes. Descriptions of the twelve GOC-agreed evidence codes, which describe the broad categories of evidence found to support a GO term-gene product association. Apart from the 'IEA' code, all others are used in manual annotation

Evidence code	Full name of evidence code	Examples of usage
IDA	Inferred from Direct Assay	Enzyme assays *In vitro* reconstitution (transcription) Immunofluorescence or cell fractionation (for cellular component) Physical interaction/binding assay
IEP	Inferred from Expression Pattern	Transcript levels (e.g. Northern blotting, microarray data) Protein levels (e.g. Western blotting)
IGI	Inferred from Genetic Interaction	'Traditional' genetic interactions such as suppressors, synthetic lethals Functional complementation Rescue Experiments Inference from the phenotype of a mutation in a different gene
IMP	Inferred from Mutant Phenotype	Any gene knockout/mutation Overexpression/ectopic expression of a wild-type or mutant gene Anti-sense experiments Specific protein inhibitors

(Continued)

Table 2
Continued

Evidence code	Full name of evidence code	Examples of usage
IPI	Inferred from Physical Interaction	Immunoprecipitation Binding assay (ion/protein) Two-hybrid interactions Co-purification Co-immunoprecipitation
ISS	Inferred from Sequence Similarity	Sequence or structural similarity Southern blotting Recognised domains
RCA	Inferred from reviewed computational analysis	Predictions from large-scale experiments (e.g. genome-wide two-hybrids, genome-wide synthetic interactions) or text-based computation (e.g. text mining)
IGC	Inferred from Genomic Context	Operon structure Syntenic regions Pathway analysis Genome-scale analysis of processes
TAS	Traceable Author Statement	Referenced statement on experimental findings which is traceable to a primary paper (e.g. a review paper)
NAS	Nontraceable Author Statement	Author statement that is not supported by any references, such as a hypothesis by the author or information from a paper's abstract
ND	No Biological Data Available	Where no biological data found (this is only used with 'unknown' GO terms)
IEA	Inferred from Electronic Annotation	Computational techniques (e.g. GO mappings, function prediction based on sequence similarity or pattern matching)

Descriptions of the 12 GO Consortium (GOC)-agreed evidence codes, which describe the broad categories of evidence found to support a GO term–gene product association. Apart from the 'IEA' code, all others are used in manual annotation.

an electronic method, 'IDA', which is used to indicate that a direct assay to determine a function or subcellular location was described in the cited paper, or an assertion by the author(s) that does not possess any reference to a primary paper (NAS). All the codes apart from 'IEA' are used in manual annotation. All the codes apart from 'IEA' are used in manual annotation. Users can choose to filter annotations by evidence code, for instance some may prefer only to use annotations having experimental evidence by selecting only those annotations with 'IDA', 'IGI', 'IMP', 'IEP', and 'IPI' evidence codes.

2.2.2. Gene Association Files

All annotations are supplied to the GOC in a text file called a gene association file. This file has a simple fifteen column tab-delimited format (*see* **Table 3**). Each row of the file contains information for one GO term-gene association, including details on the sequence being annotated, the GO term, and the reference used. A sequence identifier may be annotated to many different GO terms and possibly to the same GO term or a term within the same node by different references, evidence codes or annotating databases. Therefore there can be many lines within an association file providing functional information for one gene. The simple structure of the gene association file means that annotations can be easily downloaded into different database structures.

Table 3 details the information held within a gene association file, along with examples from the TAIR and GOA files. An annotation consists of four essential pieces of information. Firstly column 2 of a gene association file contains the identifier of the sequence and column 5 contains the identification number of the GO term being applied. Column 6 provides either the literature reference (normally a PubMed identifier) that an annotator used (for manual annotations) or the method applied (for electronic annotations). Finally an evidence code is included to indicate the type of evidence found to support the assertion that the GO term describes the activity of the gene product (column 7). Further interpretation of GO terms within an annotation can additionally be indicated within the 'Qualifier' column (column 4, *see* **Subsection 2.3.1.**). Information on the gene or gene product being annotated is also included (columns 2, 3, 10, 11, 12 and 13), allowing users to extract accession numbers, gene symbols, names and/or taxonomic identifiers.

Information within column 8 (the 'with' column) is used to hold an additional identifier for annotations using certain evidence codes ('IC', 'IEA', 'IGI', 'IPI' or 'ISS'). This field can contain a GO identifier (for 'IC'-evidenced annotation), a gene or protein identifier (for 'IGI', 'IPI' or 'ISS'-evidenced annotation), or a domain/motif identifier (e.g. INTERPRO: IPR000002). The content of this

Table 3
Information Contained Within Each Column of the Gene Association File,
Showing an Example from GOA and TAIR

Column	Content	Example: GOA	Example: TAIR
1*	Acronym of Database contributing the file	UniProt	TAIR
2	Database identifier of gene/gene product being annotated	O22870	gene:2058217
3	Symbol of the gene/gene product	TL15_ARATH	AT2G43560.1
4*	Qualifier (NOT, colocalizes_with or contributes_to)		
5	GO identifier for the term being used to describe the gene/gene product	GO:0005515	GO:0009543
6	Reference identifier supporting the annotation (Pubmed IDs recommended for manual annotation)	PMID:15352244	TAIR:Publication: 501681963|PMID: 11826309
7	Evidence code (*see* **Table 2**)	IPI	IDA
8*	Additional ('With') identifier	Q42403	
9	Ontology of the GO term (P – Biological Process, M – Molecular Function, C – Cellular Component)	M	C
10*	Name of gene/gene product	Probable FKBP-type peptidyl-prolyl cis-trans isomerase 2, chloroplast precursor	
11*	Synonym, gene symbol or other text e.g. International Protein Index (IPI) identifier	IPI00553170	T1O24.30

12	Indication of the type of entity annotated (gene, transcript or protein)	protein	gene
13	Taxonomic identifier for the species encoding the gene product (usually NCBI taxon id)	taxon:3702	taxon:3702
14	Date of last annotation update (YYYYDDMM)	20020620	20030930
15	Database providing the annotation	UniProt	TAIR

* indicates columns that are not mandatory.

field provides further information about the supporting evidence. For instance, if a paper provides experimental evidence that protein A binds to protein B, the annotation for protein A would contain the 'protein binding' (GO:0005515) term with the Inferred from Physical Interaction (IPI) evidence code, and the identifier for interacting protein B in the 'with' column. The GOA group integrates manually-curated annotations for binary protein-protein interactions from the EBI's IntAct group *(7)* using the term 'protein binding' (GO:0005515) or one of its child terms (e.g. enzyme binding). This collaboration has provided approximately 828 annotations to plant proteins.

2.3. Manual Annotation

As manual annotation is a time-consuming and expensive activity for any database to undertake, many groups have annotation priorities that reflect their users requirements. In addition some databases aim towards annotating as many genes as fully as possible (a gene-based approach) (e.g. GOA), whilst other databases have a paper-by-paper annotation approach (e.g. TAIR).

Manual annotation involves annotators reading the most recent scientific literature and converting this information into sequence identifier-GO term associations. Such work needs to be carried out by experienced biologists as it requires a high-level of domain knowledge to create accurate and detailed annotations. GO annotation follows the guidelines set down by the GOC, and publicly available annotations must conform to these standards. Many of these rules are described in this chapter, however further information is available from the Annotation Guide on the GOC Web site.

Annotators are encouraged to read the entire text of any article they curate so that a complete set of annotations are provided for a sequence, along with the most appropriate evidence codes. To provide maximum detail in an annotation, the most specific term appropriate for the evidence available in the literature should be selected. Annotators capture all relevant data within annotations, including conflicting views so that users can evaluate the evidence supporting the annotation themselves. When terms required by annotators are not available within the GO, they are encouraged to submit requests to the GOC (*see* **Note 1**).

2.3.1. Manual Annotation Qualifiers

In addition to the four main pieces of information required in an annotation (*see* **Subsection 2.2.2.**), extra information can be added into the 'qualifier column' (column 4) which modifies the interpretation of the chosen GO term. When filled, this column can contain one of three values – 'contributes_to', 'co-localizes_with' or 'NOT'. Of these, 'NOT' produces the most drastic change in the interpretation of an annotation. It indicates that whilst the GO term might have been expected to apply to a gene product, an experiment or sequence analysis has found otherwise. For instance the 'NOT' qualifier should be applied where a protein has high sequence similarity to an enzyme but was experimentally found not to possess the activity. The 'NOT' value is also used to document conflicting claims in the literature.

The 'colocalizes_with' qualifier is used with terms from the Cellular Component ontology to identify gene products that may only be transiently or peripherally associated with an organelle or complex. Finally, the 'contributes_to' qualifier is used with terms from the Molecular Function ontology, where it distinguishes the activity of a subunit from the activity of its complex. When annotating a gene product, GO terms that describe the function of the whole complex as well as the contribution made by the individual gene product can be used. To distinguish between these two sets of activities, the qualifier 'contributes_to' value is used where a subunit does not possess the activity of the whole complex.

2.3.2. Annotations to Sequences Whose Function is Unknown

Not all genes, even those that have been described in the literature, have been characterized in the detail necessary to allow annotation to all three GO categories. When this situation occurs, curators are able to annotate the gene or protein identifier to the root GO terms ('molecular_function' (GO:0003674), 'biological_process' (GO:0008150) and 'cellular_component' (GO:0005575)) in combination with the 'ND' evidence code (*see* **Table 2**). Annotations using any of these three terms mean that an attempt to find functional information on the gene product was made, but as of the date of the curation effort, no information was available.

2.3.3. Maintenance of Manual Annotation Standards

To maintain high standards of manual annotation, the GOC provides support by hosting annotation camps and annotation e-mail discussion lists. Extensive training within groups is also recommended. One recent study comparing annotations of the same proteins based on the same papers by different annotators found that there was some variability in the granularity of GO terms chosen *(8)*. Though all annotations were correct, differences resulted from the different domains of specialisation of the different annotators. To ensure consistency of annotation, different groups that annotate highly similar species, such as MGI (mouse), RGD (rat), and GOA (human) also apply annotation consistency checks between orthologs *(9)*.

2.3.4. Software Available for Creating Manual Annotations

Many GOC members have created their own annotation tools for annotators that connect directly to their databases. This allows annotations to be updated immediately, basic checks to be carried out, and records to be maintained on when an annotation was last modified, and the name of the annotator(s). Some of these databases are online and have been made accessible to external contributors on a case-by-case basis. However as the format of the gene association file is so simple and easily read, smaller groups are able to produce annotations using a spreadsheet (e.g. Excel). TAIR has developed a simple spreadsheet (*see* **Table 1**) that can be downloaded and completed by members of the research community and submitted back to the database for loading. All submitted entries are reviewed to ensure data consistency.

The Manatee editor (*see* **Table 1**) is provided by the TIGR group and is a web-based gene evaluation and genome annotation tool that allows users to view, modify and store annotation for prokaryotic and eukaryotic genomes, including GO annotation.

PubSearch (*see* **Table 1**) is a literature curation management system developed by TAIR as part of the Generic Model Organism Database (GMOD) project to facilitate literature-based annotation. It stores gene, paper and controlled vocabulary data, automatically indexes the literature against genes and controlled vocabulary terms, and provides a user-friendly web interface for manual verification of matches and curation.

2.3.5. Manual Annotation Aids

To generate and update annotations, annotators have to navigate through an ever-increasing number of publications and biological databases. This is particularly difficult for well-studied genes, for instance those involved in tumour development or infection. The bioinformatics research community has responded to this surge in data by developing a range of systems to help researchers navigate through these interrelated information resources. Text mining methods in particular have started to play a key role in locating information stored in original literature sources *(10)*.

One of the first generation of freely available text mining systems is Information Hyperlinked over Proteins (iHOP) (*see* **Table 1**) *(11)*. This online service helps users locate information on genes and proteins in journal abstracts archived in PubMed by hyperlinking between the relevant sentences extracted and their corresponding abstracts. Similarly, GOPubMed *(12)* provides a categorization of scientific literature based on either the full or a subset of GO terms.

Another text mining tool called Whatizit (*see* **Table 1**) *(13)* is being developed at the EBI for the annotators of the GOA, IntAct and UniProtKB databases, and will in time be fully applied to the EBI's citation database, CiteXplore (*see* **Table 1**). The Whatizit tool provides a box to upload articles in either html or pdf formats. Depending on the task you request it can highlight gene and protein names (hyperlinked to UniProtKB records) and exact matches to GO terms (hyperlinked to the QuickGO browser).

In addition, in the QuickGO browser (*see* **Table 1**), annotators can use the 'Often Annotated With' section provided in a GO term's entry, where GO terms that are often assigned in tandem are displayed. The assignments are suggested along with statistics calculated on UniProtKB's existing manual and electronic GO annotations *(14)*.

2.4. Computational Annotation Methods

While manual annotation provides high-quality, detailed annotations, it is a slow and expensive activity and as the number of proteins requiring annotation continues to increase exponentially, manual annotation methods alone cannot provide sufficient coverage. This phenomenon is exemplified by UniProtKB/TrEMBL where the number of protein entries have increased by more than thirteen times between 2002 and 2004, reaching a total of 1,400,776 records *(15)*. Computational annotation methods are used by databases to cope with this backlog as they can quickly populate a database, and with conservative usage, can produce reliable although less detailed annotation. With respect to the *Arabidopsis* genome, use of computational techniques by TAIR has provided 75% of the genome with at least one assignment to a known GO term *(16)*. Of the 31,808 plant species that have GO annotation, 99.5% contain only electronic annotation.

After sequencing a species that does not have 'model organism status', computational techniques provide a useful first round of annotation. There are a wide variety of techniques that can be used, including the translation of external terms to GO terms (GO mappings) and function prediction based on sequence similarity or pattern matches (e.g. subcellular localization prediction). All electronic annotations are identified by the Inferred from Electronic Annotation (IEA) evidence code and column 6 in the gene association file should indicate the method used (*see* **Table 3**). Different computational methods will of course provide sets of annotations with different levels of accuracy, coverage and reliability.

2.4.1. GO Mappings

GO mappings are used to exploit other well-established systems of functional annotation and allow the quick transfer of GO terms to a dataset. The GOC Web site (*see* **Table 1**) provides manually-curated translation tables between concepts in external resources and analogous GO terms. For each set of mappings GO does not try to supersede the external system but to compliment it. As the majority of such external concepts were developed for different purposes, the mappings are neither complete nor exact.

Swiss-Prot keywords and Enzyme Commission (EC) numbers have been manually curated into UniProtKB entries for many years and the Swiss-Prot Keyword and EC to GO mappings allow GO terms to be retrofitted to corresponding protein records. This method complements UniProtKB's Plant Proteome Annotation Program whose manual annotation of plant species has greatly improved the number of cross-references available to translate to GO terms *(17)*. Similar alternative controlled vocabularies that possess GO mappings include MIPS FunCat *(18)*, MultiFun *(19)* and MetaCyc biochemical pathways *(20)*.

GO mapping files can also convert annotation provided by tools that analyse sequence features to predict membership of a particular protein family. This includes mappings to InterPro domains and subfamilies *(21)*, HAMAP families *(22)*, TIGR families *(23)* and COG families *(24)*.

An evaluation carried out on the Swiss-Prot keyword, EC, and InterPro to GO mappings found that whilst the GO terms predicted were likely to be less detailed than those chosen manually, all strategies predicted the corrected GO term 91–100% of the time *(8)*.

All GO mapping files have the following line syntax:

database:<identifier> > GO:<term>; GO:<GO_id>

e.g. TIGR_role:11030 73 Amino acid biosynthesis Glutamate family > GO:glutamine family amino-acid biosynthesis; GO:0009084.

2.4.1.1. INTERPRO2GO

The InterPro2GO mapping provides the highest electronic annotation coverage *(8,21)*, supplying 80% of UniProtKB proteins with electronic GO annotation *(21)*. InterPro integrates the different protein signature recognition methods of the InterPro member databases (ProDom, PRINTS, SMART, TIGRFAMs, Pfam, PROSITE, PIRSF, SUPERFAMILY, Gene3D and PANTHER) *(24)* and uses their combined methods to assign proteins into InterPro domains and families. As each method has different strengths and weaknesses this combined approach overcomes the drawbacks of individual techniques and provides highly accurate categorizations. InterPro curators create GO mappings by assigning GO terms to an InterPro identifier that correctly describes the function of all manually-annotated proteins in the UniProtKB/Swiss-Prot database that possess this InterPro domain. GO annotations are then automatically applied to all UniProtKB proteins possessing the same InterPro identifier.

2.4.1.2. EXAMPLE OF USING INTERPROSCAN

The InterProScan service *(25)* (*see* **Table 1**) applies the protein signature recognition methods from InterPro member databases to user-provided genomic or protein sequences. This service is free to all academic and commercial users and offers interactive or e-mail job submissions. The different methods available to analyse an input sequence can be seen on the job submission page (*see* **Fig. 3**). DNA sequences are translated to hypothetical protein sequences produced in all six translation frames which are then queried across the database. InterProScan translation rules are specified in the 'TRANSLATION TABLE' and 'MIN OPEN READING FRAME SIZE' menus on the submission page. **Figure 3** shows the InterProScan job submission page and results produced from querying with an /*Arabidopsis*/ASK1 (*Arabidopsis*/Serine/threonine Kinase 1) protein sequence. The results show that three protein domains and an active site were recognised. Clicking on either the 'Table View', 'Raw Output', or 'XML Output' buttons shows the GO annotation suggested by each match. Some GO tools also allow uploading of InterProScan output files for graphical views of GO coverage.

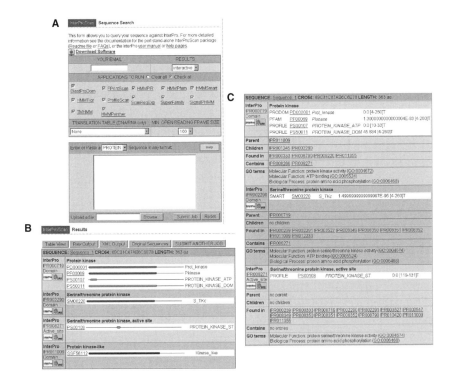

Fig. 3. (**A**) EBI's InterProScan job submission page. (**B**) Graphical results view from InterProScan when queried with the protein sequence of *Arabidopsis* serine/threonine protein kinase ASK1. (**C**) Table results view, showing the GO terms assigned to the queried protein.

2.4.2. Predicting GO Terms Based on Sequence Similarity

A powerful method for annotating an uncharacterised dataset is to transfer annotation from a similar, highly annotated species. The *Brassica* ASTRA database has done exactly this, incorporating the TAIR GO annotation from *Arabidopsis* sequences that had highly similar BLASTx matches to *Brassica* sequences. Using this method 43% of *Brassica* sequences were supplied with electronic GO annotation *(26)*.

Different software teams have released tools to provide this type of service, which often uses BLAST algorithms *(27)* to find homologs to GO-annotated sequences. Matching sequences are scored (often using the BLAST E-value), then GO terms for the hit sequences are retrieved and applied to the input sequence (sometimes together with original evidence codes). Tools allow submission of either single or multiple cDNA or protein sequences (e.g. GoFigure *(28)*, GOtcha *(29)*, GOblet *(30)* and OntoBlast *(31)*), or stretches of genomic sequence (Blast2GO, *(32)*). Many tools have the option of either using the tool online or as a local installation and results can be returned either directly in a graphical form or by e-mail. Such tools tend to have an accuracy of 65–70% *(28,30,32)*. Results can be improved if users have the option of restricting the set of sequences searched against those from a similar taxonomic group *(32)*. In addition, tools such as SIFTER *(33)* and GOAnno *(34)* can transfer annotations between protein sequences based on phylogenetic relationships. For all these tools, as annotations are being transferred automatically between homologs, the resulting evidence code must be 'IEA'.

2.4.3. Electronic Subcellular Localization Prediction

Cellular Component GO terms are difficult to annotate using sequence similarity transfer methods, as often the BLAST algorithm is unable to recognise the targeting information encoded in signal and transit peptides *(29)*. TargetP is a pattern-recognition program used by the TAIR database that detects consensus targeting sequences within a proteome (*see* **Table 1**) *(35)*. The subcellular locations determined by this analysis can be mapped to the GO terms: 'chloroplast', 'mitochondrion' or 'extracellular space'. Parameters are set to use the organism group 'plant' and to the default parameters, with no cutoffs and winner-takes-all.

2.5. Accessing Publicly Available GO Annotations

The GOC maintains a central repository where GO annotations that have been contributed by member databases are stored. Files placed here are passed through a control script that checks that they possess the minimum standard format and partially checks the data. All annotations based on electronic analyses deposited at this site must be updated regularly as those older than one year are removed. In addition, files are filtered so that annotations to particular model species can only be included in an association file provided by a specific group. For example, annotations to *Arabidopsis* (taxon 3702) are limited to the file provided by the TAIR project. The GOC has defined a set of authoritative groups responsible for annotations to the major model organisms.

In addition to the files available from the GO Web site, gene association files can also be downloaded from member database sites which can sometimes contain different groupings of annotations. For example, users can download the GOA-UniProt gene association file from the GOA Web site (*see* **Table 1**) where manual and electronic GOA annotations have been supplemented with manual annotations from twelve other model organism databases (including TAIR, TIGR and Gramene). This data provides a comprehensive multi-species GO resource for the UniProtKB database *(14)*.

Non-standard formats of GO annotations are also available from member databases. Unlike gene association files that can contain many lines of similar gene associations, the gene2GO file from the National Center for Biotechnology Information (NCBI) (*see* **Table 1**) provides annotations to Entrez Gene identifiers in a non-redundant format, which can be useful for high-throughput users. Within this file each line contains a Gene identifier, GO ID, representative GO evidence code, pipe-delimited set of PubMed identifiers, taxonomy identifier, qualifiers of the assignment of the GO term, and the GO term name. An alternative TAIR gene association file is also available from their FTP site (*see* **Table 1**), which additionally includes the *Arabidopsis* Genome Initiative (AGI) locus name, the name of the GO term and its associated GO slim term(s) (*see* **Note 2**).

GOC member databases annotate to a number of different gene or protein identifiers. For instance the GOA and Gramene groups annotate to UniProtKB accessions (e.g. P43291) whereas the TAIR and TIGR groups annotate to AGI gene model codes (e.g. AT1G10940.1). Therefore, when applying annotations to databases, a mapping between identifiers may need to be carried out (*see* **Note 3**).

2.6. Functional Analysis of Datasets and GO Slims

GO has been widely used in the analysis of high-throughput experiments *(1)*, where specific gene or protein sets can be easily clustered and compared with respect to common functional features. Many tools (e.g. FatiGO *(36)*, OntoTools *(37)*) have been developed to allow users to carry out a bulk query of GO using a list of gene, protein or probe identifiers that have been identified from microarray or other high-throughput experiments. Commercial oligo chips are often provided with corresponding GO annotation for the probes (e.g. Affymetrix *(38)*).

To provide a functional classification of genes or proteins in a dataset, especially when the input gene list and/or the number of associated GO terms is large, many analysis tools apply a 'GO slim'. GO slims are cut-down versions of the whole GO, containing a subset of terms that can provide a broad overview of either all or part of the GO content. Tools allow users either to map the terms onto an existing GO slim or to fit the terms to a given level (depth) of the GO hierarchy. Generic and plant-specific GO slims have been created and are maintained by the GOC (*see* **Table 1**), however, users can also create their own

GO slim using OBO-Edit, an ontology editor developed by the GOC (freely available from the GOC Web site). Perl scripts available from the GOC Web site enable users to independently 'map' annotations up to their corresponding GO slim terms (*see* **Note 4**).

2.7. Submission of Annotations to the GOC

The GOC encourages the submission of annotations from external groups and actively seeks to make links with groups that would be interested in annotating *(5)*. To contribute annotations, groups should first contact the GOC directly via the mailing list (go@geneontology.org) and individual annotators should contact the database that most closely represents their species (e.g. *Arabidopsis* annotators should directly contact TAIR). It is important that any annotations submitted are maintained by a member database so that these can be updated as GO develops over time.

There is also a central requirement that any annotation submitted to GO should be to a stable, unique identifier, therefore annotations to transient expressed sequence tag (EST) clusters that only exist until the next clustering analysis run, cannot be deposited at the moment.

3. Notes

1. The GO structure is continually being developed and modified in response to requests from the biological community (both within and outside of the GOC). Requests for new terms and alterations to existing terms are very much welcomed. Users are asked to use the curator request tracker within Gene Ontology's Source-Forge site, which allows requests to be entered, viewed and monitored (*see* **Table 1**). This site, in addition to the GO e-mail archive and meeting reports, allows users to follow developments within GO. New terms are shown in a monthly report, and a daily e-mail of changes, which are available to all users. Databases within the GOC are also very responsive to their user community and welcome feedback concerning any erroneous annotations. Comments should be directed to the member database involved.

2. Whilst annotations submitted to the GOC repository use only the twelve evidence codes, some GOC groups, notably TAIR (*see* **Table 1**), provide annotations from their own site which contain an expanded set of evidence codes describing in further detail the type of evidence found within the literature *(16)*. GO annotations produced by TAIR (available from their site, *see* **Table 1**) also contain an additional annotation field-called 'Relationship type', which can contain one of 21 different values and provide further qualification of the GO term used in the annotation. For instance, such annotations can qualify whether a gene is 'involved in', 'represses' or 'not required for' a certain GO process *(16)*.

3. As mentioned in **Subsection 2.5.**, as different GOC member databases annotate to different gene/protein identifiers (for instance UniProtKB accessions or AGI locus codes), it may be necessary for users to convert the identifier referred to in a gene association. To aid in this the majority of databases supply a 'gp2protein' file (available from the GOC site, *see* **Table 1**) that maps the sequence identifiers to UniProtKB accessions. Additionally the GOA group provides a number of model organism gene association files (including *Arabidopsis*) that are supplied alongside a file of cross-references for conversion to a range of alternative identifiers, such as EMBL/GenBank/DDBJ, Ensembl, RefSeq, TAIR gene symbols and locus identifiers.

4. As the 'NOT' qualifier when present in an annotation breaks the ontology's 'true path rule', it is best to exclude such annotations when using GO slims.

Acknowledgements

This work was supported by Grants QRLT-2001-00015 and QLRI-2000-00981 of the European Commission and a supplementary NIH grant, 1R01HGO2273-01. We would like to thank Jennifer Clark, Linda Hannick, Suzanna Lewis, Jane Lomax and Duncan Legge for their advice.

References

1. Lee, V., Camon, E., Dimmer, E., Barrell, D. and Apweiler, R. (2005) Who tangos with GOA? – use of Gene Ontology Annotation (GOA) for biological interpretation of '-omics' data and for validation of automatic annotation tools. *In Silico Biol.* 5, 5–8.

2. Bada, M., Stevens, R., Goble, C., Gil, Y., Ashburner, M., Blake, J.A., Cherry, J.M., Harris, M. and Lewis, S. (2004) A short study on the success of the Gene Ontology. *J Web Semantics.* 1, 235–240.

3. Lewis, S.E. (2005) Gene Ontology: looking backwards and forwards. *Genome Biol.* 6, 103.

4. Lomax, J. (2005) Get ready to GO! A biologist's guide to the Gene Ontology. *Brief Bioinform.* 6, 298–304.

5. Clark, J.I., Brooksbank, C. and Lomax, J. (2005) It's All GO for Plant Scientists. *Plant Physiol.* 138, 1268–1278.

6. Plant Ontology Consortium (2002) The Plant Ontology Consortium and Plant Ontologies. *Comp Funct Genomics.* 3, 137–142.

7. Hermjakob, H., Montecchi-Palazzi, L., Lewington, C., Mudali, S., Kerrien, S., Orchard, S., Vingron, M., Roechert, B., Roepstorff, P., Valencia, A., Margalit, H., Armstrong, J., Bairoch, A., Cesareni, G., Sherman, D. and Apweiler, R. (2004) IntAct – an open source molecular interaction database. *Nucl. Acids. Res.* 32, D452–D455.

8. Camon, E.B., Barrell, D.G., Dimmer, E.C., Lee, V., Magrane, M., Maslen, J., Binns, D. and Apweiler, R. (2005) An evaluation of GO annotation retrieval for BioCreAtIvE and GOA. *BMC Bioinformatics.* 6, Suppl 1: S17.

9. Dolan, M.E., Ni, Li, Camon, E. and Blake, J.A. (2005) A procedure for assessing GO annotation consistency. *Bioinformatics.* 2, i136–i143.

10. Rebholz-Schuhmann, D., Kirsch, H. and Couto, F. (2005) Facts from text-is text mining ready to deliver? *PloS Biol.* 3, e65.

11. Hoffman, R. and Valencia, A. (2005) Implementing the iHOP concept for navigation of biomedical literature. *Bioinformatics.* 21, Suppl 2:ii252–ii258.

12. Doms, A. and Schroeder, M. (2005) GoPubMed: exploring PubMed with the Gene Ontology. *Nucleic Acids Res.* 33, W783–W786.

13. Rebholz-Schuhmann, D. and Kirsch, H. (2004) Extraction of biomedical facts – a modular Web server at the EBI (Whatizit). Proceedings HDL 2004, Bath, UK.

14. Camon, E., Magrane, M., Barrell, D., Lee, V., Dimmer, E., Maslen, J., Binns, D., Harte, N., Lopez, R. and Apweiler, R. (2005) The Gene Ontology Annotation (GOA) Database: sharing knowledge in UniProt with Gene Ontology. *Nucleic Acids Res.* 32, D262–D266.

15. Bairoch, A., Apweiler, R., Wu, C.H., Barker, W.C., Boeckmann, B., Ferro, S., Gasteiger, E., Huang, H., Lopez, R., Magrane, M., Martin, M.J., Natale, D.A., O'Donovan, C., Redaschi, N. and Yeh, L.S. (2005) The Universal Protein Resource (UniProt). *Nucleic Acids Res.* 33, D154–D159.

16. Berardini, T.Z., Mundodi, S., Reiser, L., Huala, E., Garcia-Hernandez, M., Zhang, P., Mueller, L.A., Yoon, J., Doyle, A., Lander, G., Moseyko, N., Yoo, D., Xu, I., Zoeckler, B., Montoya, M., Miller, N., Weems, D. and Rhee, S.Y. (2004) Functional annotation of the Arabidopsis Genome Using Controlled Vocabularies. *Plant Physiol.* 135, 745–755.

17. Schneider, M., Bairoch, A., Wu, C.H. and Apweiler, R. (2005) Plant Protein Annotation in the UniProt knowledgebase. *Plant Physiol.* 138, 59–66.

18. Mewes, H.W., Amid, C., Arnold, R., Frishman, D., Güldener, U., Mannhaupt, G., Münsterkötter, M., Pagel, P., Strack, N., Stümpflen, V., Warfsmann, J. and Ruepp, A. (2004) MIPS: analysis and annotation of proteins from whole genomes. *Nucleic Acids Res.* 32, D41–D44.

19. Serres, M.H. and Riley, M. (2000) MultiFun, a multifunctional classification scheme for Escherichia coli K-12 gene products. *Micro. Comp. Genomics.* 5, 205–222.

20. Krieger, C.J., Zhang, P., Mueller, L.A., Wang, A., Paley, S., Arnaud, M., Pick, J., Rhee, S.Y. and Karp, P.D. (2004) MetaCyc: a multiorgansim database of metabolic pathways and enzymes. *Nucleic Acids Res.* 23, D438–D442.

21. Biswas, M., O'Rourke, J., Camon, E., Fraser, G., Kanapin, A., Karavidopoulou, Y., Kersey, P., Kriventseva, E., Mittard, V., Mulder, N., Phan, I., Servant, F. and Apweiler, R. (2002) Applications of InterPro in protein annotation and genome analysis. *Brief Bioinform.* 3, 285–295.

22. Gattiker, A., Michoud, K., Rivoire, C., Auchincloss, A.H., Coudert, E., Lima, T., Kersey, P., Pagni, M., Sigrist, C.J., Lachaize, C., Veuthey, A.L., Gasteiger, E. and Bairoch, A. (2003) Automated annotation of microbial proteomes in SWISS-PROT. *Comput. Biol. Chem.* 27, 49–58.

23. Haft, D.H., Selengut, J.D. and White, O. (2003) The TIGRFAMs database of protein families. *Nucleic Acids Res.* 31, 371–373.

24. Tatusov, R.L., Fedorova, N.D., Jackson, J.D., Jacobs, A.R., Kiryutin, B., Koonin, E.V., Krylov, D.M., Mazumder, R., Mekhedov, S.L., Nickolskaya, A.N., Rao, B.S., Smirnov, S., Sverdlov, A.V., Vasudevan, S., Wolf, Y.I., Yin, J.J. and Natale, D.A. (2003) The COG database: an updated version includes eukaryotes. *BMC Bioinformatics.* 4, 41.

25. Quevillon, E., Silventoinen, V., Pillai, S., Harte, N., Mulder, N., Apweiler, R. and Lopez, R. (2005) InterProScan: protein domains identifier. *Nucleic Acids Res.* 33, W116–W120.

26. Love, C.G., Robinson, A.J., Lim, G.A.C., Hopkins, C.J., Bately, J., Barker, G., Spangenberg, G.C. and Edwards, D. (2005) *Brassica* ASTRA: an integrated database for *Brassica* genome research. *Nucleic Acids Res.* 33, D656–D659.

27. Altschul, S.F., Gish, W., Miller, W., Myers, E.W. and Lipman, D.J. (1990) Basic local alignment search tool. *J Mol Biol.* 215, 403–410.

28. Khan, S., Situ, G., Decker, K. and Schmidt, C.J. (2003) GoFigure: Automated Gene Ontology annotation. *Bioinformatics.* 19, 2484–2485.

29. Martin, D.M.A., Berriman, M. and Barton, G.J. (2004) GOtcha: a new method for prediction of protein function assessed by the annotation of seven genomes. *BMC Bioinformatics.* 5, 178.

30. Groth, D., Lehrach, H. and Hennig, S. (2004) GOblet: a platform for Gene Ontology annotation of anonymous sequence data. *Nucleic Acids Res.* 32, W313–W317.

31. Zehetner, G. (2003) OntoBlast function: from sequence similarities directly to potential functional annotations by ontology terms. *Nucleic Acids Res.* 31, 3799–3803.

32. Conesa, A., Götz, S., García-Gómez, J.M., Terol, J., Talón, M. and Robles, M. (2005) Blast2GO: a universal tool for annotation, visualization and analysis in functional genomics research. *Bioinformatics.* 21, 3674–3676.

33. Engelhardt, B.E., Jordan, M.I., Muratore, K.E. and Brenner, S.E. (2005) Protein Molecular Function Prediction by Bayesian Phylogenomics. *PLoS Comput. Biol.* 1, e45.

34. Chalmel, F., Lardenois, A., Thompson, J.D., Muller, J., Sahel, J.-A., And Léveillard, T. and Poch, O. (2005) GOAnno: GO annotation based on multiple alignment. *Bioinformatics.* 21, 2095–2096.

35. Emanuelsson, O., Nielsen, H., Brunak, S. and von Heijne, G. (2000) Predicting Subcellular Localization of Proteins Based on their N-terminal Amino Acid Sequence. *J Mol Biol.* 300, 1005–1016.

36. Al-Shahrour, F., Diaz-Uriarte, R. and Dopazo, J. (2004) FatiGO: a web tool for finding significant associations of Gene Ontology terms with groups of genes. *Bioinformatics.* 20, 578–580.

37. Khatri, P., Sellamuthu, S., Malhotra, P., Amin, K., Done, A. and Draghici, S. (2005) Recent additions and improvements to the Onto-Tools. *Nucleic Acids Res.* 33, W762–W765.

38. Liu, G., Loraine, A.E., Shigeta, R., Cline, M., Cheng, J., Valmeekam, V., Sun, S., Kulp, D. and Siani-Rose, M.A. (2003) NetAffx: Affymetrix probesets and annotations. *Nucleic Acids Res.* 31, 82–86.

25

Gene Structure Annotation at PlantGDB

Volker Brendel

Summary

The accurate identification of exons and introns that comprise a complete plant gene structure can be a time-consuming and challenging task. Novel Web-based tools facilitate the process by providing a convenient interface to current transcript evidence, and portals to relevant bioinformatics software. With a few keystrokes, the user can explore alternative transcript assemblies and, for example, select for annotation those that are clearly supported by transcript evidence and similarity to known genes. The implementation of the tool at the PlantGDB resource also allows immediate communication of the novel annotations to the community through Web display.

Key Words: Exon; intron; alternative splicing; untranslated region; spliced alignment; community annotation.

1. Introduction

Accurate genome annotation is essential to the full utilization of completed and emerging genomes. Because of the speed of DNA sequencing in modern genome projects, fast computational approaches are employed in the early stages of a genome project to provide a first indication of the gene content of the assembled sequences. These fast approaches mostly rely on "*ab initio*" gene prediction methods, which only use the genomic sequences and statistical parameters from prior training on related sequences. For recent reviews of such methods, see **refs. *1*** and ***2***. However, the precision of such approaches remains unsatisfactory. Apart from intrinsic difficulties with the prediction of

From: *Methods in Molecular Biology, vol. 406: Plant Bioinformatics: Methods and Protocols*
Edited by: D. Edwards © Humana Press Inc., Totowa, NJ

biological sequence properties from merely statistical estimates, the complexities of non-coding exons, non-canonical splice sites, alternative splicing, overlapping genes, and anti-sense transcription cause problems for these types of software programs. Therefore, continuing annotation efforts for plant genomes have concentrated on incorporating experimental transcript evidence (expressed sequence tags [ESTs] and full-length cDNAs) into the annotation *(3–6)*. Comprehensive annotation platforms for *Arabidopsis (7)* and rice *(8)* seek to include all available evidence into the more mature releases of plant genome annotation.

Currently, most of this refined annotation is provided by dedicated curators using specialized interactive annotation tools, e.g., Apollo *(9)*, although there is now increasing emphasis on including larger segments of the biological community in the annotation efforts *(10,11)*. While the tools described in this chapter are relevant to these efforts, our exposition here concentrates on yet another annotation need. Even in the age of pervasive high-throughput, whole-genome biology, there are still numerous biologists working on particular genes, gene families, or pathways, and they require accurate gene structures for these genes of interest. The purpose of this chapter is to present an easy-to-use, Web-accessible interface for (protein-coding) gene structure prediction that should help with these needs, without requiring local installation of bioinformatics software suites or particular bioinformatics expertise.

2. Materials

2.1. Sequence Data

The tools described apply to protein-coding gene prediction in eukaryotic (plant) genomes. While the example given below assumes publicly available sequence data, this is not a technological restriction. We assume availability of a genomic sequence—a region of a completed chromosome, an assembled bacterial artificial chromosome (BAC), or some other source of continuous genomic sequence. Furthermore, our tool relies heavily, although not exclusively, on transcript evidence—ESTs or full-length cDNAs that have been or can be mapped onto the genomic sequence using spliced alignment algorithms such as GeneSeqer *(12)*.

2.2. PlantGDB Resources

The annotation tool to be described has been implemented for various species at the PlantGDB resource (http://www.plantgdb.org) *(13)*. **Table 1** lists relevant links at PlantGDB.

Table 1
PlantGDB Links

Site name	Web address	Content
PlantGDB	http://www.plantgdb.org/	Resources for plant comparative genomics
AtGDB	http://www.plantgdb.org/AtGDB/	*Arabidopsis* genome browser
OsGDB	http://www.plantgdb.org/OsGDB/	Rice genome browser
ZmGDB	http://www.plantgdb.org/ZmGDB/	Maize genome browser
GeneSeqer	http://www.plantgdb.org/PlantGDB-cgi/ GeneSeqer/PlantGDBgs.cgi	Spliced alignment Web server
Tutorial	http://www.plantgdb.org/tutorial/annotatemodule/	Annotation tutorial for class room use

2.3. Other Resources

Table 2 lists a number of other resources that are particularly relevant to plant gene structure annotation.

3. Methods

There are many different starting points for a gene structure annotation project (*see* **Note 1**). Here, we assume we are interested in a particular gene family and wish to explore whether recent data from genome sequencing projects have yielded further sequences belonging to the same family. To be specific, let us assume interest in the branched-chain amino acid transaminase (BCAT) gene family. Our starting point is the set of known BCAT genes in *Arabidopsis (17)*, and we are interested in finding maize homologs of these genes. Maize genome sequencing is under way, and the most up to date sequencing data are deposited as phase 1 and 2 sequences in GenBank (corresponding respectively to sets of unordered and ordered segments, separated by gaps, that were assembled from maize BACs). The following stepwise approach illustrates the use of PlantGDB resources to find a maize BCAT gene.

1. Pick one of the known *Arabidopsis* BCAT genes, say *AtBCAT-5* (At5g65780 product), and select its sequence in FASTA format. There are different ways to implement this simply put instruction (and several ways to get lost in cyberspace

Table 2
Resource Links

Site name	Web address	Content
NCBI	http://www.ncbi.nlm.nih.gov/	Sequences and lots more (*see* Chapters 2 and 3)
BLAST	http://www.ncbi.nlm.nih.gov/BLAST/	Sequence-based search tool *(14)*
TAIR	http://www.arabidopsis.org/	*Arabidopsis* information resource (*see* Chapter 8)
TIGR Rice Database	http://www.tigr.org/tdb/e2k1/osa1/	Rice genome annotation database (*see* Chapter 5)
Gramene	http://www.gramene.org/	Resource for comparative grass genomics (*see* Chapter 15)
GeneMark	http://opal.biology.gatech.edu/GeneMark/	Gene structure prediction Web server *(15)*
GENSCAN	http://genes.mit.edu/GENSCAN	Gene structure prediction Web server *(16)*

doing this . . .). One way is to plug the accession number AJ293804 given in **ref.** *17* into the National Center for Biotechnology Information (NCBI) home page search field, pull up the nucleotide sequence match, display the report, and follow the "protein_id" link. The protein page has a display option "FASTA", from which we can copy the header line and the sequence into our clipboard buffer.

2. Go to ZmGDB (http://www.plantgdb.org/ZmGDB) and select BLAST@ZmGDB from the top toolbar. Paste the query sequence and select to search in the ZMbac database using the tBLASTn program. This program will translate the maize BAC sequences into the amino acid alphabet in all frames using the standard genetic code, and then search for matches with our protein query (*see* **Note 2**).

3. "Run BLAST" results in the output shown in **Fig. 1**. A 61-amino-acid stretch of our query sequence matches very well (and identically) to regions from two recently deposited maize BACs. The first is a phase 2 BAC, consisting of 11

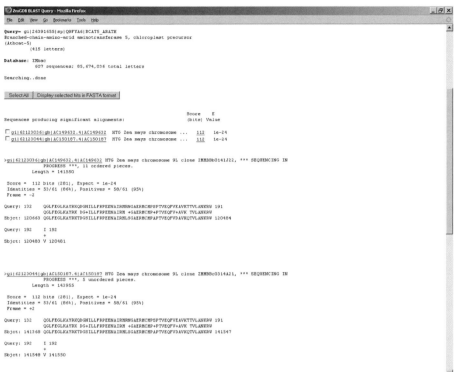

Fig. 1. A tBLASTn search at ZmGDB provides a starting point for gene discovery and annotation. The query protein matches translated sequences from two maize bacterial artificial chromosomes (BACs) deposited in GenBank. For the described case study, an expanded region around the match in the first BAC was chosen for further exploration.

ordered pieces of sequence, whereas the second BAC is phase 1 and consists of unordered pieces. We have to try our luck. The first BAC should have any exons we can deduce in the right order, but there may of course be additional exons in the unsequenced gaps.

4. To proceed, we note the gi number of the BAC (62123036) and roughly the coordinates of the match on the BAC (around 120,500 on the minus strand). Put the gi number into the BAC/Clone gi field on the beige toolbar on top of the page and select start and end positions as, for example, 120,000 and 121,000. Hit "Go".

5. The resulting page shows lots of EST evidence for transcriptional activity (and more; *see* **Note 3**). By playing with the scale factors and start/end fields a bit, we quickly arrive at the region 112,001–122,000 (about) as a good segment to hope for a full-length gene structure (*see* **Fig. 2**).

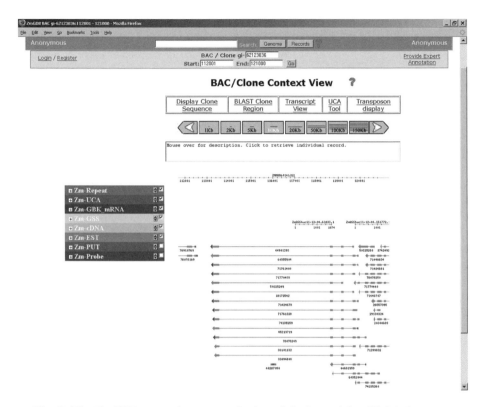

Fig. 2. The ZmGDB genomic context display of the bacterial artificial chromosome (BAC) region selected for annotation. The red box-line-and-arrow schematics represent spliced alignments of maize expressed sequence tags (ESTs) against the BAC sequence. Boxes indicate exons, and lines indicate introns. The presumed direction of transcription is in the direction of the arrows. Green and blue tips represent annotated 5′- and 3′-clones, respectively. The yellow GSS track shows the regions of mapped genome survey sequence contigs. Color is as seen on the Web page.

6. The annotation task now consists of using the EST evidence to try and build a putative full-length mRNA transcript, then to search this transcript for an open reading frame (ORF) corresponding to the gene product. Click the "UCA Tool" button (now labeled "yrGATE" for "your Gene structure Annotation Tool for Eukaryotes"; see Acknowledgments) on top of the display to get to the screen shown in **Fig. 3**.

7. This screen is at the core of our annotation method. The Evidence Plot repeats the display of the calling window (*see* **Fig. 2**). However, here each exon and each gi label are clickable with right mouse clicks, resulting in selection of the exon or

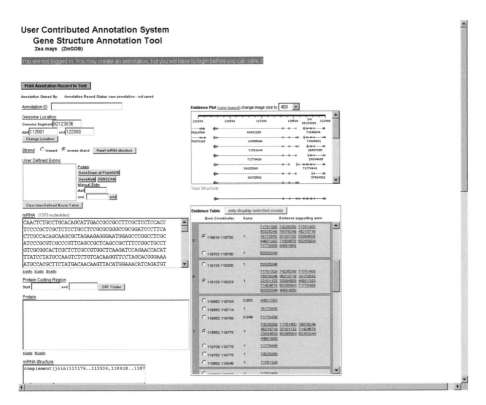

Fig. 3. Gene structure annotation at ZmGDB. This page is loaded from the button labeled "UCA" at the genomic context display shown in **Fig. 2**. The various functions available to the annotator are explained in the text. Shown here is the assembly of 10 exons into a putative mRNA of 1570 nucleotides (exon–intron structure shown in green on the Web page). The Evidence Table shows some of the selected exons, their similarity score (normalized such that 1 indicates perfect identity) and the gi numbers of the transcripts supporting the selected exons. Alternative selections are shown with unfilled radio buttons in groups of mutually incompatible choices (only one exon per group can be in a gene structure).

the entire structure of the labeled transcript into a pseudo-transcript appearing in green in the "Your Structure" field. We leave it to the reader to enjoy playing with this. Note a couple of further things happening upon each selection: the pseudo-transcript sequence shows up in the left-hand mRNA box, and the Evidence Table lists the selected exons with their match score and the EST/cDNA transcripts supporting these particular exons. To view the complete evidence for a particular exon combination, click on the links in the Evidence Table. This will bring up the detailed GeneSeqer alignments *(12)* that are schematically displayed in the graphic.

Looking at the alignment, you can decide for yourself whether or not to trust it (in this example, everything looks very good indeed; however, because PlantGDB calculates and displays tens of thousands of alignments, there may be other cases that are not entirely clear because of sequence mismatches and non-canonical splice sites, for example). See **Note 4** for additional UCA Tool functions.

8. After a bit of trial and error, you should arrive at the pseudo-transcript displayed in **Fig. 3** as the longest transcript spanning this region. Because there is no full-length cDNA corresponding to our pseudo-transcript, we must still carefully evaluate its likelihood as a real transcript. We take the attitude that if the transcript contains a long ORF spanning most of the exons, then we can be quite confident that this is a correctly assembled transcript. Moreover, if the translation product closely matches homologous proteins, then our case should be complete.

9. To facilitate the aforementioned considerations, the UCA Tool provides the ORF Finder portal on the left below the mRNA box. Click there, and you will arrive at what is shown in **Fig. 4**. The portal simply accesses the NCBI ORF Finder, but it allows the selection of an ORF from that tool to be passed back to the UCA Tool. In our example, the longest ORF encodes a 403-amino-acid protein—very close in length to our At-BACT5 query protein (415 amino acids). Thus, we should confidently pick this ORF and go back to the UCA Tool (*see* **Fig. 5**).

10. Now you will see the protein sequence pasted into the Protein field. The ORF start and stop codons are indicated with triangles on the pseudo-transcript (green for start, red for stop—no jaywalking, please!). What remains to be done?

11. Again, the UCA Tool anticipates your next question. How close is the predicted gene product to known proteins? In our case, did we really identify a maize BACT gene? The "BLASTp" button below the Protein field facilitates the answer to these questions by invoking a BLASTp search at NCBI. In our example, the predicted protein matches over its entire length to several rice genes that are already annotated as BACT family genes (*Os03g01600*, *Os05g48450*, *Os10g40200*, *Os04g47190*, and *Os03g12890*, in order of similarity score; *see* **Note 5**). Thus, we can conclude that we have found a maize BACT gene (possibly the ortholog of rice *Os03g01600*) on chromosome 9L.

So far, good for us. But it would be nice to be a bit more altruistic and share our results freely. PlantGDB provides a speedy avenue for this. All we need to do is to do what we have done above as a registered user at PlantGDB. There is nothing more to this than simply providing a name and email address that allows PlantGDB staff to identify the source of an annotation. If we use the UCA Tool after having logged into the system with our user name, the tool provides the additional functions "Save for Editing", "Submit", and "Remove Annotation". The user is asked to provide an annotation label and description

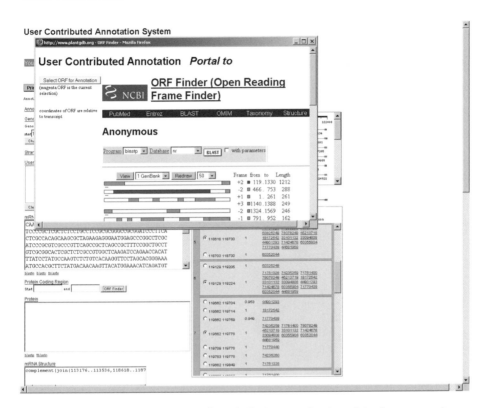

Fig. 4. The open reading frame (ORF) finder portal selected in the screen shown in **Fig. 3** by clicking the "ORF Finder" button shows an open reading frame of 1212 nucleotides in the assembled putative transcript. The "Select ORF for Annotation" button picks the magenta highlighted ORF as the presumed translation product of the assembled transcript. Color is as seen on the Web page.

of the gene and gene product. If the user invokes "Save for Editing", she or he can view the annotation in the genome browser (*see* **Fig. 2**) while logged in with her or his user name. Upon "Submit", PlantGDB staff will review the annotation for legitimacy and then make legitimate annotations viewable to the entire Web community.

Although the PlantGDB resource and the UCA software are continuously evolving, in our view this is already a great tool that should enable rapid and accurate annotation of emerging genomes, as well as refinement of current annotations (*10*). We hope that you, the reader, will agree and prove it by contributing your own annotations!

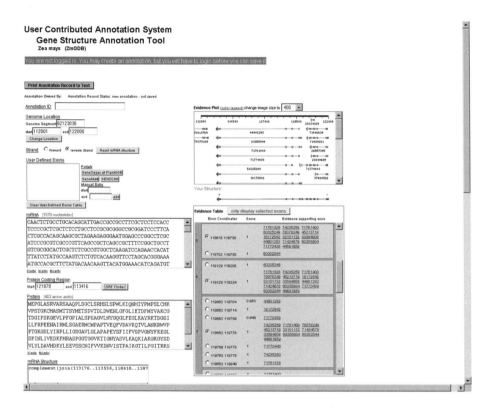

Fig. 5. After the selection of the open reading frame (ORF) as shown in **Fig. 4**, the gene structure annotation is complete. The green and red triangles above the green transcript structure indicate the positions of the start and stop codons. The predicted 403-amino-acid protein sequence has been pasted into the Protein text box and can be compared against the GenPept protein database using the BLASTp link to NCBI's BLAST server. Color is as seen on the Web page.

4. Notes

1. Another common starting point is an EST sequence as a tag for a transcribed gene. Similarly to the example described in the text, a simple BLASTn search might provide approximate matching genome locations for further exploration. The PlantGDB genome browsers provide lists of problematic or novel gene structures that should be looked at by an annotator (e.g., http://www.plantgdb.org/AtGDB-chtml/GAEVAL.php for *Arabidopsis*). Furthermore, the annotation tool also lends itself to entire BAC annotation by systematically scanning the entire sequence for transcript evidence using the genome browser. This could be especially useful for community annotation of emerging genomes such as maize.

2. Because the newly deposited BACs, as in our example, are typically deposited without annotation, tBLASTn provides the only search option to scan for matching protein-coding regions at the amino acid level. See http://www.ncbi.nlm.nih.gov/blast/producttable.shtml for help with selecting appropriate search options.

3. The display also has tracks for assembled EST contigs (PlantGDB-assembled unique transcripts, or PUTs) and mapped probe sequences that link to microarray gene expression data. The display can be customized by the user. The associated tool buttons facilitate detailed analysis of the region. In particular, the "Transcript View" button goes to a genome sequence-based multiple sequence alignment of all transcripts in the region. Scrolling along the alignment quickly identifies potential polymorphisms and alternative splice forms.

4. In less fortunate cases, the transcript evidence may not completely cover the presumed gene structure. For example, ESTs may be available for the 5′- and 3′-ends of the transcript but not for any internal exons. The tool provides several options for this case. The "GeneSeqer at PlantGDB" button goes to the PlantGDB spliced alignment server, with the current genomic region pre-selected. In many cases, non-native ESTs or full-length cDNAs from homologous genes of related species can be used to fill out the evidence gaps, as their spliced alignments will show enough sequence conservation to suggest the corresponding native gene structure. Furthermore, the "GeneMark" and "GENSCAN" buttons link to portals to these *ab initio* gene prediction programs (*see* **Table 2**). The predicted exons can be imported to the User Defined Exons table and be selected for the user's transcript assembly.

5. By now the user will be entirely enthused about gene structure prediction and wish to go further. The possibilities are (nearly) endless. Initially, the homologs identified in other species with genome sequences should be checked for their annotation accuracy. Phylogenetic studies, including comparison of exon–intron structures of the genes, might reveal gene family history and functional differentiation. The genome browsers also allow detection of syntenic relationships and associate available gene expression data.

Acknowledgments

The UCA tool was designed and developed by Matthew Wilkerson and inspired by the initial work of Shannon Schlueter. I am immensely grateful to be able to work with such talented graduate students. Technical details of the tool (now renamed "yrGATE") have been published elsewhere (*18*). Dr. Qunfeng Dong managed PlantGDB until 2006 and continually found ways to improve the utility of the resource for the community and in our own research work.

References

1. Mathé, C., Sagot, M.F., Schiex, T., and Rouzé, P. (2002) Current methods of gene prediction, their strengths and weaknesses. *Nucleic Acids Res.* 30, 4103–4107.

2. Brent, M.R., and Guigó, R. (2004) Recent advances in gene structure prediction. *Curr. Opin. Struct. Biol.* 14, 264–272.

3. Seki, M., Narusaka, M., Kamiya, A., Ishida, J., Satou, M., Sakurai, T., Nakajima, M., Enju, A., Akiyama, K., Oono, Y., et al. (2002) Functional annotation of a full-length *Arabidopsis* cDNA collection. *Science* 296, 141–145.

4. Zhu, W., Schlueter, S.D., and Brendel, V. (2003) Refined annotation of the *Arabidopsis thaliana* genome by complete EST mapping. *Plant Physiol.* 132, 469–484.

5. Haas, B.J., Delcher, A.L., Mount, S.M., Wortman, J.R., Smith, R.K. Jr., Hannick, L.I., Maiti, R., Ronning, C.M., Rusch, D.B., Town, C.D., et al. (2003) Improving the *Arabidopsis* genome annotation using maximal transcript alignment assemblies. *Nucleic Acids Res.* 31, 5654–5666.

6. Yamada, K., Lim, J., Dale, J.M., Chen, H.M., Shinn, P., Palm, C.J., Southwick, A.M., Wu, H.C., Kim, C., Nguyen, M., et al. (2003) Empirical analysis of transcriptional activity in the Arabidopsis genome. *Science* 302, 842–846.

7. Wortman, J.R., Haas, B.J., Hannick, L.I., Smith, R.K. Jr, Maiti, R., Ronning, C.M., Chan, A.P., Yu, C., Ayele, M., Whitelaw, C.A., et al. (2003) Annotation of the Arabidopsis genome. *Plant Physiol.* 132, 461–468.

8. Yuan, Q., Ouyang, S., Wang, A., Zhu, W., Maiti, R., Lin, H., Hamilton, J., Haas, B., Sultana, R., Cheung, F., et al. (2005) The Institute for Genomic Research Osa1 rice genome annotation database. *Plant Physiol.* 138, 18–26.

9. Lewis, S.E., Searle, S.M., Harris, N., Gibson, M., Lyer, V., Richter, J., Wiel, C., Bayraktaroglir, L., Birney, E., Crosby, M.A., et al. (2002) Apollo: a sequence annotation editor. *Genome Biol.* 3 (12), 1–14.

10. Schlueter, S.D., Wilkerson, M.D., Huala, E., Rhee, S.Y., and Brendel, V. (2005) Community-based gene structure annotation for the *Arabidopsis thaliana* genome. *Trends Plant Sci.* 10, 9–14.

11. Aubourg, S., Brunaud, V., Bruyère, C., Cock, M., Cooke, R., Cottet, A., Couloux, A., Déhais, P., Déleage, G., Duclert, A., et al. (2005) GeneFarm, structural and functional annotation of Arabidopsis gene and protein families by a network of experts. *Nucleic Acids Res.* 33, D641–D646.

12. Brendel, V., Xing, L., and Zhu, W. (2004) Gene structure prediction from consensus spliced alignment of multiple ESTs matching the same genomic locus. *Bioinformatics* 20, 1157–1169.

13. Dong, Q., Lawrence, C.J., Schlueter, S.D., Wilkerson, M.D., Kurtz, S., Lushbough, C., and Brendel, V. (2005) Comparative plant genomics resources at PlantGDB. *Plant Physiol.* 139, 610–618.

14. Altschul, S.F., Madden, T.L., Schäffer, A.A., Zhang, J., Zhang, Z., Miller, W., and Lipman, D.J. (1997) Gapped BLAST and PSI-BLAST: a new generation of protein database search programs. *Nucleic Acids Res.* 25, 3389–3402.
15. Lomsadze, A., Ter-Hovhannisyan, V., Chernoff, Y.O., and Borodovsky, M. (2005) Gene identification in novel eukaryotic genomes by self-training algorithm. *Nucleic Acids Res.* 33, 6494–6506.
16. Burge, C.B., and Karlin, S. (1997) Prediction of complete gene structures in human genomic DNA. *J. Mol. Biol.* 268, 78–94.
17. Diebold, R., Schuster, J., Daschner, K., and Binder, S. (2002) The branched-chain amino acid transaminase gene family in Arabidopsis encodes plastid and mitochondrial proteins. *Plant Physiol.* 129, 540–550.
18. Wilkerson, M.D., Schlueter, S.D., and Brendel, V. (2006) yrGATE: a web-based gene-structure annotation tool for the identification and dissemenation of eukaryotic genes. *Genome Biol.* 7, R58.

26

An Introduction to BioPerl

Jason E. Stajich

Summary

The BioPerl toolkit provides a library of hundreds of routines for processing sequence, annotation, alignment, and sequence analysis reports. It often serves as a bridge between different computational biology applications assisting the user to construct analysis pipelines. This chapter illustrates how BioPerl facilitates tasks such as writing scripts summarizing information from BLAST reports or extracting key annotation details from a GenBank sequence record.

Key Words: Perl; bioinformatics; BioPerl; BLAST parsing; sequence parsing.

1. Introduction

Automating data processing is essential in bioinformatics. Genomic data can be in heterogeneous file formats and scattered across databases and Web sites. In order to easily take a biologically relevant question and interrogate the available data, one requires tools that can process and manage this complex information. The BioPerl toolkit *(1)* is a programming library of subroutines for accessing sequence, annotation, and alignment data, from file formats and databases. BioPerl is written in the Perl programming language and provides hundreds of modules for processing the output from sequence analysis programs and querying local and Web-based sequence databases.

From: *Methods in Molecular Biology, vol. 406: Plant Bioinformatics: Methods and Protocols*
Edited by: D. Edwards © Humana Press Inc., Totowa, NJ

2. Materials

2.1. Obtaining and Installing the Toolkit

The BioPerl toolkit is a series of Perl modules freely available under the Perl artistic license. The software package can be obtained from the Perl Comprehensive Perl Archive Network (CPAN; http://cpan.org/) or directly from the BioPerl Web site (http://www.bioperl.org). Installation instructions are included in the distributed package and it follows like most Perl modules.

There is a set of additional modules that BioPerl depends on to perform various functions such as accessing data over the world-wide-Web or processing data files in an XML format. These extra modules are not required for basic operations of the toolkit, and the reader is not encouraged to install all the possible external dependencies.

2.2. Organization of the Modules

The Perl language convention is to organize module directories so the module names map to the directory structure they are stored in. For example, the module Bio::Tools::Genewise is located in the *Tools* directory inside the *Bio* directory of the toolkit. The BioPerl toolkit has a lot of modules, which can be overwhelming to a new user, so there are a number of different documentation resources organized on the project Web site to guide the user. This chapter will provide insight into two of these topics, (1) processing and summarizing BLAST and other sequence similarity search reports, and (2) processing DNA and protein sequence files to extract annotations.

3. Methods

3.1. Processing Sequence and Annotation Data

Much of the information gleaned from gene and protein sequences has been deposited in the sequence databases. These databases include both curated data entered by a database curator, and by automated analyses, such as similarity search algorithms for finding conserved protein domains. The information stored in sequence files often represents a great deal of what is known about a gene, protein, RNA, or genomic sequence, and so mining this data represents an important task in bioinformatics.

3.2. The BioPerl Toolkit

3.2.1. Parsing Sequence Files

The Bio::SeqIO module is used for parsing sequence files. In BioPerl, we have separated the process of parsing sequence and annotation data from the objects that store the data. Thus, there is a Bio::SeqIO object that drives the parsing, whereas another type of object, Bio::Seq, is returned from the parser and represents the data in the sequence file.

Only a few arguments are required to initialize the parser: the format of the sequence file and the filename to read. The code in **Fig. 1** demonstrates opening a sequence file called *dna.fas* that is stored in FASTA format. Line 1 tells Perl to operate in a strict manner and require that all variables be declared. This is why all the variables are first prefixed with the *my* directive. Line 2 tells Perl to load the Bio::SeqIO module that is used for parsing sequence files. Lines 3 and 4 initialize a new Bio::SeqIO object that will read a file in *FASTA* format from the file *dna.fas*.

Line 5 shows how to parse a sequence from the file. Calling next_seq on the $seqio object is where the actual parsing step happens. It returns a sequence object that is of type Bio::Seq. You can always check to see what type of object Perl has returned by calling the ref function, so on line 6 the script will print out "Bio::Seq."

Lines 7 through 12 are demonstrating some of the methods one can call on a sequence object. Line 7 shows the display_id method that provides access to the sequence name. The length method returns how many bases long the sequence is, and the alphabet method returns the type of the sequence BioPerl guessed, which can be *dna*, *rna*, or *protein*. Line 10 shows how to get the actual sequence as a string. Lines 11 and 12 show methods that can only be called on

```
1   use strict;
2   use Bio::SeqIO;
3   my $seqio = Bio::SeqIO->new(-format => 'fasta',
4                               -file   => 'dna.fas');
5   my $seq = $seqio->next_seq;
6   print ref($seq), "\n";
7   print $seq->display_id," is the sequence name\n";
8   print "The sequence is ",$seq->length,
9     " bases long and is a ",$seq->alphabet," sequence\n";
10  print "The sequence letters are ",$seq->seq,"\n";
11  print "Translation is ",$seq->translate->seq,"\n";
12  print "Reverse complement is ",$seq->revcom->seq,"\n";
13  print "bases 5 -> 20 are ", $seq->subseq(5,20), "\n";
```

Fig. 1. Reading a FASTA sequence file with Bio::SeqIO.

DNA sequences as the translate function returns a new sequence object that is the translation of the sequence into amino acids (*see* **Note 1**). Line 12 prints a new sequence object that is the reverse complement of the original sequence. In lines 11 and 12, there are actually two methods being called, the first, a call to translate or revcom and the second to seq. This is because those functions actually return a new sequence object instead of a simple value, so the seq method is called (just like in line 9) to retrieve the sequence data as a string suitable for printing out. Finally, line 13 prints out a portion of the sequence, as the subseq method returns a string that is the subsequence from bases 5 to 20 (inclusive).

Notice the arguments used to initialize the Bio::SeqIO object. The dashes in front of the *-format* and *-file* are because these are named parameters. In Perl, the "=>" operator is part of using hashes and is specifying that the value on the left is the key and the value on the right is value. Hashes are then made up of pairs of things. In this case, the pair is a parameter name, *-format*, and its value, "*FASTA*." This is very useful because you only have to remember the parameter name, not the order of the arguments as is typically done to call a function.

3.2.2. Extracting the Features

The previous example shows a simple case of reading a FASTA-formatted sequence file and printing out some basic information. However, BioPerl can also read data from more complicated sequence data files like GenBank, EMBL, and Swissprot records. These files contain sequence data, and in addition, feature annotations, such as where protein domains are located in a protein or where exons are located in a genomic contig (*see* Chapters 1, 2 and 3). The file formats also typically encode additional information such as the species, bibliographic references, cross references to other databases, keywords, and dates of creation and update. Access to these fields will be covered next after focusing on sequence features.

The annotation of features on a sequence provides an exact description of each base in a sequence. A feature is typically described as a contiguous location in a sequence, with starting and ending position and an indication of whether it is in the forward or reverse orientation. For features that are not a single contiguous sequence location, annotations are composed of multiple sub locations. This is needed, for example, when annotating a DNA sequence with Coding Sequences (CDS) that are interrupted by introns.

To better illustrate the utility of these modules, I solve an example relevant problem through the use of these modules. One goal is to extract intron sequence from an annotated sequence record and create a new FASTA file containing all the intron sequences as separate entries.

```
source            1..7315
                  /organism="Zea mays"
                  /mol_type="genomic DNA"
                  /cultivar="A188"
                  /db_xref="taxon:4577"
                  /chromosome="9"
                  /map="9S"
gene              1..7315
                  /gene="MRP3"
promoter          1..686
                  /gene="MRP3"
mRNA              join(687..874,1223..1293,1469..4536,
                  4773..4937,5130..5424,6015..7315)
                  /gene="MRP3"
5'UTR             join(687..874,1223..1293,1469..1479)
                  /gene="MRP3"
CDS               join(1480..4536,4773..4937,5130..5424,
                  6015..6940)
                  /gene="MRP3"
                  /note="MRP-type ABC transporter; ZmMRP3"
                  /codon_start=1
                  /protein_id="AAT37905.1"
                  /db_xref="GI:47717725"
ORIGIN
        1 ctgcgtactg tagggctttg accgtccggt tagcccgccc cgcccccgcg gcgagcacgt
       61 gacgacgacg gtgcgttcgt gggcgccgat cgccgagggg aggcgctatc tgccgtcgtt
      121 cgttcatcat cgggctatat atatagatat cggaggacgg aggcctctca attaagcatt
      181 ctccgcgcgc gcgcgcgccc agcagcagcc cgcgcttggc cttccttctc ctcccgaccc
      241 gacgtcgtcg gtcgcgccgc tgcctgcccc gtcccgtttc ggtgtcgtct ctcgccggcc
```

Fig. 2. GenBank feature table for accession number AY609318.

An example of the GenBank format, encoding feature data, is shown in **Fig. 2**. The *gene* feature is a simple contiguous feature, and the CDS feature is a more complicated feature with 14 sub-locations that represent where the coding exons are in the genomic sequence. Our goal is to write a general script to extract intron sequences from this or similar files.

The BioPerl Bio::SeqIO parser processes a GenBank file and builds sequence feature objects from the feature table. This creates an object called Bio::SeqFeature::Generic that captures all the information about a feature. The method 'primary_tag' provides access to the feature type such as *source* or *gene* (*see* **Note 2**). **Figure 3** demonstrates the parsing of a GenBank file and the printing of a summary of the features that are present.

Because feature locations can be complicated and not simply "start..end", we further separate the location concept from the actual annotation. This means, if you want to actually query the feature for the locations, you need to get access to the location object. This is accessible through the location method in the sequence feature. The location method returns a location object that can be any one of the following: Bio::Location::Simple (for simple contiguous start..end),

```perl
1    #!/usr/bin/perl -w
2    use strict;
3    use Bio::SeqIO;
4    use Bio::Seq;
5    my $in = Bio::SeqIO->new(-format => 'genbank',
6                             -file   => 'AY609318.gbk');
7    my $out = Bio::SeqIO->new(-format => 'fasta',
8                             -file   => '>AY609318.introns.fa');
9    while( my $seq = $in->next_seq ) {
10    my $id = $seq->display_id;
11    for my $feature ( $seq->get_SeqFeatures ) {
12     if( $feature->primary_tag eq 'CDS' ) {
13       my $count = 0;
14       my @locs = $feature->location->each_Location;
15       my $lastloc;
16       for my $subloc ( @locs ) {
17        if( $count > 0 ) {
18        # we skip the first one because introns are
19        # between exons so we need to have seen the 1st exon
20         my $iseq;
21         if( $subloc->strand > 0 ) {
22         # process forward strand features
23         # intron is from the end of last feature +1
24         # to the start of the current feature -1
25          $iseq = $seq->subseq($lastloc->end +1, $subloc->start-1);
26         } else {
27         # process reverse strand features
28         # intron is from start last feature -1
29         # to end of current feature + 1
30          my $intron = $seq->trunc($subloc->end + 1,
31                                  $lastloc->start -1);
32         # we have to ask for sequence in forward coordinates
33         # and then reverse complement.
34          $iseq = $intron->revcom->seq;
35         }
36         my $desc = "$id:".$subloc->to_FTstring;
37         my $intronseq = Bio::Seq->new(-display_id => "$id.intron$count",
38                                      -description => $desc,
39                                      -seq => $iseq);
40        $out->write_seq($intronseq);
41        }
42        $count++;
43        $lastloc = $subloc;
44       }
45      }
46     }
47    }
```

Fig. 3. Extract intron sequences from a GenBank file based on annotated CDS.

Bio::Location::Split (for locations that contain other locations to capture multiple sub-locations), or Bio::Location::Fuzzy (for locations that have an undefined start or end). All locations have three methods to keep in mind, start, end, and strand. The Split locations are container objects that hold a list of sub-locations to represent complicated non-contiguous locations like the CDS on genomic sequence in **Fig. 2**. Split locations have an additional method, 'each_Location', that returns a list of all the sub-locations within the container. By iterating over this list, one has access to the CDS locations in a particular feature.

Putting it all together to build a script that extracts the sequence of these features is illustrated in **Fig. 3**. The example shows how to extract intron sequences from between the annotated exons in a CDS feature. The code on lines 5–7 first opens a genbank file for reading and prepares a second file handle for writing FASTA sequence format to store the intron sequences. The "while loop" on line 9 iterates through all the sequences in the file, and upon each iteration, the $seq variable will have a sequence object. The features stored in the sequence object can be accessed as a list in line 12. Only processing the features that are of type "CDS" (line 12), the code gets all the sub-locations for the feature in line 14. These sub-locations are exons, and introns are defined as the sequence between them. The loop starting at line 16 iterates over the exon locations, and lines 21–35 extract the DNA sequence of the location between all the exons. A new sequence object is initialized on line 37 with this intron sequence and then written out to the sequence file in FASTA format in line 40.

Additional information from the sequence file is also stored in the sequence object. This includes the species and taxonomy information, the GI or GenBank Identifier number, bibliographic references, and comments (*see* **Note 3**).

3.2.3. Capturing Just the Features

Simple file formats exist for reading and writing sequence features. One of these is the GFF format (http://www.sanger.ac.uk/Software/formats/GFF). This is a simple tab-delimited format with columns that correspond to the name of the sequence, and location of a feature on the sequence. The GFF format has matured to some extent and there are now several versions, GFF1, GFF2, GTF or GFF 2.5, and GFF3. The only difference between these versions is to do with the structuring of the ninth column that was initially just the name of the feature or gene. In GFF3, the ninth column is structured with fields that clearly identify the name or ID of the feature and, where appropriate, the parent of the feature when hierarchies of features are necessary (e.g., exons are the children of a gene). Recent work to produce an ontology describing sequence

```
1    use strict;
2    use Bio::SeqIO;
3    use Bio::Tools::GFF;
4    my $seqio = Bio::SeqIO->new(-format => 'genbank',
5                                -file   => ' AY609318.gbk');
6    my $gffout = Bio::Tools::GFF->new(-gff_version => 2,
7                                      -file => ">AY609318.gff");
8    while( my $seq = $seqio->next_seq ) {
9    for my $feature ( $seq->get_SeqFeatures ) {
10      $gffout->write_feature($feature);
11    }
12   }
```

Fig. 4. Extract sequence features from a genbank format file and write them to a GFF2 format file.

features *(2)* (Sequence Ontology; http://song.sourceforge.net) has been applied to the GFF3 standard, so that feature relationships are unambiguous. GFF is also one representation of feature data that can be viewed through the Generic Genome Browser *(3)* (http://www.gmod.org/ggb/).

BioPerl provides several tools for generating GFF from sequence features. The module Bio::Tools::GFF will read and write sequence features in GFF versions 1, 2, 2.5 (also known as Gene Transfer Format or GTF), and 3. Code in **Fig. 4** provides an example of how to write out the features in GFF after extracting them from a genbank sequence file.

The generated GFF from the code in **Fig. 4** can be loaded into a Gbrowse database using scripts in BioPerl called bulk_load_gff.pl and load_gff.pl.

3.3. Processing BLAST and Sequence Similarity Search Reports

BLAST *(4)* is the single most used tool in bioinformatics, as it quickly identifies sequence similarities. Visual inspection of reports is useful for small sets of sequences but becomes impractical when dealing with hundreds or thousands of queries. In addition, the ability to automate analyses allows different hypotheses to be tested and compared in a uniform manner while data sets change or are improved.

Processing BLAST reports is enabled through the Bio::SearchIO system. This module is very much analogous to Bio::SeqIO that processed sequences. In this case, SearchIO stands for Input and Output of database search results from programs such as BLAST, FASTA *(5)*, Smith-Waterman, HMMER *(6)*, and BLAT *(7)*. The system provides a uniform interface to the data in these reports so that the programming interface stays the same, even when the search

```
use strict;
use Bio::SearchIO;
my $report = Bio::SearchIO->new(-format => 'blast',
                                -file   => 'report.bls');
while( my $result = $report->next_result ) {
 print "query name is ", $result->query_name, "\n";
 while( my $hit = $result->next_hit ) {
  print "hit name is ", $hit->name, "\n";
  while( my $hsp = $hit->next_hsp ) {
   print "HSP query start,end = ",
   $hsp->query->start, ",", $hsp->query->end, "\n";
   print "HSP hit start,end = ",
   $hsp->hit->start, ",", $hsp->hit->end, "\n";
  }
 }
}
```

Fig. 5. Example showing Bio::SearchIO use to parse a BLAST report.

algorithm is changed, much like how SeqIO provides sequences and features regardless of whether the data is in Swissprot, Genbank, or XML formats.

BLAST reports are broken down into three main components, the "Result", "Hit", and "HSP". The "Result" is the representation of the overall result of the search; it includes information about the query sequence that was used, in addition to the search parameters (scoring matrix, database name) and overall statistics from the result. The "Hit" represents a sequence that was found to have significant similarity. There can be many hits for a single query sequence. The "HSP" is the High Scoring Pair of sequences and represents the actual alignment segments of the query and hit pair of sequences. For algorithms such as BLAST, there can be multiple HSPs for a single query and hit pair, whereas Smith-Waterman and FASTA results will only have a single HSP per pair. The three domains of the report are parsed out by the modules in the Bio::SearchIO system and stored in three corresponding types of objects (*see* **Fig. 5**). The Bio::Search::Result::ResultI interface describes the Result objects, the Bio::Search::Hit::HitI interface describes Hits, and the Bio::Search::HSP::HSPI interface describes Hits. Each of these modules has documented methods for getting the information stored about their particular type of data. To use the Bio::SearchIO system, we use a simple code framework that iterates through the Result, Hit, and HSP objects.

3.3.1. Access to the Result Data

The Result, Hit, and HSP objects have methods for accessing specific data from the report. All the methods for these objects are outlined in the module documentation and the HOWTO document for SearchIO, but a summary of

Table 1
Bio::Search::Result::ResultI Methods

Method name	Description
query_name	Name of the query sequence
query_description	Query description
query_accession	Query accession number (where available)
query_length	Length of the query sequence
database_name	Name of the database(s) searched
database_letters	Number of residues in the database
database_entries	Number of sequences in the database
num_hits	Number of significant hits found in the search
next_hit	Iterate through the list of Hits for the next Result
available_parameters	A list of the available parameters used in the search

them are listed in **Table 1**. Some of the key methods include query_name, query_accession, and query_description, to get the name, accession, and description of the query sequence. Access to the BLAST parameters used is available from the available_parameters method, and the statistics of the search are available through available_statistics.

3.3.2. Access to the Hit Data

The Hit object provides access to the significant hits found in the database search. Some of the methods include name, description, and accession, used to get information about the sequence. In addition, the method "significance" returns the overall E-value, and raw_score returns the overall score for the hit. A list of the most commonly used methods and their description can be found in **Table 2**.

3.3.3. Access to the HSP Data

The HSP object provides access to the alignment pair of the query and hit sequences. Information about the alignment coordinates relative to the query sequence is available through the query method, which returns a feature representing the location of the alignment on the query coordinate system. Similarly, the hit method returns a feature of hit coordinates for the alignment. The methods "evalue" and "bits" report the HSP E-value and bit score, respectively. The fraction of identical bases and similar bases (similarity is only relevant for protein alignments) is available through frac_identical and frac_conserved methods. Additional methods and their description are listed in **Table 3**.

Table 2
Bio::Search::Hit::HitI Methods

Method name	Description
name	Name of the hit or subject sequence
description	Hit description
accession	Hit accession number (if available)
locus	Hit Locus name (if available)
significance	The E-value significance for the Hit
raw_score	The score for the Hit
bits	The bit score for the Hit
next_hsp	Iterate through all the HSPs for this Hit

Table 3
Bio::Search::HSP::HSPI methods

Method name	Description
query	Get the alignment coordinates for the query as a feature object
hit	Get the alignment coordinates for the hit as a feature object
evalue	The E-value for this alignment HSP
bits	The bit score for this alignment HSP
significance	The E-value significance for the Hit
matches	Get the total number of identical or similar matches
frac_identical	Get the fraction of residues aligned that are identical
frac_similar	Get the fraction of residues aligned that are similar (only relevant for proteins)
gaps	Get the number of gaps inserted in either the query or the hit
seq_str	Get the query, hit, or homology sequence as a string
get_aln	Get the HSP as an alignment object Bio::SimpleAlign

Because the SearchIO system was designed to be generic, results can be parsed from the output of many applications, including BLAST, FASTA, BLAT, HMMER, Genewise, Exonerate, and Sim4. The *flavors* of BLAST output supported include WU-BLAST, NCBI-BLAST, BLAST-XML, tabular format, bl2seq, and megablast.

Using these modules, it is possible to process results and build bioinformatics pipelines, because all the data available in the reports are available for use in a program.

3.4. Other Aspects of the Toolkit

3.4.1. Sequence Databases

When processing a genome or a database of many sequence files, it is useful to be able to efficiently retrieve a sequence based on accession number or sequence identifier. Several sequence databases exist on the Web, including EMBL, GenBank, and Swissprot. However, the latency involved in accessing hundreds of sequences can make certain tasks prohibitively slow. To achieve quick access to sequence data, the entire database must be first downloaded to the local computer and then indexed by sequence identifiers for fast random access. There are several modules in BioPerl to build these local indexes. These include Bio::Index::Fasta, Bio::Index::GenBank, Bio::Index::Swissprot, and Bio::Index::EMBL to index sequence databases in the Fasta, GenBank, Swissprot, and EMBL formats, respectively. In addition, the Bio::DB::Fasta module is a very fast indexed system for FASTA-formatted data which allows easy sub-sequence retrieval, so that one can quickly retrieve a subsection of a large chromosome sequence without having to bring the entire sequence string into memory.

Access to GenBank, EMBL, and Swissprot databases from their respective Web sites is also possible in BioPerl with the Bio::DB::GenBank, Bio::DB::EMBL, and Bio::DB::SwissProt modules.

3.4.2. Processing Multiple Sequence Alignments

BioPerl provides modules for reading, writing, and manipulating multiple sequence alignment data from programs such as ClustalW and T-Coffee. Most of the standard multiple alignment formats are supported including ClustalW, FASTA, MSF/GCG, PHYLIP, PSI-BLAST, Pfam, Prodom, Stockholm, and Selex.

3.4.3. Processing Output from Gene Prediction Programs

Another common task in bioinformatics and genome analyses is processing the output from gene prediction programs. Although GFF is a standard format, not all programs fully support this simple format. Some of the prediction tools supported include Genscan (*8*), Genewise (*9*), GeneID (*10*), Genemark (*11*), Glimmer (*12*), ESTscan (*13*), and MZEF (*14*). The parsers for the output from these applications are all named Bio::Tools::NAME, where NAME is the name of the prediction program. The parsers produce Gene, Transcript, and Exon objects that can be queried for their genomic location, spliced mRNA sequence, or translated protein sequence.

3.4.4. Running and Processing Output from Primer3

The automated design of oligonucleotide primers can improve the throughput of genomic research projects. Primer3 (15) is one tool that can automatically design primers based on a requested sequence region and annealing temperatures. The Bio::Tools::Run::Primer3 module is used to automate the running of the Primer3 program, and Bio::Tools::Primer3 can parse the predicted oligos from the report file.

3.5. Learning More

Many other bioinformatics tools are supported by the toolkit. Most of the parsers are located in the Bio/Tools directory. The list of all modules can be browsed on the documentation Web site http://doc.bioperl.org, which provides a Web version of the module synopsis and functions. In addition, several other resources for using the toolkit are located at the project Web site http://www.bioperl.org/. A set of contributed documentation is provided from this main Web site where users and developers alike can contribute code samples and descriptions on how to use a module. This community-driven documentation project will continue to grow providing a rich resource for novice users to get acquainted with the toolkit.

Notes

1. The "translate" function can be called with many more arguments including specifying the codon table, reading frame, and termination characters.
2. The full documentation about this module provides a complete list of available methods.
3. A more thorough treaty of the capabilities of the object are covered in the Feature-Annotation HOWTO available at the BioPerl Web site (http://bioperl.org/wiki/HOWTO:Feature-Annotation).

References

1. Stajich, J. E., Block, D., Boulez, K., Brenner, S. E., Chervitz, S. A., Dagdigian, C., Fuellen, G., Gilbert, J. G., Korf, I., Lapp, H., Lehvaslaiho, H., Matsalla, C., Mungall, C. J., Osborne, B. I., Pocock, M. R., Schattner, P., Senger, M., Stein, L. D., Stupka, E., Wilkinson, M. D., and Birney, E. (2002) The Bioperl toolkit: Perl modules for the life sciences. *Genome Res* 12, 1611–18.
2. Eilbeck, K., Lewis, S. E., Mungall, C. J., Yandell, M., Stein, L., Durbin, R., and Ashburner, M. (2005) The Sequence Ontology: a tool for the unification of genome annotations. *Genome Biol* 6, R44.
3. Stein, L. D., Mungall, C., Shu, S., Caudy, M., Mangone, M., Day, A., Nickerson, E., Stajich, J. E., Harris, T. W., Arva, A., and Lewis, S. (2002) The generic genome

browser: a building block for a model organism system database. *Genome Res* 12, 1599–610.

4. Altschul, S. F., Madden, T. L., Schaffer, A. A., Zhang, J., Zhang, Z., Miller, W., and Lipman, D. J. (1997) Gapped BLAST and PSI-BLAST: a new generation of protein database search programs. *Nucleic Acids Res* 25, 3389–402.

5. Pearson, W. R., and Lipman, D. J. (1988) Improved tools for biological sequence comparison. *Proc Natl Acad Sci USA* 85, 2444–48.

6. Eddy, S. R. (1998) Profile hidden Markov models. *Bioinformatics* 14, 755–63.

7. Kent, W. J. (2002) BLAT—the BLAST-like alignment tool. *Genome Res* 12, 656–64.

8. Burge, C., and Karlin, S. (1997) Prediction of complete gene structures in human genomic DNA. *J Mol Biol* 268, 78–94.

9. Birney, E., Clamp, M., and Durbin, R. (2004) GeneWise and Genomewise. *Genome Res* 14, 988–95.

10. Parra, G., Blanco, E., and Guigo, R. (2000) GeneID in *Drosophila*. *Genome Res* 10, 511–15.

11. Borodovsky, M., and McIninch, J. (1993) Recognition of genes in DNA sequence with ambiguities. *Biosystems* 30, 161–71.

12. Delcher, A. L., Harmon, D., Kasif, S., White, O., and Salzberg, S. L. (1999) Improved microbial gene identification with GLIMMER. *Nucleic Acids Res* 27, 4636–41.

13. Iseli, C., Jongeneel, C. V., and Bucher, P. (1999) ESTScan: a program for detecting, evaluating, and reconstructing potential coding regions in EST sequences. *Proc Int Conf Intell Syst Mol Biol*, 7, 138–48.

14. Zhang, M. Q. (1997) Identification of protein coding regions in the human genome by quadratic discriminant analysis. *Proc Natl Acad Sci USA* 94, 565–68.

15. Rozen, S., and Skaletsky, H. (2000) Primer3 on the WWW for general users and for biologist programmers. *Methods Mol Biol* 132, 365–86.

Index

Printed in the United States of America